Lecture Notes in Artificial Intelligence 10415

Subseries of Lecture Notes in Computer Science

More information about this series at http://www.springer.com/series/1244

Kamil Ekštein · Václav Matoušek (Eds.)

Text, Speech, and Dialogue

20th International Conference, TSD 2017
Prague, Czech Republic, August 27–31, 2017
Proceedings

 Springer

Editors
Kamil Ekštein
University of West Bohemia
Pilsen
Czech Republic

Václav Matoušek
University of West Bohemia
Pilsen
Czech Republic

ISSN 0302-9743   ISSN 1611-3349   (electronic)
Lecture Notes in Artificial Intelligence
ISBN 978-3-319-64205-5   ISBN 978-3-319-64206-2   (eBook)
DOI 10.1007/978-3-319-64206-2

Library of Congress Control Number: 2017947498

LNCS Sublibrary: SL7 – Artificial Intelligence

Printed on acid-free paper

This Springer imprint is published by Springer Nature
The registered company is Springer International Publishing AG
The registered company address is: Gewerbestrasse 11, 6330 Cham, Switzerland

Preface

The annual International Conference on Text, Speech and Dialogue (TSD), which emerged in 1997, constitutes a recognized platform for the presentation and discussion of state-of-the-art technology and recent achievements in computer processing of natural language. It has become a broad interdisciplinary forum, interweaving the topics of speech technology and language processing. The conference attracts researchers not only from Central and Eastern Europe but also from other parts of the world. Indeed, one of its goals has always been bringing together NLP researchers with various interests from different parts of the world to promote their mutual cooperation. One of the ambitions of the conference is, as its name suggests, not only to deal with dialogue systems but also to improve dialogue among researchers in areas of NLP, i.e., among the "text" and the "speech" and the "dialogue" people.

Moreover, the TSD 2017 Conference celebrated its 20-year anniversary. There was a lot to celebrate: The two teams taking turns in organizing the conference – the team from the Faculty of Applied Sciences of the University of West Bohemia in Plzeň and the team from the Faculty of Informatics of the Masaryk University in Brno – have endured and kept the wheel spinning over the last 20 years even though the situation was sometimes rather complicated, especially because of the economic crisis of 2007–2008. Not only has the conference always been vital, it has evolved from its humble beginnings into a renowned event registered in respected and widely monitored lists of conferences in the field of NLP.

We wanted to make the 20th year somehow special and therefore decided to move the conference for this year from its traditional venues in Plzeň (and surroundings) and Brno to the capital city. It is symbolic in several ways: Prague lies between Plzeň and Brno, it is the largest and the most important city of the country, there is a very important scientific and research partner facility of both organizing institutions, the Institute of Formal and Applied Linguistics, and other prominent research facilities and by this the symbolic value of Prague as the TSD venue is far from exhausted.

The TSD 2017 conference was held on the campus of the co-organizing institution, the Faculty of Mathematics and Physics of the Charles University, called the "House of Professed," during August 27–31, 2017. The conference schedule and the keynote topic were again co-ordinated with the Interspeech conference and TSD 2017 was listed as an Interspeech 2017 satellite event. Like its predecessors, TSD 2017 highlighted the importance of language and speech processing to both the academic and scientific world and their most recent breakthroughs in current applications. Both experienced researchers and professionals, as well as newcomers in the field, found in the TSD conference a forum to communicate with people sharing similar interests.

This volume contains a collection of submitted papers presented at the conference. Each of them was thoroughly reviewed by three members of the conference reviewing team consisting of more than 60 top specialists in the conference topic areas. A total of 56 accepted papers out of 117 submitted, altogether contributed by 141 authors and co-authors, were selected by the Programme Committee for presentation at the conference and for publication in this book. Theoretical and more general contributions were presented in common (plenary) sessions. Problem-oriented sessions as well as panel discussions then brought together specialists in narrower problem areas with the aim of exchanging knowledge and skills resulting from research projects of all kinds.

Last but not least, we would like to express our gratitude to the authors for providing their papers on time, to the members of the conference reviewing team and the Program Committee for their careful reviews and paper selection, and to the editors for their hard work preparing this volume. Special thanks go to the members of both Local Organizing Committees for their tireless effort and enthusiasm during the conference organization. We hope that you benefit from the event and that you these proceedings.

August 2017

Kamil Ekštein
Miloslav Konopík
Václav Matoušek

Organization

The 20th International Conference on Text, Speech and Dialogue TSD 2017 was organized by the NTIS (New Technologies for the Information Society) P2 Research Centre of the Faculty of Applied Sciences, University of West Bohemia in Plzeň (Pilsen) in co-operation with the Institute of Formal and Applied Linguistics of the Faculty of Mathematics and Physics, Charles University in Prague and co-organized by the Faculty of Informatics, Masaryk University in Brno, Czech Republic.

The conference website is located at http://www.kiv.zcu.cz/tsd2017/ or http://www.tsdconference.org/.

Programme Committee

Elmar Nöth, Germany (General Chair)
Eneko Agirre, Spain
Vladimír Benko, Slovakia
Paul Cook, Australia
Jan Černocký, Czech Republic
Simon Dobrišek, Slovenia
Kamil Ekštein, Czech Republic
Karina Evgrafova, Russia
Darja Fišer, Slovenia
Eleni Galiotou, Greece
Björn Gambäck, Norway
Radovan Garabík, Slovakia
Alexander Gelbukh, Mexico
Louise Guthrie, UK
Tino Haderlein, Germany
Jan Hajič, Czech Republic
Eva Hajičová, Czech Republic
Yannis Haralambous, France
Hynek Hermansky, USA
Jaroslava Hlaváčová, Czech Republic
Aleš Horák, Czech Republic
Eduard Hovy, USA
Maria Khokhlova, Russia
Daniil Kocharov, Russia
Miloslav Konopík, Czech Republic
Ivan Kopeček, Czech Republic
Valia Kordoni, Germany
Pavel Král, Czech Republic
Siegfried Kunzmann, Germany
Natalija Loukachevitch, Russia

Bernardo Magnini, Italy
Václav Matoušek, Czech Republic
France Mihelič, Slovenia
Roman Mouček, Czech Republic
Agnieszka Mykowiecka, Poland
Hermann Ney, Germany
Karel Oliva, Czech Republic
Karel Pala, Czech Republic
Nikola Pavešić, Slovenia
Maciej Piasecki, Poland
Josef Psutka, Czech Republic
James Pustejovsky, USA
German Rigau, Spain
Leon Rothkrantz, The Netherlands
Anna Rumshisky, USA
Milan Rusko, Slovakia
Pavel Rychlý, Czech Republic
Mykola Sazhok, Ukraine
Pavel Skrelin, Russia
Pavel Smrž, Czech Republic
Petr Sojka, Czech Republic
Stefan Steidl, Germany
Georg Stemmer, Germany
Marko Tadić, Croatia
Tamas Varadi, Hungary
Zygmunt Vetulani, Poland
Pascal Wiggers, The Netherlands
Yorick Wilks, UK
Marcin Wolinski, Poland
Victor Zakharov, Russia

Local Organizing Committee (Plzeň Team)

Václav Matoušek (Chair)   Miloslav Konopík
Romana Strapková (Secretary)   Roman Mouček
Kamil Ekštein   Tomáš Hercig

Local Organizing Committee (Prague Team)

Eva Hajičová (Chair)   Anna Kotěšovcová
Markéta Lopatková

Keynote Speakers

The organizers would like to thank the following respected scientists and researchers for delivering their keynote talks:

Eva Hajičová   Rico Sennrich
Tomáš Mikolov   Lucia Specia
Michael Picheny

Acknowledgments

The organizers would like to give special thanks to the following reviewers who substantially contributed to the successful completion of the TSD 2017 review process by voluntarily agreeing to deliver reviews beyond their duties:

Tomáš Brychcín   Ladislav Lenc
Christophe Cerisara   Boris Lobanov
Eugene Fedorov   Natalia Loukachevitch
Ivan Habernal   Oleksandr Marchenko
Tomáš Hercig   Dirk Schnelle-Walka
Ivo Ipsic   Josef Steinberger
Jozef Ivanecký   Lukáš Svoboda
Michal Konkol   Jan Trmal

Sponsoring Institutions

The organizers would like to express their immense gratitude to the following organization for providing extra funding that helped to keep the conference fees reasonable:

Springer-Verlag GmbH, Heidelberg, Germany

About the Venue – Prague (Praha)

The capital and the largest city of the Czech Republic Praha (*Prague* in English) is also called The City of a Hundred Spires or The Heart of Europe. It is situated in the very centre of the historic region Bohemia on the banks of the river Vltava (German *Moldau*). There live more than 1.2 million inhabitants. Thus, Praha is naturally considered the centre of science, higher education, culture, economy and authorities. The city has a rich history of more than 1,000 years which is projected into its multiple architectural and cultural treasures.

The city is divided into ten districts. Each of them offers its own charming atmosphere predicated upon it's rich history. A good example can be the Jewish Quarter (Josefov) known especially for the legend of Golem and famous writer Franz Kafka. Then, walking the Pařížská street (said to be the most luxurious street in the city), there is the Old Town Square. One of the most important squares of the city renowned for the rare Prague Astronomical Clock (Orloj), number of galleries, Bethlehem Chapel and a monument of religious reformer Jan Hus.

The next place of interest can be found in the area of the New Town. The Wenceslas square with the monument of St. Wenceslas, the patron saint of the Czech state, is the longest square of the republic. Its capacity is fully used by various shops, restaurants, clubs and street artists. Also the renaissance revival-styled building of National Museum, which is now under reconstruction, is situated on the upper end of the square.

Modern art and architecture together with technical mastery demonstration are represented by the Žižkov Television Tower, the Dancing House (Fred and Ginger Building) or the Štefánik's Observatory on the Petřín hill located in the neighbourhood of the quarter Hradčany. Also Křižík's light fountain or Industrial Palace in the area of the Holešovice Showground are worth seeing.

However, the dominant feature of the skyline is still created by the Prague Castle and the gothic St. Vitus Cathedral spires. The Golden Lane heading down to the Lesser Town shows the tiny and colorful medieval houses.

There are many bridges connecting the banks of the Vltava River. However, only one of them is well known in the whole world – the Charles bridge. Czech King and Holy Roman Emperor Charles IV promoted its construction in the 14th century. The bridge is 520 metres long and stands for a connection between the Lesser Town and the Old Town. It was built in the gothic style as well as the St. Vitus Cathedral.

Charles IV was also the founder of the University, which now proudly bears his name – The Charles University. It is one the world's oldest universities and with 17 faculties, 3 institutes, 6 centres of teaching, research and development it is also the largest and best rated university in the Czech Republic. The students can choose some of the 642 courses within 300 of accredited degree programmes in the field of medicine, law, theology, pharmacy, arts, science, mathematics and physics, education, social sciences, physical education and sports, and humanities.

We are justifiably very proud of the fact that the campus of the Charles University hosted the TSD 2017 conference.

Keynote Talks

A Glimpse Under the Surface: Language Understanding May Need Deep Syntactic Structure

Eva Hajičová

Faculty of Mathematics and Physics, Institute of Formal and Applied Linguistics,
Charles University in Prague, Malostranské nám. 25,
11800 Prague 1, Czech Republic
hajicova@ufal.mff.cuni.cz

Abstract. Language understanding is one of the crucial issues both for the theoretical study of language as well as for applications developed in the domain of natural language processing. As Katz (1966, p.100) puts it "to understand the ability of natural languages to serve as instrument to the communication of thoughts and ideas we must understand what it is that permits those who speak them consistently to connect the right sounds with the right meanings." The proper task of linguistics consists then in the description (and) explanation of the relation between the set of the semantic representations and that of the phonetic forms of utterances; at the same time, among the principal difficulties there belongs "a specification of the set of semantic representations" (Sgall and Hajičová 1970, p.5). In our contribution, we present arguments for the approach that follows the tradition of European structuralism which attempted at an account of linguistic meaning the elements of which are understood as "points of intersection" of conceptual contents (as a reflection of reality) and the organizing principle of the grammar of the individual language (Dokulil and Daneš 1958). In other words, we examine how "deep" the sematic representations have to be in order (i) to give an appropriate account of synonymy, and (ii) to help to distinguish semantic differences in cases of ambiguity (homonymy).

Towards Building Intelligent Machines
That We Can Communicate With

Tomáš Mikolov

Facebook Research, AI Research Group
http://research.fb.com/
tmikolov@gmail.com

Abstract. In the recent years, there has been growing interest in development of general artificial intelligence systems. I will describe some of our recent attempts to build such intelligent system, starting by specifying the end goal: A machine that accomplishes tasks via natural language. Further, I will discuss several simplifications we are considering to make this research project more manageable: Teaching the AI to communicate at first in simulated environments, and early focus on incremental learning to efficiently use small number of training examples. I will present a dataset that we recently published to help the research community in achieving these goals.

Neural Machine Translation – What's Linguistics Got to Do with It?

Rico Sennrich

University of Edinburgh, School of Informatics, Institute for Language,
Cognition and Computation, 10 Crichton Street, Edinburgh, EH8 9AB, UK
rico.sennrich@ed.ac.uk

Abstract. Neural machine translation has obtained impressive results in the last few years, establishing itself as the new state of the art. This has been achieved by learning from raw text, without explicit linguistic knowledge. In this talk, I will discuss the capability of sequence-to-sequence models to learn linguistic phenomena from text, and will present recent research on incorporating explicit linguistic structure into the models.

A Picture Is Worth a Thousand Words: Towards Multimodal, Multilingual Context Models

Lucia Specia

University of Sheffield, Department of Computer Science, NLP Research Group
Regent Court, 211 Portobello, Sheffield, S1 4DP, UK
l.specia@sheffield.ac.uk

Abstract. In Computational Linguistics, work towards understanding or generating language has been primarily based solely on textual information. However, when we humans process a text, be it written or spoken, we also take into account cues from the context in which such a text appears, in addition to our background and common sense knowledge. This is also the case when we translate text. For example, a news article will often contain images and may also contain a short video and/or audio clip. Users of social media often post photos and videos accompanied by short textual descriptions. The additional information can help minimise ambiguities and elicit unknown words. In this talk I will introduce a recent area of research that addresses the automatic translation of texts from rich context models that incorporate multimodal information, focusing on visual cues from images. I will cover some of our recent work analysing how humans perform translation in the presence/absence of visual cues and then move on to datasets and computational models proposed for this problem.

Contents

Keynote Talk

A Glimpse Under the Surface: Language Understanding May Need Deep Syntactic Structure

Eva Hajičová[(✉)]

Faculty of Mathematics and Physics Institute of Formal and Applied Linguistics,
Charles University in Prague,
Malostranské Nám. 25, 11800 Praha 1, Czech Republic
hajicova@ufal.mff.cuni.cz

Abstract. Language understanding is one of the crucial issues both for the theoretical study of language as well as for applications developed in the domain of natural language processing. As Katz (1969, p. 100) puts it "to understand the ability of natural languages to serve as instrument to the communication of thoughts and ideas we must understand what it is that permits those who speak them consistently to connect the right sounds with the right meanings." The proper task of linguistics consists then in the description (and) explanation of the relation between the set of the semantic representations and that of the phonetic forms of utterances; at the same time, among the principal difficulties there belongs "a specification of the set of semantic representations" (Sgall and Hajičová 1970, p. 5). In our contribution, we present arguments for the approach that follows the tradition of European structuralism which attempted at an account of linguistic meaning the elements of which are understood as "points of intersection" of conceptual contents (as a reflection of reality) and the organizing principle of the grammar of the individual language (Dokulil and Daneš 1958). In other words, we examine how "deep" the sematic representations have to be in order (i) to give an appropriate account of synonymy, and (ii) to help to distinguish semantic differences in cases of ambiguity (homonymy).

Synonymy can be understood as a relation between two sentences differing in a given opposition but having the same truth conditions, i.e. there does not exist a situation when one sentence would be true while the other sentence would not be true. A proof of non-existence, of course, is not possible, so that the statement of synonymy has always a nature of a hypothesis; this criterion helps us to decide for two suspicious sentences whether they are synonymous or not. Thus e.g. the sentences *Pavel sold Jirka a car. – Jirka bought a car from Pavel.* are not synonymous because if they are used in the context ... *with enthusiasm*, their meanings differ (in the first of them, the enthusiasm is on the side of Pavel, the seller, in the other on the side of Jirka, the buyer). Similar considerations hold for such pairs as Cz. *Jan si vzal Marii* (Jan married Mary) and *Jana si vzala Marie* (E. Jan-Acc married Mary-Nom.), if inserted into the context ... *for money*, or

© Springer International Publishing AG 2017
K. Ekštein and V. Matoušek (Eds.): TSD 2017, LNAI 10415, pp. 3–7, 2017.
DOI: 10.1007/978-3-319-64206-2_1

for the non-synonymous sentences *I have read a letter about a quarrel of parents* and *I have read a letter about quarrelling parents*. because the continuation "... *but about their quarrel there was no mention in the letter*" is possible only with the second sentence. On the other hand, the pairs such as *He promised to do it in time* and *He promised he would do it in time* are considered to be synonymous: a context in which they would have different truth conditions has not yet been found (Panevová 1980).

Ambiguity is a notorious problem for both theoreticians and NLP researchers. The sources for ambiguity may be either lexical or they may lie in morphemics (e.g. the ambiguity between Nominative and Accusative in Cz: *Slepice honí kuřata* 'Hens chase chicken' where *slepice* and *kuřata* may be either Nom. or Acc. and the order is not decisive (who runs after whom?)), or in syntax (the well-known example *the criticism of the Polish delegate*: who criticized whom, or *the warning of the driver* – who warns?). The sources of ambiguities may accumulate in a single sentence – Cz. *Loví tlouště na višni* 'Catch(es) fish on morello/morello-tree': *loví* "catch" he-she-it-they, *tloušť* "(kind of) fish/" Acc sg/pl., *višni*: Loc. of *višeň* "morello" (fruit) or "morello-tree"; from the syntactic point of view there are several possible structural interpretations: Subj – Verb – Object – Loc: Subj is in the Location? Object is in the Location? (multiplied by the lexical ambiguity), or: Subj – Verb – Object – Instrument (?Manner) (= Morello as a bait put on a hook to catch fish).

A special attention in the paper will be paid to the case of synonymy/ambiguity/semantic differences related to **the information structure of sentences**. Among the examples discussed there are pairs of sentences such as *Everybody in this room knows at least two languages* vs. *At least two languages are known by everybody in this room*, or *Russian is spoken in Siberia* vs. *In Siberia one speaks Russian*, or *Tom only introduced Mary to Jane* vs. *Tom introduced Mary only to Jane*, or *Dogs must be CARRIED* (with the normal placement of intonation center at the end of the sentence, as denoted by the capitals) vs. *DOGS must be carried* (with the intonation center on DOGS) vs. *Carry dogs (CARRY dogs* vs. *Carry DOGS)*. Examples such as those document that if one wants to account in a consistent way for the semantic differences between sentences that on the surface look the same, it is necessary to postulate some kind of underlying structure (for a more detailed discussion of a formal account of information structure, see e.g. Hajičová et al. 1998).

Another support for this claim is the phenomenon of **surface deletions** (see Hajič et al. 2015; Hajičová et al. 2015). There belongs e.g. the phenomenon known recently in theoretical linguistics as a pro-drop parameter (called sometimes zero subject or null-subject). Czech belongs to the pro-drop type of language: the subject is often deducible from the morphology of the verb (*Přišel*-Masc. *domů* 'He came home' vs. *Přišla*-Fem. *domů* 'She came home') but due to the ambiguity of some verb endings this is not always the case (see above the sentence *Loví tlouště na višni* 'He-she-it-they catch(es) fish on morello-tree'). Other examples of surface deletions are infinitival constructions of the type: *John decided to leave Prague* (synonymous with *John decided that he would* *leave Prague*) vs. *John recommended his friend to move to a better flat* (synonymous with *John recommended his friend that (he/she) moves to a better flat*), structures with comparison (*Paul knows a better lawyer than John:* meaning either ... *a better lawyer than John (is a lawyer)*, or ... *a better lawyer than John knows (a lawyer)*, or structures with the word '*kromě*' (besides): *Kromě Jany pozveme celou rodinu* (Besides Jane we will invite the whole family) which may mean either an addition (Jane will be invited (too)), or an exclusion (Jane will not be invited). Special problems are connected with deletions in structures with coordination (see Popel et al. 2013): it is not always clear which sentence elements are coordinated/deleted (cf. examples *red and white wine* vs. *Polish flag is white and red*, or *Romulus and Remus founded Rome* vs. *Michelangelo and Dante celebrated Rome*, or the ambiguity of the structure *sick and old people*: sick people [need not be old] and old people [need not be sick] vs. (both: sick and old) people.

The inclusion of an underlying (deep) level into the theoretical description of a language has led the research team of Prague theoretical and computational linguists to the postulation of a multilevel scheme in the theory of **Functional Generative Description** as proposed by Petr Sgall in the late sixties and developed since then by him and his pupils (for a most comprehensive treatment, cf. Sgall et al. 1986). This approach is also reflected in the proposal and build-up of the so-called **Prague Dependency Treebank** (PDT) for Czech, and the same scenario for the parallel annotation of the Prague Czech-English Dependency Treebank (PCEDT, with a two-level annotation of Czech and English; the original English texts are taken from the Penn Treebank, translated to Czech, see Hajič et al. 2011). The work on PDT started as soon as in the mid-nineties and the overall scheme was published already in 1998 (see e.g. Hajič 1998; for a detailed study on the treatment of some particular linguistic issues in PDT see Hajič et al. 2016). The basic idea was to build a corpus annotated not only with respect to the part-of-speech tags and some kind of (surface) sentence structure but capturing also the syntactico-semantic, underlying structure of sentences. The annotation is manual, and the "deep" syntactic dependency structure (with several semantically-oriented features, called "tectogrammatical" level of annotation) has been conceptually and physically separated from the surface dependency structure and its annotation, with full alignment between the elements (tree nodes) of both annotation levels being kept. The Prague Dependency Treebank consists of continuous Czech texts mostly of the journalistic style analyzed on three levels of annotation (morphological, surface syntactic and deep syntactic structure, including the annotation of the information structure of sentences, see Hajičová 2012). At present, the total number of documents annotated on all the three levels is 3,168, amounting to 49,442 sentences and 833,357 (occurrences of) nodes. The PDT version 1.0 (with the annotation of only morphology and the surface dependencies) is available from the Linguistic Data Consortium, as is the PDT version. Pronominal coreference is also annotated. Other additions (such as discourse annotation) appeared in PDT 2.5 and in PDT 3.0, which are both available from the LINDAT/CLARIN repository (Bejček et al. 2013).

The annotated corpus has a multifold exploitation. It is an indispensable resource for the study of particular linguistic phenomena in the given language, and, when a parallel corpus is available, also in comparison with other languages. The annotated material may serve as a basis for the compilation of lexicons (e.g. the VALLEX lexicon of Czech verbs with added information on valency of verbs, Lopatková et al. 2016 and the PDT-based lexicon of Czech verbs, Urešová 2011) and for the build-up of grammars (cf. Panevová et al. 2014). The annotation on the underlying, tectogrammatical level has also served as invaluable inspiration and data support for the build-up of some NLP applications (e.g. the Tecto-MT system, or the project Companions).

In our contribution, we will document that one of the basic features of this resource is its importance not only for the representation of the surface shape of the sentence but even more for the underlying sentence structure: it elucidates phenomena hidden on the surface but unavoidable for the representation of the meaning and functioning of the sentence.

Acknowledgement. This work has been supported by the LINDAT/CLARIN project of the Ministry of Education, Youth and Sports of the Czech Republic (project LM2015071) and by the project No. GA17-07313S of the Grant Agency of the Czech Republic.

References

Bejček, E., et al.: Prague Dependency Treebank 3.0. Data/Software (2013)

Dokulil, M., Daneš, F.: K tzv. významové a mluvnické stavbě věty [On the so-called semantic, grammatical structure of the sentence]. In: O vědeckém poznání soudobých jazyků, Praha: Nakladatelství Československé akademie věd, pp. 231–246 (1958)

Hajič, J.: Building a syntactically annotated corpus: the Prague dependency treebank. In: Hajičová, E. (ed.) Issues of Valency and Meaning. Studies in Honour of Jarmila Panevová, pp. 106–132. Karolinum, Prague (1998)

Hajič, J., Hajičová, E., Mírovský, J., Panevová, J.: Linguistically annotated corpus as an invaluable resource for advancements in linguistic research: a case study. Prague Bull. Math. Linguist. **106**, 69–124 (2016)

Hajič, J., Hajičová, E., Mikulová, M., Mírovský, J., Panevová, J., Zeman, D.: Deletions and node reconstructions in a dependency-based multilevel annotation scheme. In: Gelbukh, A. (ed.) CICLing 2015. LNCS, vol. 9041, pp. 17–31. Springer, Cham (2015). doi:10.1007/978-3-319-18111-0_2

Hajič, J., Hajičová, E., Panevová, J., et al.: Prague Czech-English Dependency Treebank 2.0 (2011)

Hajičová, E.: What we have learned from complex annotation of topic-focus articulation in a large Czech corpus. Echo des Etudes Romanes **8**(1), 51–64 (2012)

Hajičová, E., Mikulová, M., Panevová, J.: Reconstructions of deletions in a dependency-based description of Czech: selected issues. In: Proceedings of the Third International Conference on Dependency Linguistics, Uppsala, pp. 131–140 (2015)

Hajičová, E., Partee, B., Sgall, P.: Topic-Focus Articulation, Tripartite Structures and Semantic Content. Kluwer, Dordrecht (1998)

Katz, J.J.: The Philosophy of Language. Harper, New York (1969)

Lopatková, M., Kettnerová, V., Bejček, E., Vernerová, A., Žabokrtský, Z.: Valenční slovník českých sloves VALLEX. [Valency Dictionary of Czech Verbs]. Karolinum, Praha (2016)

Panevová, J.: Formy a funkce ve stavbě české věty [Forms and Functions in the Structure of Czech Sentence]. Academia, Praha (1980)

Panevová, J., et al.: Mluvnice současné češtiny 2. Syntax češtiny na základě anotovaného korpusu. [Syntax of Czech on the Basis of an Annotated Corpus]. Karolinum, Prague (2014)

Popel, M., Mareček, D., Štěpánek, D., Zeman, D., Žabokrtský, Z.: Coordination Structures in dependency treebanks. In: Proceedings of the 51st Annual Meeting of the Association for Computational Linguistics, pp. 517–527. Association for Computational Linguistics, Sofija (2013)

Sgall, P., Hajičová, E.: A "functional" generative description (background and framework). Prague Bull. Math. Linguist. **14**, 3–38 (1970). Revue roumaine de linguistique **16**, 9–37 (1971)

Sgall, P., Hajičová, E., Panevová, J.: The Meaning of the Sentence in Its Semantic and Pragmatic Aspect. Reidel Publishing Company and Academia, Dordrecht and Prague (1986)

Urešová, Z.: Valence sloves v Pražském závislostním korpusu [Valency of verbs in the Prague Dependency Treebank]. UFAL, Prague (2011)

Conference Papers

Robust Automatic Evaluation of Intelligibility in Voice Rehabilitation Using Prosodic Analysis

Tino Haderlein[1]([✉]), Anne Schützenberger[2], Michael Döllinger[2], and Elmar Nöth[1]

[1] Friedrich-Alexander-Universität Erlangen-Nürnberg (FAU), Lehrstuhl für Informatik 5 (Mustererkennung), Martensstraße 3, 91058 Erlangen, Germany
Tino.Haderlein@cs.fau.de
[2] Klinikum der Universität Erlangen-Nürnberg, Phoniatrische und pädaudiologische Abteilung in der HNO-Klinik, Waldstraße 1, 91054 Erlangen, Germany
http://www5.cs.fau.de
http://www.hno-klinik.uk-erlangen.de/phoniatrie/

Abstract. Speech intelligibility for voice rehabilitation has been successfully evaluated by automatic prosodic analysis. In this paper, the influence of reading errors and the selection of certain words for the computation of prosodic features (nouns only, nouns and verbs, beginning of each sentence, beginnings of sentences and subclauses) are examined. 73 hoarse patients (48.3 ± 16.8 years) read the German version of the text "The North Wind and the Sun". Their intelligibility was evaluated perceptually by 5 trained experts according to a 5-point scale. Eight prosodic features showed human-machine correlations of $r \geq 0.4$. The normalized energy in a word-pause-word interval, computed from all words ($r = 0.69$ for the full speaker set), the mean of jitter in nouns and verbs ($r = 0.67$), and the pause duration before a word ($r = 0.66$) were the most robust features. However, reading errors can significantly influence these results.

Keywords: Intelligibility · Automatic assessment · Prosody · Reading errors

1 Introduction

In speech therapy and rehabilitation, a patient's voice is usually evaluated by the therapist. Automatically computed, objective measures can support this task. However, established methods for objective evaluation, that analyze only sustained vowels, cannot evaluate speech criteria, like intelligibility. For this study, the test persons read a given standard text that underwent prosodic analysis afterwards. Earlier studies showed the suitability of this approach [3–5]. However, each prosodic feature was averaged over all words in the text and then used for further computation. Hence, content and function words, long and short words, and words at different positions in sentences, were all put together with the risk of losing information. Additionally, the influence of errors made during

© Springer International Publishing AG 2017
K. Ekštein and V. Matoušek (Eds.): TSD 2017, LNAI 10415, pp. 11–19, 2017.
DOI: 10.1007/978-3-319-64206-2_2

reading has not been analyzed in detail. When the automatic system expects the exact reproduction of a given text, then repetitions or out-of-vocabulary words have to be mapped to the pre-defined word sequence. As a consequence, the word identities and boundaries assigned by the speech recognizer are wrong. Using them for the word-based prosodic analysis leads to erroneous prosodic feature values. This problem could be solved by replacing the text reference by a transliteration of the respective speech sample. However, this method is not applicable in clinical practice. It was shown that the influence of reading errors is negligible for the average patient [5], but for smaller patient groups, the effects are unclear. Two main questions are addressed in this paper:

- How does the position and type of words that are selected from a read-out text influence the reliability of the automatic analysis of intelligibility?
- In what way is the automatic analysis influenced by the number of reading errors?

This work is organized as follows: Sect. 2 introduces the test data and the perceptual evaluation reference. The computation of the prosodic features is described in Sect. 3. The results of the experiments (Sect. 4) will be discussed in Sect. 5.

2   Test Data and Subjective Evaluation

73 German subjects with different severity of chronic hoarseness participated in this study (Table 1). Patients suffering from cancer were excluded. Each person read the text "Der Nordwind und die Sonne" ("The North Wind and the Sun", [7]), a phonetically rich standard text which is frequently used in clinical speech evaluation in German-speaking countries. It contains 108 words (71 distinct) with 172 syllables. The data were recorded with a sampling frequency of 16 kHz and 16 bit amplitude resolution using an AKG C 420 microphone (AKG Acoustics, Vienna, Austria). They were recorded in a quiet room at our university and digitally stored on a server by a client/server-based system [11, Chap. 4]. The study respected the principles of the World Medical Association (WMA) Declaration of Helsinki on ethical principles for medical research involving human subjects and has been approved by the ethics committee of our clinics.

Five voice professionals (one ear-nose-throat doctor, four speech therapists) evaluated the intelligibility of each original recording perceptually. The samples were played to the experts once via loudspeakers in a quiet seminar room without disturbing noise or echoes. Rating was performed on a five-point Likert scale. For computation of average scores for each patient, the grades were converted to integer values (1 = 'very high', 2 = 'rather high', 3 = 'medium', 4 = 'rather low', 5 = 'very low'). For each patient, an intelligibility mark, expressed as a floating point value, was calculated as the arithmetic mean of the single scores. These marks served as ground truth in our experiments.

Due to reading errors, repetitions, and additional remarks, such as "read now?", the recordings did not only contain words appearing in the text reference but also additional words and word fragments. The topic of this paper is

Table 1. The test speakers (entire set, group with few and group with many reading errors)

Group	Persons			Age				Reading errors			
	All	Men	Women	μ	σ	Min	Max	μ	σ	Min	Max
Overall	73	24	49	48.3	16.8	19	85	3.10	3.50	0	17
Low-error	32	9	23	48.5	13.7	26	76	0.34	0.47	0	1
High-error	41	15	26	48.1	18.9	19	85	5.24	3.34	2	17

not a full linguistic analysis of the reading errors, since the automatic analysis of intelligibility used here does not work on the linguistic level of speech. In order to describe the errors, a manual word-based counting of errors was adopted instead [5]. The three basic error classes are substitutions, deletions, and insertions. We also distinguished between substitutions of a word by another word or a word fragment, and between insertions of a full word or word fragment, respectively (Tables 1 and 2). We consider this word-based method sufficient since most of the errors affect one word only: the rate of single-word errors on newspaper and magazine articles among healthy speakers has been reported to be almost 70% [8].

The problem of reading errors has been addressed in two ways. In order to study the effect of errors on the evaluation on subsets of reasonable size, the overall data set was divided into an age-matched 'low-error' group with at most one reading error per speaker and a 'high-error' group with 2 to 17 errors per speaker (Tables 1 and 3). In order to determine the influence of errors within one particular data subset, a second version of the audio files was created by removing the speech parts containing additional words and fragments. Deletions,

Table 2. Number of reading errors (in parentheses: percental per speaker)

	All files		Low-error reading		High-error reading	
	Orig. files	Error-treat	Orig. files	Error-treat	Orig. files	Error-treat.
All	226 (3.10)	149 (2.04)	11 (0.34)	9 (0.28)	215 (5.24)	140 (3.41)
Substitutions	80 (1.09)	77 (1.05)	8 (0.25)	8 (0.25)	72 (1.76)	69 (1.68)
Deletions	7 (0.09)	7 (0.09)	0 (0.00)	0 (0.00)	7 (0.17)	7 (0.17)
Inserted words	55 (0.78)	3 (0.04)	2 (0.06)	0 (0.00)	53 (1.29)	3 (0.07)
Fragments	64 (0.88)	62 (0.84)	1 (0.03)	1 (0.03)	63 (1.54)	61 (1.49)
Inserted fragments	20 (0.27)	0 (0.00)	0 (0.00)	0 (0.00)	20 (0.49)	0 (0.00)

Table 3. Number of recordings with a certain number of reading errors

Errors	0	1	2	3	4	5	6	7	8	11	15	17
Original files	21	11	8	6	6	8	3	3	3	2	1	1
Error-treated files	29	13	11	3	7	3	2	2	1	1	0	1

however, cannot be repaired as the correct word was not spoken in the sample. For substitutions, the situation is similar. The text flow was supposed to be preserved, so misread single words without corrections were not removed. For instance, the repetition "einst stri- einst stritten" was reduced to the correct "einst stritten" while the word "Nordwand" instead of "Nordwind" without correction was left unchanged. The data set created in this way will further be denoted as 'error-treated'.

3   Prosodic Features

The speech recognition system used for the experiments [3] is based on semi-continuous Hidden Markov Models (HMM). For each 16 ms frame, a 24-dimensional feature vector is computed. It contains short-time energy, 11 Mel-frequency cepstral coefficients, and the first-order derivatives of these 12 static features. The recognition vocabulary of the recognizer was changed to the 71 words of the standard text. Only a unigram language model was used so that the results mainly depend on the acoustic models.

In order to find counterparts for intelligibility, a 'prosody module' was used to compute features based upon frequency, duration, and speech energy (intensity) measures. This is common in automatic speech analysis on normal voices [12–14]. The module processes the output of the word recognition module and the speech signal itself.

'Local' prosodic features are computed for each word position. Originally, there were 95 of them. After several studies on voice and speech assessment, however, a relevant core set of 33 features has been defined for further processing [6]. The components of their abbreviated names are given in parentheses:

- Length of pauses (Pause): length of silent pause before (–before) and after (–after), and filled pause before (Fill-before) and after (Fill-after) the respective word
- Energy features (En): regression coefficient (RegCoeff) and the mean square error (MseReg) of the energy curve with respect to the regression curve; mean (Mean) and maximum energy (Max) with its position on the time axis (MaxPos); absolute (Abs) and normalized (Norm) energy values
- Duration features (Dur): absolute (Abs) and normalized (Norm) duration
- F_0 features (F0): regression coefficient (RegCoeff) and mean square error (MseReg) of the F_0 curve with respect to its regression curve; mean (Mean), maximum (Max), minimum (Min), voice onset (On), and offset (Off) values as well as the position of Max (MaxPos), Min (MinPos), On (OnPos), and Off (OffPos) on the time axis; all F_0 values are normalized.

The last part of the feature name denotes the context size, i.e. the interval of words on which the features are computed (see Table 4). They can be computed on the current word (W) or in the interval that contains the second and first word before the current word and the pause between them (WPW). A full description of

Table 4. Local prosodic features; the context size denotes the interval of words on which the features are computed (W: one word, WPW: word-pause-word interval).

Features	Context size	
	WPW	W
Pause: before, Fill-before, after, Fill-after		•
En: RegCoeff, MseReg, Abs, Norm, Mean	•	•
En: Max, MaxPos		•
Dur: Abs, Norm	•	•
F0: RegCoeff, MseReg	•	•
F0: Mean, Max, MaxPos, Min, MinPos, Off, OffPos, On, OnPos		•

the features used is beyond the scope of this paper; details and further references are given in [1,3].

Besides the 33 local features, 15 **'global' features** were computed for intervals of 15 words length each. They were derived from jitter, shimmer, and the number of detected voiced and unvoiced sections in the speech signal [1]. They covered the means and standard deviations of jitter and shimmer, the number, length, and maximum length of voiced and unvoiced sections, the ratio of the numbers of voiced and unvoiced sections, the ratio of the length of the voiced sections to the length of the signal, and the same for unvoiced sections. The last feature was the standard deviation of the F_0.

The human listeners gave ratings for the entire text. In order to receive also one single value for each feature that could be compared to the human ratings, the average of each prosodic feature over all selected words served as final feature value. Pearson's correlation coefficient was computed between the respective values of all speakers and the according average human intelligibility ratings.

4   Experiments

Earlier experiments averaged each prosodic feature over the entire read-out text. For this study, we examined whether the restriction to certain subsets might be beneficial:

- averaging over *all words* (108 words; as in earlier studies, i.e. the reference)
- *nouns only* (24 words)
- *nouns and verbs* (44 words)
- *beginnings of sentences*, i.e. the first 3 words of each of the 6 sentences (18 words)
- *beginnings of sentences and subclauses*, i.e. the first 3 words of each of the 6 sentences and 10 subclauses (48 words)

Nouns and verbs were chosen because content words generally show less predictability and hence intelligibility than function words, such as articles, prepositions, and conjunctions [15]. The beginnings of sentences and subclauses, without the regard of the word classes, were chosen with respect to the medical application. Many voice and speech patients show higher speaking effort and shorter phonation time, so they will have to pause more often and fragment the paragraph to be read into shorter sections. These breaks usually occur at syntactic boundaries.

5   Results

Table 5 shows the features that for at least one of the experiments reached a human-machine correlation of $r \geq 0.4$.

The pause duration before a word (Pause–before) is only a robust indicator when it is measured before nouns. Although other scenarios, except for the beginning of sentences, also show correlations up to $r = 0.70$, the results for low-error reading are rather poor. This is supported by the correlations on error-treated files, which drop slightly when the additional utterances are removed.

The regression coefficient of the energy in a word-pause-word interval (EnReg-CoeffWPW) works best when it is measured at the beginning of sentences and subunits. On the average, its human-machine correlation is $r = 0.59$, in low-error reading it decreases to $r = 0.48$; in high-error reading, $r = 0.62$ was achieved. The difference to the values on the error-treated files is not significant.

The normalized energy in a word-pause-word interval (EnNormWPW) has been reported to be a good indicator for intelligibility [3,5]. The results in this study confirm this with $r = 0.59$ on low-error reading, $r = 0.70$ on high-error reading, and $r = 0.69$ for the entire database, computed on the full text (Fig. 1, left). Especially for low-error reading, a selection of words from the text lowers the correlation to the perceptual scores.

The normalized duration of a word-pause-word interval (DurNormWPW) has also been a good indicator for intelligibility in earlier studies and could on the average mostly replace the energy EnNormWPW [5]. Here, it shows about the same results as the energy, but the drop for the low-error reading is much more remarkable. Only for the nouns+verbs scenario, the correlation exceeds $r = 0.40$. Both DurNormWPW and Pause–before reveal the overall speaking rate.

MeanJitter shows the highest correlation of all in this study, namely $r = 0.73$ for low-error reading and computation on nouns and verbs (Fig. 1, right). The other computation scenarios in this case are by $\Delta r \approx 0.15$ lower; for high-error reading, the correlation is stable at $r \approx 0.63$. In the error-treated files, only one single significant drop of correlation appears when the prosodic features are computed on the beginnings of sentences. StandDevJitter shows the same trend as MeanJitter, but with lower correlations.

The durations of the unvoiced sections in the recording (Dur–Voiced) and the longest unvoiced section (DurMax–Voiced), that contain information about the voice quality, exceed $r = 0.40$ in a few cases, but, in general, they are too

Table 5. Human-machine correlation r for single prosodic features ($r \geq 0.4$), depending on the words used for computation: all, nouns only, nouns and verbs (n+v), beginnings of sentences (sent_i) or of sentences and subclauses (s+s_i); bold-face: best results of each line

Type	Feature name	all	nouns	n+v	sent_i	s+s_i	all	nouns	n+v	sent_i	s+s_i
		all original files					all error-treated files				
Local	Pause–before	0.64	0.65	**0.66**	0.35	0.51	0.62	0.64	0.65	0.32	0.47
Local	EnRegCoeffWPW	0.51	0.37	0.52	0.45	**0.59**	0.48	0.31	0.49	0.46	0.58
Local	EnNormWPW	**0.69**	0.64	0.59	0.59	0.66	0.68	0.62	0.59	0.59	0.65
Local	DurNormWPW	0.65	**0.66**	0.63	0.43	0.56	0.64	0.65	0.62	0.43	0.55
Global	MeanJitter	0.63	0.65	**0.67**	0.61	0.63	0.61	0.64	0.66	0.54	0.60
Global	StandDevJitter	0.55	0.58	**0.60**	0.48	0.53	0.52	0.57	0.59	0.43	0.49
Global	Dur–Voiced	0.31	0.18	0.21	0.36	0.41	0.34	0.20	0.21	0.34	**0.44**
Global	DurMax–Voiced	0.36	0.21	0.24	0.32	0.42	0.38	0.21	0.23	0.30	**0.46**
		Original low-error files					Error-treated low-error files				
Local	Pause–before	0.36	**0.62**	0.36	0.16	0.20	0.35	0.61	0.33	0.14	0.19
Local	EnRegCoeffWPW	0.38	0.44	0.44	0.36	**0.48**	0.36	0.38	0.40	0.36	**0.48**
Local	EnNormWPW	**0.59**	0.43	0.48	0.48	0.52	0.57	0.44	0.47	0.46	0.50
Local	DurNormWPW	0.39	0.38	**0.43**	0.30	0.32	0.37	0.39	**0.43**	0.28	0.31
Global	MeanJitter	0.60	0.61	**0.73**	0.57	0.57	0.60	0.61	**0.73**	0.57	0.57
Global	StandDevJitter	0.50	0.53	**0.64**	0.46	0.46	0.50	0.53	0.63	0.44	0.46
Global	Dur–Voiced	0.41	0.42	0.39	0.31	0.38	0.41	**0.43**	0.39	0.26	0.37
Global	DurMax–Voiced	0.41	0.43	0.38	0.31	0.37	0.42	**0.44**	0.38	0.27	0.36
		Original high-error files					Error-treated high-error files				
Local	Pause–before	0.69	0.66	0.70	0.44	0.60	0.69	0.65	**0.71**	0.46	0.57
Local	EnRegCoeffWPW	0.51	0.30	0.51	0.46	**0.62**	0.48	0.24	0.49	0.46	0.61
Local	EnNormWPW	**0.70**	**0.70**	0.63	0.60	0.68	**0.70**	0.67	0.63	0.61	0.67
Local	DurNormWPW	0.70	**0.71**	0.66	0.50	0.62	0.70	0.70	0.65	0.52	0.62
Global	MeanJitter	0.62	0.62	**0.63**	0.62	**0.63**	0.60	0.62	0.60	0.51	0.59
Global	StandDevJitter	**0.58**	**0.58**	0.57	0.50	0.55	0.54	0.56	0.55	0.42	0.50
Global	Dur–Voiced	0.30	0.09	0.14	0.41	0.47	0.35	0.12	0.17	0.39	**0.53**
Global	DurMax–Voiced	0.36	0.13	0.21	0.36	0.47	0.40	0.14	0.21	0.34	**0.55**

unreliable to be recommended for the evaluation of intelligibility. There is a large variation among the computation scenarios: for nouns in high-error reading, only $r = 0.09$ was reached for Dur–Voiced.

We are aware of the problem arising when standard texts are used for measuring intelligibility. However, our listeners were well-trained speech therapists who were instructed to evaluate intelligibility and not voice quality. It is obvious that spontaneous speech would be the best choice for this task, and the kind of stimulus presented to the listener has an influence on the perceptual results [9]. However, spontaneous speech causes other problems. There may be a mismatch in the vocabulary of speaker and listener, the sentence structure and distribution

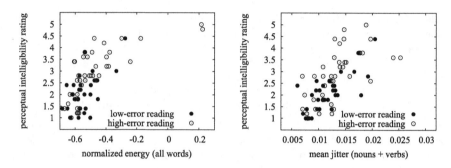

Fig. 1. Human-machine agreement: left side: for the normalized energy in a word-pause-word interval, computed on all words; right side: for the mean jitter, computed on nouns and verbs

of vowels and consonants may vary among the speakers, etc. [2]. This affects also the speech recognizer underlying the prosodic analysis. Furthermore, the prosodic evaluation of different persons is not comparable any more due to different word lengths, ratios of voiced and unvoiced sections, etc. This complexity cannot be handled properly at the moment. On the other hand, it has been shown that the text-based evaluation performed by trained listeners is as reliable as an inverse intelligibility test, where naïve raters write down a previously unknown sequence of words that was read by the test person [4]. Nevertheless, independence of a given text is a long-term goal of our work. For instance, an existing German sentence test for speech intelligibility assessment [10] contains different sentence lists which introduce more variety and complexity and represent every-day communication better.

In summary, EnNormWPW computed from all words of the text, MeanJitter of nouns and verbs, and Pause–before computed from nouns as an indicator of speaking rate, are the most robust single features for evaluation of intelligibility in this study, i.e. they show the least variability among data with different numbers of reading errors. The combination of all features and computation scenarios may reveal some more beneficial interrelations. This has been shown for features, that were averaged over the entire text, and for the average patient without regarding the reading errors. There is also room for improvement concerning the regression method, etc. This is part of future work. With additional preprocessing steps of out-of-vocabulary detection and word class identification, the automatic prosodic analysis will gain even more reliability.

Acknowledgments. Dr. Döllinger's contribution was supported by the German Research Foundation (Deutsche Forschungsgemeinschaft; DFG), grant no. DO1247/8-1.

References

1. Batliner, A., Buckow, J., Niemann, H., Nöth, E., Warnke, V.: The prosody module. In: Wahlster, W. (ed.) Verbmobil: Foundations of Speech-to-Speech Translation, pp. 106–121. Springer, Berlin (2000). doi:10.1007/978-3-662-04230-4_8
2. Ellis, L., Fucci, D.: Magnitude-estimation scaling of speech intelligibility: effects of listeners' experience and semantic-syntactic context. Percept. Mot. Skills **73**, 295–305 (1991)
3. Haderlein, T., Moers, C., Möbius, B., Rosanowski, F., Nöth, E.: Intelligibility rating with automatic speech recognition, prosodic, and cepstral evaluation. In: Habernal, I., Matoušek, V. (eds.) TSD 2011. LNCS, vol. 6836, pp. 195–202. Springer, Heidelberg (2011). doi:10.1007/978-3-642-23538-2_25
4. Haderlein, T., Nöth, E., Batliner, A., Eysholdt, U., Rosanowski, F.: Automatic intelligibility assessment of pathologic speech over the telephone. Logoped. Phoniatr. Vocol. **36**, 175–181 (2011)
5. Haderlein, T., Nöth, E., Maier, A., Schuster, M., Rosanowski, F.: Influence of reading errors on the text-based automatic evaluation of pathologic voices. In: Sojka, P., Horák, A., Kopeček, I., Pala, K. (eds.) TSD 2008. LNCS, vol. 5246, pp. 325–332. Springer, Heidelberg (2008). doi:10.1007/978-3-540-87391-4_42
6. Haderlein, T., Schwemmle, C., Döllinger, M., Matoušek, V., Ptok, M., Nöth, E.: Automatic evaluation of voice quality using text-based laryngograph measurements and prosodic analysis. Comput. Math. Methods. Med. **2015**, 11p. (2015)
7. International Phonetic Association (IPA): Handbook of the International Phonetic Association. Cambridge University Press, Cambridge (1999)
8. Kaufmann, R., Obler, L.: Classification of normal reading error types. In: Leong, C., Joshi, R. (eds.) Developmental and Acquired Dyslexia, pp. 149–157. Kluwer Academic Publishers, Dordrecht (1995)
9. Kempler, D., van Lancker, D.: Effect of speech task on intelligibility in dysarthria: a case study of Parkinson's disease. Brain Lang. **80**, 449–464 (2002)
10. Kollmeier, B., Wesselkamp, M.: Development and evaluation of a German sentence test for objective and subjective speech intelligibility assessment. J. Acoust. Soc. Am. **102**, 2412–2421 (1997)
11. Maier, A.: Speech of Children with Cleft Lip and Palate: Automatic Assessment, Studien zur Mustererkennung, vol. 29. Logos Verlag, Berlin (2009)
12. Nöth, E., Batliner, A., Kießling, A., Kompe, R., Niemann, H.: Verbmobil: the use of prosody in the linguistic components of a speech understanding system. IEEE Trans. Speech Audio Process. **8**, 519–532 (2000)
13. Origlia, A., Alfano, I.: Prosomarker: a prosodic analysis tool based on optimal pitch stylization and automatic syllabification. In: Calzolari, N., et al. (ed.) Proceedings of 8th International Conference on Language Resources and Evaluation (LREC 2012), pp. 997–1002 (2012)
14. Rosenberg, A.: Automatic detection and classification of prosodic events. Ph.D. thesis, Columbia University, New York (2009)
15. Rubenstein, H., Pickett, J.: Intelligibility of words in sentences. J. Acoust. Soc. Am. **30**, 670 (1958)

Personality-Dependent Referring Expression Generation

Ivandré Paraboni[✉], Danielle Sampaio Monteiro, and Alex Gwo Jen Lan

School of Arts, Sciences and Humanities, University of São Paulo, São Paulo, Brazil
{ivandre,daniellemonteiro,alex.lan}@usp.br

Abstract. This paper addresses the issue of how Big Five personality traits may influence the content selection task in Referring Expression generation (REG.) To this end, we build a corpus of referring expressions annotated with personality information, and then use it as the input to a machine learning approach to REG that takes the personality of the target speakers into account. Results show that personality-dependent REG outperforms standard REG algorithms, and that it may be a viable alternative to speaker-dependent approaches that require examples of descriptions produced by every individual under consideration.

Keywords: Text generation · Referring expressions · Big five personality traits

1 Introduction

In Natural Language Generation (NLG) systems, Referring Expression Generation (REG) is concerned with the production of uniquely identifying descriptions of a given target object such that the generated descriptions resemble those produced by human speakers. For instance, given the goal of describing the target highlighted in the top-left corner of Fig. 1[1], we may produce descriptions such as 'the guy with short hair, on the left', 'the one who is frowning', etc.

Given an input context of this kind, most existing approaches to REG would produce a fixed output description, that is, subsequent executions based on the same input scene would generally produce the same output. Descriptions produced by human speakers, by contrast, show much greater variation: under identical circumstances (i.e., given the same input context), different speakers will often produce different descriptions. In the above example, these may include, for instance, 'the Asian guy who looks upset', 'the man with thick eyebrows', 'the only guy who is not smiling', etc.

Human variation plays a central role in language production and, accordingly, has received a considerable amount of attention in REG and related fields.

---

[1] Stimulus images courtesy of Michael J. Tarr, Center for the Neural Basis of Cognition and Department of Psychology, Carnegie Mellon Univ. Funding provided by NSF award 0339122.

© Springer International Publishing AG 2017
K. Ekštein and V. Matoušek (Eds.): TSD 2017, LNAI 10415, pp. 20–28, 2017.
DOI: 10.1007/978-3-319-64206-2_3

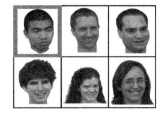

Fig. 1. An example of referential context built from Face Place images [1].

This is particularly the case when the goal of the system is to produce descriptions that are plausible or human-like. In systems of this kind, speaker-dependent information is generally modelled as a tailor-made list of preferred attributes for each individual speaker [2,3] and, in some cases, speaker's identifiers [4,5].

Speaker-dependent REG has been consistently shown to outperform standard REG algorithms that produce the same, fixed output for a given input. This method however relies on training data in the form of examples of how every individual may produce language (e.g., available from an annotated corpus of referring expressions and input contexts), and resources of this kind are known to be costly and generally unavailable for practical REG.

As an alternative to speaker-dependent REG, we may consider the generation of descriptions according to a more general speaker profile model, that is, by grouping together individuals with similar referential behaviour. This strategy may not be able to outperform algorithms that generate tailor-made descriptions for every individual but, if speaker profiles are sufficiently easy to obtain (i.e., if the need to collect examples of referring expressions can be avoided), this alternative may be worth of further investigation.

One possible way of implementing a more general REG strategy that still pays regard to differences across speakers, and which will be the basis of the present work, is by assuming that these differences are at least partially influenced by *personality traits*. Knowing the personality traits of a target speaker, we may be able to generate descriptions that resemble those produced by other speakers *with similar personality* and, in doing so, we may be able to outperform at least standard REG algorithms that produce a fixed output without the costs of collecting linguistic examples from every target speaker under consideration.

Based on these observations, this paper introduces a study on personality-dependent REG that we believe to be the first of its kind. We build a corpus for REG that is further annotated with personality information, and then use this dataset as the input to a machine learning approach that takes not only context features into account, but also the personality traits of the target speaker. Results show that personality-dependent REG outperforms standard REG algorithms, and that this may be a viable alternative to speaker-dependent approaches that rely on pre-recorded linguistic examples.

2   Background

2.1   Referring Expression Generation

The content selection task for REG takes as an input a target object r to be distinguished from a set of distractors within a given context C. Objects are generally modelled as sets of atomic or relational properties [6] represented as (*attribute*-value) pairs, as in (*gender*-male.) The goal is to produce a set L of properties of r such that L distinguishes r from every distractor in C [7]. The output description L may be subsequently realised as a definite or indefinite description. For instance, an output description $L = \{gender$-male, *emotion*-upset$\}$ could be realised as 'the guy who looks upset'.

One of the best-known approaches to REG is the Incremental algorithm in [8]. In this approach, attributes are considered for selection according to a domain-dependent list of preferences P, and provided that they are discriminatory. When an attribute a is selected, the corresponding distractors are removed from C. For instance, selecting a property (*gender*-male) rules out all distractors whose *gender* is female. The algorithm terminates when C becomes empty, or when all attributes in P have been attempted.

The Incremental approach and many of its successors are generally concerned with the generation of a single, fixed description for the given input. By contrast, more recent approaches as in [4,5] have considered the issue of human variation as well. These methods, however, rely on a set of pre-recorded examples of referring expression produced by every speaker under consideration, and are of limited use when suitable (linguistic) training data is unavailable.

2.2   Personality and Language Generation

As an alternative to speaker-dependent REG, in this work we consider the use of information about the *personality* of the target speaker. The computational treatment of human personality has become widespread in Computer science and many of its fields, and it is particularly linked to the growing popularity of the Big Five model of personality [9]. Big Five personality traits - *Extraversion, Agreeableness, Conscientiousness, Neuroticism,* and *Openness to experience* - are sufficiently straightforward to compute using a wide range of methods. These include the use of short self-report inventories and a number of less intrusive, automatic techniques (e.g., by analysing the speaker's publications on blogs etc. [10].)

Big Five personality traits have also been considered in a few NLG projects, most notably in personalised text generation as in the PERSONAGE system [11]. Systems of this kind make a large number of low- and high-level generation decisions based on personality information but, to the best of our knowledge, have not addressed the issue of using personality information in REG content selection directly.

3   Data Collection

In order to investigate possible relations between referential behaviour and personality, we built a corpus of referring expressions annotated with Big Five scores obtained from self-report inventories of every speaker. The corpus was produced by 152 speakers (86 or 56.6% female) on average 25.8 years old (min 18 and max 59). All subjects were native speakers of Brazilian Portuguese and had normal or corrected vision.

Referring expressions were elicited from a set of 12 stimulus images built from the Face Place image database [1] as in the previous Fig. 1. For each image, participants were requested to uniquely identify a particular target within the given visual context by completing a sentence in the form 'The person highlighted in red is the...'. This procedure is similar to [12–14] and other data collection tasks for REG.

Each context image contained six human faces displaying different physical and emotional features. The situations of reference under consideration involve contexts with different degrees of ambiguity (i.e., requiring shorter or longer descriptions) and physical and affective properties with different degrees of salience (e.g., a scene containing only one smiling person, or a single woman, etc.)

A dataset comprising 1810 descriptions was collected. Both images and descriptions were annotated by two judges with 27 attributes representing the most frequent information observed in the elicited data. For certain attributes, the corresponding values were obtained directly from the image annotations available from Face Place [1]. These include some of the most frequent attributes in the domain, such as gender, race and emotion. For others, annotation was also relatively straightforward (e.g., deciding which characters were smiling or not.) For more subjective cases (e.g., deciding which characters had short or long hair in a particular context), the value chosen by the majority of speakers was selected.

Table 1 summarises the seven most frequent attributes in the corpus. These correspond to 74% of all referential attributes in this domain.

Table 1. Most frequent referential attributes found in the corpus descriptions.

Attribute	Possible values	Instances	%
gender	{male, female}	1707	23.7%
race	{asian, black, caucasian}	794	11.0%
smile	{yes, no}	784	10.9%
isyoung	{yes}	705	9.8%
hair.colour	{dark, blonde}	633	8.8%
hair.length	{short, long }	434	6.0%
emotion	{positive, negative, neutral}	266	3.7%

In the present annotation scheme, we notice that some attributes such as *isyoung* are assigned the same value (e.g., 'yes') for all objects in the domain (e.g., because all stimuli images depicted reasonably young people.) Thus, attributes of this kind do not have any discriminatory power, although we would still expect our REG algorithms to handle them appropriately. Moreover, we notice that target objects in this domain are in principle identifiable by making use of fairly short descriptions (from one to three properties each.) In practice, however, the elicited descriptions are slightly longer, with an average of four properties each. This means that referential overspecification plays a significant role in this domain, an issue that will influence the design of our REG algorithms in Sect. 4.

4   Personality-Dependent REG

Since target objects in the current domain are easily described by making use of short descriptions conveying highly discriminatory properties, as in 'the smiling guy', the main differences across speakers (and which may or may not be determined by personality) are to be found in *referential overspecification*, that is, in the kinds of information that a speaker may add *beyond what is strictly necessary* for disambiguation [15]. Thus, in our work we will focus on the effects of personality on the use of such additional information. More specifically, we will take as a starting point a basic, uniquely identifying description, and we will attempt to overspecify it (up to a certain limit, as discussed below) using personality-related information or not.

Our approach consists of two stages: first, a uniquely identifying descriptions is produced by a standard implementation of the Incremental algorithm [8]. Next, additional (i.e., overspecified) properties may be added by considering two competing strategies: a baseline model called *DT-scene*, and our proposed *DT-b5* model.

In both *DT-scene* and *DT-b5* models, we follow [16] and others and implement attribute selection as a classification task, namely, by making use a set of decision-tree binary classifiers *Class*[] for each attribute under consideration. Thus, for instance, we define a binary classifier to decide whether to select the *gender* attribute or not, a binary classifier for *race*, and so on.

The *DT-scene* model is intended to select additional properties as in standard REG, that is, by taking only scene (i.e., contextual) features into account. These features represent frequency estimates and discriminatory power of the most frequent attributes (cf. Table 1) found in a set of training descriptions. The *DT-b5* classifier, by contrast, selects additional properties by considering both scene[2] and personality-related features as scalar values representing the Big Five

---

[2] The use of frequency estimates in *DT-b5* may in principle defeat the purpose of not relying on pre-recorded examples of referring expressions. In the current *DT-b5* implementation, however, these features were included only as a means to provide a meaningful comparison with *DT-scene*, and could in principle be replaced by a more realistic account of salience.

personality scores of each speaker. These five personality-related features are the only difference between *DT-scene* and *DT-b5*.

The input to the algorithm is the target object r, the context scene C, the list P of most frequently referred attributes for the context C, the predictions made by the appropriate (*scene* or *b5*) set of classifiers $Class[]$, and the average description *length* found in the training data (which in the current domain is set to 4, and which will prevent the generation of overly long descriptions.) The output of the algorithm is a list L of properties that are true of r, and which could be subsequently realised as a definite description. The general approach is presented in Algorithm 1.

```
1  Algorithm describe(r, C, P, Class[], length)
2  |   L ← MakeReferringExpression(r, C, P)
3  |   for  A_i ∈ P do
4  |   |   if |L| <= length then
5  |   |   |   if Class[A_i] == true then
6  |   |   |   |   v ← ⟨A_i, value(r, A_i)⟩
7  |   |   |   |   if ⟨A_i, v⟩ ∉ L then
8  |   |   |   |   |   L ← L ∪ ⟨A_i, v⟩
9  |   |   |   |   end
10 |   |   |   end
11 |   |   end
12 |   end
13 |   return L
```
Algorithm 1. Referential overspecification strategy

Before the main overspecification procedure in lines 03..13, a standard implementation of the Incremental algorithm [8] is invoked (line 02) to produce a uniquely identifying description L of the target r. This description may then be modified by adding (hence overspecified) attributes as predicted by the relevant set of classifiers $Class[]$. As discussed above, recall that *DT-scene* makes predictions based on scene information alone, whereas *DT-b5* takes personality information into account as well.

Overspecification is performed by considering every attribute a (line 03) within the given *length* limit (04). Provided that a is predicted by the corresponding classifier (05) and that it has not been previously selected (07), a is added to the output L regardless of being discriminatory or not (08). This procedure is repeated until the output description reaches the maximal description *length*.

5   Evaluation

In this section we describe the evaluation work of our two main REG strategies described in the previous section - *DT-scene* and *DT-b5* - and baseline systems using training and test portions of the corpus described in Sect. 3.

DT-scene and *DT-b5* are to be evaluated alongside two baseline strategies: the *Greedy* approach in [17] and the *Incremental* algorithm [8]. *Greedy* selects the referential attributes of highest discriminatory power, leading to very short descriptions (in the current domain, usually one- or two-properties long.) The *Incremental* baseline follows the list of preferred attributes ordered by frequency in each context as seen in the training data, and produces descriptions that are slightly longer than those produced by *Greedy* (but, in our corpus, never longer than three properties).

In the present evaluation, both *Greedy* and *Incremental* are solely intended to provide a lower boundary for the task, and are not expected to outperform the *DT* strategies since these baselines have access to more limited information. Moreover, unlike the *DT* models, these algorithms do not explicitly attempt to produce overspecified descriptions of the kind found in the corpus. Thus, the main focus of the evaluation work is the comparison between *DT-b5* and *DT-scene*, that is, between REG models with and without access to personality information.

Evaluation was carried out as follows. Corpus descriptions were randomly split into training (1400 instances) and test (410 instances) sets. From the training portion of the data, we obtained frequency estimates for the *Incremental* baseline and for both *DT* strategies. Next, the test dataset was taken as the input to each of the four systems (*Greedy, Incremental, DT-scene* and *DT-b5*) while computing overall Accuracy (Acc), Dice and MASI scores obtained by each strategy.

Results from the baseline methods and the two *DT* strategies are summarised in Table 2. We notice that *DT-b5* outperforms all alternatives - including its counterpart without personality information, *DT-scene* - as measured by Accuracy, Dice and MASI scores alike. A Wilcoxon's signed rank test shows that the differences between *DT-b5* and the second best alternative (i.e., *DT-scene*) are significant both in terms of Dice ($W = -9959, n = 263, Z = -4.03, p = 0.0001$) and MASI scores ($W = -6491, n = 261, Z = -2.66, p = 0.0078$.) Taking personality information into account allows the generation of descriptions that are closer to those produced by human speakers. This outcome provides support to our main research hypothesis.

Table 2. REG results.

Strategy	Acc.	Dice	MASI
Greedy	0.05	0.30	0.15
Incremental	0.16	0.59	0.34
DT-scene	0.17	0.62	0.36
DT-b5	0.19	0.66	0.38

Further analysis of the output descriptions produced by each strategy under consideration showed that, generally speaking, the *DT* strategies make more accurate attribute selection decisions than *Greedy* and *Incremental*. This outcome was to be expected since machine-learning REG is known to outperform standard algorithmic approaches when sufficient training data is available [16].

As for the main comparison between *DT-scene* and *DT-b5*, that is, the question of whether to use personality information or not, we notice that the difference between the two strategies becomes more evident in the case of non-discriminatory attribute usage. Non-discriminatory attributes such as *isyoung*, although ubiquitous in our data, are rarely selected by *DT-scene*, but are much more common in *DT-b5* descriptions. The ability to use personality information to select less discriminatory attributes when a standard context model would not predict them is precisely why *DT-b5* outperforms *DT-scene* in our experiments.

6   Final Remarks

This paper presented the initial results of an investigation on personality-dependent REG. We collected a corpus of referring expressions elicited from photographic face images conveying both physical and affective referential attributes, and presented a REG algorithm that implements attribute selection as a classification task based on contextual and personality features alike. Results suggest that taking personality information into account outperforms both standard REG algorithms and a similar classification method based on contextual features alone.

The present work leaves a number of opportunities for improvement. First, we notice that our study was limited to the use of the most frequent attributes found in the corpus, and that these attributes represent, to a large extent, discriminatory physical features (e.g., gender, race, etc.) Focusing on the most frequent attributes allowed us to meaningfully compare our results to well-known algorithms in which attribute selection is mainly driven by discriminatory power. However, we notice that attributes of this kind may be arguably less influenced by differences in personality than others (e.g., facial expressions), and that more significant results may be considered by focusing on more emotionally-charged or affective attributes, which were presently disregarded due to their low frequency in the corpus.

Finally, we notice that affective attributes may provide useful correlations with personality traits but, in order to observe their full effect on personality-dependent REG, it may be necessary to consider polarity as well. This means that content selection may need to consider using *properties* (i.e., attribute-value pairs, as in *emotion*-happy), and not simply attributes. Implementing fine-grained decisions of this kind in a personality-dependent REG algorithm is also left as future work.

Acknowledgements. This work has been supported by FAPESP grant 2016/14223-0.

References

1. Righi, G., Peissig, J.J., Tarr, M.J.: Recognizing disguised faces. Vis. Cogn. **20**(2), 143–169 (2012)
2. Bohnet, B.: The fingerprint of human referring expressions and their surface realization with graph transducers. In: INLG-2008, Stroudsburg, USA, pp. 207–210 (2008)
3. Fabbrizio, G.D., Stent, A.J., Bangalore, S.: Trainable speaker-based referring expression generation. In: CoNLL 2008, Stroudsburg, PA, USA, pp. 151–158 (2008)
4. Viethen, J., Dale, R.: Speaker-dependent variation in content selection for referring expression generation. In: Australasian Language Technology Association WS, pp. 81–89 (2010)
5. Ferreira, T.C., Paraboni, I.: Referring expression generation: taking speakers' preferences into account. In: Sojka, P., Horák, A., Kopeček, I., Pala, K. (eds.) TSD 2014. LNCS, vol. 8655, pp. 539–546. Springer, Cham (2014). doi:10.1007/978-3-319-10816-2_65
6. dos Santos Silva, D., Paraboni, I.: Generating spatial referring expressions in interactive 3D worlds. Spat. Cogn. Comput. **15**(03), 186–225 (2015)
7. Krahmer, E., van Deemter, K.: Computational generation of referring expressions: a survey. Comput. Linguis. **38**(1), 173–218 (2012)
8. Dale, R., Reiter, E.: Computational interpretations of the Gricean maxims in the generation of referring expressions. Cogn. Sci. **19**(2), 233–263 (1995)
9. John, O.P., Donahue, E.M., Kentle, R.L.: The big five inventory - versions 4a and 54. University of California, Berkeley, CA, Technical report (1991)
10. Iacobelli, F., Gill, A.J., Nowson, S., Oberlander, J.: Large scale personality classification of bloggers. In: D'Mello, S., Graesser, A., Schuller, B., Martin, J.-C. (eds.) ACII 2011. LNCS, vol. 6975, pp. 568–577. Springer, Heidelberg (2011). doi:10.1007/978-3-642-24571-8_71
11. Mairesse, F., Walker, M.A.: Controlling user perceptions of linguistic style: trainable generation of personality traits. Comput. Linguist. **37**(3), 455–488 (2011)
12. Gatt, A., van der Sluis, I., van Deemter, K.: Evaluating algorithms for the generation of referring expressions using a balanced corpus. In: Proceedings of ENLG-07 (2007)
13. Teixeira, C.V.M., Paraboni, I., da Silva, A.S.R., Yamasaki, A.K.: Generating relational descriptions involving mutual disambiguation. In: Gelbukh, A. (ed.) CICLing 2014. LNCS, vol. 8403, pp. 492–502. Springer, Heidelberg (2014). doi:10.1007/978-3-642-54906-9_40
14. Paraboni, I., Galindo, M., Iacovelli, D.: Stars2: a corpus of object descriptions in a visual domain. Lang. Res. Eval. **51**(2), 495–524 (2016)
15. Paraboni, I., Lan, A.G.J., de Sant'Ana, M.M., Coutinho, F.L.: Effects of cognitive effort on the resolution of over specified descriptions. Comput. Linguist. **43**(2), 273–310 (2017)
16. Ferreira, T.C., Paraboni, I.: Classification-based referring expression generation. In: Gelbukh, A. (ed.) CICLing 2014. LNCS, vol. 8403, pp. 481–491. Springer, Heidelberg (2014). doi:10.1007/978-3-642-54906-9_39
17. Dale, R.: Cooking up referring expressions. In: Proceedings of the 27th Annual Meeting of the Association for Computational Linguistics, pp. 68–75 (2002)

Big Five Personality Recognition from Multiple Text Genres

Vitor Garcia dos Santos, Ivandré Paraboni$^{(\boxtimes)}$,
and Barbara Barbosa Claudino Silva

School of Arts, Sciences and Humanities, University of São Paulo, São Paulo, Brazil
{vitor.garcia.santos,ivandre,barbara.barbosa.silva}@usp.br

Abstract. This paper investigates which Big Five personality traits are best predicted by different text genres, and how much text is actually needed for the task. To this end, we compare the use of 'free' Facebook text with controlled text elicited from visual stimuli in descriptive and referential tasks. Preliminary results suggest that certain text genres may be more revealing of personality traits than others, and that some traits are recognisable even from short pieces of text. These insights may aid the future design of more accurate models of personality based on highly focused tasks for both language production and interpretation.

Keywords: Big Five · Personality recognition

1 Introduction

Knowing the personality traits of an individual (e.g., a social network user) enables a broad range of content personalisation strategies, such as presenting customised results in a search engine, or generating more effective advertisement. Accordingly, computational models of personality recognition have been the focus of a number of NLP studies in recent years [1], and are also the focus of the present work.

Essential personality traits are observable in many instances of human behaviour and, of particular interest for NLP, in language use. This observation has led to a framework that is now a standard in Psychology and related fields - the Big Five model - comprising five dimensions of human personality: *Extraversion, Neuroticism, Agreeableness, Conscientiousness* and *Openness to experience.*

The recognition of Big Five traits from text is now an established computational task [1], and its results tend to vary according to both the text genre under consideration (e.g., some personality traits are more evident in certain text genres than others), and according to the amount of text made available for the task (e.g., a system that attempts to recognise personality from Facebook text may be unhelpful if the intended individual is not a very active Facebook user.) Despite these variations, however, best-known approaches to personality recognition are usually based on free (or non-topic specific) text obtained from essays [2,3], personal blogs [4–6], Facebook [7,8] or Twitter [9,10].

© Springer International Publishing AG 2017
K. Ekštein and V. Matoušek (Eds.): TSD 2017, LNAI 10415, pp. 29–37, 2017.
DOI: 10.1007/978-3-319-64206-2_4

The use of free text for personality recognition is arguably well motivated both in psychological and computational terms. Free text may however be inadequate for the task when more fine-grained mappings from linguistic forms to personality traits are needed (e.g., in order to build personality-dependent models of language production, cf. [3]), or simply when such text sources are not available. Situations of this kind give rise to the question of which text genres are more suitable for the recognition of each personality trait, and how much text is actually needed.

As a means to investigate possible alternatives to free text, this work compares supervised models of personality recognition built from Facebook text with models built from controlled text elicited in a descriptive and in a referential task. Preliminary results suggest that some text genres are more revealing of certain personality traits than others, and that some traits may be recognisable even from short pieces of text. These insights may aid the future design of more accurate models of personality based on highly focused tasks for both language production and interpretation.

2   Related Work

Big Five personality recognition from text has become a popular research topic in recent years [1]. Existing work usually makes use of machine learning methods to model the problem as a classification task (e.g., to decide whether an individual is an extrovert or not), regression (e.g., to determine the scalar value of the Big Five Extraversion class) or ranking (e.g., in order to rank individuals based on their degree of Extraversion) [3]. In what follows we briefly discuss some of these alternatives.

One of the first studies on Big Five personality recognition from text is the work in [2]. The study presented an experiment focused on the English language involving 2263 essays written by 1200 students who had responded a Big Five inventory. The experiment was limited to the extreme of the scale for Extraversion and Neuroticism. Words were clustered in four categories: function, conjunctive, modality and appraisal words. Individual texts were represented by the relative frequency of each category, and the binary classes representing Extraversion and Neuroticism were estimated using SVMs. Results show overall accuracy of up to 58%.

The work in [3] addresses the issue of personality recognition by making use of 88 LIWC [11] word categories and additional knowledge sources (e.g., psycholinguistic attributes such as concreteness, age of acquisition of words etc.) in order to recognise Big Five personality traits from an expanded version of the set of essays considered in [2], and also from a speech corpus. Personality recognition is modelled as classification, regression and ranking tasks for all Big Five dimensions, with accuracy ranging from 50% to 62% when using SVM models.

Studies as in [2,3] make use of lexical knowledge to estimate personality traits, usually provided by the LIWC dictionary or similar resources. By contrast, studies as in [4,6] make use of n-gram statistics. The goal in [4,6] is to classify individuals of extreme personality scores for four of the Big Five traits (the exception is Openness, which was disregarded.) The classification task makes use of both Naive-Bayes and SVM models. The work in [4] makes use of 71 personal blogs and achieves accuracy from 45% (random baseline) to 100% (depending on the n-gram model under consideration and on how the classes are defined). In [6] this procedure is repeated by using a much larger data set (1672 blogs) with maximum accuracy of 65%.

In addition to the above early approaches, a large number of recent studies of personality recognition from Twitter, based on a shorter (10-questions) version of the Big Five inventory, have been discussed in the light of the CLEF author profiling shared task in [1]. These include the work in [12], which combines second order representation with latent semantic analysis, the work in [13], which makes use of char and POS n-grams, and the work in [14], which makes use of TF-IDF n-grams and stylistic features.

3   Current Work

Existing work in personality recognition from text as discussed in the previous section usually makes use of free (non-topic specific) text such as essays [2,3], personal blogs [4–6], Facebook [7,8] or Twitter [1,9,10]. However, when this kind of text source is not available, or when more fine-grained mappings from text features to personality traits are required (e.g., to build language production models), it may be necessary to consider the use of controlled text instead. To investigate this, we designed an experiment to recognise Big Five personality traits from a number of alternative text genres. This involved collecting data (Big Five personality inventories and text in different genres) from human subjects, and then building a series of computational models of personality recognition from each text genre.

3.1   Data Collection

Data collection involved requesting Facebook users to respond a standard 44-item self-report Big Five inventory [15] in Portuguese, and then collecting two kinds of text: free text published by the subjects on Facebook, and controlled text obtained from descriptive and referential language production tasks in a closed domain.

Big Five personality inventories and Facebook text were produced by 110 subjects who acted as volunteers, and who gave us explicit consent for the use of their data. Subjects were on average 25.4 years-old and nearly half (58, or 52.7%) were female. All participants were native speakers of Brazilian Portuguese. The distribution of personality traits among the 110 subjects is illustrated in Fig. 1.

Both personality inventories and free text were obtained from Facebook with the aid of a purpose-built application intended to present the 44-item Big Five Inventory, and to save the participant's status updates (up to 1,000 per user) to a

Fig. 1. Personality distribution (110 subjects).

text database. Controlled texts in both descriptive and referential tasks were collected through in-person experiments involving the same group of participants. The scene description task made use of 10 emotionally-charged images (which are arguably more likely to reveal possible differences across personality types) obtained from the GAPED image database [16]. Figure 2 shows an example of stimulus image for this task.

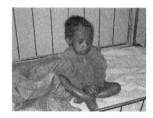

Fig. 2. A stimulus image for the description task, taken from [16].

Stimulus images ranged from 3 to 54° of valence selected at regular intervals. Subjects were requested to describe each scene as if they were talking to a hypothetical visually-impaired colleague. As a means to compare the use of long and short texts in a same, closed domain, the task required the scenes to be described in two ways: first by providing as much detail as possible to produce a multisentential text description, and then by summarising it as a single sentence (as if producing a picture caption.)

In addition to the descriptive task, subjects were also requested to perform a referential task to uniquely identify a certain target object within a context containing a number of distractor objects, as in standard data collection tasks for, e.g., corpus-based referring expression generation [17–19].

The task made use of scenes displaying photographic human faces obtained from the Face Place image database [20] interleaved with scenes displaying abstract Greeble entities[1]. Figure 3 shows examples of the two kinds of stimulus images.

---

[1] Face Place and Greeble images are courtesy of Michael J. Tarr, Center for the Neural Basis of Cognition and Department of Psychology, Carnegie Mellon University.

Fig. 3. Stimulus images for human face (left) and Greeble (right) referential tasks. (Color figure online)

Human faces were selected so as to depict positive and negative emotions, and showing multiple ethnic or racial features. Greebles, on the other hand, are abstract objects studied in face recognition, and are notoriously difficult to describe (let alone unambiguously, as required in the present task.) In both cases, subjects were shown stimulus images representing a referential context with the target highlighted in a red frame and one or more distractor objects, and were requested to inform which person (or object) was highlighted. Examples of possible responses include "the blonde girl who is not smiling" and 'the red thing with pointy beaks', among many others.

Table 1 provides descriptive statistics for the five genres of collected data - free text, descriptive (multi- and monosentential) and referential (human faces and Greebles).

Table 1. Text genre descriptive statistics

Text genre	Sentences	Words
Free text (Facebook posts)	21,392	241,522
Descriptive (monosentential)	9,898	160,084
Descriptive (multisentential)	13,637	255,195
Referential (human faces)	12,537	194,261
Referential (Greebles)	11,218	182,266

3.2 Personality Recognition

Using the five text genres and accompanying Big Five scores obtained from the personality inventories described in the previous section, we would like to investigate which text genres are more revealing of each Big Five personality trait. More specifically, we would like to shed light on the following questions:

$q1$ Whether free (Facebook) text outperforms controlled text.
$q2$ Whether multisentential descriptive text outperforms monosentential text.
$q3$ Whether descriptive text outperforms referential text.
$q4$ Whether Greeble reference outperforms face reference.

Question $q1$ is motivated by existing work on personality recognition, which generally make use of free text from blogs, social networks, etc. In particular, we would like to determine what alternatives to free text may be useful for the task.

Question $q2$ is motivated by our interest in determining how much text is actually needed for personality recognition. Although making use of a larger amount of text is likely to improve results, we would like to know whether it may be still possible to obtain comparable results from shorter pieces of text as well.

Questions $q3$ and $q4$ are motivated by our interest in finding ideal text sources for personality recognition. Question $q3$ is based on the assumption that the greater variety of visual elements to be considered in the picture description task (cf. previous Fig. 2) should be more revealing of personality than the somewhat simpler referential task (cf. previous Fig. 3). Moreover, question $q4$ hypothesises that abstract Greeble objects are in principle more difficult to describe and, as a result, may be more revealing of differences of personality than standard face descriptions.

In order to investigate these questions, we built supervised models of personality recognition from each individual text genre. We notice however that these models are not intended to provide state-of-art results for the task (which in any case would not be available for our target language - Brazilian Portuguese), and that the present work is solely focused on the comparison between alternative text genres based on their surface form, and not on semantics [21].

A set of 155 learning features was considered, including both standard feature definitions found in the literature and additional definitions motivated by the particular language style of Facebook status updates. These comprise 31 features computed during text normalisation, 64 LIWC features [11] and 60 features obtained from a lexicon (e.g., word classes, gender, number and tense information etc.) All features were computed as word counts and normalised by document size.

Taking the 155-feature set as an input, personality recognition is presently modelled as a binary classification task to determine whether an individual speaker shows positive or negative tendency towards each trait. To this end, we assigned positive/negative class labels based on the average score for each trait, namely, individuals with above-average score for a personality trait t make positive instances of t, and the other individuals make negative instances.

4 Results

We built recognition models for each of the five personality traits and for each of the five text genres. As a means to compare free and controlled text in general, further five models (one for each trait) were built from the combination of all sources of controlled text as well. This makes 30 models (5 personality dimensions $*$ 6 text sources) in total.

All models use decision-tree induction with 10-fold cross-validation over the entire dataset. Resulting F1 scores are summarised in Table 2. Since all classes

Table 2. F1 score results. Highest scores for each personality trait are highlighted.

Model	Extrav.	Agreeabl.	Conscient.	Neurot.	Openness
1. Free text (Facebook posts)	**0.64**	0.50	0.45	0.45	**0.54**
2. Descriptive (monosentential)	0.59	0.45	0.43	0.45	0.48
3. Descriptive (multisentential)	0.54	0.47	0.49	0.42	0.43
4. Referential (faces)	0.55	0.50	**0.63**	0.47	**0.54**
5. Referential (Greebles)	0.53	**0.59**	0.46	**0.51**	0.53
6. All controlled text (2..5)	0.56	0.44	0.40	0.45	0.46

are reasonably well-balanced, we notice that simply selecting the majority class would obtain considerable lower F1 scores than any of the current classifiers.

From these results the following observations are warranted. First, we notice that different personality traits are best predicted by different text genres. In particular, Extraversion and Openness were best predicted by models built from Facebook texts (#1), although in the latter case there is a tie with #3 (reference to faces.) For the other traits, best results were obtained by models built from face referring expressions (#4 and #5). Interestingly, models built from descriptive text (#2 and #3) never produced top results.

Regarding our research question $q1$ (the use of free versus controlled text), we notice that models built from Facebook text (#1) outperform those built from controlled text (#6) in the recognition of all traits but Neuroticism, in which case there is a tie.

Regarding $q2$ (the use of mono- versus multisentential descriptive text) we notice that despite the larger amount of text available from multisentential descriptions (#3), these models were actually outperformed by the use of monosentential descriptions (#2) in the recognition of Extraversion, Neuroticism and Openness.

For $q3$ (the use of descriptive versus referential text) we consider the mean results (not shown) for descriptive text #2 and #3 compared to the mean results (not shown) for referential text #4 and #5. Referential text outperforms descriptive text in all cases except for Extraversion (with top results divided between face and Greeble genres).

Finally, for $q4$ (reference to faces versus Greebles) we notice that Greebles (#5) outperform face descriptions (#4) in the recognition of Agreeableness and Neuroticism only, and that face descriptions provide better predictions for the other traits.

5   Discussion

This paper has presented an investigation of Big Five personality recognition from multiple text genres, focusing on four main questions: the use of free versus controlled text, monosentential versus multisentential description, descriptive versus referential text, and human face versus Greebles reference.

Some of our current results confirm findings from previous work. In particular, the use of free (Facebook) text in the personality recognition task generally outperforms the controlled text alternatives under consideration. The difference is however relatively small, and in some cases best results were observed when using controlled text. This was particularly the case when using face and/or Greeble referring expressions. As future work, we intend to refine these experiments and investigate which particular aspects of referential behaviour may help predict personality.

An interesting outcome of our experiments is the overall positive results obtained by the use of referring expressions as a text source for personality recognition. Descriptions of both human faces and Greebles were reasonably good predictors of Agreeableness and Conscientiousness, which were precisely the two weakest results obtained from Facebook text. This may be explained by the fact that our stimuli for this task was emotionally-charged in a consistent fashion, whereas Facebook text combines many kinds of emotionally-charged and neutral contents. A study on automatic personality recognition based on short text elicited in highly focused tasks is left as future work.

Other results, on the other hand, may seem less expected given the literature on personality recognition. This is the case of the model based on monosentential text, which turned out to outperform the model based on multisentential text for the same (descriptive) genre. The use of larger text seems to add considerable noise to the model, an observation that is also partially supported by the comparison between descriptive and referential text. Since text sizes in both genres were approximately equal, the fact that referential text always outperforms descriptive text should be explained by differences in content. Further investigation on this issue is also left as future work.

Acknowledgements. This work has been supported by FAPESP grant 2016/14223-0.

References

1. Rangel, F., Celli, F., Rosso, P., Potthast, M., Stein, B., Daelemans, W.: Overview of the 3rd author profiling task at PAN 2015. In: CLEF 2015 Evaluation Labs and Workshop (2015)
2. Argamon, S., Dhawle, S., Koppel, M., Pennebaker, J.W.: Lexical predictors of personality type. In: The Joint Annual Meeting of the 37th Interface Symposium and the CSNA (2005)
3. Mairesse, F.: Learning to adapt in dialogue systems: data-driven models for personality recognition and generation. Ph.D. thesis, University of Sheffield (2008)
4. Oberlander, J., Nowson, S.: Whose thumb is it anyway? Classifying author personality from weblog text. In: COLING/ACL 2006 Poster Sessions, Sydney, Australia, pp. 627–634 (2006)
5. Gill, A.J., Nowson, S., Oberlander, J.: What are they blogging about? Personality, topic and motivation in blogs. In: ICWSM-2009, pp. 18–25. The AAAI Press (2009)
6. Nowson, S., Oberlander, J.: Identifying more bloggers: towards large scale personality classification of personal weblogs. In: International Conference on Weblogs and Social Media (2007)

7. Celli, F.: Adaptive personality recognition from text. Ph.D. thesis, University of Trento (2012)

8. Farnadi, G., Zoghbi, S., Moens, M.F., de Cock, M.: Recognising personality traits using Facebook status updates. In: Workshop on Computational Personality Recognition (2013)

9. Plank, B., Hovy, D.: Personality traits on Twitter - or - how to get 1,500 personality tests in a week. In: Proceedings of WASSA-2015, pp. 92–98 (2015)

10. Iacobelli, F., Gill, A.J., Nowson, S., Oberlander, J.: Large scale personality classification of bloggers. In: D'Mello, S., Graesser, A., Schuller, B., Martin, J.-C. (eds.) ACII 2011. LNCS, vol. 6975, pp. 568–577. Springer, Heidelberg (2011). doi:10.1007/978-3-642-24571-8_71

11. Pennebaker, J.W., Francis, M.E., Booth, R.J.: Inquiry and Word Count: LIWC. Lawrence Erlbaum, Mahwah (2001)

12. Álvarez-Carmona, M., López-Monroy, A., Montes-y-Gómez, M., Villaseñor-Pineda, L., Escalante, H.: INAOE's participation at PAN 2015: author profiling task. In: CLEF 2015 (2015)

13. González-Gallardo, C., et al.: Tweets classification using corpus dependent tags, character and POS N-grams. In: CLEF 2015 (2015)

14. Şulea, O.M., Dichiu, D.: Automatic profiling of Twitter users based on their tweets. In: CLEF 2015 (2015)

15. John, O.P., Donahue, E.M., Kentle, R.L.: The big five inventory - versions 4a and 54. University of California, Berkeley, CA, Technical report (1991)

16. Dan-Glauser, E.S., Scherer, K.R.: The Geneva affective picture database (GAPED): a new 730-picture database focusing on valence and normative significance. Behav. Res. Methods **43**(2), 468–477 (2011)

17. Teixeira, C.V.M., Paraboni, I., da Silva, A.S.R., Yamasaki, A.K.: Generating relational descriptions involving mutual disambiguation. In: Gelbukh, A. (ed.) CICLing 2014. LNCS, vol. 8403, pp. 492–502. Springer, Heidelberg (2014). doi:10.1007/978-3-642-54906-9_40

18. Paraboni, I., Galindo, M., Iacovelli, D.: Stars2: a corpus of object descriptions in a visual domain. Lang. Resour. Eval. **51**(2), 49–62 (2016)

19. Ferreira, T.C., Paraboni, I.: Generating natural language descriptions using speaker-dependent information. Nat. Lang. Eng. 1–22 (2017). doi:10.1017/S1351324917000079

20. Righi, G., Peissig, J.J., Tarr, M.J.: Recognizing disguised faces. Vis. Cogn. **20**(2), 143–169 (2012)

21. de Lucena, D.J., Paraboni, I., Pereira, D.B.: From semantic properties to surface text: the generation of domain object descriptions. Inteligencia Artificial. Revista Iberoamericana de. Inteligencia Artificial **14**(45), 48–58 (2010)

Automatic Classification of Types of Artefacts Arising During the Unit Selection Speech Synthesis

Jiří Přibil[1,2(✉)], Anna Přibilová[3], and Jindřich Matoušek[1]

[1] Department of Cybernetics, Faculty of Applied Sciences,
University of West Bohemia, Univerzitní 8, 306 14 Plzeň, Czech Republic
Jiri.Pribil@savba.sk, jmatouse@kky.zcu.cz
[2] Institute of Measurement Science, Slovak Academy of Sciences,
Dúbravská cesta 9, 841 04 Bratislava, Slovakia
[3] Faculty of Electrical Engineering and Information Technology,
Institute of Electronics and Photonics, Slovak University of Technology
in Bratislava, Ilkovičova 3, 812 19 Bratislava, Slovak Republic
Anna.Pribilova@stuba.sk

Abstract. The paper describes an experiment with automatic classification of the basic types of artefacts in the synthetic speech produced by the Czech text-to-speech system using the unit selection synthesis method. The developed classifier based on the Gaussian mixture models (GMM) is solved finally as the open-set classification task due to a limited database of speech artefacts resulting from incorrectly chosen or exchanged speech units during the synthesis process. The realized experiments prove principal impact of the accuracy of determination of the speech artefact section on the final precision of the artefact type classification. From the auxiliary investigations follows a relatively great influence of the number of mixtures and the type of a covariance matrix on the output artefact classification error rate as well as on the computational complexity.

Keywords: Quality of synthetic speech · Text-to-speech system · GMM classification · Statistical analysis

1 Introduction

At present, various methods of speech synthesis are implemented in the text-to-speech (TTS) systems. In each of them, the aim is to obtain the maximum quality, comprehensibility, intelligibility, and naturalness of the synthetic speech as much as possible. But during the synthesis process artefacts of different origin

The work was supported by the Czech Science Foundation GA16-04420S (J. Matoušek, J. Přibil), by the Grant Agency of the Slovak Academy of Sciences 2/0001/17 (J. Přibil), and by the Ministry of Education, Science, Research, and Sports of the Slovak Republic VEGA 1/0905/17 (A. Přibilová).

K. Ekštein and V. Matoušek (Eds.): TSD 2017, LNAI 10415, pp. 38–46, 2017.
DOI: 10.1007/978-3-319-64206-2_5

can occur in the finally generated speech. In the often used concatenative synthesis of the output speech production [1], any concatenation point is a potential source of an audible artefact. Such artefacts may be caused by the wrong annotation and/or segmentation of the built speech corpus of the natural speech and are related mainly to the fundamental frequency (F0) and the energy discontinuities [2].

The artefact detection, localization, and classification is usually performed by the manual determination of the artefact type, and visual identification of the beginning and ending frames of the artefact, and by comparing the clean original speech signal and the same sentence with the artefact to determine the type and the localization of the artefact. In this way obtained results must be subsequently confirmed by several independent annotators using the listening tests. This technique is very challenging and time consuming, and in addition, it is heavily affected by the human subjectivity during the evaluation process. Therefore, we try to develop an automatic method using the objective criterion that works with good accuracy and without any human interaction. Among the objective methods, the automatic speech recognition system yielding the final evaluation in the form of the recognition score can be used [3]. These systems are often used for speaker recognition or identification, and are usually based on hidden Markov models [4], support vector machines (SVM) [5], or Gaussian mixture models (GMM) [6,7].

The experiments described in this paper are, first of all, performed to verify functionality of the proposed automatic GMM-based classifier for artefact type determination in the synthetic speech produced by the unit selection (USEL) method [8]. Next, the correctness of selection of a region of interest (ROI) with the artefact inside the tested sentence was checked for its influence on accuracy and stability of the classification results. In addition, the dependence of error rate of artefact classification (ERAC) on the number of used GMM mixture components and the method of the covariance matrix calculation was analysed. Finally, the computational complexity (CPU processing time) was evaluated and compared.

2   Method

The process of analysis of the input sentence starts with determination of speech spectral and prosodic features (see the block diagram in Fig. 1) which are subsequently used for the first-step ANOVA-based detection whether there is any speech artefact. In the next step, if the sentence is marked as an artefacted one, the second type of speech features analysis is performed yielding another types of spectral and prosodic parameters. These features are then used for artefact localization with the help of the trained GMM models of the starting/ending parts and the bodies of the artefacts. After localization of the artefact position, the nearest ROI is determined for further processing. In the third step, the united GMM models of the starting, ending, and body parts are used for final classification of the artefact type which is processed only inside the selected ROI of the input tested sentence. The artefact detection as well as the artefact localization problem was successfully solved in [9,10], so the present work is focused on the last phase – the artefact type classification.

Our previous experience has shown that the TTS system based on the USEL synthesis method can principally generate up to six types of speech artefacts:

1. local increase of the signal RMS (energy) – $Artf_{C1}$,
2. local decrease of the signal RMS – $Artf_{C2}$,
3. local increase of F0 – $Artf_{C3}$,
4. local decrease of F0 – $Artf_{C4}$,
5. superposition of the local energy and F0 increase/declination – $Artf_{C5}$,
6. incorrectly chosen or exchanged speech units from the database – $Artf_{C6}$.

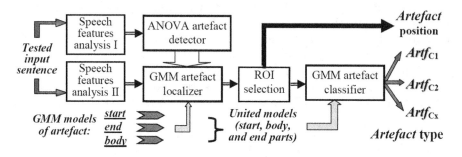

Fig. 1. Block diagram of the whole system for artefact detection, localization, and classification.

While the artefact classes 1–5 occur relatively often, the corresponding change in prosodic and spectral parameters is well defined and consequently the classification can be carried out with a relatively high precision, in the last class the principal problem of a different context of the artefact each time results in insufficient amount of reference data for GMM model creation and training, etc. Therefore, it was necessary to create of the GMM-based classifier in the open set with the last 6th class containing all artefacts that had not been classified as the types 1–5. The modified structure of the GMM classifier is depicted in Fig. 2.

Principally, the GMMs represent a linear combination of multiple Gaussian probability distribution functions of the input data vector. The covariance matrix and the vector of means together with the weighting parameters must be determined from the input training data [11]. In general, covariance matrices of three types may be used: spherical, diagonal, and full. For correlated elements of the feature vectors there is necessity for a relatively high number of components and a sufficient approximation can be achieved only with the full covariance matrix. However, the lower computational complexity of the diagonal covariance matrix justifies its usage in the speaker identification.

In our experiment, the united GMM models are trained on the speech signal consisting of the starting part, the body of the artefact, and the ending part including also $\pm i$ frames in the left/right vicinity of the ROI part. During the expectation-maximization iteration process, the maximum likelihood function of GMM is found for every artefact type and voice (male/female). This process is controlled by the number of used mixtures (N_{MIX}) and the number of iteration steps. In the classification phase, the input feature vectors from the tested sentence are compared in parallel with the three trained GMM models, so we obtain three output vectors of the normalized score. These output scores are next analysed to determine the maximum overall probability in the discriminator block performing basic classification to M output classes. One of them is assigned to each of the processed speech feature vectors, then the class distribution based on histograms is constructed, and the maximum occurrence is determined as shown in Fig. 3. The final classification block works with $M + 1$ output classes – the virtual class *"Exchange"* is added to the basic closed set of M artefact types to create the open-set speaker identifier. The classification strategy is based on the consideration that when the class distribution is relatively flat (without a dominant class), the final classification of the whole tested sentence is set to the *"Exchange"* class – see an example in Fig. 3b. Practically, the maximum occurrence is compared with the threshold $Tresh_0$ given as a ratio between the number of the currently processed frames (length of ROI) and the number of basic classes (M).

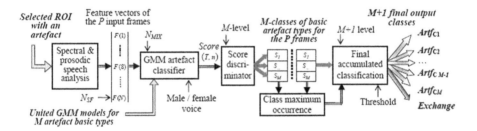

Fig. 2. Block diagram of the proposed GMM-based open-set classifier for identification of the artefact type after its detection and localization.

In speaker recognition tasks the use of spectral features like mel-frequency cepstral coefficients (MFCC) together with prosodic parameters is the most common [3,5–7,11]. In this experiment, first, the fundamental frequency and the speech signal energy are determined in each segment of the input sentence and the prosodic parameters like microintonation, jitter, shimmer, etc. are derived. Next, the smoothed spectral envelope and the power spectral density are computed for determination of the spectral features. As the relative position of the formants and the formant trajectories can be used as the main indicator for voiced speech description, the first two formant positions and their ratios are also used. Every vector of P speech features is subsequently processed to obtain N_{SF} representative statistical values (mean, relative minimum/maximum, skewness, kurtosis, etc.) that are used in the process of GMM training and classification.

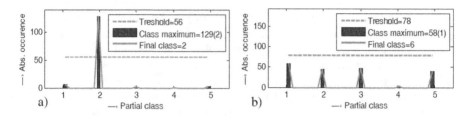

Fig. 3. Histograms of absolute occurrence of the GMM scores for 5 basic output classes including the threshold and the finally determined class: with one class maximum (a), without any dominant class - the virtual 6th class *"exchange"* is set (b).

3   Material, Experiments, and Results

The synthetic speech produced by the Czech TTS system using the USEL synthesis method [8] was used in this artefact classification experiments. The main speech corpus was divided into two groups of 40 declarative sentences of male/female voices generated by the synthesizer. The first group comprises the sentences without any audible artefact designated as *"clean"*; the second group consists of another 40 sentences of the same voices with just one speech artefact in each sentence. The groups in the database are parallel – the clean ones and those with artefacts correspond to the exactly same input text that was generated by the TTS system using the male and female voices. All the sentences were sampled at 16 kHz and their duration was from 2.5 to 5 s. The derived database consisting of ROI parts of the artefacted sentences was used for training of the GMM models to classify the artefact type. The k-fold cross-validation method [11] was employed during the training and the testing processes to obtain independence within the male/female voice. Practically $k = 4$, so the groups of sentences were divided using the ratio of 3:1 – three for training and one for testing/classification.

The main performed experiment consists in testing and verifying whether the proposed automatic GMM-based classifier of the artefact type is principally correct and produces values of sufficiently low ERAC – see the results in the form of 3D confusion matrices in Fig. 4. In the next step, the comparison was realized to show the influence of correct selection of the ROI part with the artefact inside the tested sentence – see the numerical results in Table 1. To obtain high accuracy among seven artefact type classes together with high stability of the results for this open-set classification task, our further analysis was aimed at investigation of:

– influence of different methods of calculation of the covariance matrix for the GMM training $G_{CMtype} = \{\text{'Spherical', 'Diagonal', and 'Full'}\}$ – see detailed values per class type and voice gender in Fig. 5 and numerical results in Table 2,
– influence of the number of applied Gaussian mixtures on the mean ERAC values for $N_{MIX} = \{4, 8, 16, \text{and } 32\}$ – see the left part of Table 2,

– comparison of the computational complexity: CPU times of the GMM training and classification phases for different number of used mixtures and different types of covariance matrices summarized for both voice genders in Fig. 6.

For speech feature analysis, the artefact processing neighbourhood of $\pm i$ frames (before the beginning part and after the ending part) was set to $i = 11$ in correlation with the results presented in [9]. According to the research results published in [9,12] the length of the input feature vector was set to $N_{SF} = 16$ and for determination of the dominant class inside the open-set classification process the threshold was set experimentally to $1.2 * Tresh_0$ (i.e. adding 20 % to the basic level given by the calculated P/M ratio). In all cases, the ROI was selected manually for further comparison and evaluation of the ERAC calculated from the number X_C of the sentences with the correctly determined artefact class and the total number N_T of the tested sentences as $ERAC = (1 - X_C/N_T) * 100$ [%].

The described analysis and speech signal processing were currently realized in the Matlab environment (ver. 2012a) and the Ian T. Nabney *"Netlab"* pattern analysis toolbox [13] was used for implementation of the basic functions for the proposed GMM classifier. The computational complexity was determined using the UltraBook with the following configuration: processor Intel(R) Intel i5-4200U at 2.30 GHz, 8 GB RAM, and Windows 10 (64-bit) OS.

Table 1. Comparison of dependence of the mean ERAC [%] on correctness of the ROI artefact part selection inside the tested sentence; summarized for both genders, $N_{MIX} = 16$, $G_{CMtype} = Full$.

Selection/ERAC	$Artf_{C1}$	$Artf_{C2}$	$Artf_{C3}$	$Artf_{C4}$	$Artf_{C5}$	C6-Exc	Summary
ROIs-correct	0	0	0	0	25	20	7.5
Full sentences	62	51	57	48	72	85	62.5
ROIs-incorrect	100	50	100	100	80	50	80

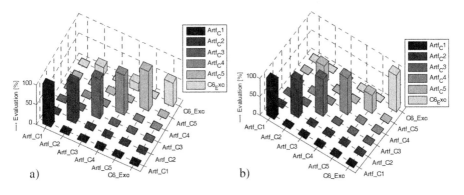

Fig. 4. Visualization of the obtained artefact type classification results: 3D confusion matrix for a male (a) and a female voice (b); $N_{MIX} = 16$, $G_{CMtype} = Full$.

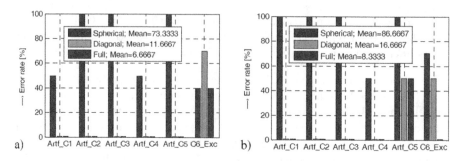

Fig. 5. Detailed comparison of the influence of the used covariance matrix type on the resulting ERAC [%]: for a male voice (a), for a female voice (b); $N_{MIX} = 16$ (diagonal and full covariance matrices yield ERAC of 0% in the first five/four classes for male/female voices).

Table 2. Comparison of the mean ERAC in [%] and its standard deviation (in parentheses) depending on the number of used GMM mixtures and on the type of the covariance matrix.

Voice/ERAC	$N_{\mathrm{MIX}}(G_{\mathrm{CMtype}} = \text{Full})$				$G_{\mathrm{CMtype}}(N_{\mathrm{MIX}} = 16)$		
	4	8	16	32	Spherical	Diagonal	Full
Male	20.4 (24.7)	8.1 (13.3)	6.7 (16.3)	6.7 (16.3)	63.3 (29.4)	11.7 (28.5)	6.7 (16.3)
Female	21.6 (34.8)	16.7 (25.8)	8.3 (20.4)	11.7 (20.4)	76.7 (21.6)	16.7 (25.8)	8.3 (20.4)
Summary	**21**	**12.4**	**7.5**	**9.2**	**70**	**14.2**	**7.5**

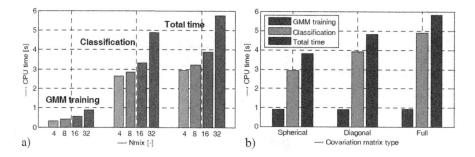

Fig. 6. Comparison of the computational complexity (CPU time [s]): for different number of Gaussian mixtures using $G_{CMtype} = Full$ (a), for different types of covariance matrices when $N_{MIX} = 16$ (b); summarized for both genders.

4   Discussion and Conclusions

The performed experiments have confirmed that the proposed automatic GMM-based classifier of the artefact type is principally correct and produces the results comparable with those attained by manual determination method. Next, from the realized analysis follows that there exist a principal influence of the accuracy setting of the ROI containing the whole classified speech artefact on the precision of the artefact type classification. If the ROI with the artefact is set incorrectly, the output error rate rapidly increases up to 100% making the whole artefact detection, localization, and classification system useless. In the case that the ROI is not set, i.e. the full sentence is analyzed, the error rate is unacceptable mostly for the last virtual sixth class – compare the ERAC values in Table 1. Therefore, the detailed analysis for correct setting of the threshold for the sixth class determination is necessary.

The further obtained results show significant dependence of the computational complexity for the GMM creation and training as well as for the classification phase on the used number of mixtures. For the maximum of 32 tested mixtures the total CPU time increases more than 2.5 times when compared with 4 mixtures (see the left bar-graph in Fig. 6). As regards the influence of the type of the covariance matrix in the GMM models on the overall CPU time consumption, the maximum value was measured for the 'Full' type and the minimum one for the 'Spherical' matrix – compare the results in the right bar-graph in Fig. 6). Contrary to general expectations, the achieved mean ERAC values do not fall for $N_{MIX} > 16$ – they may even rise as documented in the left part of Table 2. Considering the fact that the 'Spherical' covariance matrix yields practically even 100% error rate in the majority of tested artefact classes (over 70% in total), this type of matrix is not suitable for this classification task. The full covariance matrix and 16 mixtures were finally applied in our experiments to obtain the acceptable computational complexity and good results of the ERAC parameter for both male and female voices.

In the near future, the analysis of dependence of the obtained results on different types and different numbers of speech parameters used in the input feature vectors must be performed. Finally, the computation complexity analysis of the current realization in the Matlab environment revealed that optimization and implementation in a higher programing language is necessary for real-time processing – particularly in the classification phase where the CPU times of about 5 s for the tested sentence are unacceptable.

References

1. Tiomkin, S., Malah, D., Shechtman, S., Kons, Z.: A hybrid text-to-speech system that combines concatenative and statistical synthesis units. IEEE Trans. Audio Speech Lang. Proces. **19**(5), 1278–1288 (2011)
2. Legát, M., Matoušek, J.: Identifying concatenation discontinuities by hierarchical divisive clustering of pitch contours. In: Habernal, I., Matoušek, V. (eds.) TSD 2011. LNCS, vol. 6836, pp. 171–178. Springer, Heidelberg (2011). doi:10.1007/978-3-642-23538-2_22
3. Bello, C., Ribas, D., Calvo, J.R., Ferrer, C.A.: From speech quality measures to speaker recognition performance. In: Bayro-Corrochano, E., Hancock, E. (eds.) CIARP 2014. LNCS, vol. 8827, pp. 199–206. Springer, Cham (2014). doi:10.1007/978-3-319-12568-8_25
4. Bapat, O.A., Fastow, R.M., Olson, J.: Acoustic coprocessor for HMM based embedded speech recognition systems. IEEE Trans. Consum. Electron. **59**(3), 629–633 (2013)

5. Campbell, W.M., Campbell, J.P., Reynolds, D.A., Singer, E., Torres-Carrasquillo, P.A.: Support vector machines for speaker and language recognition. Comput. Speech Lang. **20**(2–3), 210–229 (2006)

6. Matza, A., Bistritz, Y.: Skew Gaussian mixture models for speaker recognition. IET Sign. Process. **8**(8), 860–867 (2014)

7. Dileep, A.D., Sekhar, C.C.: Class-specific GMM based intermediate matching kernel for classification of varying length patterns of long duration speech using support vector machines. Speech Commun. **57**, 126–143 (2014)

8. Tihelka, D., Kala, J., Matoušek, J.: Enhancements of Viterbi search for fast unit selection synthesis. In: Proceedings of Interspeech 2010, Makuhari, Japan, pp. 174–177 (2010)

9. Přibil, J., Přibilová, A., Matoušek, J.: Detection of artefacts in Czech synthetic speech based on ANOVA statistics. In: Proceedings of the 37th International Conference on Telecommunications and Signal Processing, TSP 2014, Berlin, Germany, pp. 414–418 (2014)

10. Přibil, J., Přibilová, A., Matoušek, J.: Experiment with GMM-based artefact localization in Czech synthetic speech. In: Král, P., Matoušek, V. (eds.) TSD 2015. LNCS, vol. 9302, pp. 23–31. Springer, Cham (2015). doi:10.1007/978-3-319-24033-6_3

11. Reynolds, D.A., Rose, R.C.: Robust text-independent speaker identification using Gaussian mixture speaker models. IEEE Trans. Speech Audio Process. **3**, 72–83 (1995)

12. Přibil, J., Přibilová, A.: Evaluation of influence of spectral and prosodic features on GMM classification of Czech and Slovak emotional speech. EURASIP J. Audio Speech Music Process. **2013**(8), 1–22 (2013)

13. Nabney, I.T.: Netlab pattern analysis toolbox, Release 3.3. http://www.aston.ac.uk/eas/research/groups/ncrg/resources/netlab/downloads. Accessed 15 Oct 2015

A Comparison of Lithuanian Morphological Analyzers

Jurgita Kapočiūtė-Dzikienė[1][(✉)], Erika Rimkutė[2], and Loic Boizou[2]

[1] Department of Applied Informatics, Vytautas Magnus University,
Vileikos str. 8, 44404 Kaunas, Lithuania
jurgita.kapociute-dzikiene@vdu.lt
[2] Centre of Computational Linguistics, Vytautas Magnus University,
Putvinskio str. 23, 44243 Kaunas, Lithuania
{erika.rimkute,loic.boizou}@vdu.lt
http://if.vdu.lt/
http://tekstynas.vdu.lt/

Abstract. In this paper we present the comparative research work disclosing strengths and weaknesses of two the most popular and publicly available Lithuanian morphological analyzers, in particular, *Lemuoklis* and *Semantika.lt*. Their lemmatization, part-of-speech tagging, and fined-grained annotation of the morphological categories (as case, gender, tense, etc.) performance was evaluated on the morphologically annotated gold standard corpus composed of four domains, in particular, administrative, fiction, scientific and periodical texts. *Semantika.lt* significantly outperformed *Lemuoklis* by ~1.7%, ~2.5%, and ~8.1% on the lemmatization, part-of-speech tagging, and fine-grained annotation tasks achieving ~98.0%, ~95.3% and, ~86.8% of the accuracy, respectively.

Semantika.lt was also superior on the administrative, fiction, and periodical texts; however, *Lemuoklis* yielded similar performance on the scientific texts and even bypassed *Semantika.lt* in the fine-grained annotation task.

Keywords: Lithuanian morphological analysers · Gold-standard corpus · Experimental evaluation · The Lithuanian language

1 Introduction and Related Work

If excluding so-called isolating languages as Mandarin Chinese which do not have grammatical categories (as case, gender, number, tense, etc.), languages show a varied degree of inflection (and derivation), starting from weakly inflected as English and going to highly inflected as Spanish, Czech, Lithuanian, Turkish, Arabic or Hebrew. In this paper we focus on the fusional Lithuanian language, which has the rich inflectional morphology even complex to Latvian or Slavic languages [20]. Different morphological categories are defined with various endings attached to the stable parts of words (i.e., to a root or to a root with affixes). In highly inflectional languages hundreds of word's forms can be generated from

© Springer International Publishing AG 2017
K. Ekštein and V. Matoušek (Eds.): TSD 2017, LNAI 10415, pp. 47–56, 2017.
DOI: 10.1007/978-3-319-64206-2_6

a single root (e.g., ~2–3 million grammatical forms can be used for a dictionary of ~100 thousand words [9]); moreover, these forms often match other grammatical categories or parts-of-speech. Thus, a rate of ambiguous morphological forms for the Lithuanian language reaches even ~47 [18].

Morphological analysis has experienced a great success since the invention of the Two-Level morphology and the development of the finite-state technology that Two-level formalism is based on [14]. All existing morphological analyzers according to their creation method can be divided into knowledge-based (sometimes called rule-based and/or lexicon-based), supervised, and unsupervised. Despite that unsupervised approaches (segmenting the raw text into morphs as, e.g., *un+fail+ing+ly*) have become very attractive recently (because do not require gold morphological labels and for any language there is an unlimited number of text resources), they, however, are more suitable for agglutinative languages [3]. Knowledge-based approaches rely on rules/lexicons prepared by linguist-experts and do not require additional resources. Probably due to this reasons, this approach is still the most widely spread, thus, used for many different languages: English [11,19], French [7], Russian and Spanish [14], Urdu and Hindi [1,5], Tamil [2], etc.

Corpus-based morphological analyzers are the closest alternative to knowledge-based approaches. Although such systems are already built automatically in the supervised manner, induced rules are based on gold morphological annotations found in the training data. The annotation process itself is very laborious and requires deep language expertise, but such analyzers can be easily redeveloped and improved after adding more annotated texts. Analyzers of this type are used for many languages: Dutch [6], Swahili [17], Hindi [15], Kazakh [12], Arabic [13], Polish [10], etc.

Morphological analysis is important in such NLP applications as information retrieval, parsing or machine translation (especially when translating direction points from/to the morphologically rich language). Each module of such complex system has to be as accurate and reliable as possible (because the overall accuracy depends on cumulative accuracies of separate modules), including the morphological analyzer. The priority is its accuracy, no matter if the analyzer is developed using rule-based or corpus-based approach. The aim of this research is to evaluate, to compare and to determine the most accurate morphological analyzer for the Lithuanian language.

2   The Lithuanian Morphological Analyzers

The Lithuanian language is spoken by only ~3.2 million people world-wide; therefore it is not very attractive for big companies. Nevertheless this field of research has a rather long history. The first prototype of the Lithuanian morphological analyzer was created ~30 years ago and ever since there were several attempts towards creation of the accurate tool coping with the complex Lithuanian morphology. Despite all of those attempts, there are only two reliable morphological analyzers and lemmatizers which are still maintained, updated, and publicly available on-line:

1. *Lemuoklis*[1] at the beginning was purely rule and lexicon-based approach (described in detail by it's founder V. Zinkevičius in [22]), later extended with the statistical approach for the disambiguation of morphological homoforms. In *Lemuoklis* the knowledge about the Lithuanian language is stored in the lexical and grammar database, which contains 6 lexicons with various Lithuanian lexical groups; proper nouns, in the forms they as found in the corpus (that is without lemmatization); the stems of these proper nouns; obsolete and dialectal word forms; forms with the shortened endings which appear in literary and colloquial styles; abbreviations and acronyms. Since the database also contains word stems, each stem can be augmented with the affixes (prefixes, suffixes, endings) which, in turn, are determined according to the word's morphological type (i.e., its morphemic structure). Each analyzed word is divided on the basis of the various scheme options using prefix + stem + postfix pattern, therefore the implemented inflectional models not only recognize different inflectional word forms (including obsolete or dialectal as e.g., illative) of the existent words, but also synthesize some derivatives in their various inflected forms. The lexical database contains ~91 thousand different headwords in total; however, the number of theoretically possible grammatical word forms can reach even several billions.

   Lemuoklis has been used for many practical tasks. One of its first versions was used in preparing the first frequency world list for the Lithuanian language [21], and it was later integrated into the Microsoft Office package and Information Base components and used for the automatic spell checking [22]. In 2000–2005 *Lemuoklis* was applied on ~1 million word corpus, which afterwards was manually corrected by a linguist-expert (for more information see [18]) and led to the creation of the first lemmatized and morphologically annotated gold-standard corpus for the Lithuanian language. This research showed that within the ~89% of all automatically recognized Lithuanian words no less than ~47% are ambiguous. The disambiguation problem was solved out by complementing the rule and lexicon based approach with the statistical trigram Hidden Markov Model method (described in [8]): this version reached ~94% of accuracy for annotation and ~99% for lemma assignment.

2. *Semantika.lt*[2] morphological analyzer was created with the ambition to outperform its ancestor *Lemuoklis*, which still has not got rid of such shortcomings as rather low performance on the proper nouns. The main reason for designing a new tool, was the fact that *Lemuoklis* data was hard coded, which makes difficult to enrich the lexical database. *Semantika.lt* is also based on the hybrid approach: it is based on the Hunspell open source platform (consisting of the lexicon and the affixes) supplemented with the statistical method for the disambiguation task. The information included into the lexicon was taken from the following sources: from the 6th edition of the *Modern*

---

[1] At http://tekstynas.vdu.lt/page.xhtml;jsessionid=C27B0743101187E540CD32D049 8C9887?id=morphological-annotator.

[2] At http://www.semantika.lt/TextAnnotation/Annotation/Annotate.

Lithuanian Dictionary; from the *Corpus of the Contemporary Lithuanian Language* at the Centre of Computational Linguistics of Vytautas Magnus University (∼100 million tokens; ∼600 thousand unique); from the database of the Lithuanian Parliamentary documents (∼400 million tokens; ∼1 million unique); from various public Internet sources. The created analyser resulted in 429 groups of rules; 1,518 explicit tags for flexing/non-flexing properties; 5,832 rules for suffix and affix alternation in 16,734 alternation cases. The total number of headwords in *Semantika.lt* is ∼146 thousand of which ∼38 thousand are common nouns, ∼67 thousand proper nouns, ∼12 thousand adjectives, ∼23 thousand verbs, ∼4 thousand words from other classes. The disambiguation problem as in *Lemuoklis* is solved using statistical trigram Markov model + Viterbi algorithm.

Thus, according to the number of headwords, *Semantika.lt* obviously outperforms *Lemuoklis*; but on the other side, *Lemuoklis* uses the synthesis method in order to handle some frequent derivation patterns (e.g. some regular agentive and diminutive forms). Besides, *Lemuoklis* had been updated in the past, but since 2007 it has not been experimentally evaluated. Moreover, *Semantika.lt* has never been fully evaluated on the basis of a gold-standard corpus. In general, there was no evaluation with explicit methodology. Therefore currently it is not clear which one is more accurate and whether difference in their accuracy is statistically significant.

The contribution of this research is to evaluate both of these analyzers and to compare their results following standard up-to-date methods for tool evaluation. However, that the research would be carried out correctly it is important (1) to equalize experimental conditions for the both analyzers (to evaluate them on the same gold-standard corpus; to equalize their annotation tags; to use the same evaluation metrics); (2) to test them on the unseen corpus which was neither used in the rule or lexicon creation nor in training for the disambiguation problem solving. Besides, we anticipate that the publicly available morphologically annotated gold-standard corpus (presented in this paper) could be treated as the benchmark corpus and used for evaluation and comparison purposes of other existing or forthcoming morphological analyzers.

3   The Comparative Evaluation

The experimental comparison (described in Sect. 3.2) of both Lithuanian morphological analyzers (presented in Sect. 2) was performed on a morphologically annotated gold-standard corpus (described in Sect. 3.1). The issue of the annotation format discordance (in the gold corpus and texts produced by both analyzers) was solved out by converting all formats to one based on the Leipzig glossing rules [4] used in the Universal Dependencies Project[3].

---

[3] More about the Universal Dependencies Project is presented in http://universaldependencies.org/.

3.1 The Gold-Standard Corpus

The first morphologically annotated gold-standard corpus (called MATAS[4]) was prepared by the Centre of Computational Linguistics at Vytautas Magnus University. It contains 1,641,263 words and covers 4 domains, in particular, administrative, fiction, scientific and periodical texts. MATAS was prepared in a semi-automatic manner: the initial annotations were obtained with *Lemuoklis* and afterwards manually verified and corrected by one linguist-expert.

Unfortunately for our experiments we could not take the entire corpus, because some parts of it have already been used in training of the *Semantika.lt* morphological analyzer. Thus, we had to select and annotate additional texts taking into account two important factors: (1) they must not have been used in creation/training of *Lemuoklis* and *Semantika.lt*; (2) the obtained gold-standard corpus has to be balanced (in terms of words) that results would not be biased towards the largest domains. Hence, for experiments we selected texts that contain ~5 thousand words in each domain, resulting ~20 thousand in totals. The statistics about the gold-standard corpus is presented in Table 1.

Table 1. The distribution of total and distinct (in brackets) words over different parts-of-speech and domains in the gold-standard corpus. The *unrecognized words* caption defines foreign language or misspelled Lithuanian words.

Part-of-speech	Administrative	Fiction	Scientific	Periodicals	All domains
Noun	2,102 (911)	1,305 (1,034)	2,158 (1,092)	1,768 (1,105)	7,333 (3,492)
Verb (all forms)	834 (513)	1,098 (869)	773 (506)	1,045 (754)	3,750 (2,306)
Adjective	469 (317)	372 (334)	557 (416)	313 (279)	1,711 (1,234)
Conjunction	364 (14)	497 (26)	355 (22)	340 (24)	1,556 (33)
Pronoun	186 (80)	754 (193)	179 (84)	398 (140)	1,517 (273)
Adverb	159 (74)	411 (175)	153 (85)	252 (120)	975 (302)
Proper noun	151 (34)	29 (27)	362 (190)	422 (239)	964 (462)
Preposition	155(15)	250 (25)	135 (15)	253 (25)	793 (31)
Particle	50 (16)	333 (48)	36 (9)	151 (34)	570 (64)
Numeral	179 (114)	25 (19)	165 (96)	137 (91)	506 (239)
Unrecognized word	89 (22)	11 (10)	105 (37)	114 (38)	319 (94)
Interjection	0 (0)	3 (3)	0 (0)	2 (2)	5 (4)
In total	4,738 (2,110)	5,088 (2,763)	4,978 (2,552)	5,195 (2,851)	19,999 (8,534)

3.2 The Experimental Set-Up and Results

In our experiments we compared the gold annotations with the automatic annotations produced by *Lemuoklis* and *Semantika.lt* and calculated the *accuracy* and *f-score* values.

Moreover, we evaluated if the differences between the results obtained by different morphological analyzers are statistically significant. The evaluation was

---

[4] The annotated corpus can be downloaded from https://clarin.vdu.lt/xmlui/handle/20.500.11821/9.

done using the McNemar test [16] with one degree of freedom at the significance level of $\alpha = 0.05$, meaning that the differences are considered statistically significant if calculated probability density function $p < \alpha$.

The obtained lemmatization accuracies are presented in Fig. 1. The *micro-accuracy* (or *micro-f-score*) values for the parts-of-speech (i.e., coarse-grained morphological information) and the morphological categories (i.e., fine-grained information as case, gender, number, voice, tense, etc.) are summarized in Fig. 2. The *f-score* values distributed over the different parts-of-speech are presented in Table 2.

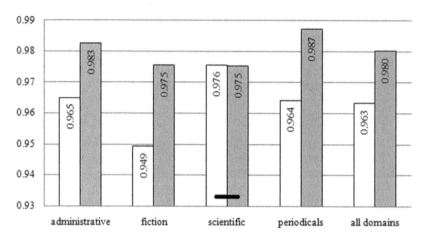

Fig. 1. The lemmatization *accuracies* in white and gray columns for *Lemuoklis* and *Semantika.lt*, respectively. The results which differences are not statistically significant are connected with a solid black line (see the *scientific* domain).

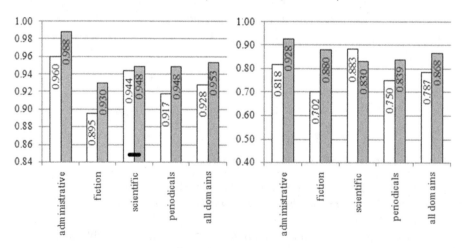

Fig. 2. The *micro-accuracy/micro-f-score* values for parts-of-speech (the left diagram) and morphological categories (the right diagram) in white and gray columns for *Lemuoklis* and *Semantika.lt*, respectively.

Table 2. The calculated *f-score* values for various parts-of-speech in different domains. *Lem* and *Sem* stands for *Lemuoklis* and *Semantika.lt*, respectively.

Part-of-speech	Administr.		Fiction		Scientific		Periodicals		All domains	
	Lem	Sem	Lem	Sem	Lem	Sem	Lem	Sem	Lem	Sem
Noun	0.995	0.997	0.976	0.983	0.989	0.993	0.983	0.989	0.987	0.992
Verb (all forms)	0.974	0.992	0.966	0.976	0.993	0.971	0.987	0.992	0.979	0.983
Adjective	0.955	0.982	0.919	0.948	0.986	0.953	0.939	0.944	0.954	0.958
Conjunction	0.957	0.992	0.823	0.893	0.966	0.948	0.872	0.923	0.896	0.934
Pronoun	0.943	0.984	0.907	0.964	0.977	0.969	0.908	0.959	0.920	0.966
Adverb	0.862	0.953	0.806	0.860	0.923	0.872	0.825	0.868	0.837	0.879
Proper noun	0.873	0.969	0.462	0.711	0.706	0.889	0.754	0.902	0.750	0.903
Preposition	0.926	0.997	0.982	0.982	0.989	1.000	0.986	0.992	0.973	0.991
Particle	0.472	0.585	0.575	0.608	0.740	0.196	0.359	0.459	0.532	0.543
Numeral	0.632	1.000	0.800	0.894	1.000	0.778	0.929	0.763	0.892	0.818
Unrecognized word	0.725	0.889	0.032	0.032	0.451	0.305	0.398	0.369	0.472	0.382
Interjection	-	-	0.667	1.000	-	-	0.333	0.667	0.188	0.889

4   Discussion

As it can be seen from the Fig. 1 the best lemmatization results are obtained with the *Semantika.lt* morphological analyzer. Although the difference is very small (i.e., only ~1.7% points on entire gold-standard corpus), it is still statistically significant. The superiority of *Semantika.lt* over *Lemuoklis* is especially apparent on fiction and periodical texts. The fiction is usually characterized by a high abundance of words, whereas periodical texts are full of neologisms and specific terminology. Thus, a larger number of headwords incorporated into *Semantika.lt* has an obvious advantage over *Lemuoklis*. Surprisingly *Semantika.lt* slightly underperformed *Lemuoklis* on the scientific texts, but the difference is not statistically significant. The terminology used in the scientific texts is not completely settled: some Anglicisms are more popular than their Lithuanian equivalents, some equivalents in Lithuanian sometimes even does not exist, thus are not recorded in the dictionary.

The left diagram in Fig. 2 presents the coarse-grained annotation results. The difference between the results on the entire gold-standard corpus is ~2.5%: the largest gap is again on the fiction (~3.5%) and periodicals (~3.1%), the smallest – on scientific texts (~0.4%). In the lemmatization task, the *Semantika.lt* morphological analyzer outperforms *Lemuoklis*, but the difference again is not statistically significant. In the right diagram of Fig. 2, which already presents fine-grained morphological categorization results (determined cases, genders, tenses, etc.), the robustness of *Lemuoklis* over *Semantika.lt* on the scientific texts is already statistically significant (the difference is ~5.3%). However, on the entire gold-standard corpus (the difference is ~8.1%) and on the other domains, in particular, fiction (~17.8%), administrative (~11.0%), periodicals

(~8.9%), the superiority of *Semantika.lt* is apparent. The lower accuracy in the fine-grained annotation is due to the complicated disambiguation problem and out-of-vocabulary words. The main drawback of both morphological analyzers is due to out-of-the-vocabulary words: i.e., if analyzer cannot recognize the word and indicate its lemma (leaving in original untouched form), it cannot recognize any other morphological information. Thus, the errors in the first lemmatization stage cause errors in the following morphological annotation stages: part-of-speech recognition and afterwards in the morphological categorization.

The detailed error analysis (see Table 2) reveals some major mistakes. The most complicated issue for both analyzers is the auxiliary words (i.e., conjunctions and particles) which can be assigned to the different parts-of-speech without absolutely clear criteria (by the way, some numerals also face this problem). However, *Semantika.lt* analyzer demonstrates significant improvement for the proper nouns compared to *Lemuoklis*. A very specific mistake of *Lemuoklis* is due to the confusion of one letter abbreviations with one letter interjections (e.g., despite interjections in the upper-case at the end of a direct sentence are very rare).

5   Conclusion and Future Work

This comparative research work disclosed strengths/weaknesses of two the most popular and publicly available Lithuanian morphological analyzers: *Lemuoklis* and *Semantika.lt*. Both analyzers were evaluated on 4 domains of the same gold-standard corpus.

The morphological analyzers *Lemuoklis/Semantika.lt* achieved ~96.3%/ ~98.0%, ~92.8%/~95.3%, ~78.7%/~86.8% of accuracy on the lemmatization, part-of-speech tagging, and annotation of the morphological categories, respectively. Despite *Semantika.lt* was superior over *Lemuoklis* on the entire gold-standard corpus and on the administrative, fiction, and periodical texts; *Lemuoklis* yielded equal performance on the scientific texts and even outperformed *Semantika.lt* on the annotation task of the morphological categories.

The experiments with *Lemuoklis* and *Semantika.lt* were carried out on the normative Lithuanian texts. In the future research we are planning to test their robustness on the challenging types of texts: forum posts, Internet comments, tweets, etc.

Acknowledgments. The authors thank the researchers from LLC Fotonija, especially Virginijus Dadurkevičius, for providing information about the *Semantika.lt* morphological analyzer.

References

1. Agarwal, A., Pramila, Singh, S.P., Kumar, A., Darbari, H.: Morphological analyser for Hindi - a rule based implementation. Int. J. Adv. Comput. Res. **4**(1), 19–25 (2014)
2. Akilan, R., Naganathan, E.R.: Morphological analyzer for classical Tamil texts: a rule-based approach. IJISET - Int. J. Innovative Sci. Eng. Technol. **1**(5), 563–568 (2014)
3. Baisa, V., Suchomel, V.: Large corpora for Turkic languages and unsupervised morphological analysis. In: Proceedings of the Eighth Conference on International Language Resources and Evaluation (LREC) (2012)
4. Bickel, B., Comrie, B., Haspelmath, M.: Leipzig Glossing Rules: Conventions for Interlinear Morpheme-by-Morpheme Glosses (2008)
5. Bögel, T., Butt, M., Hautli, A., Sulger, S.: Developing a finite-state morphological analyzer for Urdu and Hindi. In: The 6th International Workshop on Finite-State Methods and Natural Language Processing (FSMNLP 2007), pp. 86–96 (2007)
6. den Bosch, A.V., Daelemans, W.: Memory-based morphological analysis. In: Proceedings of the 37th Annual Meeting of the Association for Computational Linguistics on Computational Linguistics (ACL 1999), pp. 285–292 (1999)
7. Byrd, R.J., Tzoukermann, E.: Adapting an English morphological analyzer for French. In: Proceedings of the 26th Annual Meeting on Association for Computational Linguistics (ACL 1988), pp. 1–6 (1988)
8. Daudaravičius, V., Rimkutė, E., Utka, A.: Morphological annotation of the Lithuanian corpus. In: Proceedings of the Workshop on Balto-Slavonic Natural Language Processing: Information Extraction and Enabling Technologies (ACL 2007), pp. 94–99 (2007)
9. Gelbukh, A., Sidorov, G.: Approach to construction of automatic morphological analysis systems for inflective languages with little effort. In: Gelbukh, A. (ed.) CICLing 2003. LNCS, vol. 2588, pp. 215–220. Springer, Heidelberg (2003). doi:10.1007/3-540-36456-0_21
10. Jęrzejowicz, P., Strychowski, J.: A neural network based morphological analyser of the natural language. In: Proceedings of the International Conference on Intelligent Information Processing and Web Mining (IIPWM 2005), pp. 199–208 (2005)
11. Karp, D., Schabes, Y., Zaidel, M., Egedi, D.: A freely available wide coverage morphological analyzer for English. In: Proceedings of the 14th Conference on Computational Linguistics, vol. 3, pp. 950–955 (1992)
12. Kessikbayeva, G., Cicekli, I.: A rule based morphological analyzer and a morphological disambiguator for Kazakh language. Linguist. Lit. Stud. **4**(1), 96–104 (2016)
13. Khoufi, N., Boudokhane, M.: Statistical-based system for morphological annotation of Arabic texts. In: Recent Advances in Natural Language Processing (RANLP 2013), pp. 100–106 (2013)
14. Koskenniemi, K.: Two-level model for morphological analysis. In: Proceedings of the International Joint Conferences on Artificial Intelligence Organization (IJCAI 1983), pp. 683–685 (1983)
15. Malladi, D.K., Mannem, P.: Statistical morphological analyzer for Hindi. In: International Joint Conference on Natural Language Processing (IJCNLP 2013), pp. 1007–1011 (2013)
16. McNemar, Q.M.: Note on the sampling error of the difference between correlated proportions or percentages. Psychometrika **12**(2), 153–157 (1947)

17. Pauw, G.D., de Schryver, G.M.: Improving the computational morphological analysis of a Swahili corpus for lexicographic purposes. Lexikos **18**, 303–318 (2008)
18. Rimkutė, E.: Morfologinio daugiareikšmiškumo ribojimas kompiuteriniame tekstyne [The Limitation of the Morphological Disambiguation in the Digitalized Corpus] (in Lithuanian). Ph.D. thesis, Vytautas Magnus University (2006)
19. Russell, G.J., Pulman, S.G., Ritchie, G.D., Black, A.W.: A dictionary and morphological analyser for English. In: Proceedings of the 11th Conference on Computational Linguistics (COLING 1986), pp. 277–279 (1986)
20. Savickienė, I., Kempe, V., Brooks, P.J.: Acquisition of gender agreement in Lithuanian: exploring the effect of diminutive usage in an elicited production task. J. Child Lang. **36**, 477–494 (2009)
21. Žilinskienė, V.: Lietuvių kalbos dažninis žodynas [The Frequency Dictionary of the Lithuanian Language] (1990). (in Lithuanian)
22. Zinkevičius, V.: Lemuoklis - morfologinei analizei [Morphological analysis with Lemuoklis]. In: Gudaitis, L. (ed.) Darbai ir Dienos, vol. 24, pp. 246–273 (2000) (in Lithuanian)

Constrained Deep Answer Sentence Selection

Ahmad Aghaebrahimian[(⊠)]

Faculty of Mathematics and Physics, Institute of Formal and Applied Linguistics,
Charles University in Prague, Malostranske nam. 25, 11800 Praha 1, Czech Republic
ebrahimian@ufal.mff.cuni.cz

Abstract. In this paper, we propose Constrained Deep Neural Network (CDNN) a simple deep neural model for answer sentence selection. CDNN makes its predictions based on neural reasoning compound with some symbolic constraints. It integrates pattern matching technique into sentence vector learning. When trained using enough samples, CDNN outperforms regular models. We show how using other sources of training data as a mean of transfer learning can enhance the performance of the network. In a well-studied dataset for answer sentence selection, our network improves the state of the art in answer sentence selection significantly.

Keywords: Deep neural network · Sentence selection · Transfer learning

1 Introduction

A typical Question Answering (QA) system consists of three basic components; passage retrieval, sentence selection, and answer extraction [21]. Given a question, different possible Information Retrieval (IR) methods can be used to extract a relevant passage that hopefully contains the answer.

Given a question and all sentences in its passage, the job of sentence selection component is to choose a subset of sentences as the answer of the question. If a finer answer is expected, the third component (i.e. answer extraction) extracts a word or a span of words from the selected sentences.

In the literature, the last two components are referred to as sentence-level [23] and word-level [18] QA systems. The sentence selection component helps the answer extraction component by eliminating non-relevant sentences and reducing the search space. Moreover, it provides a sentence which can be used as an evidence for the final answer. This formulation of QA has also other applications such as paraphrase detection [27] or Recognizing Textual Entailment (RTE) [16].

Deep Neural Networks (DNN) have been shown to outperform traditional machine learning algorithms in many Natural Language Processing (NLP) tasks. A natural choice for sentence selection using DNNs is the hinge approximation approach in which the weights associated to positive question-answer couples are increased and those associated to negative couples are decreased [19] through training. We extend this idea to integrate the number of shared patterns of question-answer couples into the model.

© Springer International Publishing AG 2017
K. Ekštein and V. Matoušek (Eds.): TSD 2017, LNAI 10415, pp. 57–65, 2017.
DOI: 10.1007/978-3-319-64206-2_7

Such a large DNN has millions of parameters that should be trained properly. This requires a large number of samples which are not available in regular datasets. In order to provide the network with enough training samples, we used SQuAD [18] to train our model and we show that this out-sourced trained model performs remarkably better than previous best models.

The main contributions of this paper are a DNN model for sentence selection and integrating learning transfer into DNNs for this and other similar tasks.

2   Related Work

The emergence of DNNs in recent years has remarkably helped to improve the performance of NLP applications including different kinds of QA systems. From QA systems based on semantic parsing [4,13] to IR-based systems [24], from cloze-type [9,10] to free-text systems [18] and from factoid [1–3] to reasoning-type systems [11] all has been benefited from DNNs.

Before the introduction of DNNs, feature engineering based on knowledge base data, n-grams or syntactic rules [6,14,22] was a common yet cumbersome practice in QA systems. DNNs have helped QA systems in at least two ways; first, to get rid of many time-consuming intermediate feature engineering processes, and second, to reduce the need for domain-specific knowledge and to make transfer learning possible and easier.

In DNN-based QA systems, instead of manual and hard-coded features, DNNs learn an internal representation of both questions and sentences. One common approach in learning the internal representation is to define the problem as a point-wise ranking problem [7,8,26,28]. In point-wise ranking approach, a DNN is trained on tuples of questions and their correct sentences. In this way, it learns a probability distribution over all sentences given each question.

A more intuitive approach to QA systems is to train a DNN on positive and negative sentences at the same time [19]. Our model is similar to this one with two differences. First, we defined the loss function to enforce the number of shared patterns between questions and sentences as a hard constraint. Second, instead of working with triplets (question, positive sample, negative sample), we used tuples (question, positive sample) and (question, negative sample). Our approach is more manageable especially when working with big datasets. Moreover, instead of a convolution, we used an attention mechanism which is much faster and more flexible with the length of sentences.

DNNs are very good at domain adaptation and transfer learning. There are numerous successful cases for transfer learning such as object detection [29], leaning word vectors (embeddings) [15] and most recently Question Answering [16]. Our work is similar to the latter experiment. However, we used exact match and overlap measures to map SQuAD [18] questions to their answer sentences. Besides, the use of the attention mechanism on sentences and our pattern constraint on the hinge approximation helped our model to outperform their models.

3   The Datasets

We tried our model on two widely used datasets namely TrecQA [25] and SQuAD [18].

TrecQA is compiled using the data in TREC 8–13 QA tracks. There is a modified version of TrecQA available in which unanswered questions and questions with only one positive and negative sentences are removed from the development set and the test set divisions. The training questions in both the original and the modified versions are the same.

SQuAD is a dataset for QA in the context of machine comprehension. It includes 107,785 question-answer pairs posed by crowd workers on 536 Wikipedia articles. The answers in SQuAD can be any span of consecutive words in a paragraph which comes with each question. SQuAD is not designed for sentence selection models. However, it is a good source of data for transfer learning experiments [16].

Although SQuAD is not designed originally for answer sentence selection, it can be used for these experiments with some trivial modifications. For this purpose, we first extracted sentences that contain the answer given each question using exact matching.

Each question in SQuAD is mapped to a paragraph. Some paragraphs contain tens of different sentences, and hence by exact matching a question can be mapped to more than one sentence. We assumed that each question has just one positive answer sentence. In order to make sure that each question is mapped to the best possible sentence, in the next step, among sentences which contain the answer, we selected the sentence that has the maximal n-gram overlap with the question. These n-grams were kept limited to uni, bi and trigrams.

To obtain negative sentences, we used two different settings, which seem to cause no real difference in the final results. In the first setting, given a question, we detected the positive sentence as explained above and used the other sentences in the same paragraph as negative samples. In the other setting, we sampled negative sentences for a given question from all the paragraphs disregarding witch paragraph the question belongs to. At the end, we had one positive and up to five negative samples for each question.

The original test set of SQuAD is unseen and is not available online. Hence, for our experiments, we used the original training set for both training and development purposes. For testing purpose, we used the original development set. Table 1 shows the number of the questions in each dataset. Since this is the first time that SQuAD is being tested in a sentence selection experiment, we established a simple baseline for sentence selection in our test set by counting the number of shared uni, bi and trigrams in each question and sentence and assigning the sentence with the highest number of shared patterns to the question.

Table 1. Datasets statistics

	Train set	Dev. set	Test set
Original TrecQA	1229	82	100
Modified TrecQA	1229	65	68
SQuAD	68983	17245	10778

4   The Approach

The task in a sentence selection experiment is to estimate a probability distribution over all sentences given each question and to get the sentence with the highest probability.

$$s_{best} = argmax_s \quad p(\mathbf{s}|\mathbf{q}) \tag{1}$$

For learning the probability associated with each question and sentence or, in other words, for learning the association between questions and their correct sentences, training on negative sentences are informative as training on positive ones. In order to feed both negative and positive sentences into the network at the same time, we defined a loss function that not only considers the similarity between questions and their sentences but also enforces their pattern similarity as a hard constraint.

$$loss = \max\{0, \mathbf{m} - \alpha * \mathrm{Sim}(\mathbf{q}, \mathbf{s}^+) + \beta * \mathrm{Sim}(\mathbf{q}, \mathbf{s}^-)\} \tag{2}$$

This loss decreases with the similarity between a question and its positive samples and increases with the similarity between a question and its negative samples.

Parameter $\mathbf{m}$ is the margin between positive and negative sentences. There is a trade off between the margin and the number of mistakes in positive and negative classification in the model. A margin between 0.01 and 0.1 usually gives good results in this model. $\mathbf{s}^+$ are correct sentences and $\mathbf{s}^-$ are wrong ones. α and β are the numbers of shared patterns between a question and its positive and negative sentences, respectively. Finally, Sim is the similarity function that computes the similarity between questions and sentences.

Among various techniques for measuring the similarity between two sentences, we adopted Geometric mean of Euclidean and Sigmoid Dot product (GESD) [5] which seems to outperform other similarity measures in this model. GESD linearly combines two other similarity measures called L2-norm and inner product. L2-norm is the forward-line semantic distance between two sentences and inner product measures the angle between two sentences vectors.

$$GESD(\mathbf{q}, \mathbf{s}) = \frac{1}{1 + \exp\left(-(\mathbf{q} \cdot \mathbf{s})\right)} * \frac{1}{1 + \|\mathbf{q}, \mathbf{s}\|} \tag{3}$$

In order to provide the loss function with its inputs, we should train three vectors; question, positive sample and negative sample vectors. Similar to [20],

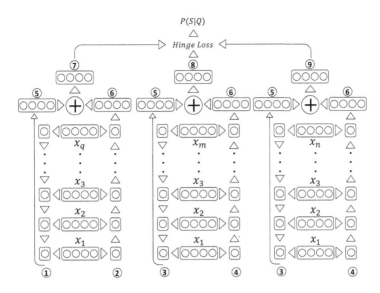

Fig. 1. Abstract model architecture. Numbered components in the figure are: 1-Forward LSTM, 2-Backward LSTM, 3-Forward Attentive LSTM, 4-Backward Attentive LSTM, 5-Backward max-pooling, 6-Forward max-pooling, 7-Question vector, 8-Positive sample vector, 9-Negative sample vector

we defined an embedding, an LSTM with attention layer, a max pooling and a loss function in four separate layers as illustrated in Fig. 1.

Question vectors are generated by passing through these four layers. Embedding layer in our model is a static layer. Word vectors in this layer are initialized with pre-trained 100-dimensional Glove vectors [17]. During training, they were held fixed as we observed no improvements in the performance of the model on the validation set. We also observed no noticeable effect on the performance by increasing the dimensionality of the embeddings to 200 or 300-dimensional vectors.

In the next layer, two LSTM cells are allocated to each token in questions to form a forward and a backward LSTM. Given a question, in forward LSTM, each LSTM cell receives the output of its previous cell as the input layer. It happens from left to right to cover all the tokens. The first LSTM receives the embedding layer output as its input and the output of the last LSTM is fed into the max-pooling layer.

In the backward LSTM, the same question is fed into the same LSTM, but in reverse order (i.e. from right to left). The outputs of forward and backward LSTMs then go through a column-wise max-pooling layer. Similar to [20], the max-pooling layer is a column-wise max operator that estimates the importance of each token given the surrounding context. Max-pooling is applied on both forward and backward LSTMs and the resulting vectors are concatenated to form the final vectors.

For generating positive and negative vectors, positive and negative samples are passed through the same network, where instead of regular LSTM, attention LSTM cells over corresponding questions are used. For this part we used the attentive LSTM model proposed by [20].

Having a couple of questions and sentences, we first compute their vectors and then compute their similarity, which is parameterized by the parameters of the network. Now, by putting the similarity of positive and negative couples into the loss function which was mentioned above, we can train the network to maximize the similarity between questions and the corresponding correct sentences and to minimize the similarity between questions and wrong sentences.

At the end, we rearrange the scores generated by the model by a coefficient of question-sentence shared patterns similar to what we used in the loss function.

5   Experimental Results

For training, we used Adam [12] to optimize the parameters using SGD. We used LSTM with 128 layer size and set the margin of the loss to 0.05. Since the number of the samples in our training set is very large, a couple of iterations over all the samples is enough for training the model.

The questions in TrecQA has more than one positive and negative samples and the output of the model is an ordered list of sentences. Therefore, for evaluation purposes the most suitable measures are Mean Average Precision (MAP) and Mean Reciprocal Rank (MRR).

However, the questions in SQuAD are limited only to one positive answer and a couple of negative ones. For this reason, we used Accuracy to evaluate the performance of the model for SQuAD dataset. It should be mentioned that Accuracy measure is the floor measure for the performance of this system[1]. Table 2 shows the experimental results.

This research demonstrated that transferring learned models which are trained on large datasets to be test on small ones can be beneficial. However,

Table 2. Experimental results. TrecQA-org is the original TrecQA, TrecQA-mod is the modified TrecQA and state-of-the-art refers to [19]

	Trained on	Test set	MRR	MAP	ACC
State-of-the-art	TrecQA	TrecQA-mod	87.7	80.1	-
State-of-the-art	TrecQA	TrecQA-org	83.4	78.8	-
Current paper	SQuAD + TrecQA	TrecQA-mod	86.2	73.2	-
Current paper	SQuAD + TrecQA	TrecQA-org	**89.5**	**79.5**	-
Baseline	SQuAD	SQuAD	-	-	69
Current paper	SQuAD	SQuAD	93.4	90.1	**86**

---

[1] Compared to MRR and MAP, Accuracy is a pessimistic measure for this experiment, because it just considers the first selected answer and disregards the others.

the opposite does not seem to be valid. It seems that adding training data to a large dataset distorts its true distribution. In our experiment, we get significantly better results when we included SQuAD in TrecQA for training, but including TrecQA in SQuAD for training decreased the model performance on SQuAD.

6 Conclusion

We proposed CDNN as a network which is able to enforce hard constraints on the parameters of a deep neural network. We also showed that applying transfer learning technique on a model with enough learning capacity can obtain state-of-the-art results without dependence on any external resource or doing any time-consuming feature designing procedure.

Acknowledgments. This research was partially funded by the Ministry of Education, Youth and Sports of the Czech Republic under SVV project number 260 453, core research funding, and GAUK 207-10/250098 of Charles University in Prague.

References

1. Aghaebrahimian, A., Jurčíček, F.: Constraint-based open-domain question answering using knowledge graph search. In: Sojka, P., Horák, A., Kopeček, I., Pala, K. (eds.) TSD 2016. LNCS (LNAI), vol. 9924, pp. 28–36. Springer, Cham (2016). doi:10.1007/978-3-319-45510-5_4
2. Aghaebrahimian, A., Jurčíček, F.: Open-domain factoid question answering via knowledge graph search. In: Proceedings of the Workshop on Human-Computer Question Answering, The North American Chapter of the Association for Computational Linguistics (NAACL) (2016)
3. Bordes, A., Usunier, N., Chopra, S., Weston, J.: Large-scale simple question answering with memory networks. arXiv preprint arXiv:1506.02075 (2015)
4. Clarke, J., Goldwasser, D., Chang, M.W., Roth, D.: Driving semantic parsing from the worlds response. In: Proceedings of the Conference on Computational Natural Language Learning (2010)
5. Feng, M., Xiang, B., Glass, M.R., Wang, L., Zhou, B.: Applying deep learning to answer selection: a study and an open task. In: Proceedings of IEEE ASRU Workshop (2015)
6. Fern, S., Stevenson, M.: A semantic similarity approach to paraphrase detection. In: Proceedings of the 11th Annual Research Colloquium of the UK Special-Interest Group for Computational Linguistics (2008)
7. He, H., Gimpel, K., Lin, J.: Multi-perspective sentence similarity modeling with convolutional neural networks. In: Proceedings of the Conference on Empirical Methods in Natural Language Processing (EMNLP) (2015)
8. He, H., Lin, J.: Pairwise word interaction modeling with deep neural networks for semantic similarity measurement. In: The North American Chapter of the Association for Computational Linguistics (NAACL) (2016)

9. Hermann, K.M., Kocisky, T., Grefenstette, E., Espeholt, L., Kay, W., Suleyman, M., Blunsom, P.: Teaching machines to read and comprehend. In: Advances in Neural Information Processing Systems (2015)
10. Kadlec, R., Vodolan, M., Libovicky, J., Macek, J., Kleindienst, J.: Knowledge-based dialog state tracking. In: 2014 IEEE Spoken Language Technology Workshop (SLT) (2014)
11. Khashabi, D., Khot, T., Sabharwal, A., Clark, P., Etzioni, O., Roth, D.: Question answering via integer programming over semi-structured knowledge. In: Proceedings of International Joint Conference on Artificial Intelligence (IJCAI) (2016)
12. Kingma, D., Ba, J.: Adam: a method for stochastic optimization. arXiv:1412.6980 (2014)
13. Kwiatkowski, T., Zettlemoyer, L., Goldwater, S., Steedman., M.: Inducing probabilistic CCG grammars from logical form with higher-order unification. In: Proceedings of the Conference on Empirical Methods in Natural Language Processing (2010)
14. Madnani, N., Tetreault, J., Chodorow, M.: Re-examining machine translation metrics for paraphrase identification. In: Proceedings of the 2012 Conference of the North American Chapter of the Association for Computational Linguistics: Human Language Technologies (2012)
15. Mikolov, T., Chen, K., Corrado, G., Dean, J.: Efficient estimation of word representations in vector space. In: Proceedings of Workshop at ICLR (2013)
16. Min, S., Seo, M., Hajishirzi, H.: Question answering through transfer learning from large fine-grained supervision data. arXiv:1702.02171 (2017)
17. Pennington, J., Socher, R., Manning, C.D.: Glove: Global vectors for word representation. In: Proceedings of the Empirical Methods in Natural Language Processing (2014)
18. Rajpurkar, P., Zhang, J., Lopyrev, K., Liang, P.: Squad: 100,000+ questions for machine comprehension of text. arXiv preprint arXiv:1606.05250 (2016)
19. Rao, J., He, H., Lin, J.: Noise-contrastive estimation for answer selection with deep neural networks. In: Proceedings of the 25th ACM International on Conference on Information and Knowledge Management, CIKM (2016)
20. Santos, C.D., Tan, M., Xiang, B., Zhou, B.: Attentive pooling networks. arXiv:1602.03609 (2016)
21. Tellex, S., Katz, B., Lin, J., Fernandes, A., Marton, G.: Quantitative evaluation of passage retrieval algorithms for question answering. In: SIGIR (2003)
22. Xu, W., Ritter, A., Callison-Burch, C., Dolan, W.B., Ji, Y.: Extracting lexically divergent paraphrases from twitter. Trans. Assoc. Comput. Linguist. 2, 435–448 (2014)
23. Yang, Y., Yih, W.T., Meek, C.: WikiQA: a challenge dataset for open-domain question answering. In: Proceedings of the Conference on Empirical Methods in Natural Language Processing (EMNLP) (2015)
24. Yao, X., Durme, B.V.: Information extraction over structured data: question answering with freebase. In: Proceedings of Association for Computational Linguistics (2014)
25. Yao, X., Van Durme, B., Callison-Burch, C., Clark, P.: Answer extraction as sequence tagging with tree edit distance. In: HLT-NAACL (2013)
26. Yih, W.T., Chang, M.W., Meek, C., Pastusiak, A.: Question answering using enhanced lexical semantic models. In: Proceedings of Association for Computational Linguistics (ACL) (2013)
27. Yin, W., Schtze, H., Xiang, B., Zhou, B.: ABCNN: attention-based convolutional neural network for modeling sentence pairs. arXiv:1512.05193 (2015)

28. Yu, L., Moritz Hermann, K., Blunsom, P., Pulman, S.: Deep learning for answer sentence selection. In: NIPS Deep Learning Workshop (2014)
29. Zeiler, M.D., Fergus, R.: Visualizing and understanding convolutional networks. In: Fleet, D., Pajdla, T., Schiele, B., Tuytelaars, T. (eds.) ECCV 2014. LNCS, vol. 8689, pp. 818–833. Springer, Cham (2014). doi:10.1007/978-3-319-10590-1_53

Quora Question Answer Dataset

Ahmad Aghaebrahimian[(✉)]

Faculty of Mathematics and Physics, Institute of Formal and Applied Linguistics,
Charles University in Prague, Malostranske nam. 25, 11800 Praha 1, Czech Republic
ebrahimian@ufal.mff.cuni.cz

Abstract. We report on a progressing work for compiling Quora Question Answer dataset. Quora dataset is composed of questions which are posed in Quora Question Answering site. It is the only dataset which provides sentence-level and word-level answers at the same time. Moreover, the questions in the dataset are authentic which is much more realistic for Question Answering systems. We test the performance of a state-of-the-art Question Answering system on the dataset and compare it with human performance to establish an upper bound.

Keywords: Dataset · Question answering · Sentence-level answer · Word-level answer

1 Introduction

As one of the oldest applications of Natural Language Processing (NLP), Question Answering (QA) is one of the most interesting research areas with lots of commercial potentials. Given a question and a passage in which the question's answer is mentioned explicitly, QA is the task of providing the question with its sentence-level or word-level answers. As it is shown in Table 2 in Sect. 3, given a question, we may want to get the answer either as a sentence [13] or a span of words [4,8].

Through the last couple of recent years, QA has been subjected to many big achievements due to some advancements in machine learning as well as the emergence of powerful processing hardwares known as Graphics Processing Unit (GPU). Through these years, variety of QA systems have been emerged for different purposes such as sentence selection [9], entity selection [2,4] or machine comprehension [5,7,8].

A major problem with the datasets which are used in these experiments is that all of them are either sentence-level or word-level datasets. It means that they are suitable for testing either a sentence-level QA system or a word-level one. These datasets can not be used to test QA systems with a pipe-line architecture in which the sentence selection and the answer extraction components are different modules.

We present Quora dataset which is compiled from the questions in Quora. Quora is a Question Answering site where people ask their questions and other

© Springer International Publishing AG 2017
K. Ekštein and V. Matoušek (Eds.): TSD 2017, LNAI 10415, pp. 66–73, 2017.
DOI: 10.1007/978-3-319-64206-2_8

people answer them according to their expertise or experience. The answers in Quora enjoy a high degree of variance since each question is answered by variety of people with different perspectives.

The compilation of Quora dataset is in progress and at the moment it contains 300 questions. However, we intend to expand it using a crowd sourcing platform in the near future[1].

In contrast to previous datasets in which the questions are synthetic or answer-oriented [4,8], Quora dataset is totally authentic. It means that real people asked these questions for obtaining real information. Moreover, Quora dataset makes sentence-level and word-level QA possible at the same time. Finally, Quora is the only dataset for answering questions with multiple-part answers. In contrast to other datasets where the answers are either one entity [4] or one span of consecutive words [8], the answers in Quora are normally multiple-part answers from different parts of their accompanying passage.

In order to establish a baseline for Quora dataset, we tested it using a state-of-the-art sentence selection system [1]. We also used human annotation to establish an upper bound for the dataset. Our results show a wide margin between machine and human performance on the dataset which demonstrates the difficulty of the task.

In the rest of this paper, we elaborate on recent datasets of QA in Sect. 2. Then, we explain the process of dataset compilation in Sect. 3. Finally, we evaluate the dataset in Sect. 4 before we conclude in Sect. 5.

2   Question Answering Datasets

We can roughly recognize all QA datasets as either sentence-level or word-level datasets. Sentence-level QA datasets provide one or multiple correct sentences [9, 11] for each question. Word-level datasets provide one answer in the form of a single word [10] or a span of consecutive words [8] for each question.

Single word QA algorithms usually are tested on cloze-type datasets. Cloze-type QA systems are word-level systems where the system is trained to select the answer among four or five answer choices. There are a wide range of datasets for cloze-type [5,6] QA.

Cloze-type QA datasets are easy to compile because they can be generated automatically. CBT [6] and CNN [5] are among the biggest cloze-type datasets which contain more than half a million and one and half a million questions respectively.

Level of understanding in these datasets are adjustable by changing the type of missing elements in the questions. For instance, while looking for prepositions are rather easy in a cohesive text, looking for name entities is not always straightforward.

In order to put our work in an appropriate context, we focus on WikiQA [13] and TrecQA [12] as two sentence-level datasets and SQ [4] and SQuAD [8] as two word-level datasets. Table 1 summarizes some statistics of these datasets.

---

[1] Quora dataset is available at https://github.com/Q2AD.

WikiQA [13] is a small dataset for sentence-level QA. It is compiled from the query logs of Bing search engine. The Wikipedia page selected for each query is used as the passage for that question. All sentences in the summary paragraph of the selected passage are used as the candidate sentences which in turn are presented to crowd workers for sentence selection.

TrecQA [12] is a standard and well-studied benchmark for answer sentence selection experiments. It is compiled from the data in TREC 8–13 QA tracks. Like WikiQA, the source of questions in TrecQA is users' log files. In both WikiQA and TrecQA, each question is mapped to more than one correct and several wrong sentences and QA systems are expected to return an ordered list of correct sentences.

The mapping between the questions to paragraphs in both WikiQA and TrecQA is not accurate due to indeterministic retrieval methods used for data retrieval in the first place. This is the reason why some of the questions in both datasets contain no correct answer.

SimpleQuestions (SQ) [4] is a collection of 108,442 questions composed in natural language. Each question in the dataset is mapped to a triple of subject-predicate-object (a.k.a assertion) in Freebase knowledge graph [3]. The questions in this dataset are synthesized by crowd workers. They are posed to Freebase facts and are asked to synthesize a question. SQ is a dataset for entity selection where the entities are limited to the entities in the system's knowledge graph.

Stanford Question Answering Dataset (SQuAD) [8] is a dataset for QA in the context of machine comprehension. It includes 107,785 question-answer pairs synthesized by crowd workers on 536 Wikipedia articles. SQuAD is a word-level QA dataset. The answers in SQuAD can be any span of words in a paragraph which comes with each question.

Table 1. Datasets statistics

	Train set	Dev. set	Test set	Type
TrecQA	1229	82	100	Sentence-level
WikiQA	873	126	243	Sentence-level
Simple Questions	75910	10845	21687	Entity-level
SQuAD	86228	10778	10778	Word-level
Quora	210	30	60	Word-level Sentence-level

There are three differences between Quora and other QA datasets like SQuAD or SQ.

The first difference lies on the type of Quora questions. The questions in SQuAD and SQ are crowd sourced and synthetic while Quora questions are authentic and original.

The second difference is that the answers of Quora questions are multi-part answers where each part is located in a different part of the accompanying passage. A full answer to a question in Quora dataset is the one which contains all the separate parts.

Finally, the last difference is that Quora dataset accommodates sentences-level answers with word-level ones in the same dataset which makes it an ideal dataset for open domain QA pipelines for testing sentence selection and answer exaction components at the same time.

3   Dataset Compilation

Quora dataset contains 300 questions accompanied with their long and short answers. It is divided between train set with 80% and test set with 20% of the questions[2]. Each question in the dataset is provided with a full-text passage which contains the answer sentences and short answers of the question.

The answer sentences are complete sentences in full-text passages and the short answers are any possible span of consecutive words in correct sentences.

As seen in Table 2 the question *How do I push myself to study in the afternoon?* is answered by three full sentences and three short answers. The short answers are a span of consecutive words from the full sentences. However, the choice of boundary is totally arbitrary and each answer is from different parts of the passage.

Table 2. A sample question with its answers from Quora dataset. Correct sentences are full sentences from full-text passages and short answers are spans of words taken from correct sentences

Question	How do I push myself to study in the afternoon?
Full text passage	Many people have a down cycle after lunch. You can eat a light lunch which will help. That way your body isn't processing a heavy load of carbs and not as much energy is spent in the digestion of your meal. You can also build in a small amount of exercise: stretching, a short walk. Exercise helps digestion and helps improve your overall energy level. You can also deliberately set a timer for a work period. There is a method called the Pomodoro method you can check out which lets you set a clock and at the end of that time, you stop your work, rest for 5 min and then reset the clock. This method has been scientifically proven to help people be more productive
Correct sentences	You can eat a light lunch which will help You can also build in a small amount of exercise: stretching, a short walk You can also deliberately set a timer for a work period
Short answers	Eat a light lunch Exercise Set a timer for a work period

---

[2] The choice of development size is given to the preference of researchers and the attributes of their experiments.

As shown in Table 2, there is no any common words between the question and its answers. This feature makes answer selection difficult for QA systems which works based on the overlap between questions and answers.

Quora dataset is compiled in three steps; question screening, passage selection and answer annotation. These steps are explained in details in the following subsections.

3.1   Question Screening

In the question screening step, we compile a list of questions which are suitable for the purpose of the dataset. Questions for inclusion in the dataset should be straightforward, well stated and eloquent. Besides, they should be answerable in at least one and at most three explicit sentences. To satisfy these conditions, to maintain a good variance in the dataset and to make sure that we choose our questions randomly enough, we went through the following processes.

At the time of dataset compilation, Quora had hosted around 200000 answered questions among which we randomly sampled 20000 ones. In order to make sure we would collect the most popular questions, we did a weighted sampling based on the views number of the questions and we selected the first 5000 questions.

At this step, a group of 10 annotators went through the questions one by one to eliminate questions which ask about more than one thing (e.g. which computer system should I buy? apple or Asus, why and preferably where?).

Afterward, the annotators are asked to eliminate opinionated or subjective (e.g. what is the most nasty things happen to you recently?) and descriptive or procedural questions (e.g. how can I reinstall windows on my pc?) from the list of questions. The answer to these questions are usually very long. Besides, there is usually, no way to answer them explicitly.

Finally, the annotators chose questions which were self-explanatory and could be answered without any clarification[3]. At this point we had 4000 question for the next step; passage selection.

3.2   Passage Selection

Having a list of questions from previous step, in this step, the annotators chose a passage for each question.

Each question in Quora is answered by different number of people. Some of the questions are answered more than 100 times. It means that there are a large number of different answers to each question.

The passages of questions should be lengthy enough to convey the message fully and clearly. However, we do not like them to be very lengthy. Too short passages do not provide enough context for answering and too lengthy passages add undesirable noise to the dataset. Hence, the first step in the passage selection

---

[3] Some users in Quora provides their questions with a comment which helps to clarify the question better.

is to eliminate too short or too long passages. We eliminated questions with full-text passages less than 100 and more than 1000 words in length.

Full-text passages should answer their questions explicitly and it can not be decided merely based on the up-vote statistic associated to them. The best full-text passage is not necessarily the one which has the highest up-votes. It is the one which contains correct answer sentences and short answers explicitly. Hence, given a question and its remaining full-text passages after the screening above, we asked our annotators to choose the best passage which satisfies the above mentioned criteria.

Full-text passage selection is done on the merit of answering questions explicitly and providing enough context for answering them correctly. At the end of this step, we had our list of questions mapped to their correct full-text passages.

3.3   Answer Annotation

In this step, we extract answer sentences and short answers from full-text passages. The answers in Quora dataset are multifaceted. It means that an answer to a question may contain different aspects which are expressed in multiple sentences. For this reason, the number of correct answer sentences and short answers is different for each question. The answers in Quora dataset may contain at least one and at most three different aspects (Table 3).

Given a list of questions accompanied with their best full-text passages, our annotators were asked to extract correct sentences and short answers. They were asked, first to choose the right sentences as sentence-level answers, and then to choose the right spans of words among the right sentences as word-level answers.

Table 3. Quora dataset statistics

	Train dataset	Test data
Number of sentences	1212	302
Number of questions	240	60
One sentence	81	22
Two sentences	108	30
Three sentences	51	8
One answer	76	16
Two answers	104	33
Three answers	60	11

4   Evaluation

We make use of a state-of-the-art sentence-level QA system [1] to measure the difficulty of the dataset and to establish a baseline. In contrast to SQuAD, the answers in Quroa dataset have multiple parts, hence the questions have multiple correct sentences and multiple short answers. Since the questions in Quora dataset are supposed to be answered in multiple segments, the short answers as well as their correct sentences are supposed to be in ordered lists. Therefore, similar to TrecQA and WikiQA, we use Mean Average Precision(MAP) and Mean Reciprocal Rank (MRR) as our metric of performance.

To establish an upper bound for the dataset, we asked 5 of our annotators to select correct sentences for the questions in Quora test set. Then, we computed MRR and MAP scores for each and reported their average.

Table 4 summarizes the results of Quora dataset for random guess, upper bound and baseline experiments. In this work, we only report sentence selection performance on the dataset.

Table 4. Experimental results. State-of-the-art refers to [1]

	MRR	MAP
Random guess	35.6	31.4
State-of-the-art	61.2	45.9
Upper bound	94.6	92.8

The wide margin between the random guess and the state-of-the-art results in Table 4 suggests the effectiveness of the approach used in [1]. However, there is a wide gap between machine and human performances as well which suggests there is still a lot of room for improvement.

Error analysis shows that by increasing the number of answers in the answer set of each question the error rate decreases. It shows that most errors are attributed to single answer questions while least errors are attributed to triple answer ones.

As it is mentioned in the original paper [1], training the integrated pre-trained word vectors do not change the performance of the system significantly. However, in case of Quora dataset, enabling word vector training during training phase decreases the performance of the system by 4% which seems normal due to the small size of current dataset.

5   Discussion and Future Work

Unlike WikiQA and TrecQA, the connection between the questions and their passages in Quora database is deterministic and direct. This connection is strengthen by answer up-voting statistic associated to each question. So the answers in Quora dataset are totally tailored toward the questions.

In contrast to SQ and SQuAD, the questions in Quora are totally authentic and original. In Quora, people ask questions to elicit useful information. This is similar to the way a real-world QA system works.

Each question in Quora is tagged with one or several related topic. The number of topics in Quora is very large. Questions about different countries, people, places, civilizations, experiences in life, philosophy and many other topics are asked and answered in Quora. There are more than 600 different topics in Quora train set and more than 150 different topics in the test set. There are numerous notions to ask about in Quora and it requires much more questions to provide a sufficient representation that we hope to provide in the near future.

All of these attributes make Quora an ideal source for experimentation on open domain Question Answering. In the next step, we intend to expand the number of the questions in our dataset to a number which makes more intensive data-driven approaches possible.

Acknowledgments. This research was partially funded by the Ministry of Education, Youth and Sports of the Czech Republic under SVV project number 260 453, core research funding, and GAUK 207-10/250098 of Charles University in Prague.

References

1. Aghaebrahimian, A.: Constrained deep answer sentence selection. In: Proceedings of the 20th International Conference on Text, Speech and Dialogue (TSD) (2017)
2. Aghaebrahimian, A., Jurčíček, F.: Open-domain factoid question answering via knowledge graph search. In: Proceedings of the Workshop on Human-Computer Question Answering, The North American Chapter of the Association for Computational Linguistics (NAACL) (2016)
3. Bollacker, K., Tufts, P., Pierce, T., Robert, C.: A platform for scalable, collaborative, structured information integration. In: Proceedings of the Sixth International Workshop on Information Integration on the Web (2007)
4. Bordes, A., Usunier, N., Chopra, S., Weston, J.: Large-scale simple question answering with memory networks. arxiv:1506.02075 (2015)
5. Hermann, K.M., Kocisky, T., Grefenstette, E., Espeholt, L., Kay, W., Suleyman, M., Blunsom, P.: Teaching machines to read and comprehend. In: Advances in Neural Information Processing Systems (2015)
6. Hill, F., Bordes, A., Chopra, S., Weston, J.: The goldilocks principle: reading children's books with explicit memory representations. arxiv:1511.02301 (2015)
7. Kadlec, R., Schmid, M., Bajgar, O., Kleindienst, J.: Text understanding with the attention sum reader network. In: Proceedings of the Association for Computational Linguistics (2016)
8. Rajpurkar, P., Zhang, J., Lopyrev, K., Liang, P.: Squad: 100,000+ questions for machine comprehension of text. arxiv:1606.05250 (2016)
9. Rao, J., He, H., Lin, J.: Noise-contrastive estimation for answer selection with deep neural networks. In: Proceedings of the 25th ACM International on Conference on Information and Knowledge Management (2016)
10. Richardson, M., Burges, J.C., C., Erin, R.: MCTest: a challenge dataset for the open-domain machine comprehension of text. In: Empirical Methods in Natural Language Processing (EMNLP) (2013)
11. Santos, C.D., Tan, M., Xiang, B., Zhou, B.: Attentive pooling networks. arXiv:1602.03609v1 (2016)
12. Voorhees, E.M., Tice, D.M.: Building a question answering test collection. In: ACM Special Interest Group on Information Retreival (SIGIR) (2000)
13. Yang, Y., Yih, S.W.T., Meek, C.: WikiQA: a challenge dataset for open-domain question answering. In: Empirical Methods in Natural Language Processing (EMNLP) (2015)

Sentiment Analysis with Tree-Structured Gated Recurrent Units

Marcin Kuta[(✉)], Mikołaj Morawiec, and Jacek Kitowski

Department of Computer Science, AGH University of Science and Technology,
Al. Mickiewicza 30, 30-059 Krakow, Poland
`mkuta@agh.edu.pl`

Abstract. Advances in neural network models and deep learning mark great impact on sentiment analysis, where models based on recursive or convolutional neural networks show state-of-the-art results leaving behind non-neural models like SVM or traditional lexicon-based approaches. We present Tree-Structured Gated Recurrent Unit network, which exhibits greater simplicity in comparison to the current state of the art in sentiment analysis, Tree-Structured LSTM model.

Keywords: Sentiment analysis · Recursive neural network · Gated Recurrent Unit · Tree-Structured GRU · Long Short-Term Memory

1 Introduction

Sentiment analysis is the problem of assigning sentiment to a document, sentence or a phrase. A document may be a movie review or an opinion about a particular product. Finding an accurate solution to the sentiment analysis problem has strong economic justification, as it allows companies to find opinions about their products and recognize their characteristics.

Currently neural network based models achieve the most competitive results in sentiment analysis. One challenge for neural network architectures is handling an input of variable length. Recurrent Neural Networks (RNNs) easily process sequences of variable length, unfortunately are hard to train, due to vanishing or exploding gradient problems. Long Short Term Memory (LSTM) units were proposed [6] as a remedy to vanishing gradient, the most ubiquitous problem encountered during RNNs training. Over the years, numerous extensions and refinements of the original LSTM, including adding peephole cells, have been developed [5].

Recently, significant simplification over LSTM, known as Gated Recurrent Unit (GRU), was introduced [2]. GRUs contain less subcells and are described by much simpler set of equations, thus require less computational power. Relation between GRU and LSTM effectiveness is an open issue and an area of research. Evaluation of GRU-based neural networks on sequence modelling [3] showed effectiveness similar to those build from LSTMs, while in [8] GRU outperformed the LSTM on nearly all tasks except language modelling with the naive initialization.

K. Ekštein and V. Matoušek (Eds.): TSD 2017, LNAI 10415, pp. 74–82, 2017.
DOI: 10.1007/978-3-319-64206-2_9

This paper proposes the Tree-Structured Gated Recurrent Unit (TS-GRU) model for sentiment analysis and compares it with the state-of-the-art Tree-Structured Long Short Term Memory (TS-LSTM) [16] and other models in terms of effectiveness. The model is evaluated on the Stanford Sentiment TreeBank (SST) for the binary (number of classes, $C = 2$) and fine-grained ($C = 5$) sentiment classification problems.

2 Related Work

The promising avenue of research in sentiment analysis opened up with expansive growth of deep learning. A broad range of neural networks was already harnessed to the sentiment analysis problem, including recurrent and recursive neural networks, convolutional neural networks, autoencoders and Restricted Boltzmann Machines (RBMs). The common aspect of these models is that they do not rely on man-made features.

Recursive autoencoders (RvAEs), introduced into sentiment analysis by Socher [14], work over a parse tree of a sentence, build by an external parser. RvAEs were further extended into Matrix-Vector Recursive Neural Networks (MV-RNNs) [13]. The original idea of this model is that an embedding is not represented by a vector, but by a (vector, matrix) pair. This increases representational power of phrases, but very large number of additional parameters introduced with matrices makes the model more prone to overfitting. The next model, Recursive Neural Tensor Network (RNTN) [15], tries to alleviate this problem. Embeddings are represented back with vectors only, but equations define the model using tensors instead of matrices. Tensor multiplications give the model enough representational power, at the same time solving the problem with overfitting.

Current state of the art in the fine-grained sentiment analysis, measured on the Stanford Sentiment TreeBank, is achieved by Tree-Structured LSTMs [16]. Similarly to other recursive neural networks, this approach uses a parse tree as a backbone for sentiment signal propagation. An alternative to Tree-Structured LSTMs way of combining in a parent node sentiment signals from children, named S-LSTMs, was proposed in [19].

Document level sentiment can be determined by hierarchically building document neural representation on the basis of neural representations of its sentences. Gated recurrent neural networks with LSTM or convolutional components achieve here state-of-the-art results, as measured on Yelp and IMBD datasets [18].

3 Tree-Structured GRU Model

3.1 LSTM and GRU

Let x be an input sequence of length T, i.e., $x = (x^{(1)}, x^{(2)}, \ldots, x^{(T)})$, and let N be dimensionality of its elements, i.e., each $x^{(t)} \in \mathbb{R}^N$.

LSTMs are able to remember long-term dependencies due to the presence of memory cell $s^{(t)}$. The flow of a signal is controlled by three gates: input gate $i^{(t)}$, forget gate $f^{(t)}$ and output gate $o^{(t)}$. Cell $g^{(t)}$ denotes candidate hidden state on the basis of which hidden state $h^{(t)}$ is computed. Dimensionality of $h^{(t)}$ and other LSTM components equals M. Values of LSTM cells and gates are determined according to the following equations ($\odot$ denotes the element-wise multiplication – the Hadamard product):

$$
\begin{bmatrix} g^{(t)} \\ i^{(t)} \\ f^{(t)} \\ o^{(t)} \end{bmatrix} = \begin{bmatrix} \tanh \\ \sigma \\ \sigma \\ \sigma \end{bmatrix} W \begin{bmatrix} x^{(t)} \\ h^{(t-1)} \end{bmatrix} + b
$$

$$
s^{(t)} = g^{(t)} \odot i^{(t)} + s^{(t-1)} \odot f^{(t)}
$$

$$
h^{(t)} = \tanh(s^{(t)}) \odot o^{(t)}
$$

(1)

where $W \in \mathbb{R}^{4M \times (N+M)}$, bias $b \in \mathbb{R}^{4M}$ and σ is the sigmoid logistic function.

Structure of GRU, shown in Fig. 1, is simpler than the LSTM one. GRU contains only two gates: reset gate $r \in \mathbb{R}^M$ and update gate $z \in \mathbb{R}^M$. Compared to LSTM, GRU does not have an output gate and is defined by a simpler set of equations:

$$
z = \sigma\left(U^z x^{(t-1)} + V^z h^{(t-1)}\right)
$$

$$
r = \sigma\left(U^r x^{(t-1)} + V^r h^{(t-1)}\right)
$$

$$
\bar{h} = \tanh\left(U^h x^{(t-1)} + V^h(h^{(t-1)} \odot r)\right)
$$

$$
h^{(t)} = (\mathbf{1} - z) \odot \bar{h} + z \odot h^{(t-1)}
$$

(2)

Before computing hidden state $h^{(t)} \in \mathbb{R}^M$, candidate hidden state $\bar{h} \in \mathbb{R}^M$ must be determined. Vector $\mathbf{1}$ denotes M-dimensional vector composed of ones.

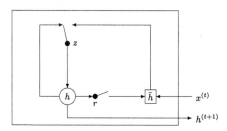

Fig. 1. Structure of Gated Recurrent Unit. Source: [2]

3.2 Model

GRU is designed for work with linear structures like sequences coped with RNNs. GRU version adjusted to operations on branching structures, called grConv, was proposed and applied to machine translation [1].

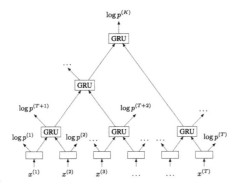

Fig. 2. Tree-Structured GRUs network over a parse tree

Tree-Structured GRU extension proposed in the paper is able to cope with recursive structures like parse trees (Fig. 2). The model is determined by parameters θ: $W^{\mathrm{x}} \in \mathbb{R}^{M \times N}$; $U^{\mathrm{z}1}, V^{\mathrm{z}1}, U^{\mathrm{z}2}, V^{\mathrm{z}2}, U^{\mathrm{r}}, V^{\mathrm{r}}, U^{\mathrm{h}}, V^{\mathrm{h}} \in \mathbb{R}^{M \times M}$; $W^{\mathrm{y}} \in \mathbb{R}^{C \times M}$; and $F \in \mathbb{R}^{N \times \#W}$, where $\#W$ is the number of different words in the training set.

Tree-Structured GRU is governed by the following equations:

$$
\begin{aligned}
z^i &= \sigma\big(U^{\mathrm{z}i} h_1 + V^{\mathrm{z}i} h_2\big), \qquad i = 1, 2 \\
r &= \sigma\big(U^{\mathrm{r}} h_1 + V^{\mathrm{r}} h_2\big) \\
\bar{h} &= \tanh\big(U^{\mathrm{h}}(h_1 \odot r) + V^{\mathrm{h}}(h_2 \odot r)\big) \\
h &= \Big(1 - \sum_{i=1}^{2} z^i\Big) \odot \bar{h} + \sum_{i=1}^{2} z^i \odot h_i
\end{aligned}
\tag{3}
$$

States h_1 and h_2 denote hidden states of the left and right child unit of a parent GRU. Gate z was implemented as two binary gates z^1 and z^2.

The model works as follows. At first, each element $x \in \mathbb{R}^N$ of input sequence $\boldsymbol{x}$, represented with a GloVe vector [11], is projected onto the input hidden state $h \in \mathbb{R}^M$:

$$
h = \tanh\big(W^{\mathrm{x}} x\big).
\tag{4}
$$

Hidden states are propagated up the tree according to TS-GRU definition (3). On the basis of the hidden state, h, each GRU computes its output signal, $\log p \in \mathbb{R}^C$, with the logsoftmax function according to (6). To improve network effectiveness regularization dropout layer (5) is applied in-between h and p. In experiments dropout ratio was set to $1/2$.

$$
\tilde{h} = \mathrm{dropout}(h)
\tag{5}
$$

$$
p = \mathrm{softmax}(W^{\mathrm{y}} \tilde{h}) = \frac{\exp(W^{\mathrm{y}} \tilde{h})}{\sum_{j'} \exp(W^{\mathrm{y}}_{j'.} \tilde{h})}
\tag{6}
$$

$$
\hat{c} = \arg\max_k p_k
\tag{7}
$$

Output signal is compared with the true sentiment and error is backpropagated down a parse tree. At the input, error signal is backpropagated through fine-tuning matrix F to pre-input storing fine-tuned version of the input vectors.

The cost function $J(\theta)$ used as the training criterion is defined for one sentence as follows:

$$J(\theta) = -\frac{1}{K} \sum_{i=1}^{K} \log p_c^{(i)} + \frac{\lambda}{2} ||\theta||_2^2, \tag{8}$$

where sum goes over all nodes of a parse tree. The number of such nodes, K, equals $2T - 1$, as the network is spanned over a binary constituency tree. Weight decay parameter, λ, determines the importance of the regularization term $||\theta||_2^2$.

Vector $p^{(i)}$ (6) contains probabilities a word or phrase belongs to each of C classes, and $p_c^{(i)}$ is just the probability corresponding to the true class $c \in \{1, \ldots, C\}$, denoting the correct word or phrase sentiment read from the SST gold standard. The sentiment class $\hat{c}$ of a word or phrase returned by the model is determined according to (7).

4   Experiments and Results

Experiments were conducted on the Stanford Sentiment TreeBank (SST) [15], containing movie reviews. SST consists of 11 855 sentences parsed with the Stanford Parser into 239 232 phrases. Each sentence and phrase goes with the assigned sentiment, s, being a real number in interval [0,1]. Using appropriate cutoffs, this number can be mapped into one of 5 fine-grained sentiment classes: very negative, negative, neutral, positive and very positive. Removing neutral opinions and merging together very negative and negative classes into one class, and similarly merging positive and very positive classes we obtain coarse-grained, binary sentiment classification version of the problem. The mapping is done with functions f_V and f_{II}:

$$f_V(s) = \begin{cases} 1, & \text{for } s \in [0, 0.2) \\ 2, & \text{for } s \in [0.2, 0.4) \\ 3, & \text{for } s \in [0.4, 0.6) \\ 4, & \text{for } s \in [0.6, 0.8) \\ 5, & \text{for } s \in [0.8, 1] \end{cases} \quad \text{and} \quad f_{II}(s) = \begin{cases} 1, & \text{for } s \in [0, 0.4) \\ 2, & \text{for } s \in [0.6, 1] \end{cases} \tag{9}$$

SST comes with the predefined split to the training, optimization, and test set, containing 8544, 1101 and 2210 sentences respectively for the fine-grained version. In the binary sentiment classification 6920, 872 and 1821 sentences from SST were used, respectively.

There are 24860 different words in SST, of which 15665 were initialized with GloVe vectors and remaining words, not present in the GloVe dictionary, got random initialization. The model parameters, θ, were initialized from the uniform distribution $\mathcal{U}(-\frac{1}{\sqrt{D}}, \frac{1}{\sqrt{D}})$, where D is a dimension of the given layer input, i.e., $D = N$ for W^x, $D = M$ for U^r, etc. The size of GloVe embedding was fixed to $N = 300$.

The best model was selected through the full grid search of the following hyperparameters: learning rate $\varepsilon \in \{0.005, 0.05, 0.1\}$, size of the hidden layer $M \in \{40, 60, 80, 100, 150, 200\}$, and weight decay $\lambda \in \{10^{-4}, 2 \cdot 10^{-4}, 10^{-3}\}$. Both in the binary and fine-grained problem the highest effectiveness on the optimization set was achieved for hyperparameters $\varepsilon = 0.05$, $M = 100$, and $\lambda = 10^{-4}$.

The network was trained with the AdaGrad algorithm in mini-batches of 25 samples. The optimization algorithm itself was model selected from 3 algorithms: SGD, AdaGrad and Adam. The error signal was propagated with the Backpropagation Through Structure (BTS) algorithm [4]. The parameters of the optimal model, θ, were found for the network trained 4 epochs (the fine-grained problem) and 6 epochs (the binary problem), when the highest accuracy was achieved on the optimization set.

Accuracy of various approaches to sentiment classification is compared in Table 1. For sentences TS-GRU achieved 49.28% accuracy in the fine-grained classification and 76.16% accuracy in the binary classification. Analysis of sentiment of phrases is always simpler, TS-GRU classified them with 66.50% (fined-grained sentiment) and 77.80% (binary sentiment) accuracy.

TS-GRUs revealed significant impact of proper initialization of input vectors – initialization with subsequently fine-tuned GloVe vectors showed over 6% improvement over a random initialization. This behaviour is similar to TS-LSTMs, where 7% improvement was achieved. Exact impact of input vector initialization on sentiment accuracy is shown in Table 2.

Table 1. Accuracy of sentiment classification of sentences on the test set of SST, [%]

Method	Fine-grained	Binary
TS-LSTM [16]	**51.0**	88.0
TS-GRU	49.2	76.1
S-LSTM [19]	48.0	n/a
Bidirectional LSTM [16]	49.1	87.5
CNN-multichannel [10]	47.4	**88.1**
DCNN [9]	48.5	86.8
CharSCNN [12]	48.3	85.7
Deep RvNN [7]	49.8	86.6
RNTN [15]	45.7	85.4
MV-RNN [15]	44.4	82.9
RvAE [15]	43.2	82.4
Naïve Bayes [15]	41.0	81.8
SVM [15]	40.7	79.4

Table 2. Impact of initialization of input tokens on sentiment classification accuracy, [%]

Gate	Representation	Fine-grained	Binary
GRU	GloVe fine-tuned	49.2	76.1
GRU	GloVe fixed	48.9	74.1
GRU	Random	42.8	73.3
LSTM	GloVe fine-tuned	51.0	88.0
LSTM	GloVe fixed	49.7	87.5
LSTM	Random	43.9	82.0

The number of parameters of the TS-GRU network for $C = 5$ equals $8M^2 + MN + CM + N \cdot \#W$, i.e., 7568500 parameters when fine-tuning is applied and $8M^2 + MN + CM$, i.e., 110500 parameters when fixed vectors are used. The corresponding TS-LSTM network [16] needed 316800 parameters for the input represented with $N = 300$ dimensional GloVe vectors.

5 Conclusions

In this paper we proposed the TS-GRU network, which adapts GRU to recursive networks, spanned over a parse tree. The model was inspired by the TS-LSTM network and the gated recursive convolutional neural network (grConv).

TS-GRU network achieved high accuracy on the binary sentiment classification, ranking third, behind TS-LSTM and Deep RvNN, although it performed badly on the fine-grained sentiment classification. Without input vectors fine-tuning, the TS-GRU network needed, however, three times less parameters than the TS-LSTM network.

Obtained accuracies are also comparable with human judgments, which vary between 70%–90% effectiveness, in particular [17] reports 83.6%–87.9% human accuracy on a different evaluation set.

Acknowledgments. This research was supported in part by PL-Grid Infrastructure. The research was also partially financed by AGH University of Science and Technology Statutory Fund.

References

1. Cho, K., van Merrienboer, B., Bahdanau, D., Bengio, Y.: On the properties of neural machine translation: encoder-decoder approaches. In: Wu, D., Carpuat, M., Carreras, X., Vecchi, E.M. (eds.) Proceedings of SSST-8, Eighth Workshop on Syntax, Semantics and Structure in Statistical Translation, pp. 103–111 (2014)
2. Cho, K., van Merriënboer, B., Gülçehre, Ç., Bahdanau, D., Bougares, F., Schwenk, H., Bengio, Y.: Learning phrase representations using RNN encoder-decoder for statistical machine translation. In: Proceedings of the 2014 Conference on Empirical Methods in Natural Language Processing, pp. 1724–1734 (2014)

3. Chung, J., Gülçehre, Ç., Cho, K., Bengio, Y.: Empirical evaluation of gated recurrent neural networks on sequence modeling. CoRR abs/1412.3555 (2014)
4. Goller, C., Küchler, A.: Learning task-dependent distributed representations by backpropagation through structure. In: Proceedings of the International Conference on Neural Networks, ICNN 1996, pp. 347–352 (1996)
5. Greff, K., Srivastava, R.K., Koutník, J., Steunebrink, B.R., Schmidhuber, J.: LSTM: a search space Odyssey. arXiv preprint arxiv:1503.04069 (2015)
6. Hochreiter, S., Schmidhuber, J.: Long short-term memory. Neural Comput. **9**(8), 1735–1780 (1997)
7. Irsoy, O., Cardie, C.: Deep recursive neural networks for compositionality in language. In: Ghahramani, Z., Welling, M., Cortes, C., Lawrence, N.D., Weinberger, K.Q. (eds.) Advances in Neural Information Processing Systems, vol. 27, pp. 2096–2104 (2014)
8. Józefowicz, R., Zaremba, W., Sutskever, I.: An empirical exploration of recurrent network architectures. In: Bach, F.R., Blei, D.M. (eds.) Proceedings of the 32nd International Conference on Machine Learning, pp. 2342–2350 (2015)
9. Kalchbrenner, N., Grefenstette, E., Blunsom, P.: A convolutional neural network for modelling sentences. In: Toutanova, K., Wu, H. (eds.) Proceedings of the 52nd Annual Meeting of the Association for Computational Linguistics, pp. 655–665 (2014)
10. Kim, Y.: Convolutional neural networks for sentence classification. In: Moschitti, A., Pang, B., Daelemans, W. (eds.) Proceedings of the 2014 Conference on Empirical Methods in Natural Language Processing, pp. 1746–1751 (2014)
11. Pennington, J., Socher, R., Manning, C.: GloVe: global vectors for word representation. In: Moschitti, A., Pang, B., Daelemans, W. (eds.) Proceedings of the 2014 Conference on Empirical Methods in Natural Language Processing, pp. 1532–1543 (2014)
12. dos Santos, C., Gatti, M.: Deep convolutional neural networks for sentiment analysis of short texts. In: Tsujii, J., Hajič, J. (eds.) Proceedings of COLING 2014, The 25th International Conference on Computational Linguistics, Technical Papers, pp. 69–78 (2014)
13. Socher, R., Huval, B., Manning, C.D., Ng, A.Y.: Semantic compositionality through recursive matrix-vector spaces. In: Tsujii, J., Henderson, J., Pașca, M. (eds.) Proceedings of the 2012 Joint Conference on Empirical Methods in Natural Language Processing and Computational Natural Language Learning, EMNLP-CoNLL 2012, pp. 1201–1211 (2012)
14. Socher, R., Pennington, J., Huang, E.H., Ng, A.Y., Manning, C.D.: Semi-supervised recursive autoencoders for predicting sentiment distributions. In: Barzilay, R., Johnson, M. (eds.) Proceedings of the 2011 Conference on Empirical Methods in Natural Language Processing, pp. 151–161 (2011)
15. Socher, R., Perelygin, A., Wu, J., Chuang, J., Manning, C.D., Ng, A.Y., Potts, C.: Recursive deep models for semantic compositionality over a sentiment treebank. In: Proceedings of the 2013 Conference on Empirical Methods in Natural Language Processing, pp. 1631–1642 (2013)
16. Tai, K.S., Socher, R., Manning, C.D.: Improved semantic representations from tree-structured long short-term memory networks. In: Zong, C., Strube, M. (eds.) Proceedings of the 53rd Annual Meeting of the Association for Computational Linguistics and the 7th International Joint Conference on Natural Language Processing, pp. 1556–1566 (2015)

17. Takala, P., Malo, P., Sinha, A., Ahlgren, O.: Gold-standard for topic-specific sentiment analysis of economic texts. In: Proceedings of the Ninth International Conference on Language Resources and Evaluation, LREC, vol. 2014, pp. 2152–2157 (2014)
18. Tang, D., Qin, B., Liu, T.: Document modeling with gated recurrent neural network for sentiment classification. In: Màrquez, L., Callison-Burch, C., Su, J. (eds.) Proceedings of the 2015 Conference on Empirical Methods in Natural Language Processing, pp. 1422–1432 (2015)
19. Zhu, X., Sobhani, P., Guo, H.: Long short-term memory over recursive structures. In: Bach, F.R., Blei, D.M. (eds.) Proceedings of the 32nd International Conference on Machine Learning, pp. 1604–1612 (2015)

Synthetic Speech in Therapy of Auditory Hallucinations

Kamil Sorokosz[1], Izabela Stefaniak[2], and Artur Janicki[1(✉)]

[1] Institute of Telecommunications, Warsaw University of Technology,
Nowowiejska 15/19, 00-665 Warsaw, Poland
{K.Sorokosz,A.Janicki}@tele.pw.edu.pl
[2] Institute of Psychiatry and Neurology, Sobieskiego 9, 02-957 Warsaw, Poland
blaszczuk@poczta.onet.pl

Abstract. In this article we propose using speech synthesis in the therapy of auditory verbal hallucinations, which are sometimes called "voices". During a therapeutic session a patient converses with an avatar, which is controlled by a therapist. The avatar, based on the XFace model and commercial text-to-speech systems, uses a high quality synthetic voice synchronized with lip movements. A proof-of-concept is demonstrated, as well as the results of preliminary experiments with six patients. The initial results are highly encouraging – all the patients claimed that the therapy helped them, and they also highly assessed the quality of the avatar's speech and its synchronization with the animations.

Keywords: Speech synthesis · Auditory hallucinations · Assistive technologies · Avatar · Visual speech

1 Introduction

Speech processing in the context of assisting people with various impairments has been researched for decades. The oldest and simplest devices, hearing aids, were just used to amplify the speech signal, and in this way they helped hearing impaired people. Next, when speech processing technologies advanced, speech synthesis and speech recognition started to be applied for visually impaired people. Along with the progress of speech, image and video processing, new, even more advanced systems were proposed, which were used in therapy for autism or dementia, for example.

In the current research we propose a system that uses speech synthesis, accompanied with facial animation, to be used in the therapy of auditory verbal hallucinations. The system was inspired by initial research by Huckvale et al. [13], who for that purpose proposed an avatar with voice conversion. In contrast to their approach, we proposed the use of synthetic speech. Thanks to this change, the therapist can accompany their patient during a therapeutic session and the therapy can take place in a single room.

© Springer International Publishing AG 2017
K. Ekštein and V. Matoušek (Eds.): TSD 2017, LNAI 10415, pp. 83–91, 2017.
DOI: 10.1007/978-3-319-64206-2_10

This paper is organized as follows: first, we will present the theoretical background, briefly describing the problem of voice hallucinations, treatment of psychological diseases using audio-visual techniques and the use of speech synthesis in the context of assistive technologies. Next, in Sect. 3, we describe our proposed solution. Section 4 will present the results of initial experiments using the proposed therapy. Finally, Sect. 5 will summarize the article.

2 Theoretical Background

2.1 Auditory Verbal Hallucinations

In psychopatology, hallucinations are the primary perception disorder in which qualitative changes occur [15]. They are false sensory perceptions, accompanied by a sensation of reality [3], which occur without the involvement of any external stimulus. According to the diagnostic criteria of the International Classification of Diseases (ICD-10), auditory verbal hallucinations are the most commonly encountered and, at the same time, the most characteristic symptoms of schizophrenia – a disease of the central nervous system in which abnormal neurotransmission in the area of the dopamine and serotonine system is observed. The hallucinations interact with the patient, comment on their behavior and provoke them to various actions, etc.

Medications applied for the treatment of schizophrenia symptoms (antipsychotic drugs) lead to, i.a., dopamine receptors blocking (D2). Thanks to that, a decrease in the intensification of productive symptoms (which include auditory hallucinations) is obtained [21]. Auditory verbal hallucinations, often called "voices," occur with 75% of people diagnosed with schizophrenia. What is more, they are also symptoms of other disorders, such as borderline personality disorder, posttraumatic stress disorder, epilepsy, Parkinson's disease, as well as dissociative, psychotic and affective disorders. They can also be observed in people without any clinical diagnosis [16].

Clinical practice often shows that medical procedures are not able to help the patients with voice hallucinations in a sufficient way. Half of the hallucinations progress into a chronic form so they are retained for a period of several months or even years despite the application of pharmacological treatment. This symptom often becomes the reason for the psychiatric hospitalization as well as the patient's prolapse from social functions. Such a situation requires a search for new therapeutic solutions. One of them is cognitive behavioral therapy that allows for the intensification of symptom understanding. It is included in the Schizophrenia PORT Treatment Recommendation [5] and acknowledged as one of the most effective therapeutic approaches.

Auditory hallucinations are often accompanied by feeling difficult emotions by the patient as well as the occurrence of non-adaptive behavior. In the cognitive model of auditory hallucinations [6], stressful emotions and actions do not result from the contents coming from the heard auditory hallucinations but from the meaning and belief that the person gives them. In this conception, work on the relationship between the patient and the symptom is important. In the proposed method, the use of an avatar within the cognitive behavioral therapy enables the extraction and the modification of this relationship.

2.2 Treating Psychological Disorders Using Audio-Visual Techniques

Using audio-visual techniques to treat psychological disorders, including voice hallucinations, has been tested in the past. One such method was described in 2013 by Huckvale and his colleagues from University College London [13]. The authors developed and evaluated a therapy based on an audio visual dialog system. In this method, first, the patient was asked to create an avatar by choosing a face resembling the visual hallucinations in their mind, as well as modifying the voice timbre to resemble their voice hallucinations. Next, during a therapeutic session, such an avatar was driven by the therapist's voice and the patient ran a conversation with the avatar. Voice conversion was used to change the timbre of the therapist's voice to sound more similar to the voice in the patient's hallucinations. A visual speech synthesis (VSS) system with a real time lip synchronization algorithm was used. During the session the therapist was located in another room and was able to talk with the patient with their natural voice by means of a video conference or, depending on a switch position, with their converted voice through the avatar.

The pilot studies confirmed that the application helped patients to control their hallucinations in real life after a series of short sessions. A clinical trial was started in 2012 and it is still in progress [7].

There are also other examples of successfully using audio-visual techniques in assisting patients with psychological disorders. One of them is the therapy of patients suffering from depression [10,17], and other examples are described for patients with different kinds of phobias [4,20].

2.3 Synthetic Speech in Assistive Context

Speech synthesis, apart from classical applications in man-machine interfaces, has been known to assist people with various disabilities. One such application is using synthetic speech to support patients who lost their ability to speak, either due to surgery (e.g., a laryngectomy) or neurological damage (e.g., caused by a stroke). Such solutions using text-to-speech (TTS) technology are often called Voice Output Communication Aids (VOCAs).

There has been research on creating VOCAs with personalized voices. Such an approach was proposed for individuals with dysarthria [8]. They proposed the adaptation of a statistical model of dysarthric speech extracted from an individual's voice, using the HTS toolkit. During the study, personalized synthetic voices for two participants with dysarthria were built and evaluated. Participants assessed the technique as promising and convincing.

TTS technology undoubtedly provides enormous support to visually impaired people. There are a lot of free and commercialized reading assistants that read aloud the specified content, e.g., e-mails, books, messages, news articles, bus schedules, temperature and weight, etc. – and an overview of TTS-based software and hardware for visually impaired people can be found in [11].

So far, however, we have found no information about using synthetic speech in therapy for psychological disorders. Our work proposes the use of this technique for the therapy of auditory verbal hallucinations.

3 Proposed Approach

In the proposed approach we decided to use synthetic speech instead of converted speech, as described in [13]. There were several reasons for that change:

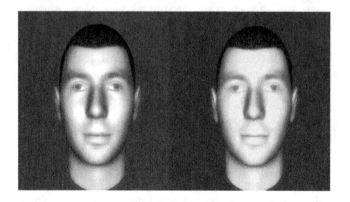

Fig. 1. Two faces used by avatar.

- we wanted the therapist to sit next to the patient and assist him/her during interactions with the avatar, so that the therapist can control the relationship between the patient and their symptom. This was hardly possible when the therapist was in another room;
- we did not want to rely on the oral skills of the therapist. When using synthetic speech the vocal effect is fully controlled by the system;
- voice conversion alters to some extent voice timbre and pitch; however, it usually has no impact on duration nor other speaking habits, based on which a speaker can be recognized. Therefore, there is a risk that the patient will associate the avatar's voice with the therapist, which would be highly unwanted;
- using converted speech required two separate rooms, which can sometimes pose logistical problems.

Following these presumptions we decided to set up a proof-of-concept implementation and ran initial experiments with the patients. The system was developed to use the Polish language as the patients were Polish native speakers.

3.1   Proof-of-Concept Setup

We proposed that the therapeutic session took place in a single room, so that the therapist was able to talk with the patient face to face. The patient sat in front of the screen with the avatar displayed and watched the animations, which were discretely controlled by the therapist. The therapy session was divided into two phases:

1. In the first (offline) one, the therapist prepared an individualized set of prompts for a given patient (i.e., with content of hallucinations) and the video files with animations were generated.
2. In the second one, the patient, accompanied by the therapist, interacted with the avatar, which played back a required animation in a way that was controlled by the therapist.

3.2   Speech and Video Generation

For our proof-of-concept implementation we used a VSS engine developed for Polish [14] with two faces (Fig. 1). The system was based on the XFace toolkit [2] – an open source tool for the development of 3D talking heads implemented by FBK-irst in Trento, Italy, that supported both MPEG-4 muscle deformation and keyframe interpolation based animations. In our approach the VSS engine was driven by synthetic speech sampled at 16 kHz. We decided to use a number of different commercial TTS systems (unit selection-based or HTS-based) to get a naturally sounding voice that would be fully intelligible to the patients. XFace as the input required a sequence of visemes with the timestamps provided for each lip movement in the form of a .pho file. The overall system design is presented in Fig. 2.

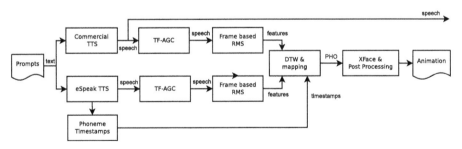

Fig. 2. Avatar's system design.

3.3   Speech and Video Synchronization

The drawback of the commercial TTS software was the lack of phoneme timestamps, which are required to control the face movements. To overcome this problem we used another TTS engine – the freely available eSpeak [1], which generates poor quality speech, but with timestamps. Next, we used the dynamic time warping (DTW) algorithm [19] for voice alignment between the speech signal generated by the commercial and eSpeak TTS systems. To improve this alignment we normalized both speech signals, using a time-frequency automatic gain control (TF-AGC) algorithm implemented by D. Ellis [9].

The DTW algorithm was applied to a sequence of the root mean square (RMS) energy features calculated on 50 ms frames on the normalized signals. The result of this alignment was used for mapping the timestamps generated by eSpeak TTS to the speech generated by the commercial software. Next, phoneme to viseme mapping was performed, according to the description in [14], using an inventory of 12 visemes. Next, the XFace animation was generated.

The animation together with the synchronized voice were recorded using screen recording software called *recordmydesktop*, which is available on the Linux platform. When needed, the video file was edited using a video editor. The video modifications applied were only minor, e.g., the image was rescaled if it was know that the patient's auditory hallucination was accompanied by a slim or a corpulent face.

4   Results Assessment and Discussion

To assess the initial results of the therapy of auditory hallucinations we asked six patients, who had either finished their therapy or were in its final stage, to fill in a questionnaire. Similar to other studies (e.g., [18]), we used a 5-degree Likert scale [12], which is widely used in the assessment of patient satisfaction. We exposed the patients to 13 statements on the quality of the speech and animation as well as about how helpful the therapy was, etc. The patients were asked to decide whether they strongly disagreed with the statement (scored as 1), they disagreed (scored as 2), they had no opinion (scored as 3), they agreed (scored as 4), or they strongly agreed (scored as 5). The full list of questions with the average results is displayed in Fig. 3.

A somewhat similar set of statements was presented to the therapist – during this initial stage of the experiment the therapy was conducted by a single psychiatrist. This set of statements was extended to include statements about the therapist's experience as a user of the system: how friendly the tool was and would additional functionalities be needed, etc.

The results from the patient survey showed that the therapy using the synthetic speech was accepted by all the patients taking part in the experiment. All of them found that the therapy helped them (score: 4.7); what is more, five out of six of the patients claimed that they like talking to the avatar (score: 4.2). All of the patients said that the speech generated by the avatar was intelligible. A high score (4.7) next to the statement: "Avatar's utterances are natural and fluent" is proof of the high quality of the TTS systems used. Synchronization of the animation was also highly assessed and judged as "fluent" – the score yielded here was 4.7. In the comments the patients stressed that the therapist's presence next to them during the sessions with the avatar was very important.

In contrast, scores referring to how the avatar's face fit the hallucinations were lower – it yielded 3.8. This result was not surprising, since no advance

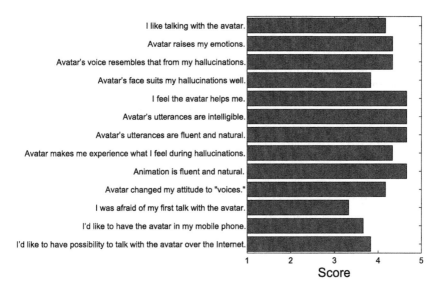

Fig. 3. Results of initial evaluation of avatar therapy by patients.

face modification was used, apart from basic operations, such as re-scaling. The answers from the therapist confirmed that the proposed technique was highly effective in the treatment of auditory hallucinations and, if developed into a fully functional system, it would be a valuable tool.

5   Conclusions and Future Work

The proposed avatar solution with the use of synthetic speech turned out to be highly promising. All of the patients that used the solution claimed that the avatar-based therapy was helpful and that after the therapy the auditory hallucinations were either less severe or the patients learned how to cope with them in a better way. One of the patients wrote it the survey: "Thanks to the avatar I understood better my hallucinations and now most of them are gone." They also commented that the therapist's presence during the sessions with the avatar was highly supportive.

All of the patients accepted the quality of the synthetic speech and found it easy to understand, and that it was "fluent and natural." We suspect that a somewhat "non-natural" origin of the synthetic speech seemed to suit well the "non-naturalness" of the voice hallucinations. To the best of our knowledge, this work is the first to show the use of synthetic speech for therapy of psychological diseases.

In future work, following the patients' and the therapist's suggestions, we plan to develop a fully functional system that would offer the therapist full control over the system (including the possibility to quickly generate new sentences), higher quality animations and, possibly, a mobile version for a patient.

References

1. eSpeak - a free TTS engine. http://espeak.sourceforge.net/
2. Balcí, K.: Xface: open source toolkit for creating 3D faces of an embodied conversational agent. In: Butz, A., Fisher, B., Krüger, A., Olivier, P. (eds.) SG 2005. LNCS, vol. 3638, pp. 263–266. Springer, Heidelberg (2005). doi:10.1007/11536482_25
3. Bilikiewicz, A., Pużyński, S., Rybakowski, J., Wciórka, J.: Psychiatry. Wydawnictwo Medyczne Urban & Partner, Wrocław II (2002)
4. Brinkman, W.P., Hartanto, D., Kang, N., de Vliegher, D., Kampmann, I.L., Morina, N., Emmelkamp, P.G.M., Neerincx, M.: A virtual reality dialogue system for the treatment of social phobia. In: Extended Abstracts on Human Factors in Computing Systems, CHI 2012, pp. 1099–1102. ACM, New York (2012)
5. Buchanan, R.W., Kreyenbuhl, J., Kelly, D.L., Noel, J.M., Boggs, D.L., Fischer, B.A., Himelhoch, S., Fang, B., Peterson, E., Aquino, P.R., et al.: The 2009 schizophrenia PORT psychopharmacological treatment recommendations and summary statements. Schizophrenia Bull. **36**(1), 71–93 (2010)
6. Chadwick, P., Birchwood, M.: The omnipotence of voices. A cognitive approach to auditory hallucinations. Br. J. Psychiatry **164**(2), 190–201 (1994)
7. Craig, T.K.J., Rus-Calafell, M., Ward, T., Fornells-Ambrojo, M., McCrone, P., Emsley, R., Garety, P.: The effects of an audio visual assisted therapy aid for refractory auditory hallucinations (avatar therapy): study protocol for a randomised controlled trial. Trials **16**(1), 349 (2015)
8. Creer, S., Cunningham, P.G.S., Yamagishi, J.: Building personalized synthetic voices for individuals with Dysarthria using the HTS toolkit. In: Mullenix, J., Stern, S. (eds.) Computer Synthesized Speech Technologies: Tools for Aiding Impairment, pp. 92–115. IGI Global press, Hershey (2010)
9. Ellis, D.: Time-frequency automatic gain control (2010). https://labrosaeecolumb iaedu/matlab/tf_agc
10. Falconer, C.J., Rovira, A., King, J.A., Gilbert, P., Antley, A., Fearon, P., Ralph, N., Slater, M., Brewin, C.R.: Embodying self-compassion within virtual reality and its effects on patients with depression. Br. J. Psychiatry **2**(1), 74–80 (2016)
11. Freitas, D., Kouroupetroglou, G.: Electronic speech processing for persons with disabilities. Technol. Disabil. **20**, 135–156 (2008)
12. Grogan, S., Conner, M., Willits, D., Norman, P.: Development of a questionnaire to measure patients' satisfaction with general practitioners' services. Br. J. Gen. Pract. **45**(399), 525–529 (1995)
13. Huckvale, M., Leff, J., Williams, G.: Avatar therapy: an audio-visual dialogue system for treating auditory hallucinations. In: Proceedings Interspeech 2013, pp. 392–396, August 2013
14. Janicki, A., Bloch, J., Taylor, K.: Visual speech synthesis for Polish using keyframe based animation. In: Pułka, A., Golonek, T. (eds.) Proceedings of International Conference on Signals and Electronics Systems, ICSES 2010, pp. 423–426. IEEE, September 2010
15. Jarema, M.: Psychiatry. In: PZWL (2016). (in Polish)
16. Larøi, F., Sommer, I.E., Blom, J.D., Fernyhough, C., Hugdahl, K., Johns, L.C., McCarthy-Jones, S., Preti, A., Raballo, A., Slotema, C.W., et al.: The characteristic features of auditory verbal hallucinations in clinical and nonclinical groups: state-of-the-art overview and future directions. Schizophrenia Bull. **38**(4), 724–733 (2012)

17. Pagliari, C., Burton, C., Mckinstry, B.H., Wolters, M.: Psychosocial implications of avatar use in supporting therapy for depression. Stud. Health Technol. Inform. **181**, 329–333 (2012)
18. Paulo, S., Oliveira, L.C., Mendes, C., Figueira, L., Cassaca, R., Viana, C., Moniz, H.: DIXI – a generic text-to-speech system for European Portuguese. In: Teixeira, A., Lima, V.L.S., Oliveira, L.C., Quaresma, P. (eds.) PROPOR 2008. LNCS, vol. 5190, pp. 91–100. Springer, Heidelberg (2008). doi:10.1007/978-3-540-85980-2_10
19. Sakoe, H., Chiba, S.: Dynamic programming algorithm optimization for spoken word recognition. IEEE Trans. Acoust. Speech Signal Process. **26**(1), 43–49 (1978)
20. Sarver, N.W., Beidel, D., Spitalnick, J.S.: The feasibility and acceptability of virtual environments in the treatment of childhood social anxiety disorder. J. Clin. Child Adolesc. Psychol. **43**, 63–73 (2013)
21. Stahl, S.M.: Stahl's Essential Psychopharmacology: Neuroscientific Basis and Practical Applications. Cambridge University Press, Cambridge (2013)

Statistical Pronunciation Adaptation
for Spontaneous Speech Synthesis

Raheel Qader[1], Gwénolé Lecorvé[1(✉)], Damien Lolive[1], Marie Tahon[1],
and Pascale Sébillot[2]

[1] IRISA/University of Rennes 1 (ENSSAT), Lannion, France
{raheel.qader,gwenole.lecorve,damien.lolive,marie.tahon}@irisa.fr
[2] IRISA/INSA Rennes, Rennes, France
pascale.sebillot@irisa.fr

Abstract. To bring more expressiveness into text-to-speech systems, this paper presents a new pronunciation variant generation method which works by adapting standard, i.e., dictionary-based, pronunciations to a spontaneous style. Its strength and originality lie in exploiting a wide range of linguistic, articulatory and prosodic features, and in using a probabilistic machine learning framework, namely conditional random fields and phoneme-based n-gram models. Extensive experiments on the Buckeye corpus of English conversational speech demonstrate the effectiveness of the approach through objective and perceptual evaluations.

Keywords: Speech synthesis · Spontaneous speech · Pronunciation modeling · Statistical adaptation · Conditional random field

1   Introduction

Modeling pronunciation variation in spontaneous speech is critical to achieve expressive Text-To-Speech (TTS) synthesis since pronunciation variants reflect the emotional state of a speaker, his/her intention, or a specific accent. However, phonetizers used by most current TTS systems fail to capture these variants as they only rely on standard pronunciations, i.e., extracted or learned from a general dictionary. Thus, the resulting synthetic speech conveys a neutral and formal style. A solution to this problem is to adapt standard pronunciations in order to reflect spontaneousness. In a machine learning perspective, this task corresponds to predicting a sequence of spontaneous phonemes from an input sequence of canonical phonemes, i.e., deciding whether input phonemes should be deleted, substituted, simply kept as is, or if new phonemes should be inserted.

Most of the early work in the area of pronunciation adaptation relied on using predefined or automatically extracted phonological rules to derive alternative pronunciations [1–3]. In the recent literature, various machine learning and statistical approaches have been proposed. Notably, decision trees [4,5], random forests [6], neural networks [7,8], hidden Markov models [9], and Conditional Random Fields (CRFs) [8,10,11] have been investigated. Regarding features, two

© Springer International Publishing AG 2017
K. Ekštein and V. Matoušek (Eds.): TSD 2017, LNAI 10415, pp. 92–101, 2017.
DOI: 10.1007/978-3-319-64206-2_11

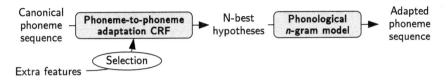

Fig. 1. Overview of the proposed pronunciation adaptation method.

categories are considered important to model pronunciation variation: linguistic-phonological features and prosodic ones. Linguistic-phonological features can be derived from textual data (POS, word predictability, lexical stress, etc.) [5,12], while prosodic features (F0, energy, duration, etc.) can be directly extracted from speech signals or predicted from text using a prosodic model [12,13]. Besides those two feature types, the benefits of using articulatory features have also been experimented [14,15]. Most of the mentioned studies have been applied in the context of Automatic Speech Recognition (ASR) and concentrated on utilizing either linguistic, articulatory or prosodic features.

In contrast, following our preliminary adaptation method proposed in [10], the method here combines a wide range of features and focuses on TTS rather than ASR. More precisely, the contributions are the following:

1. The importance and complementarity of linguistic, articulatory and prosodic information are studied w.r.t. the spontaneous style, highlighting that linguistic features are sufficient to perform good adaptations.
2. The usage of a phonological n-gram model is proposed to guarantee the *a posteriori* plausibility of the adapted pronunciations.
3. Perceptual tests demonstrate that adapted pronunciations are judged spontaneous while remaining reasonably intelligible.

In the remainder, the overall method and corpus are presented in Sect. 2. A study on feature selection and combination is provided in Sect. 3. The usage of a phonological model is exposed in Sect. 4. Finally, perceptual tests are discussed in Sect. 5.

2   Method Overview

Given a textual utterance, our fundamental idea for pronunciation adaptation is to predict the sequence of spontaneously realized phonemes from an input sequence of canonical phonemes. As shown in Fig. 1, we propose to perform this task in 2 steps. First, adapted pronunciation hypotheses are generated by a phoneme-to-phoneme CRF trained on canonical phonemes and a combination of linguistic, articulatory, and prosodic features. These features are selected offline, i.e., while setting up the method, in an automatic manner to optimize the CRF accuracy. Second, hypotheses are reranked using a phonological n-gram model of spontaneous phoneme sequences.

Table 1. List of all features. Selected features are in bold.

Linguistic features (22)

---

canonical phoneme • **word** • **is a stop word** • **syllable lexical stress** •
syllable part • **word frequency in English** • **reverse phoneme position in
syllable** • **phoneme position in syllable** • **syllable location** • stem frequency
in the interview • word frequency in the interview • syllable type • POS • number
of syllables of the word • stem frequency in English • grapheme • word length •
reverse utterance position • utterance position • word position • reverse word
position • word occurrence count in interview

Articulatory features (9)

---

vowel/consonant • **manner** • **place** • **shape** • **aperture** • **voiced** • **rounded**
• affricate • doubled

Prosodic features (10)

---

syllable energy • **syllable F0 shape** • **syllable tone** • **speech rate** • **pause
per syllable** • phone tone • distance to previous silence • distance to next silence
• distance to previous hesitation (*um/uh*) • distance to next hesitation (*um/uh*)

The method is experimented on 20 h of spontaneous American English speech
from the Buckeye corpus [16]. This represents 20 interviews with speakers from
central Ohio, USA, of various ages and both genders. Interviews are annotated
with their orthographic transcript and two phonemic transcripts: the standard
pronunciation of the words (*canonical phonemes*), and the one effectively uttered
by the speaker (*realized phonemes*). The average numbers of phonemes and words
per speaker are 22,789 and 7,354, respectively. The Phoneme Error Rate (PER)
between the canonical and realized phonemes is 28.3%. This very high rate shows
how different standard and spontaneous pronunciations are, and how difficult
adapting pronunciation to a spontaneous style is. Phone segmentation is also
available and about 40 linguistic-phonological (shortened to *linguistic* in the
remainder), articulatory, and prosodic features have been automatically added
using speech and natural language processing tools (see Table 2). Prosodic features have been directly estimated in an oracle way by processing signals of each
speaker, normalizing and strongly approximating the derived information. This
simulates a perfect prosody modeling, leading to adaptation results which are
not biased by prosody prediction errors, while remaining realistic. Finally, the
corpus has been randomly divided into a training set (60% of the utterances),
a development set (20%), and a test set (20%), with an equal representation of
each speaker in each set.

Phoneme sequences generated by our method are evaluated by PERs w.r.t.
the ground truth, i.e., the sequence realized by the speaker. Thus, the lower the
PER the better, the baseline being the PER of the canonical pronunciation, that
is before adaptation. Listening tests have also been conducted to perceptually
validate the method. All models have been learned on the training set, optimized
on the development set and evaluated on the test set. Canonical phonemes have
been automatically aligned with realized phonemes using `m2m-aligner` [17] to
train the phoneme-to-phoneme CRF.

3   Phoneme-to-Phoneme Adaptation

Phoneme-to-phoneme adaptation is performed by CRFs trained on the canonical phonemes and relevant selected features. This section briefly describes how linguistic, articulatory, and prosodic features have been selected, before presenting how the selected features have been combined to produce the final adaptation CRF.

Automatic selection is applied on each feature group separately in order to eliminate irrelevant and redundant features. The selection process relies on a greedy approach where votes are assigned to the most influential features, that is features leading to the lowest PER when training adaptation CRFs. These CRFs are trained without contextual information to avoid large training time overheads, i.e., information about the neighbors of each canonical phoneme is disregarded. Features resulting from this selection are highlighted in bold in Table 1. Linguistic and prosodic information derived from syllables is particularly valuable, as well as information about word frequencies. Regarding articulatory features, the selection has less effects (only 2 discarded features), meaning that no clear dominance can be established among them. Table 2 reports the influence of feature selection on PER for CRFs trained on the development set and on

Table 2. PERs (%) on the development set with selected features vs. all features. CRFs are trained without contextual information. In brackets, variations from the baseline (in percentage points).

Baseline (not adapted)			28.3
Adapted using	Canonical phonemes only (C)		30.7 (+2.4)
	+ Linguistic features	All (22)	26.6 (−1.7)
	(C + L)	Selected (8)	25.1 (−3.2)
	+ Articulatory features	All (9)	30.9 (+2.6)
	C + A)	Selected (7)	30.8 (+2.5)
	+ Prosodic features	All (10)	27.1 (−1.2)
	C + P)	Selected (6)	26.7 (−1.6)

Table 3. PERs (%) on the test set for all possible combinations. CRFs use contextual information.

Baseline (not adapted)			28.3				
Adapted using	C		24.2 (−4.1)		C + L + A		24.0 (−4.3)
	C + L		24.0 (−4.3)	Adapted using	C + L	+ P	21.1 (−7.2)
Adapted using	C	+ A	24.4 (−3.9)		C	+ A + P	21.4 (−6.9)
	C	+ P	21.5 (−6.8)		C + L + A + P		21.2 (−7.1)

each group of features independently. Results show that the selection is efficient for linguistic and prosodic features, whereas again almost useless for articulatory ones.

To optimize the method and search for potential complementarities, all possible combinations of selected features are tested. Moreover, adaptation CRFs are trained with contextual information, precisely 2 neighbors on the left and on the right, as this configuration has shown to lead to the best PER in preliminary tests. Table 3 reports PERs on the test set for these combination experiments. First, it appears that CRFs already perform rather well when solely relying on canonical phonemes (configuration C), thanks to contextual information. Then, when separately including the selected features, results show that linguistic features provide a small improvement (C + L), articulatory features bring worse results (C + A), and prosodic features (C + P) lead to a clear improvement with a reduction of 2.7 percentage points (pp) compared to the use of the sole canonical phonemes (C). Although prosodic features are extracted in an oracle way and thus lead to optimistic results, the latter result tends to show a strong relationship between prosody and pronunciation. When feature types are combined together, articulatory features bring worse results in all cases, definitely showing that they should not be considered in our method. On the contrary, results again demonstrate that linguistic and prosodic features are useful for pronunciation adaptation, bringing the best PER down to 21.1%. This conclusion has been validated by paired t-test and paired Wilcoxon test with confidence level $\alpha = 0.05$, whether these feature groups are considered individually or together.

Table 4. PERs (%) before and after reranking when adapting using canonical phonemes (C), linguistic (L), and prosodic features (P).

		Before reranking	After reranking
Baseline (not adapted)		28.3	
Adapted using	C	24.2 (−4.1)	23.7 (−4.6)
	C + L	24.0 (−4.3)	23.7 (−4.6)
	C + L + P	21.1 (−7.2)	**20.6 (−7.7)**

4   Phonological Rescoring and Reranking

A qualitative analysis of the adapted pronunciations shows that some phoneme sequences returned by the finally adopted CRF are very unlikely. For instance, the sequence /dt/ is rare in spontaneous English, and rather simply reduced to /d/ or /t/. To fix these imperfections, we propose as a second step to generate several pronunciation hypotheses using the adaptation CRF, and to rerank them according to scores given by a probabilistic phonological model[1]. Precisely, each hypothesis $\mathbf{h} = (p_1, \ldots, p_m)$ of m phonemes p_i is assigned a score $s(\mathbf{h})$ mixing the CRF and phonological model (PM) probabilities. This mixture is computed by a log-linear interpolation—which has been successfully used for N-best list reranking in various domains [18,19]—, and is formulated as follows:

$$s(\mathbf{h}) = \Pr{}_{\mathrm{CRF}}(\mathbf{h}) \times \Pr{}_{\mathrm{PM}}(\mathbf{h})^{\alpha} \times \beta^{m}, \tag{1}$$

where α and β are two parameters to be optimized. The parameter β is used to prevent the phonological model from favoring short hypotheses. Finally, the hypothesis with the highest score is selected as the adapted pronunciation.

In our experiments, the phonological model is a phoneme-based n-gram model estimated on the training set using a Witten-Bell smoothing. The order n of the model as well as α and β have been optimized such that they minimize PER on the development set, and consequently set to 5, 0.48 and 0.024, respectively. Training, optimization and reranking have all been conducted using SRILM [20]. Reranking is performed on the 10 best hypotheses predicted by the adaptation CRF, as tuned on the development set.

As shown in Table 4, our reranking technique always reduces PERs. The largest reduction is 0.5 pp, achieved for both canonical phonemes (C) and linguistic+prosodic configurations (C + L + P). Alongside, phonological reranking surprisingly seems to obviate linguistic features (C against C + L). However, results of the perceptual tests (Sect. 5) show that this is not true. Overall, and given the difficulty of the task, results show that our whole approach is effective as it reduces PER to a large extent with significant improvements up to 7.7 pp w.r.t. the baseline.

5   Perceptual Tests

AB tests on 40 synthesized speech samples have been conducted with 10 native English speakers. Listeners were asked to answer two questions: *"Between A and B, which sample is pronounced in the most spontaneous way?"*, and *"Which once is pronounced in the most intelligible way?"*. For both questions, listeners were also allowed to indicate that they do not have any preference. Orthographic transcripts were given along with the samples to help listeners to focus on pronunciations. Tests were set up to compare canonical and realized pronunciations to those generated by our adaptation method using either configurations C, C + L or C + L + P, all including phonological reranking.

Utterances have been selected among the 2,000 available utterances in the test set such that their PER between the canonical and realized pronunciations is high. This strategy has been designed to ensure that selected utterances reflect the difficulty of the task. Utterances were synthesized using the parametric HTS

---

[1] CRFs allow dependencies between predicted phonemes but it appeared in preliminary work that using a separate phonological model is better to avoid overfitting the training data.

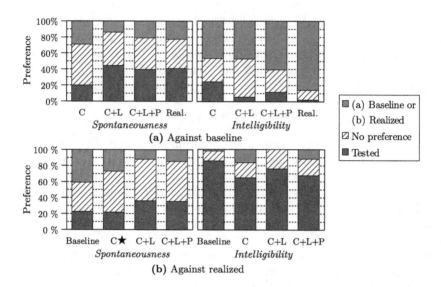

Fig. 2. Preference on spontaneousness and intelligibility between baseline, realized and adapted pronunciations. Adaptations were performed using canonical phonemes (C), linguistic features (L), and prosodic features (P). ★ stands for *"not statistically significant* (see footnote 2)".

v2.2 speech synthesis system trained with standard features [21] and on the Blizzard Challenge 2012 data [22], i.e., audiobooks with mixed speech styles and uttered by a same US male speaker. Hence, no bias toward standard or spontaneous speech can be observed. Unit selection has voluntarily been discarded since this type of system is usually sensitive to pronunciation variants, producing disturbing artefacts.

Figure 2 shows the comparison of speech samples generated using (*a*) the standard pronunciations (baseline) against adapted or realized ones and (*b*) realized pronunciations against the baseline and adapted ones, in terms of spontaneousness and intelligibility. Preference percentages are given as bar segments on the y-axis. Statistical significances of these ratios have been computed for all the tests[2].

First, Fig. 2a shows that realized pronunciations are logically judged as more spontaneous than the baseline, while being much less intelligible. Regarding adapted pronunciations, the configuration C performs poorly. Conversely, the two other adapted configurations are judged as much more spontaneous than the baseline, but again leading to intelligibility degradations. Finally, adaptation performs equally or even slightly better when using linguistic features alone, i.e., without prosodic ones. This is interesting since predicting prosodic features is difficult in TTS.

---

[2] Binomial test with $\alpha = 0.1$ and votes for "No preference" equally spread over A and B, following the methodology proposed in [23].

As for Fig. 2b, it surprisingly appears that C + L and C + L + P configurations are preferred over the realized pronunciations w.r.t. spontaneousness. This is probably correlated with the large intelligibility gap reported. Similarly, it can again be noticed that the use of the sole linguistic features performs slightly better than when also accounting for prosodic features, especially regarding intelligibility. On the one hand, these results demonstrate the effectiveness of our method in generating spontaneous pronunciations. On the other, they are in contradiction with PERs of Table 4. A deeper analysis shows that pronunciations produced using prosodic features, as well as the realized ones, seem to be more spontaneous but they are too complex for current TTS systems, especially because of strong coarticulations like /dn̩/ (like in "didn't") or /fm/ ("familiarity"). This penalizes intelligibility and, as a side effect, spontaneousness. Hence, a reasonable conclusion is that the proposed method is enough effective yet to reflect a spontaneous style while results could still be improved, on the condition that the speech data used to build the TTS voice are consistent with the desired degree of spontaneousness.

6   Conclusions and Future Work

In this paper, we have proposed a TTS-dedicated spontaneous pronunciation adaptation method which combines a phoneme-to-phoneme CRF and a phonological model. Objective and perceptual tests have shown that produced pronunciations effectively better reflect spontaneous speech than non adapted ones. The study of linguistic, articulatory and prosodic features shows that linguistic features are good predictors for spontaneous pronunciations, while articulatory ones are useless and prosodic ones tend to produce less intelligible speech. More generally, it seems that there is a tradeoff between the degree of spontaneousness and intelligibility. In the future, it would be interesting to more deeply investigate the relationship between prosody and pronunciation, and their impact on the perceived spontaneousness. Following this direction, finding out mechanisms to enable a fine control of intelligibility against spontaneousness is another interesting perspective, especially by taking into account the intrinsic phonemic and prosodic variability of the TTS system's voice. Finally, we have planed to apply the proposed method to emotional speech to generate synthetic speech samples.

Acknowledgments. This study has been realized under the ANR (French National Research Agency) project SynPaFlex ANR-15-CE23-0015.

References

1. Tajchman, G., Foster, E., Jurafsky, D.: Building multiple pronunciation models for novel words using exploratory computational phonology. In: Proceedings of Eurospeech (1995)
2. Giachin, E., Rosenberg, A., Lee, C.H.: Word juncture modeling using phonological rules for HMM-based continuous speech recognition. In: Proceedings of ICASSP (1990)

3. Oshika, B.T., Zue, V.W., Weeks, R.V., Neu, H., Aurbach, J.: The role of phonological rules in speech understanding research. IEEE Trans. Acous. Speech Signal Process. **23**, 104–112 (1975)

4. Goronzy, S., Rapp, S., Kompe, R.: Generating non-native pronunciation variants for lexicon adaptation. Speech Commun. **42**(1), 109–123 (2004)

5. Vazirnezhad, B., Almasganj, F., Ahadi, S.M.: Hybrid statistical pronunciation models designed to be trained by a medium-size corpus. Comput. Speech Lang. **23**, 1–24 (2009)

6. Dilts, P.C.: Modelling phonetic reduction in a corpus of spoken English using random forests and mixed-effects regression. Ph.D. thesis, University of Alberta (2013)

7. Chen, K., Hasegawa-Johnson, M.: Modeling pronunciation variation using artificial neural networks for English spontaneous speech. In: Proceedings of Interspeech (2004)

8. Karanasou, P., Yvon, F., Lavergne, T., Lamel, L.: Discriminative training of a phoneme confusion model for a dynamic lexicon in ASR. In: Proceedings of Interspeech (2013)

9. Prahallad, K., Black, A.W., Mosur, R.: Sub-phonetic modeling for capturing pronunciation variations for conversational speech synthesis. In: Proceedings of ICASSP (2006)

10. Qader, R., Lecorvé, G., Lolive, D., Sébillot, P.: Probabilistic speaker pronunciation adaptation for spontaneous speech synthesis using linguistic features. In: Dediu, A.-H., Martín-Vide, C., Vicsi, K. (eds.) SLSP 2015. LNCS (LNAI), vol. 9449, pp. 229–241. Springer, Cham (2015). doi:10.1007/978-3-319-25789-1_22

11. Tahon, M., Qader, R., Lecorvé, G., Lolive, D.: Improving TTS with corpus-specific pronunciation adaptation. In: Proceedings of Interspeech (2016)

12. Bell, A., Brenier, J.M., Gregory, M., Girand, C., Jurafsky, D.: Predictability effects on durations of content and function words in conversational English. J. Mem. Lang. **60**, 92–111 (2009)

13. Bates, R., Ostendorf, M.: Modeling pronunciation variation in conversational speech using prosody. In: Proceedings of ISCA Tutorial and Research Workshop (ITRW) on Pronunciation Modeling and Lexicon Adaptation for Spoken Language Technology (2002)

14. Livescu, K., Jyothi, P., Fosler-Lussier, E.: Articulatory feature-based pronunciation modeling. Comput. Speech Lang. **36**, 165–172 (2016)

15. Rasipuram, R., Doss, M.M.: Articulatory feature based continuous speech recognition using probabilistic lexical modeling. Comput. Speech Lang. **36**, 165–172 (2016)

16. Pitt, M.A., Johnson, K., Hume, E., Kiesling, S., Raymond, W.: The Buckeye corpus of conversational speech: labeling conventions and a test of transcriber reliability. Speech Commun. **45**, 89–95 (2005)

17. Jiampojamarn, S., Kondrak, G., Sherif, T.: Applying many-to-many alignments and hidden Markov models to letter-to-phoneme conversion. In: Proceedings of NAACL-HLT (2007)

18. Rosti, A.V.I., Matsoukas, S.: Combining outputs from multiple machine translation systems. In: Proceedings of NAACL-HLT (2007)

19. Huet, S., Gravier, G., Sébillot, P.: Morpho-syntactic post-processing of N-best lists for improved French automatic speech recognition. Comput. Speech Lang. **24**(4), 663–684 (2010)

20. Stolcke, A., Zheng, J., Wang, W., Abrash, V.: SRILM at sixteen: update and outlook. In: Proceedings of IEEE ASRU Workshop (2011)

21. Zen, H., Nose, T., Yamagishi, J., Sako, S., Masuko, T., Black, A.W., Tokuda, K.: The HMM-based speech synthesis system (HTS) version 2.0. In: Proceedings of SSW (2007)
22. King, S., Karaiskos, V.: The Blizzard challenge 2012. In: Proceedings of Blizzard Challenge 2012 Workshop (2012)
23. Karhila, R., Remes, U., Kurimo, M.: Noise in HMM-based speech synthesis adaptation: analysis, evaluation methods and experiments. IEEE J. Sel. Top. Signal Process. **8**(2), 285–295 (2014)

Machine Learning Approach to the Process of Question Generation

Miroslav Blšták[(⊠)] and Viera Rozinajová

Faculty of Informatics and Information Technologies, Institute of Informatics, Information Systems and Software Engineering, Slovak University of Technology in Bratislava, Ilkovicova 2, Bratislava, Slovakia
{miroslav.blstak,viera.rozinajova}@stuba.sk

Abstract. In this paper, we introduce an interactive approach to generation of factual questions from unstructured text. Our proposed framework transforms input text into structured set of features and uses them for question generation. Its learning process is based on combination of machine learning techniques known as reinforcement learning and supervised learning. Learning process starts with initial set of pairs formed by declarative sentences and assigned questions and it continuously learns how to transform sentences into questions. Process is also improved by feedback from users regarding already generated questions. We evaluated our approach and the comparison with state-of-the-art systems shows that it is a perspective way for research.

Keywords: Question generation · Machine learning · Computer assisted learning

1 Introduction

E-learning provides many opportunities how to enrich standard educational process. There are various educational tools available - like massive open online courses, online encyclopedias, online lexicons and so on. Although there is a variety of learning materials online and in electronic form, it is difficult to substitute the role of teacher. Interaction with him in the phase of knowledge verification is still hardly replaceable.

In order to contribute to solution of this problem, we propose an interactive question generation framework which creates a set of questions to verify students' knowledge and is also able to check their answers. It learns how to generate question from set of sentence-question pairs and it improves its ability by feedback from students. In many research areas of natural language processing (NLP), we can see the trend of suppression of classical linguistic approaches and prioritization of machine learning approaches. The big advantage of this approach lies in the possibility of gradual improvement of the system by adding new data (new sentence-question pairs and positive or negative feedback to generated questions). In the background of learning process, we transform unstructured text

© Springer International Publishing AG 2017
K. Ekštein and V. Matoušek (Eds.): TSD 2017, LNAI 10415, pp. 102–110, 2017.
DOI: 10.1007/978-3-319-64206-2_12

into structured data (by obtaining set of features about input sentence and its tokens). We mainly use two level of part-of-speech tags (simplified tags and full tags), named entities, multiword expressions and some semantic categories of words. The idea of our approach comes from ensemble methods: the combination of various simple approaches can be more effective in comparison to one tuned approach. That also allows us to overcome the fact that NLP tools do not give always correct results. We assume that this approach would also be beneficial in other task of NLP, where it is required to apply transformation process on text (e.g. text summarization or text simplification).

The rest of the paper is organized as follows. In the second section, we introduce some state-of-the-art ideas and question generation systems. In the third section, we describe our framework for question creation and its abilities. In the fourth section, we evaluate the framework and analyze the correctness of generated questions on two datasets. In the last section, we make conclusion and propose new research challenges for the future.

2   Related Works

There are many various works in areas of intelligent tutoring and question generation. One of the most typical approach is known as fill-the-blank. These systems firstly identify a part of the text (word or set of words) which will be an answer and remove it from text. Words for removing are usually chosen by part-of-speech (POS) tag (e.g. adjectives [9], prepositions [8], nouns [7,11] or combination of vocabulary words [5]). These systems just focus on simple questions: "fill the word in correct form into sentence". On the other side, there are systems focusing on whole text which generates more general questions (e.g. main idea of article or presentation [4]).

Somewhere in the middle, there are approaches for generating questions as interrogative sentences from input declarative sentences. This is known as sentence to question transformation and it usually uses set of rules for generation various types of questions known as wh-questions (e.g.: why, where, who, when). This approach requires generation of grammatically, syntactically and semantically correct question. In [3], they transform sentences into questions by modification of syntactic tree. At the beginning, set of rules for sentence simplification was applied on the text and then the simplified sentences were transformed into questions by another set of rules. System proposed in [6] used syntactic patterns to match the sentences and generates questions based on its syntactic structure (occurrence of subject, verb and prepositional phrase in sentences). In [2] used lexico-syntactic patterns, in [12] were proposed lexico-semantic patterns and in [1] the combination of multiple patterns (syntactic, semantic and lexical) was used.

The drawbacks of these approaches lie in coverage and extensibility of transformation rules for various sentences. Manually created rules and patterns cover only subset of possible sentences from input text. If the patterns are too general (e.g. syntactic patterns), they cover various types of questions, but the created questions are general too. If they use more detailed patterns which generate better questions, they need large number of these patterns to cover various sentence types and this is also related to system extensibility. Adding new patterns and rules into system or their derivation is time consuming. In [2] they propose a method for pattern derivation and they use google as judge about question correctness, but their overall results were weak. In [6] it is mentioned, that the system is capable to hold thousands to millions of rules inserted by user, but verification is missing and there is no idea how to search in that large number of patterns. It is also difficult to imagine that someone will create millions of syntactic rules. We try to overcome these constraints in our approach.

3 Our Approach

In this section, we will describe our framework for automatic question generation. We have used data driven approach, which is realized by extracting features from text and choosing the most appropriate class of questions (on the basis of these features). Each class of sentences has assigned set of possible questions and questions are chosen based on similarity of features. We describe features and matching process in more detail in the next section. When the questions are generated, users are asked to evaluate them or fix the incorrect ones. System uses this feedback for self-improvement (approach known as reinforcement learning). The framework consists of four modules and database of learned rules. We are working with English language as NLP tools work more reliably here in comparison to other languages and there are also better possibilities to compare our results with state-of-the-art systems.

3.1 Question Creation Pipeline

The first module serves as text preprocessing component. Its role is to divide input text into sentences and to obtain information about tokens - it transforms the unstructured text into structured data. Obtained information are used as features of the sentence. We now use these features:

- full POS tags (POS tags with grammar classes) obtained by Stanford CoreNLP,
- simplified POS tags (POS tags without grammar classes) derived from full POS,
- named entities (NE) obtained by Stanford CoreNLP,
- word classes obtained by Google Knowledge Graph (GKG),
- super sense tags (SST) obtained by Wordnet,
- multiword expressions calculated by our system and evaluated by Wordnet, Google Knowledge graph and Virtual International Authority File (VIAF): calculation based on merging the adjacent tokens with similar labels).

The second module searches for possible transformation rules stored in database which can be used for question generation. Matching criteria are based on similarity of features obtained by first module (example of similarity calculation between two sentences is shown in Table 1). Database of transformation rules is continuously built and improved as the new training data come in. Matching algorithm reflects number of consistent features and if some labels are different, it calculates the relatedness and interchangeability. Each feature has its own similarity model which calculates how related (interchangeable) are two tokens if they are not identical. For example, if we compare two tokens of which one is labelled as noun and the second as pronoun, their syntactic relatedness is higher in comparison to situation when we compare token with label noun to token with label preposition. Table 1 shows the simplified view on sentence similarity calculation between two sentences.

Table 1. Similarity calculation of two sentences: number of identical features of sentences are in the last column and number of identical tokens are in last row.

| Sentence: | The | president | of | Slovakia | is | Andrej Kiska. | Feature |
Sentence 2:	The	capital city	of	Czech Republic	is	Prague.	match
POS	DT	NN	IN	NN	VB	NN	6 of 6
POS full	DT	NN	IN	NNP	VBZ	NNP	6 of 6
NE				location		person location	1 of 2
SST		role place		country		- city	1 of 2
GKG				country		person city	1 of 2
Identical lemmas	The		of		is		3 of 6
Token match	3 of 3	2 of 3	3 of 3	4 of 5	3 of 3	2 of 4	

Sentences in real text are usually more complex and they have different length (different number of words), so we have to consider situations when the edit distance calculation is needed. We use Levenshtein distance known from string similarity [10] which reflects insertions (occurrence of extra token), deletion (some token is missing) or replacements (token with different label).

The third module, question estimator, estimates the quality of questions created by each rule. Estimated quality reflects similarity of sentence with the matched rule, number of correct questions generated in the past by this rule and ratio between correct and incorrect questions generated by this rule. In the end, there is simple grammar checker tool which checks the question correctness and decreases the total confidence score if some obstacles have been revealed (e.g. incorrect article before word). Finally, similar questions are eliminated and the rest questions are ordered by its confidence score. The questions above a certain threshold are sent to output as plausible questions.

The last module is in role of feedback. If the output question is not acceptable, it is possible to denote it and the system will take this information into account for the next time in similar questions.

3.2 Learning Process

First set of transformation rules were obtained from training dataset. At the beginning, we trained the system on small dataset of sentence-question pairs: about 50 pairs were added to system manually. The rest of pairs were obtained by interactive interface: the sentences from text were sequentially shown on interface and users were asked to write ideal questions to these sentences. System analyzed the structures of sentences from text and also structures of assigned questions written by users. Sentence structure is a set of features described in previous section. Based on these pairs it created new transformation rules. When some generated questions were marked as incorrect and the user has corrected the question, the system will adopt these changes.

This learning approach allows us to let the system build its set of rules for each type of questions and the users do not have to manage it. The framework also keeps positive and negative feedbacks about acceptance of previously generated similar questions and questions generated by rules. After the questions are generated, user can evaluate them and system updates the confidence of stored transformation rules which may change the priority of the rule for the next time (known as reinforcement machine learning approach).

4 Evaluation

In this section, we will evaluate ability of our framework to learn how to create new questions and then we compare correctness of generated questions with state-of-the art systems. Whereas, as far as we know, there is only one small dataset publicly available for question generation task [13], we decided to train the system on our datasets and leave the question generation dataset for comparison with state-of-the-art systems.

4.1 Question Generation Ability

Firstly, we trained the system on our dataset which consists of 306 articles about countries obtained from Wikipedia. We split dataset into two parts: the first quarter of articles which names start on letters from A to D (from Afghanistan to Denmark) were used for training. The sentences were sequentially shown on user interface and the students could assign the most desirable questions created from these sentences. The framework learns transformation rules based on these pairs (input sentence and question assigned by student). Then, these rules were used for question generation from the rest of this dataset and from another dataset of cities articles.

We manually evaluated the correctness of questions from countries dataset (Table 2). Whereas human evaluation takes some time and it was done for purpose of question generation verification, questions from cities dataset were evaluated only automatically (by question estimator which was described in previous section). In Table 3, correctness of questions generated from country datasets is shown. All these questions were checked by human and its correctness was proved.

Table 2. Datasets and number of created questions

	Articles	Sentences	Questions	Ques./Sent
Countries - training	51	499	1098	2,19
Countries - testing	155	1540	1466	0,95
Cities	460	3101	2426	0,78

Table 3. Correctness and sentence coverage of generated questions

	Correct questions	Incorrect questions	Total	Correctness (%)
Countries - training	1008	90	1098	91,8
Countries - testing	1288	178	1466	87,9

We also looked at types of generated questions (Tables 4 and 5). Some types of questions are underrepresented, what is typical in real texts. In this case, it is also due to the topic of articles in dataset (there are small possibilities to generate "who" questions from articles about countries and this is reflected in the number of questions and their correctness).

Table 4. Generated questions and their types from training dataset

Question type	All	Correct	Incorrect	correctness (%)
What	432	401	31	92,8
True-false	267	248	19	92,8
Where	138	129	9	93,4
How many	62	54	8	87,1
Who	51	29	22	56,8

Table 5. Generated questions and their types from testing dataset

Question type	All	Correct	Incorrect	Correctness (%)
What	701	628	73	89,6
True-false	429	398	31	92,8
Where	186	162	24	87,1
How many	39	38	1	97,4
Who	62	46	16	74,2

4.2   Comparison of Generated Questions

After learning phase, we evaluated the quality of generated questions on question generation shared task challenge (QGSTEC) dataset. It contains 81 sentences and there were manually created 180 desired questions by authors of dataset. Questions generated by participants of question generation challenge were evaluated from various points of view (correctness, ambiguity, variety) by human evaluators on scale one to four. The number of correct questions was calculated by average correctness of all questions and number of generated questions. It was done because it is difficult to assign correctness value on this scale and keep the same evaluation criteria (questions from our system was divided into two categories: correct and incorrect). When there were generated too similar questions, only one was included into results. This is also reason, why we generate less questions in comparison with participants of QGSTEC, but as it is shown in Table 6, average correctness of our questions is much higher. From our point of view, it is also interesting to analyze correctness of each type of questions whereas our dataset for training has come from one set of articles. In [13] published number of questions generated from this dataset ordered by question type and its average correctness. However, results are summarized for all participants together, so we will compare our results with average score of participants (total score divided by number of participants). As we can see in Table 7, our system generates more correct questions of types what, where and true-false. As we expected from situation above, our system generates significantly less correct questions of types which and who, because only a few occurrences of persons were in training dataset.

Table 6. Comparison of AQG approaches on QGSTEC dataset

	Correctness (%)	Q/S	Correct questions	Generated questions
Participant 1	0,48	4,37	170	354
Participant 2	0,26	2,04	42	165
Participant 3	0,23	2,58	48	209
Participant 4	0,16	2,07	27	168
Curto [2]	0,56	0,31	14	25
Our approach	0,8	1,5	98	122

4.3   Summarization of Evaluation and Comparisons

We have made several evaluations. First, we created dataset of articles from country topic and divided it into training and testing parts. Training part of articles contained pairs of sentences and questions. They were used in learning process of question generation. Then we let the framework create questions from the rest of dataset and compared overall correctness with correctness of each

Table 7. Comparison of AQG approaches on QGSTEC dataset by question type

Question type	Participants - correct questions	Our approach - correct questions
What	36,3	39
Where	7,6	10
How many	10,3	6
When	11,0	6
Who/Whom	11,1	6
Which	11,1	1
True-false	6,2	30

type of questions. We have compared question generation ability with state-of-the-art systems on standard question generation dataset from several viewpoints (number of correct questions, number of generated questions, types of created questions). Our results confirm, that the proposed approach is perspective direction in question generation task and its automation.

5   Conclusions and Future Work

There are several approaches to question generation task. Their common drawback is the inability of automatized improving and covering various types of input sentences. We proposed framework for question generation task which can continuously learn how to extract question from declarative sentence. It uses combination of supervised learning and reinforcement learning techniques. We also presented question estimator which estimates the correctness of generated question based on various features obtained from text. This allows us to create various types of questions without affecting the code of system. We successfully evaluated our approach and compared generated questions with state-of-the-art systems.

In the future, we plan to extend the list of features which are used for sentence matching. There are still problems with complex sentences (sentences with large number of words), so sentence simplification module could increase the number of sentences covered by less number of transformation rules. It would be also interesting to evaluate this approach on different language.

Acknowledgments. The work reported here was supported by the Scientific Grant Agency of Slovak Republic (VEGA) under the grants No. VG 1/0752/14, VG 1/0646/15, ITMS 26240120039 and STU Grant scheme for Support of Young Researchers.

References

1. Blšták, M., Rozinajová, V.: Automatic question generation based on analysis of sentence structure. In: Sojka, P., Horák, A., Kopeček, I., Pala, K. (eds.) TSD 2016. LNCS, vol. 9924, pp. 223–230. Springer, Cham (2016). doi:10.1007/978-3-319-45510-5_26
2. Curto, S., Mendes, A.C., Coheur, L.: Question generation based on lexico-syntactic patterns learned from the web. Dialogue Discourse 3(2), 147–175 (2012)
3. Heilman, M., Smith, N.A.: Question generation via overgenerating transformations and ranking. Technical report, DTIC Document (2009)
4. Huang, Y.T., Tseng, Y.M., Sun, Y.S., Chen, M.C.: Tedquiz: Automatic quiz generation for ted talks video clips to assess listening comprehension. In: 2014 IEEE 14th International Conference on Advanced Learning Technologies, pp. 350–354, July 2014
5. Huang, Y.T., Chen, M.C., Sun, Y.S.: Personalized automatic quiz generation based on proficiency level estimation. In: 20th International Conference on Computers in Education (2012)
6. Hussein, H., Elmogy, M., Guirguis, S.: Automatic english question generation system based on template driven scheme. Int. J. Comput. Sci. Issues (IJCSI) 11(6), 45–53 (2014)
7. Le, N.T., Pinkwart, N.: Question generation using wordnet. In: Proceedings of the 22nd International Conference on Computers in Education (2014)
8. Lee, J., Seneff, S.: Automatic generation of cloze items for prepositions. In: INTERSPEECH, pp. 2173–2176 (2007)
9. Lin, Y.C., Sung, L.C., Chen, M.C.: An automatic multiple-choice question generation scheme for English adjective understanding. In: Workshop on Modeling, Management and Generation of Problems/Questions in eLearning, The 15th International Conference on Computers in Education, ICCE 2007, pp. 137–142 (2007)
10. Navarro, G.: A guided tour to approximate string matching. ACM Comput. Surv. (CSUR) 33(1), 31–88 (2001)
11. Rakangor, S., Ghodasara, Y.: Automatic fill in the blanks with distractor generation from given corpus. Int. J. Comput. Appl. 105(9) (2014)
12. Rodrigues, H., Coheur, L., Nyberg, E.: QGASP: a framework for question generation based on different levels of linguistic information. In: The 9th International Natural Language Generation conference, p. 242 (2016)
13. Rus, V., Wyse, B., Piwek, P., Lintean, M.C., Stoyanchev, S., Moldovan, C.: A detailed account of the first question generation shared task evaluation challenge. Dialogue Discourse 3(2), 177–204 (2012)

Automatic Extraction of Typological Linguistic Features from Descriptive Grammars

Shafqat Mumtaz Virk[1]([⊠]), Lars Borin[1], Anju Saxena[2],
and Harald Hammarström[2]

[1] Språkbanken, Department of Swedish,
University of Gothenburg, Gothenburg, Sweden
virk.shafqat@gmail.com, lars.borin@svenska.gu.se
[2] Department of Linguistics and Philology, Uppsala University, Uppsala, Sweden
{anju.saxena,harald.hammarstrom}@lingfil.uu.se

Abstract. The present paper describes experiments on automatically extracting typological linguistic features of natural languages from traditional written descriptive grammars. The feature-extraction task has high potential value in typological, genealogical, historical, and other related areas of linguistics that make use of databases of structural features of languages. Until now, extraction of such features from grammars has been done manually, which is highly time and labor consuming and becomes prohibitive when extended to the thousands of languages for which linguistic descriptions are available. The system we describe here starts from semantically parsed text over which a set of rules are applied in order to extract feature values. We evaluate the system's performance on the manually curated Grambank database as the gold standard and report the first measures of precision and recall for this problem.

Keywords: Information extraction · Semantic parsing · Language typology · Typological database

1 Introduction

The area of linguistics which deals with comparison and classification of languages based on their structural features is known as linguistic typology. The objectives of this area are to find commonalities between, and to explore diversity across, the world's languages, and to explain these in historical and/or universal terms. The typical flow of information in such investigations runs from primary linguistic data (collected in the field), to analyzed linguistic data (written down in descriptive grammars), to databases of features of interest (distilled into a language × feature matrix) which form input to the computational tools. The most widely known databases of linguistic structure include the *World Atlas of Language Structures* (WALS) (wals.info), the *Atlas of Pidgin and Creole Language Structures* (APiCS) (apics.org), the *South American Indigenous Language Structures* (SAILS) (sails.clld.org), AUTOTYP

© Springer International Publishing AG 2017
K. Ekštein and V. Matoušek (Eds.): TSD 2017, LNAI 10415, pp. 111–119, 2017.
DOI: 10.1007/978-3-319-64206-2_13

(github.com/autotyp/autotyp-data), and the *Phonetics Information Base and Lexicon* (PHOIBLE) (phoible.org). A fuller listing of available linguistic databases is provided at languagegoldmine.com/.

To the best of our knowledge, all the linguistic databases published so far have been manually constructed, where human experts have turned information from field data or analyzed data into datapoints in the database. The use of human expertise guarantees a certain level of quality and robustness, but is highly labor intensive and consequently costly. There are some 6,500 languages in the world, out of which descriptive grammars – ranging from brief grammar sketches to multi-volume reference grammars – are available for over 4,000 (see glottolog.org). Manually extracting information about 200–300 features from each of them is a very ambitious – and in practice unrealistic – undertaking.

Significant amounts of analyzed language data (grammatical descriptions in discursive textual form) are increasingly being made available in digital form, and the field of natural language processing (NLP) offers tools that potentially can aid us in extracting information about linguistic features from such textual sources, at least for sources in English and some other languages. To take advantage of these advancements, and to help the linguistic community in populating the linguistic feature databases, we report on a pioneer system for automatically extracting information about selected linguistic features from descriptive grammars. The system is based on semantic parsing and a set of rules to extract and formulate the required information. We test our system on Grambank data taking it as a gold standard, and report first measures of precision and recall in this direction.

2   Data

The work presented in this paper has been conducted as part of a larger long-term research project where work is ongoing on compiling a comprehensive database of linguistic features from the *Linguistic Survey of India* (LSI, a massive text data source) [3], in order to investigate areal-linguistic features of the languages of South Asia. In part, this database is to be used for evaluating the NLP tools for automatic feature extraction developed in the project. Since the tools themselves are intended to be applicable to more than one data source, we are in the meantime drawing on another source of gold standard data, using linguistic features also relevant for the South Asian languages. In the experiments reported here, we have used a subset of the Grambank data to test and evaluate the system. Grambank is the name of a database of structural (typological) features being developed at the Max Planck Institute for the Science of Human History at Jena. It surveys 195 structural features of language drawing on and extending the set used by [6]. The scope of Grambank is worldwide and it currently contains data for over 700 languages. The main advantage of using Grambank is that the descriptive grammars consulted in compiling it are both known and available to us in digital form, a set of approximately 6,000 digital descriptive grammars.

Table 1. Semantic parse

Predicate	Semantic arguments
follow	ARG1: The_adjectives, ARG2:the_noun_they_qualify
qualify	ARG1: the_noun_they

For our experiments, we preprocessed the available set of digital grammars as follows. First, we removed all those documents whose language of description was not English (since the semantic parser that we are using in this study is for English, we have restricted our experiments to English-language documents only). From the remaining document set, we grouped all those documents which were the source of the value for a particular feature into one subset. This gave us five subsets of documents, one for each of our five target features listed in Sect. 3. Cases where there was more than one source document for a given feature in a particular language were removed from consideration as they were not corresponding to the LSI use case (with only one description per language). Further, if a document was a source of the same feature for multiple languages, it was filtered out, since for those cases an extra step is required to map the feature value to the language, which we left for future work.

3   Automatic Feature Extraction and Formulation

In this study, we have targeted the following features: (1) Apos: What is the order of adnominal property word and noun?[1] (2) NLpos: What is the order of numeral and noun in the NP? (3) AagrNum: Can an adnominal property word agree with the noun in number? (4) AagrGen: Can an adnominal property word agree with the noun in gender? (5) DefArticle: Are there definite or specific articles?

Though the focus in this article is only on these features, it is worth mentioning that the proposed methodology can also be used easily to extract information about other features too. The procedure of automatically extracting feature values is as follows: As a first step, a given reference grammar was sentence segmented using the Natural Language Tool Kit (NLTK).[2] Each of the sentences was then parsed using a Propbank based semantic parser [1]. From each parsed sentence a list of predicates and their semantic arguments were extracted. The predicates and their arguments were then further analyzed to examine if a particular predicate and its semantic arguments contain the information we are interested in (i.e. information about a particular linguistic feature). Additional analysis steps included: (1) checking for particular predicates for particular features; (2) inspecting the semantic arguments' structure and contents; and

---

[1] An *adnominal property word* corresponds to an adjective or participle in English and many other languages.

[2] http://www.nltk.org/.

Table 2. Linked predicate set for features

Feature	Linked-predicate set
Apos, NLpos	(i) follow (ii) precede (iii) come (iv) place
AagrNum, AagrGen	(i) agree (ii) inflect (iii) change
DefArticle	(i) be (ii) use (iii) lack

(3) formulating the feature values. Let's take an example to illustrate the whole procedure. Suppose we are interested in extracting information about adjective–noun order for a particular LSI language, e.g., Siyin.[3] In the descriptive grammar used for this language, the information about adjective–noun order has been conveyed through the sentence *The adjectives follow the noun they qualify*. Parsing this sentence using the semantic parser will return a set of predicates and their semantic arguments which are listed in Table 1. The predicate 'follow' is one of those predicates that were identified, as a separate process, to be linked to the adjective–noun order feature. Using a development data set, we identified a set of predicates linked to each of target features. This simply involved finding sentences in the descriptive grammars which were used to provide information about a particular feature, and then analyzing them to find the associated list of predicates. Table 2 contains the list of predicates that were identified for each of the target features.

The next step is to examine the semantic arguments of the predicate 'follow', and formulate the feature value. As per Propbank, for the predicate 'follow', ARG1 represents the thing following, while ARG2 represents the thing followed. In the analysis shown in Table 1, the string *The adjectives*, is ARG1 (i.e. the thing following), while the string *the nouns they qualify* is ARG2 (i.e. the thing followed). The substrings representing ARG1, and ARG2 can be further analyzed to formulate and return the feature value '2-N-ANM' (the fact that adjectives follow the nouns). Had ARG1 contained *noun(s)*, ARG2 contained *adjective(s)* with predicate being 'follow', or ARG1 contained *adjective(s)*, ARG2 contained *noun(s)*, and the predicate being 'precede', '1-ANM-N' (the fact that adjectives precede nouns) would have been returned as the feature value. We have used simple if-then-else condition based rules to examine predicates and their semantic argument strings for the purpose of extracting and formulating the feature values. In the future, we have plans to experiment with more advanced techniques, such as active learning, for the feature extraction and formulation from the semantic parses.

A simplified version of the algorithm that was used to extract the adjective–noun order feature values is given in Algorithm 1. As can be noted, the algorithm simply loops over the set of predicates (line 4), and for each predicate it collects the numbered and modifier arguments (lines 5–6). If the predicate is one of the linked predicate for the target feature (line 7), it loops through

---

[3] A Tibeto-Burman language of Burma with about 10,000 speakers.

the list of (argument_label, argument_string) pairs (line 12). At each iteration, the argument_label and argument_string[4] are examined and appropriate boolean variables are adjusted (lines 13–21). Once we have examined all semantic arguments of a predicate, and have adjusted the variables, different combinations of these variables are tested to adjust different order-specifying variables (lines 23–33). For example, if 'adjective' is the agent of the predicate 'follow', and 'noun' is the patient, it means 'adjective follow noun'. Similarly, if 'noun' is the agent, and 'adjective' is the patient, it means 'adjective precede noun'. Also note, how the contents of modifier arguments are tested. For example, if we have a sentence *Adjectives usually follow nouns*, it means adjective may follow or may precede the noun, and the system should be able to return 'both' as a feature value. The algorithm takes care of such cases with an extra condition analyzing the contents of modifier arguments (lines 25–27 and 30–32). Finally, the algorithm formulates the feature values in the required format and returns them back (lines 36–42). For clarity of exposition, the algorithm shows the handling of the predicate 'follow' only, but it is easy to think of a similar sort of handling for other linked predicates (given in Table 2) for each of the target features discussed in this study.

The sentence containing the description of a particular feature may have anaphoric expressions referring to an antecedent or subsequent expression. For example, the sentence *They follow nouns* may appear instead of *Adjectives follow nouns* in the description, with the antecedent expression *adjectives* appearing somewhere else. To extract feature values from such sentences properly, such anaphoric relations need to be resolved. There exist many anaphora resolution systems [2,5], and a classical solution will involve using a state-of-the art system for this purpose. At the current stage of our experiments, however, we have chosen to employ a simple rule-based strategy to resolve such co-reference relations and extract feature values/descriptions. The main idea is to investigate the context with a particular window size to resolve such co-references (if any). For example, if a semantic argument is a pronoun (e.g. *they, it*, or *them*), we just investigate the semantic arguments of one or more preceding or following sentences, and if those arguments contain the potentially linked entity (e.g. *nouns* or *adjectives*, etc.), we just assume that they are related to each other. This procedure can easily be incorporated in Algorithm 1 with an extra if-else condition, but for simplicity, we have excluded it here. It is worth mentioning here that the rule-based anaphora resolution solution was chosen not only for its simplicity, but we also observed in experiments that in many cases this simple strategy was actually able to relate the arguments, whereas the Stanford anaphora resolution system [5] failed to do that.

4   Evaluation

The evaluation results are given in Table 3. As can be seen, the system has varying precision and recall for different features, which highlights the difficulty/ease

---

[4] The argument string is split into a set of words using NLTK's word tokenizer.

Algorithm 1. Extract Adjective Noun Order

```
1: procedure EXTRACTADJECTIVENOUNORDER(parses)
2:        AdjectiveFollowNoun ← False
3:        AdjectivePrecedeNoun ← False
4:        for <every predicate in parses> do
5:            NumberedArgs ← NumberedArgumentsOfPredicate
6:            ModifierArgs ← ModifierArgumentsOfPredicate
7:            if predicate = follow then
8:                AdjectiveAgentFollow ← False
9:                NounPatientFollow ← False
10:               NounAgentFollow ← False
11:               AdjectivePatientFollow ← False
12:               for <every (ArgStr,ArgLabel) of NumberedArgs> do
13:                   if ArgLabel = A1 ∧ adjective ∈ ArgStr then
14:                       AdjectiveAgentFollow ← True
15:                   else if (ArgLabel = A2 ∧ noun ∈ ArgStr) then
16:                       NounPatientFollow ← True
17:                   else if ArgLabel = A1 ∧ noun ∈ ArgStr) then
18:                       NounAgentFollow ← True
19:                   else if ArgLabel = A2 ∧ adjective ∈ ArgStr then
20:                       AdjectivePatientFollow ← True
21:                   end if
22:               end for
23:               if AdjectiveAgentFollow ∧ NounPatientFollow then
24:                   AdjectiveFollowNoun ← True
25:                   if usually ∈ ModifierArgs ∨ sometimes ∈ ModifierArgs then
26:                       AdjectivePrecedeNoun ← True
27:                   end if
28:               else if NounAgentFollow ∧ AdjectivePatientFollow then
29:                   AdjectivePrecedeNoun ← True
30:                   if usually ∈ ModifierArgs ∨ sometimes ∈ ModifierArgs then
31:                       AdjectiveFollowNoun ← True
32:                   end if
33:               end if
34:           end if
35:       end for
36:       if AdjectiveFollowNoun = True ∧ AdjectivePrecedeNoun = True then
37:           return '3 − both'
38:       else if AdjectiveFollowNoun = True ∧ AdjectivePrecedeNoun = False then
39:           return '2 − N − ANM'
40:       else if AdjectiveFollowNoun = False ∧ AdjectivePrecedeNoun = True then
41:           return '1 − ANM − N'
42:       end if
43: end procedure
```

---

of automatically extracting the corresponding feature values using the described method. The 'NLPos' feature has the best precision while the 'Apos' feature has the best recall value. The next section contains a detailed error analysis, discussion of possible reasons for the low recall, and suggestions for how to improve both precision and recall.

As mentioned previously, there does not exist any work related to automatic linguistic feature extraction, which means we do not have any other system to compare the proposed system's performance with. Instead, we evaluate the system performance against a baseline calculated for each feature on the basis of the most frequent feature value in the entire Grambank database. As can be noted, for four out of the five features, the proposed system was able to easily beat the baseline precision values, the exception being the feature 'AagrNum'.

Table 3. Evaluation results

Feature	Precision	Recall	F-score	Baseline precision
Apos	0.76	0.40	0.52	0.41
NLpos	0.85	0.30	0.44	0.75
AagrNum	0.69	0.21	0.32	0.77
AagrGen	0.64	0.14	0.23	0.27
DefArticle	0.84	0.27	0.41	0.27

5 Error Analysis and Discussion

We did a detailed error analysis and found that there are three major sources of errors: (1) debatable gold feature values; (2) pre-processing and semantic parsing errors; (3) a grammar of one language including information about another language.

In many of the erroneous cases, the output produced by the proposed system actually seems reasonable. For example, for the feature Apos, the system determined the value to be '3-both' based on the modifier argument *usually* in the sentence *'An attributive adjective usually precedes (comes before) the noun that it modifies'* from the grammar of the Pulaar[5] language. In the gold data, the value is listed as '2-N-ANM'. Another similar case is in the grammar of the language Pech,[6] where it is stated that *Both qualitative and quantitative adjectives follow the nouns they modify; however, demonstrative adjectives precede nouns.* Based on this description the system determined the value to be '3-both', but in the gold standard, the human experts have determined the feature value to be '2-N-ANM'. Similar cases were found for other features as well. Such cases need to be investigated further. There are two possibilities: (1) either the feature value in the gold standard is debatable (or wrong) – there are studies suggesting that the accuracy of manually curated typological databases is not always 100% and it may vary from 80 to 90% [4] – or (2) the information has been formulated in a different or an indirect way in the grammar, which the system was unable to extract. This needs to be further investigated. Whatever the reason may be, one can use the system proposed in this study to do a kind of validation of the gold data, which can be considered as an additional use-case of the system.

Sentence segmentation errors may propagate all the way from semantic parses to the output produced by the proposed system. For example, consider the following sequence of sentences which was (erroneously) not segmented and passed to the semantic parser as a single sentence:

An indefinite noun phrase negated by md may also be preceded by cdd "still, etc." 5 or gad, as in : ma cad halib yirjac darrih md gad yahudi nasah muslim "milk will not return to the breast" Z.12/51 "no Jew has advised a Muslim" Indefinite adjectives are likewise negated by ma in predicand position (cf.

---

[5] An Atlantic-Congo language spoken in Africa.
[6] A Chibchan language spoken in Central America.

In the resulting semantic parse, the predicate 'precede' was identified having arguments (1) ARG0, which contained the string *adjectives*, and (2) ARG1, which contained the string *nouns*. This caused the feature formulation algorithm to return the wrong feature value '1-ANM-N' (i.e. adjective precede nouns).

It also happens that a grammar contains information about other languages than the main language being described, which may become a source of errors. For example, consider the following sentence from the reference material of a language called Tu: *Numerals in Baonan may either precede or follow a head noun.* The sentence provides information about the language Baonan, but the proposed system will extract and mark it for the language Tu. A solution to this issue could be to identify the language name Baonan first, and then link the extracted information to this language – a task that we leave for future work.

While the precision is relatively good, recall seems too low. We have identified two probable reasons for this: (1) At times, the information about a particular feature is presented indirectly in a descriptive grammar, which is easy for a human to extract but difficult for the automatic system. For example, the grammar may provide language examples with noun phrases containing adnominal property words, which could be used by a human to find about the adjective–noun order in a language, even in the absence of an explicit statement. Since the current system does not extract such indirect information, this may contribute towards a low recall. (2) Due to the small development data-set, it is very likely that we could not find out all the ways and linked predicates (Table 2) which might have been used to encode information about particular features. This, too, can be a factor contributing towards low recall. With the availability of more development data, we believe the system's recall can be considerably improved.

6   Conclusions and Future Work

Given that the manual compilation and curation of typological databases are very labor and time consuming[7] enterprises, any assistance for doing it (semi)automatically is a welcome addition. In a semi-automatic setting, such a system can be used to get pointers to the relevant text segments which may contain the required information. Once these have been pinpointed, it becomes a lot easier to manually extract the feature values from language descriptions which may be as long as a multi-volume book.

In this study, we have reported our experiments related to automatic extraction of linguistic features, and first measures of precision and recall. Since we don't have any other systems to compare our results with, we have shown improvements over an expected value based baseline system.

The most urgent next step is to improve the recall. As mentioned above, the recall can be improved given that we have more development data to find other ways in which information about features could have been encoded. As this depends on development and availability of more data, we have another

---

[7] In a separate study we found that the average cost for a salaried student assistant to extract one datapoint from a descriptive grammar is 1.53 EUR.

possibility in mind, and that is to ask the assistants, who are working in our LSI project, as well as in the Grambank or other typological database projects to record the sentence(s) which they think contain the relevant information. Later, the sentence structure and contents can be used to enhance the system's ability to automatically extract more information and hence to improve the system's recall. As a long term goal, we would like to extend the experiments to other linguistic features (ideally to all of Grambank's 195 linguistic features).

References

1. Björkelund, A., Hafdell, L., Nugues, P.: Multilingual semantic role labeling. In: Proceedings of the Thirteenth Conference on Computational Natural Language Learning: Shared Task, CoNLL 2009, pp. 43–48. Association for Computational Linguistics, Stroudsburg (2009)
2. Broscheit, S., Poesio, M., Ponzetto, S.P., Rodriguez, K.J., Romano, L., Uryupina, O., Versley, Y., Zanoli, R., Kessler, F.B.: Bart: a multilingual anaphora resolution system. In: Proceedings of the 5th International Workshop on Semantic Evaluation, SemEval 2010, pp. 104–107 (2010)
3. Grierson, G.A.: A Linguistic Survey of India, vol. I–XI. Government of India, Central Publication Branch, Calcutta (1903–1927)
4. Polyakov, V.N., Solovyev, V.D., Wichmann, S., Belyaev, O.: Using wals and jazyki mira. Linguist. Typology **13**, 137–167 (2009)
5. Raghunathan, K., Lee, H., Rangarajan, S., Chambers, N., Surdeanu, M., Jurafsky, D., Manning, C.: A multi-pass sieve for coreference resolution. In: Proceedings of the 2010 Conference on Empirical Methods in Natural Language Processing, EMNLP 2010, pp. 492–501. Association for Computational Linguistics, Stroudsburg (2010)
6. Reesink, G., Singer, R., Dunn, M.: Explaining the linguistic diversity of sahul using population models. PLoS Biol. **7**(11), 1–9 (2009)

Text Punctuation: An Inter-annotator Agreement Study

Marek Boháč[1], Michal Rott[1(✉)], and Vojtěch Kovář[2]

[1] Institute of Information Technology and Electronics, Technical University
of Liberec, Studentská 2, Liberec, Czech Republic
{marek.bohac,michal.rott}@tul.cz
[2] Natural Language Processing Centre,
Masaryk University, Botanická 68a, Brno, Czech Republic
xkovar3@muni.cz
https://www.ite.tul.cz
https://nlp.fi.muni.cz

Abstract. Spoken language is a phenomenon which is hard to be annotated accurately. One of the most ambiguous tasks is to fill in the punctuation marks into the spoken language transcription. Used punctuation
marks are often dependent on how annotators understand the transcription content. This may differ as the spoken language often lacks clear
structure (inherent to written language) due to the utterance spontane-
ity or due to skipping between ideas.

Therefore we suspect that filling commas into the spoken language
transcription is a very ambiguous task with low inter-annotator agreement (IAA). Low IAA also means that application of Gold Truth (GT)
annotations for automatic algorithm evaluation is questionable as already
discussed in [7,8].

In this paper we analyze the IAA within group of annotators and we
propose methods to increase it. We also propose and evaluate a refor-
mulation of classical GT annotations for cases with multiple annotations
available.

Keywords: Comma adding · Spoken language · Inter-annotator agreement

1 Introduction

The task of adding commas into some text is addressed in many scenarios. One
kind of scenarios process various texts written by human (grammar checking,
automatic corrections [11]) while another scenarios process automatic transcriptions made by speech-to-text (S2T) systems [3,9,10]. However both tasks have
the same goal, prepared and well-structured written language poses different
challenge than unprepared (and often discontinuous) spoken language. The dis-
continuity of transcription is usually caused by changes of topic, unclear borders
between ideas and other bad speaker habits. These differences between prepared

© Springer International Publishing AG 2017
K. Ekštein and V. Matoušek (Eds.): TSD 2017, LNAI 10415, pp. 120–128, 2017.
DOI: 10.1007/978-3-319-64206-2_14

and spontaneously created texts make it harder to add commas into the S2T transcriptions. The inter-annotator agreement (IAA) becomes dependent not only on annotators' grammatical skills but also on how they interpret ambiguities within the text.

The most common way to evaluate the accuracy of comma adding tools is to compare them against some gold truth standard [6]. These standards are often created by one expert annotator and therefore may be vulnerable to text-content understanding errors. In order to evaluate the importance of the text-content understanding and to estimate the accuracy of the gold truth (GT) standard, we propose to gather multiple annotations of the texts and to compare it in scenarios with different information available to the annotators (annotating with/without audio recording). Such experimental setup enables us to evaluate the inter-annotator agreement (IAA) and to assess how much can 1-annotator GT differ from majority agreement. We also look for the best way to utilize the available multiple annotations to create reasonable evaluation metrics.

Adding of punctuation for S2T system usually adds commas and full-stops. In this paper we focus on text-based adding of commas. Full-stops are usually added with usage of prosodic information [3,10] which is not the focus of this paper.

In the next section we describe the setup of performed experiments (data, annotators, annotation rules). Experiments are evaluated in the following Sect. 3. Discussion and conclusions follow in Sects. 4 and 5.

2   Experimental Setup

For our experiments we prepared 2 sets of data[1] (denoted X and Y). Each set consists of 125 radio Czech broadcast recordings. Data set X contains 11,590 words with 2,857 punctuation marks. Data set Y has 11,750 words with 2,734 punctuation marks. Every recording is approximately 30 s long and contains utterance of a single speaker. Recordings were manually transcribed and provided with punctuation during works on NAKI project [10]. Speakers are radio hosts, their guests and politicians.

The annotators were students of Czech, with specialization on computational linguistics. There were 22 annotators. Each annotator worked independently on others and his/her task was to add commas and full-stop marks into text. In the end, they had to write a short report with their remarks.

Three different annotation scenarios were prepared. In all the experiments annotators were asked to place commas and full-stops into the given text data (full-stops after last word are excluded from the evaluation statistics). In the first scenario (denoted *noAudio_noSlots*) annotators had access only to the text data and were allowed to place punctuation marks anywhere in the text (potential positions of punctuation marks are called slots in this paper). In the second scenario (*Audio_noSlots*) annotators had access to the text data along with the

---

[1] All the used data are accessible at http://nlp.ite.tul.cz/punctuation.

original recording and were allowed to place punctuation marks anywhere in the text. In the third scenario (*noAudio_Slots*) annotators had only the text data and were restricted to place (or not to place) the punctuation marks only into a limited set of slots. These slots were chosen as slots used by annotators in scenario *noAudio_noSlots*. To enable a slot we demanded more than 4 punctuation marks in the slot (from total of 22 annotators; slots with less marks were considered to be singular errors). From 2,819 slots used by annotators 1,947 were enabled for use (which discarded 2,269 marks from total of 39,123). This means that we banned 30.9% of slots for annotation while discarding only 5.8% of annotated punctuation marks from experiment *noAudio_noSlots*.

The data set X was used in all scenarios in the following order: *noAudio_noSlots*, *noAudio_Slots*, *Audio_noSlots* with 2 months pause between the first experiment and the rest. Data set Y was used only in the *Audio_noSlots* scenario, so we can verify that our annotators were not "over-fit" to repeatedly seen texts. The alignment of the annotated texts to the reference text was provided by algorithm proposed in [2] which is able to handle many changes annotators could make in the texts.

3   Results

Between each two consecutive words of the textual transcription a punctuation mark can be placed. Such a position is called slot in the following paragraphs. Our annotators were asked to use only commas and full-stops in the annotations. Other marks (e.g. colons, question marks, quotes and dashes) were normalized into these two categories. Hence when the annotations were aligned with the reference text, each slot was assigned 3 numbers: comma count, full-stop count and gap count. Its sum is obviously equal to the number of annotators in the annotator group. If there was at least one punctuation mark placed in the slot, we call it *used slot*.

3.1   Comparison of Inter-annotator Agreement

In this experiment we evaluated the Inter-Annotator Agreement (IAA) of all proposed annotation scenarios. The IAA was compared via two metrics. The first metric was the slot agreement (1) which shows us if the annotators place punctuation marks into the same slots. The second one was the mark agreement (2) which tells us if the annotators used the same punctuation mark (if they chose the same ends of sentences). As 85% of all slots contained no punctuation, we decided to evaluate the IAA only amongst the used slots. The *max agreement* stands for maximum count of punctuation that can be placed into the slot if all annotators agreed the same solution.

$$slot\ agreement = \frac{\sum (commas_{used\ slots} + full\ stops_{used\ slots})}{max\ agreement_{used\ slots}} \tag{1}$$

$$mark\ agreement = \frac{\sum max(commas, full\ stops)_{used\ slots}}{max\ agreement_{used\ slots}} \tag{2}$$

Table 1. Inter-annotator agreement on slots and used punctuation marks

	Data set X						Data set Y	
	noAudio_noSlots		noAudio_Slots		Audio_noSlots		Audio_noSlots	
	Slot agr.	Mark agr.	Slot agr.	Mark agr.	Slot agr.	Mark agr.	Slot agr.	Mark agr.
A	74.10%	74.10%	97.67%	89.41%	77.88%	71.96%	76.46%	70.59%
B	74.42%	68.68%	97.04%	87.49%	76.59%	71.02%	74.87%	67.60%
C	76.24%	69.40%	98.69%	87.12%	76.24%	68.54%	76.81%	69.31%
D	77.33%	70.88%	98.42%	89.66%	81.04%	75.00%	77.32%	70.49%

3.2 Majority and Proportional References

We had access to multiple annotations of the same text data. Thus it was reasonable to define accuracy metrics using all those annotations. In this paper we proposed two variations of accuracy metric. Both can be defined via (3) where formula components have following meaning: *slots* determines if we operate over all slots of the text or only over used slots (as explained above), *slot count* is the number of slots in the text (in case of evaluating multiple results it is multiplied by the number of annotators), *mark* stands for any punctuation mark that can be put in the slot (none mark included), *weight* is accuracy score assigned to the mark in concrete slot and *count* means how many punctuation marks of given type were found in the slot.

The first proposed accuracy definition (ACC_{MAJ}) simulated the 1-annotator gold-truth approach (ACC_{GT}). This means that for every slot we chose the most frequent punctuation mark and set its weight to 1 while other weights were set to 0. The second variant (ACC_{PROP}) computed the weights of punctuation marks proportionally to number of annotators who used them. The weights were normalized so that the most frequent mark (with count p_max) had weight 1, others used weight = count/p_max. This ensured that the ACC_{PROP} values were from range $<0;1>$ and enabled to take less-often punctuation into account with some reasonable weight.

$$ACC = \frac{\sum\limits_{slots} \sum\limits_{mark=[comma,gap,full\ stop]} weight_{mark} \cdot count_{mark}}{slot\ count} \tag{3}$$

Table 2 evaluates the agreement between the original 1-annotator gold-truth and the multi-annotator reference. We also show the impact of evaluating all versus used-only slots in the evaluation. Used slots are chosen according to multi-annotator reference.

3.3 Comparison of Automatic Comma-Adding Tools by Available Reference Annotations

In this experiment we compared two automatic comma-adding tools so we could evaluate the impact of different annotations. All the annotations were made

Table 2. Comma agreement between 1-annotator gold-truth and proposed annotation scenarios

	Used slots		All slots	
	ACC_{MAJ}	ACC_{PROP}	ACC_{MAJ}	ACC_{PROP}
noAudio_noSlots; data X	88.88%	83.30%	93.66%	95.58%
noAudio_Slots; data X	84.39%	74.61%	93.67%	94.80%
Audio_noSlots; data X	87.41%	84.11%	94.30%	95.95%
Audio_noSlots; data Y	90.35%	84.91%	94.35%	95.94%

Table 3. Automatic comma-adding tools compared by different references

	Used slots			All slots		
	ACC_{GT}	ACC_{MAJ}	ACC_{PROP}	ACC_{GT}	ACC_{MAJ}	ACC_{PROP}
SET; data X	54.98%	59.64%	74.21%	90.09%	91.28%	93.45%
FST; data X	53.58%	57.35%	72.78%	90.00%	90.97%	93.18%

under similar circumstances - annotators had access to both text data and audio recordings. Accuracy and experiment labels were already established in Sect. 3.2.

First compared tool was SET [4,5]. The second tool was denoted as FST [1]. Both systems take plain text data as input and employ no acoustic information. Comparison of results is shown in Table 3.

3.4   Inter-annotator Agreement on Slots Marked by Automatic Systems

Computers (nowadays) are not able to understand human language. Nevertheless they are able to add commas into the text. Hence we suppose some slots are correctly processed using some lower-level features (i.e. word N-grams). Our hypothesis states that slots labeled by computer should have higher IAA than the slots computers can't mark. To evaluate this hypothesis we used SET and FST tools to split slots into two groups. The first group are slots marked by corresponding automatic tool (easy to find). The second group contains slots computers could not mark but at least one annotator did (probably dependent on text understanding).

As both tools use only text information we compare the annotator agreement with noSlots_noAudio scenario. Texts were annotated by all 22 annotators so maximum agreement was 22 and minimal was 0 (false detections of automatic system). The results are shown in Figs. 1 and 2. The horizontal axes shows how many annotators agreed on slots. The vertical axes represent the number of slots with corresponding annotators' agreement. Every column of the figures is split to two parts. The black part represents slots marked by computer systems, the gray one represents slots marked only by annotators. For example, the second column in Fig. 1 means that 21 of 22 annotators agreed on 204 slots and automatic system was able to detect 150 of these slots.

4   Discussion

4.1   Annotators' Remarks

After annotators finished all punctuation scenarios, they were asked to write short comment about the task. In the following paragraphs we summarize their remarks.

Adding punctuation is often ambiguous for two main reasons. First, some punctuation usage is dependent on annotator's understanding to the given text (which is not always clear). Second, it is often unclear if a slot starts new sentence or embedded sentence. On the other hand, some subordinate conjunctions suggest to use commas automatically (without any analysis of text structure).

Annotators also reported differences between written and spoken language that complicated their task. Filler words and speech discontinuities (repetitions) were hard to annotate as well as identification of direct speech. Reference texts also contained some amount of inaccuracies - typing errors and reformulations of speech discontinuities.

Annotators were sometimes missing allowed slot in certain positions in the noAudio_Slots scenario. Adding punctuation with audio recording available (Audio_noSlot scenario) was easier. However some speakers had very vague intonation. Especially utterances by politicians were hard to understand and annotate in contrast with radio hosts whose utterances were well pronounced (and probably prepared).

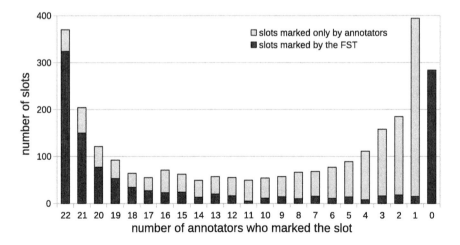

Fig. 1. Annotators' agreement on slots marked by FST system

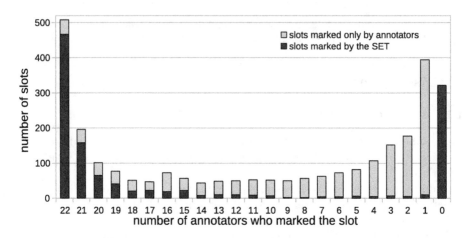

Fig. 2. Annotators' agreement on slots marked by SET system

4.2 Discussion About Performed Experiments

In the first experiment (Sect. 3.1) we evaluated the impact of different annotation scenarios on the inter-annotator agreement. Table 1 shows that audio recordings slightly increase IAA (by decreasing the ambiguity of the text interpretation). At the same time we have to admit that this impact is much smaller than we anticipated. Audio recording allowed annotators to process the recordings faster as the meaning of text was clear to them. The results for data set Y prove that annotators did not over-fit to repeatedly annotated data. The scenario which reduces the number of punctuation slots greatly increases the IAA on used slots however its effect on the choice of punctuation marks is smaller (it does not help to disambiguate the text meaning).

Second experiment (Sect. 3.2) evaluates the agreement between 1-annotator gold truth annotation and multi-annotator references. Both scenarios that allowed annotators to use any slots (noAudio_noSlots and Audio_noSlots) report very similar results (as can be seen in Table 2). In 3 of 4 cases the Audio_noSlots shows better agreement with audio-supported 1-annotator gold truth than the noAudio_noSlots scenario (audio recording helps to increase the IAA). The scenario with limited set of slots (noAudio_Slots) seems to be very volatile to cases when evaluated annotation uses forbidden slots. This result suggests that our strategy of limiting slots was too strict.

In Sect. 3.3 we use single and multiple-annotator references to compare the performance of two comma-adding tools. We can estimate how much the results vary, i.e. how reliable is one-annotator reference. Table 3 shows that by all comparisons the SET system slightly outperforms the FST system. In the most varying comparison the relative error was 8%.

Our last experiment showed that it is easier to place commas in some slots (mostly marked by automatic systems). These slots have higher inter-annotator agreement than the slots marked only by annotators (which are dependent on understanding the annotated text).

5 Conclusions

Annotators reported that some slots are easy for annotation - they use commas "automatically" with high inter-annotator agreement (IAA). Another group of slots is very ambiguous (as is shown in Sect. 3.4). Audio recordings partly help to resolve this ambiguity (as is shown by higher IAA in Table 1). If the speaker is well trained and his utterance is prepared (e.g. broadcast host has a prepared set of questions) the utterance is easy for annotation. Annotation is more challenging (and data proves lower IAA) when the speaker is not prepared, utterance is spontaneous or intentionally unclear (e.g. politicians).

The impact of audio recordings seems to be small which is caused by the following statistics: 88% of all slots are never used by annotators, approx. 6.8% of slots are "added automatically" with high IAA and only the remaining 5.2% of slots are ambiguous for annotators. The percentage of ambiguous slots explains why the audio recording has such small impact on IAA.

In the noAudio_Slots annotation scenario we tried to increase IAA by forcing annotators to use limited set of slots. The IAA increased but annotators reported "missing slots" they wanted to use. The evaluation via this reference shows that our slot-reducing strategy was too restrictive, so this approach does not provide suitable references.

Despite our expectations, 1-annotator gold-truth references proved high agreement with multi-annotator references. This can be explained by relatively small number of slots which contain ambiguous punctuation (approx. 5.2%) in which annotators most often disagree. The most important prerequisites to obtain high quality annotations are annotators' knowledge of grammar and well prepared annotation manuals. Special focus must be paid to such phenomena as filler words, speech discontinuities (repetitions, hesitation) and speaker's auto-corrections which are inherent to spoken language.

Acknowledgment. We are very grateful to the students doing the annotation work, thank you. This work was supported by the Student's Grant Scheme at the Technical University of Liberec (SGS 2016), by the Ministry of Education of CR within the LINDAT-Clarin project LM2015071 and by the Grant Agency of CR within the project 15-13277S.

References

1. Boháč, M., Blavka, K., Kuchařová, M., Škodová, S.: Post-processing of the recognized speech for web presentation of large audio archive. In: 2012 35th International Conference on Telecommunications and Signal Processing (TSP), pp. 441–445, July 2012

2. Boháč, M., Nouza, J., Blavka, K.: Investigation on most frequent errors in large-scale speech recognition applications. In: Sojka, P., Horák, A., Kopeček, I., Pala, K. (eds.) TSD 2012. LNCS, vol. 7499, pp. 520–527. Springer, Heidelberg (2012). doi:10.1007/978-3-642-32790-2_63

3. Kolář, J., Švec, J., Psutka, J.: Automatic punctuation annotation in Czech broadcast news speech. In: 9th Conference Speech and Computer (2004)

4. Kovář, V.: Partial grammar checking for Czech using the SET parser. In: Sojka, P., Horák, A., Kopeček, I., Pala, K. (eds.) TSD 2014. LNCS (LNAI), vol. 8655, pp. 308–314. Springer, Cham (2014). doi:10.1007/978-3-319-10816-2_38

5. Kovář, V., Horák, A., Jakubíček, M.: Syntactic analysis as pattern matching: the SET parsing system. In: Proceedings of 4th Language and Technology Conference, Wydawnictwo Poznańskie, Poznań, Poland, pp. 978–983 (2009)

6. Kovář, V., Machura, J., Zemková, K., Rott, M.: Evaluation and improvements in punctuation detection for Czech. In: Sojka, P., Horák, A., Kopeček, I., Pala, K. (eds.) TSD 2016. LNCS, vol. 9924, pp. 287–294. Springer, Cham (2016). doi:10.1007/978-3-319-45510-5_33

7. Kovář, V.: Evaluating natural language processing tasks with low inter-annotator agreement: the case of corpus applications. In: Recent Advances in Slavonic Natural Language Processing, RASLAN 2016, pp. 127–134 (2016)

8. Kovář, V., Jakubíček, M., Horák, A.: On evaluation of natural language processing tasks - is gold standard evaluation methodology a good solution? In: Proceedings of the ICAART 2016, vol. 2, pp. 540–545. SCITEPRESS (2016)

9. Mihajlik, P., Fegyó, T., Németh, B., Tüske, Z., Trón, V.: Towards automatic transcription of large spoken archives in agglutinating languages – Hungarian ASR for the MALACH Project. In: Matoušek, V., Mautner, P. (eds.) TSD 2007. LNCS, vol. 4629, pp. 342–349. Springer, Heidelberg (2007). doi:10.1007/978-3-540-74628-7_45

10. Nouza, J., Červa, P., Ždánský, J., et al.: Speech-to-text technology to transcribe and disclose 100, 000+ hours of bilingual documents from historical Czech and Czechoslovak radio archive. In: INTERSPEECH 2014, pp. 964–968 (2014)

11. Petkevič, V.: Kontrola české gramatiky (český grammar checker). Studie z aplikované lingvistiky-Studies in Applied Linguistics 5(2), 48–66 (2014)

PDTSC 2.0 - Spoken Corpus with Rich Multi-layer Structural Annotation

Marie Mikulová[⊠], Jiří Mírovský, Anja Nedoluzhko, Petr Pajas,
Jan Štěpánek, and Jan Hajič

Faculty of Mathematics and Physics, Institute of Formal and Applied Linguistics,
Charles University, Prague, Czech Republic
{mikulova,mirovsky,nedoluzhko,pajas,hajic}@ufal.mff.cuni.cz

Abstract. We present a richly annotated spoken language resource, the Prague Dependency Treebank of Spoken Czech 2.0, the primary purpose of which is to serve for speech-related NLP tasks. The treebank features several novel annotation schemas close to the audio and transcript, and the morphological, syntactic and semantic annotation corresponds to the family of Prague Dependency Treebanks; it could thus be used also for linguistic studies, including comparative studies regarding text and speech. The most unique and novel feature is our approach to syntactic annotation, which differs from other similar corpora such as Treebank-3 [8] in that it does not attempt to impose syntactic structure over input, but it includes one more layer which edits the literal transcript to fluent Czech while keeping the original transcript explicitly aligned with the edited version. This allows the morphological, syntactic and semantic annotation to be deterministically and fully mapped back to the transcript and audio. It brings new possibilities for modeling morphology, syntax and semantics in spoken language – either at the original transcript with mapped annotation, or at the new layer after (automatic) editing. The corpus is publicly and freely available.

Keywords: Speech · Spoken corpus · Syntax · Semantics · Coreference · Treebank · Annotation

1 Introduction

Spontaneous speech breaks many rules by which written texts are constituted. Despite most spontaneous oral communications not meeting the basic written-text standards, the mutual understanding among humans does usually not get harmed. Posing no problem for humans, spontaneous speech is yet very difficult to handle for machines. POS taggers, parsers and semantic analyzers trained on written texts cannot cope with the morphological and syntactic irregularities typical of spontaneous speech. In this paper, we describe the (manually built) Prague Dependency Treebank of Spoken Czech 2.0 aimed at automatic recognition of spontaneous speech and its "understanding". We present our annotation scheme – which includes a speech "reconstruction" layer – above a corpus of

© Springer International Publishing AG 2017
K. Ekštein and V. Matoušek (Eds.): TSD 2017, LNAI 10415, pp. 129–137, 2017.
DOI: 10.1007/978-3-319-64206-2_15

spontaneous dialogs. The reconstruction layer enables standard structural annotation, while linking the original transcript to syntax and semantics as well.
The overall scheme conforms to the complex PDT-style annotation scenario
that spans from linear text to dependency based syntax and semantics. The
annotation scheme and the internal linking allows for future machine learning
experiments using either the reconstruction layer or directly the combined links
across layers.

2   Related Work

There is a wide range of corpora with disfluency annotation and subsequent
syntax annotation, e.g., Switchboard corpus in Treebank-3 [8], Childes Database [18], the treebank of English, German, and Japanese created within the
Verbmobil project [7], Corpus Gesproken Nederlands [19], or Treebank of Spoken French [2]. All these projects aim at identifying and labeling segments of the
original audio (and transcript) for the chosen disfluencies. However, this style
of disfluency annotation (consisting only in identifying and labeling spoken phenomena) cannot, in general, arrive at grammatical, fluent and understandable
text readable for the human readers as well as appropriate for subsequent manual syntactic annotation or automatic processing. The development of a robust
speech understanding pipeline requires not only a knowledge of what is a disfluency and where the disfluencies occur in an annotated spoken language corpus,
but also how to understand them (cf. Sect. 4.1).

3   Prague Dependency Treebank of Spoken Czech

PDTSC 2.0[1] is a new release of Prague Dependency Treebank of Spoken Czech. It
is a corpus of spoken language, consisting of 742,257 tokens and 73,835 sentences,
representing 6,174 min (over 100 h) of spontaneous dialogs. The dialogs have been
recorded, transcribed and edited in several interlinked layers: audio recordings,
automatic and manual transcripts and manually reconstructed text. These layers
along with morphological annotation were part of the first version of the corpus
(PDTSC 1.0[2]; [3]). Version 2.0 is extended by annotation at the dependency
syntax layer and the "deep" syntax layer, which contains semantic roles and
relations as well as annotation of coreference. PDTSC 2.0 is freely and publicly
available. Table 1 shows the inclusion and status of layers of annotation in both
versions of the corpus.

3.1   The Data

PDTSC recordings consist of two parts covering two types of dialogs; both parts
contain mostly colloquial Czech, even though some people spoke close to the

---

[1] http://ufal.mff.cuni.cz/pdtsc2.0.
[2] http://ufal.mff.cuni.cz/pdtsc1.0/en/index.html.

Table 1. Annotation in PDTSC 1.0 and PDTSC 2.0

			Coreference	manually
PDTSC 2.0		t-layer	**Deep Syntax Annotation**	manually
		a-layer	Dependency Syntax Parsing	automatically
	PDTSC 1.0	m-layer	Tagging and Lemmatization	automatically
			Speech Reconstruction	manually
		w-layer	**Transcript**	manually
		z-layer	Speech Recognition	automatically
			Audio	

standard. First, it contains a part of the Czech portion of the Malach project corpus, i.e., lightly moderated interviews (testimonies) with Holocaust survivors, originally recorded by the Shoa Visual History Foundation.[3] The second part of the corpus consists of dialogs recorded for the Companions project.[4] The domain is also personal memories, but in a Wizard-of-Oz setting where the two dialog participants chat over a collection of personal photographs. The goal of this project was to create virtual companions that would be able to have a natural conversation with humans. Domain-identical dialogs were created also in English (corpus PDTSE 1.0[5]), allowing comparison with the Czech data, even if the English data have not yet been upgraded to version 2.0.

The markup used in PDTSC 2.0 is the language-independent Prague Markup Language (PML), which is an XML subset customized for multi-layered linguistic annotation [16].

4   Layers of Annotation

PDTSC 2.0 is a treebank from the family of PDT-style corpora developed in Prague (for more information, see [4]). The main features of this annotation style are:

- based on a well-developed dependency syntax theory which is known as the Functional Generative Description [20],
- interlinked hierarchical layers of standoff annotation,
- "deep" syntax layer.

PDTSC differs from other PDT-style corpora mainly in the "spoken" part of the corpus. The layers stack starting at the external base layer with audio files (in the Vorbis format). The bottom layer of the corpus (**z-layer**) contains automatic speech recognition output synchronized to audio. The next layer, **w-layer**, contains manual transcript of the audio, i.e. everything the speaker has said including all slips of the tongue as well as non-speech events like coughing, laugh, etc.

---

[3] http://sfi.usc.edu/collections/holocaust.
[4] http://companions-project.org.
[5] http://ufal.mff.cuni.cz/pdtse1.0/en/index.html.

W-layer is synchronized to the automatic transcript and through it thus to the original audio. The subsequent **m-layer** contains a manually "reconstructed", i.e. edited, grammatically corrected version of the transcript, including punctuation and assumed sentence boundaries. The reconstructed tokens are automatically morphologically tagged and lemmatized. From this point on, annotation on the upper layers is the same as in the other PDT-style corpora. The dependency syntax layer (**a-layer**) is parsed automatically, while the "deep" syntax layer (**t-layer**) is annotated manually. There is a one-to-one correspondence between the tokens at the m-layer and the nodes at the a-layer. The syntactic dependencies are provided with dependency relations (e.g., Subject or Adverbial). The t-layer, which is also a tree-shaped graph (with content words only), is the highest and most complex linguistic representation that combines syntax and semantics in the form of semantic labeling, coreference annotation and argument structure description based on a valency lexicon.

In order not to lose any piece of the original information, tokens (nodes) on a lower layer are explicitly referenced from the corresponding closest (immediately higher) layer. These links allow for tracing every unit of annotation all the way down to the original audio and transcript, with the exception of reconstructed

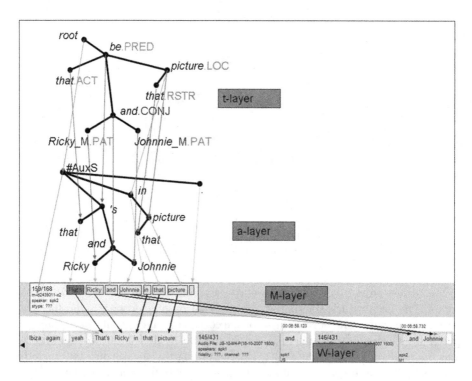

Fig. 1. Layers of annotation in PDTSC 2.0 (demonstrated on a English sentence *That's Ricky and Johnnie in that picture.*; audio not shown.)

ellipsis, which might only point in between audio segments. Figure 1 shows the relations between the layers as annotated and represented in the data.

In the following subsections, the manual annotation of the most important corpus parts (i.e., speech reconstruction and deep syntax annotation) is shortly described.

4.1 Spontaneous Speech Reconstruction

Spontaneous speech is "ungrammatical", full of a class of phenomena called disfluencies, such as false starts, repetitions, fillers, ellipses, etc. These phenomena cause problems for any subsequent processing. The purpose of speech reconstruction as defined in the present work is to "translate" the input spontaneous speech to a written text, before it is tagged and parsed. The transcript is segmented into sentence-like segments and these segments are edited to meet written-text standards, which means cleansing the text from the discourse-irrelevant and contentless material (superfluous connectives and deictic words, false starts, repetitions, etc. are removed) and re-chunking and re-building the original segments into grammatical sentences with acceptable word order and proper morpho-syntactic relations between words. The annotators are thus simulating the work of, e.g., magazine editors when preparing recorded interviews to appear in printed form. There are two basic annotation principles they have to follow:

A. The Content-Preservation Principle: the modifications of the original transcript may not affect the content.

B. The Minimal Modification Principle: modifications are only performed when it is necessary to follow written-text standards.

The annotators are also required to correctly **link the reconstructed text tokens to the original transcription** (which is, of course, then linked implicitly by using the synchronization marks to both the automatically recognized audio (z-layer) and to the audio itself). Even though the rules are relatively simple, certain conventions had to be introduced:

- source deletions: not linked (implicit links only based on order),
- word and punctuation insertions: not linked (implicit links as above),
- word substitution changes: linked to the source tokens that are the ones edited (and most similar in case of ambiguity),
- no change (identity between source and annotation): links to the source token,
- the reconstructed sentence (segment) boundaries (begin, end) are mapped onto the raw-transcript segments. These two links indicate the span of transcript that was used as the input for the given reconstructed sentence.
- word order changes are not labeled since they are deterministically extractable from the (crossing) links.

An example of linking the reconstructed text to original transcript is depicted in Fig. 1 (links between the M-layer and W-layer).

Manual annotation of speech reconstruction was the crucial part of the first version of the corpus. The annotation is described in more detail in [3] and the guidelines are also specified in the annotation manual [9]. PDTSC annotation scheme of speech reconstruction has been developed in parallel (and often in cooperation) with Fitzgerald and Jelinek [1].

4.2 Deep Syntax and Coreference Annotation

One of the important distinctive features of the PDT-style annotation is the fact that in addition to the morphological and syntactic (dependency) layer, it includes complex semantically based annotation on the highest annotation layer (t-layer).

On the t-layer, every sentence is represented as a rooted tree with labeled nodes and edges. The tree reflects the **underlying dependency structure** of the sentence. The nodes stand for content words only. Unlike at the a-layer, not all the original tokens from the edited transcript (or, in the case of text, all the word tokens) are represented at the t-layer as nodes. Function words (prepositions, auxiliary verbs, etc.) do not have nodes of their own, but their contribution to the meaning of the sentence is not lost – several attributes are attached to the t-nodes the values of which represent such a contribution (e.g. tense for verbs). Some of the t-nodes do not correspond to any morphological token; they are added in case of surface deletions (ellipses). The types of the (semantic) dependency relations are represented by the "functor" attribute attached to all t-nodes.

The core ingredient in the annotation of the t-layer is **valency** (the theoretical description of the valency theory as developed in the framework of Functional Generative Description is summarized mainly in [17]). The valency criterion divides functors into the argument functors and adjunct functors. There are five arguments: Actor (ACT), Patient (PAT), Addressee (ADDR), Origin (ORIG) and Effect (EFF). In addition, we distinguish about 50 types of adjuncts (temporal, local, casual, etc.). The valency lexicon that all the PDT-family corpora use, PDT-Vallex [6,21], was built in parallel with the annotation of sentences and it has been used for consistent annotation of valency modifications in the annotated sentences. The t-layer annotation of PDTSC extended PDT-Vallex with approximately 1,500 new lemmas and 2,500 new valency frames [14].

The PDTSC 2.0 also captures grammatical and textual **coreference** relations. Grammatical coreference is based on language-specific grammatical rules, whereas to resolve textual coreference, the context knowledge is needed. Textual coreference annotation is based on the "chain principle", the anaphoric entity always referring to the last preceding coreferential antecedent. Coreference relations are technically part of the t-layer.

Annotation principles used at the t-layer and the annotation guidelines are described in the annotation manuals [10,11]. Compared to the anchoring original project of Prague Dependency Treebank[6] [5], the t-layer annotation in PDTSC 2.0 is slightly simplified; e.g., it does not contain information structure annotation (topic-focus).

---

[6] http://ufal.mff.cuni.cz/prague-dependency-treebank.

5   Annotation Quality Checking (Inter-annotator Agreement)

There are many ways to produce correct written text from a literal transcript. To capture this fact, we provide multiple parallel annotations for each transcript, but we do not unify the individual annotation streams. We believe that it will lead to more possibilities of training and evaluation of any tools that might be developed using such data, in a similar vein to the way multiple reference translations are used for automatic machine translation evaluation (more about speech reconstruction quality checking see in [3]).

A multiple parallel annotation of the same data becomes impossible (with regard to time and work) if the treebank is large and the annotated information is complex. For measuring an inter-annotator agreement (IAA) of deep syntax annotation, only a subset of the data was annotated in parallel. Since there is no "golden" annotation, we measure the agreement of all the pairs of annotators. A system of automatic quality checking of the annotated data was developed as well (see [12]). For more detailed account how the IAA for deep syntax annotation and for coreference relations are measured, see [13] and [15]. Table 2 shows average values of IAA measurements for deep syntax annotation and for textual coreference relations. Problems of low inter-annotator agreement and ambiguity in annotation of coreference relations are also described in [15].

Table 2. IAA in deep syntax annotation and coreference relations

Syntax	Annot1	Annot2	Annot3	Annot4
Annot1	-	98.7	98.4	98.6
Annot2	98.7	-	98.4	98.5
Annot3	98.4	98.4	-	98.4
Annot4	98.6	98.5	98.4	-
Coreference	Annot1	Annot2	Annot3	Annot4
Annot1	-	87.5	-	87.0
Annot2	87.5	-	86.8	-
Annot3	-	86.8	-	89.5
Annot4	87.0	-	89.5	-

6   Conclusion: What Is the Data Good For?

With the release of PDTSC 2.0, we have to a large extent closed the gap between the full annotation of the Prague Dependency Treebank (which is a written text-based corpus) and the Prague spoken dialog corpus, the PDTSC. We are not aware of any other spoken language corpus that would have both the "disfluencies" marked and a full annotation of syntax and semantics. In addition, we have kept the unique "reconstruction" layer of annotation, which allows different views of and annotation mapping onto the original data: either the annotation can be mapped all the way to audio (or its automatic or manual transcripts), getting the usual style of speech corpora annotation with syntax built over the original transcript, or one might attempt to use the reconstruction layer - for example, one can perform the reconstruction step directly, using the upper layer annotation possibly only as a "hidden" layer (or not at all). Either way, we hope that this resource can help build automatic speech understanding and dialog systems.

As with similar projects, this release is a step towards bigger corpora, with more manual annotation. The PDTSC 2.0 will be also extended in the future, most notably by manual annotation on the m- and a-layers, and will become part of a consolidated Prague Dependency Treebanks release in 2018, which will contain four different treebanks of Czech, uniformly annotated using the scheme described in part here, with data coming from text, speech and internet sources.

Acknowledgments. The research reported in the paper was supported by the Czech Science Foundation under the projects GA16-05394S and GA17-12624S. This work has also been supported by the LINDAT/CLARIN project of Ministry of Education, Youth and Sports of the Czech Republic (project LM2015071).

References

1. Fitzgerald, E., Jelinek, F.: Linguistic resources for reconstructing spontaneous speech text. In: Proceedings of the 6th LREC, Marrakech, Moroco (2008)
2. Gerdes, K., Kahane, S., Lacheret, A., Truong, A., Pietrandrea, P.: Intonosyntactic data structures: the Rhapsodie Treebank of spoken French. In: Proceedings of the 6th Linguistic Annotation Workshop, Jeju, Korea, pp. 85–94. ACL (2012)
3. Hajič, J., Cinková, S., Mikulová, M., Pajas, P., Ptáček, J., Toman, J., Urešová, Z.: PDTSL: an annotated resource for speech reconstruction. In: Proceedings of the 2008 IEEE Workshop on Spoken Language Technology, Goa, India, pp. 93–96 (2008)
4. Hajič, J., Hajičová, E., Mikulová, M., Mírovský, J.: Prague dependency treebank. In: Handbook on Linguistic Annotation, Volume II, pp. 555–594. Springer, Dordrecht (2017)
5. Hajič, J., Panevová, J., Hajičová, E., Sgall, P., Pajas, P., Štěpánek, J., Havelka, J., Mikulová, M., Žabokrtský, Z., Ševčíková-Razímová, M., Urešová, Z.: Prague Dependency Treebank 2.0 (LDC2006T01) (2006)
6. Hajič, J., Panevová, J., Urešová, Z., Bémová, A., Kolářová, V., Pajas, P.: PDT-VALLEX: creating a large-coverage valency lexicon for treebank annotation. In: Proceedings of the 2nd Treebanks and Linguistic Theories Workshop, pp. 57–68. Vaxjo University Press, Vaxjo (2003)
7. Hinrichs, E.W., Bartels, J., Kawata, Y., Kordoni, V., Telljohann, H.: The verbmobil treebanks. In: KONVENS, pp. 107–112 (2000)
8. Marcus, M., Santorini, B., Marcinkiewicz, M.A., Taylor, A.: Penn Treebank-3. Linguistic Data Consortium, LDC99T42, University of Pennsylvania (1999)
9. Mikulová, M.: Rekonstrukce standardizovaného textu z mluvené řeči v Pražském závislostním korpusu mluvené češtiny. Manuál pro anotátory. Technical report ÚFAL TR-2008-38 (2008)

10. Mikulová, M.: Annotation on the tectogrammatical level. Additions to annotation manual (with respect to PDTSC and PCEDT). Technical report ÚFAL TR-2013-52 (2014)

11. Mikulová, M., Bémová, A., Hajič, J., Hajičová, E., Havelka, J., Kolářová, V., Kučová, L., Lopatková, M., Pajas, P., Panevová, J., Razímová, M., Sgall, P., Štěpánek, J., Urešová, Z., Veselá, K., Žabokrtský, Z.: Annotation on the tectogrammatical level in the Prague Dependency Treebank. Annotation manual. Technical report 30, Prague, Czech Republic (2006)

12. Mikulová, M., Štěpánek, J.: Annotation quality checking and its implications for design of Treebank (in Building the Prague Czech-English Dependency Treebank). In: Proceedings of 8th Treebanks and Linguistic Theories Workshop, Milano, Italy, pp. 137–148 (2009)

13. Mikulová, M., Štěpánek, J.: Ways of evaluation of the annotators in building the Prague Czech-English Dependency Treebank. In: Proceedings of the 7th LREC, Valletta, Malta, pp. 1836–1839 (2010)

14. Mikulová, M., Štěpánek, J., Urešová, Z.: Liší se mluvené a psané texty ve valenci? Korpus "gramatika" axiologie **8**, 36–46 (2013)

15. Nedoluzhko, A., Mírovský, J.: Annotators' certainty and disagreements in coreference and bridging annotation in Prague Dependency Treebank. In: Proceedings of the 2nd International Conference on Dependency Linguistics, Prague, Czech Republic, pp. 236–243 (2013)

16. Pajas, P., Štěpánek, J.: Recent advances in a feature-rich framework for treebank annotation. In: Proceedings of the 22nd International Conference on Computational Linguistics, Manchester, UK, vol. 2, pp. 673–680 (2008)

17. Panevová, J.: On verbal frames in functional generative description. Prague Bull. Math. Linguist. **22**, 3–40 (1974)

18. Sagae, K., MacWhinney, B., Lavie, A.: Adding syntactic annotations to transcripts of parent-child dialogs. In: Proceedings of the 4th LREC, Lisbon, Portugal (2004)

19. Schuurman, I., Goedertier, W., Hoekstra, H., Oostdijk, N., Piepenbrock, R., Schouppe, M.: Linguistic annotation of the spoken Dutch corpus: if we had to do it all over again. In: Proceedings of the 4th LREC, Lisbon, Portugal (2004)

20. Sgall, P., Hajičová, E., Panevová, J.: The Meaning of the Sentence and Its Semantic and Pragmatic Aspects. Academia/Reidel Publishing Company, Prague/Dordrecht (1986)

21. Urešová, Z.: Building the PDT-VALLEX valency lexicon. In: Proceedings of the 5th Corpus Linguistics Conference, pp. 1–18. University of Liverpool, Liverpool (2012)

Automatic Phonetic Segmentation Using the Kaldi Toolkit

Jindřich Matoušek$^{(\boxtimes)}$ and Michal Klíma

Department of Cybernetics, Faculty of Applied Sciences, New Technology for the Information Society (NTIS), University of West Bohemia, Plzeň, Czech Republic
jmatouse@kky.zcu.cz, klima42@students.zcu.cz

Abstract. In this paper we explore the possibilities of hidden Markov model based automatic phonetic segmentation with the Kaldi toolkit. We compare the Kaldi toolkit and the Hidden Markov Model Toolkit (HTK) in terms of segmentation accuracy. The well-tuned HTK-based phonetic segmentation framework was taken as the baseline and compared to a newly proposed segmentation framework built from the default examples and recipes available in the Kaldi repository. Since the segmentation accuracy of the HTK-based system was significantly higher than that of the Kaldi-based system, the default Kaldi setting was modified with respect to pause model topology, the way of generating phonetic questions for clustering, and the number of Gaussian mixtures used during modeling. The modified Kaldi-based system achieved results comparable to those obtained by HTK—slightly worse for small segmentation errors but better for gross segmentation errors. We also confirmed that, for both toolkits, the standard three-state left-to-right model topology was significantly outperformed by a modified five-state left-to-right topology, especially with respect to small segmentation errors.

Keywords: Automatic phonetic segmentation · HTK · Kaldi · Hidden Markov models

1 Introduction

Phonetic segmentation is a process of detecting boundaries between phones in the speech signal. The knowledge of these boundaries is beneficial in a variety of speech processing algorithms and applications, especially in those which use the speech signal directly. Such algorithms (e.g. *concatenative speech synthesis*) rely on the precise placement of boundaries between phones or other phonetic units [6]. In addition, various speech corpora intended e.g. for spoken term detection [15] or speech understanding [17] exploit this information for indexing their content on the phonetic level. In today's world of big data and large speech corpora, the automation of the phonetic segmentation process is very important.

This research was supported by the Technology Agency of the Czech Republic (TA CR), project No. TH02010307.

© Springer International Publishing AG 2017
K. Ekštein and V. Matoušek (Eds.): TSD 2017, LNAI 10415, pp. 138–146, 2017.
DOI: 10.1007/978-3-319-64206-2_16

The most successful approaches to the *automatic phonetic segmentation* are based on *hidden Markov models* (HMMs), a statistical framework adopted from the area of *automatic speech recognition*. However, instead of the recognition, so-called *forced alignment* is performed to find the best alignment between phone-level HMMs and the corresponding speech data, producing a set of boundaries which delimit speech segments belonging to each HMM.

For decades, the *Hidden Markov Model Toolkit* (HTK) [19] has been used for building and manipulating HMMs in the context of speech modeling, recognition, segmentation, and with some extension also for speech synthesis [20]. Although HTK is still perfectly usable today, it is considered rather obsolete. Therefore, another toolkit, *Kaldi* [14], is getting much more attention and popularity these days. This is also due to its supportive community—Kaldi is used and developed by researchers all over the world. As a result, Kaldi supports many modern speech-processing and feature-extraction techniques, and it contains many "recipes" for advanced training and manipulation of speech models which would be (if at all) difficult to implement in HTK. Although Kaldi is intended mainly for speech recognition [12,14], some experiments to use it for the purposes of speech synthesis has been presented recently [1,13]. However, the application of Kaldi for automatic phonetic segmentation has not been researched so much [11].

In this paper we explore the possibilities of hidden Markov model based automatic phonetic segmentation with the Kaldi toolkit. We compare Kaldi- and HTK-based phonetic segmentation systems in terms of segmentation accuracy. The well-tuned HTK-based phonetic segmentation framework that we traditionally use to segment speech is taken as the baseline and compared to a newly proposed segmentation framework built from the default examples and recipes available in the Kaldi repository. Such a default setting of the Kaldi toolkit would probably be adopted by a newcomer to Kaldi. Later, we will modify the default Kaldi setting in order to achieve better segmentation results. Although there are many aspects that can affect the performance of the HMM-based phonetic segmentation such as the usage of context-independent (*monophone*) [4,18], context-dependent (*triphone*) models or their combination [6], HMM initialization and training strategies [2,6,18], number of Gaussian mixtures used during modeling [8,18], feature extraction schemes [6], etc., we will focus on the following parameters: HMM topology (for both pause and non-pausal models), the way of generating phonetic questions to cluster contextually similar model states, and the number of Gaussian mixtures used during modeling. Since the Kaldi toolkit currently does not support to bootstrap HMMs with manually-aligned speech data (so-called *bootstrap* initialization), HMMs in all experiments described here were initialized uniformly using the *flat-start* initialization scheme [19].

2   Speech Data Description

For our experiments, we used a Czech phonetically and prosodically rich speech corpus designed primarily for the purposes of unit selection based speech synthesis [3,9]. The utterances in the corpus were carefully selected [5], spoken by a

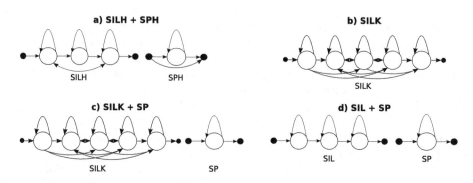

Fig. 1. Various pause models topologies: the default configuration used in HTK (a), the default one-pause model in Kaldi (b), and some combinations of these models (c) and (d). SIL and SP are modifications of the default HTK pause models.

professional male speaker in an anechoic chamber, recording at 16-bit precision with 48 kHz sampling frequency (later down-sampled to 16 kHz) and carefully annotated on the orthographic and phonetic levels. Phonetic transcripts for all utterances plus some manual segmentations from a phonetic expert were available. In order to train the segmentation systems examined in this paper, a feature vector was computed for each frame of the length $l = 25$ ms with a step $s = 6$ ms using 12 mel-frequency cepstral coefficients (MFCCs), log energy, and their delta and delta-delta coefficients (39 coefficients for each frame in total).

The corpus consists of 12,242 utterances (approx. 18 h of speech excluding pauses, 675,809 phone boundaries in total), 90 of them were segmented manually (approx. 12 min, 7,789 phone boundaries in total). As mentioned above, the manually segmented boundaries were not used for HMM initialization; they were used for evaluation purposes only.

3   Experiments and Results

In this section, we describe experiments with the automatic phonetic segmentation using the Kaldi toolkit and compare them with the well-established HTK-based segmentation framework. Firstly, experiments were made with the default Kaldi toolkit setting. Then, various modifications of the default setting were examined.

3.1   Parameters Under Consideration

In the comparison, we focused on the following parameters which were found to have an impact on the segmentation accuracy:

– **Pause models.** The pause models typically differentiate between long silence (SIL) and short pause (SP) and their various configurations (see Fig. 1).

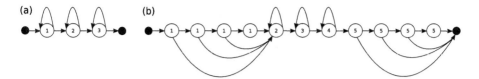

Fig. 2. Standard three-state left-to-right (a) and modified five-state left-to-right HMM topology (inspired by [16]).

Table 1. Parameters of the default and modified Kaldi settings and of the tuned HTK setting.

Setting	Kaldi-default	Kaldi-modif	HTK-tuned
Pauses	SILK	SIL+SP	SILH+SPH
Questions	AUT	AUT+EXP	EXP
Mixtures	M1k+T10k	M4+T1	M4+T1

- **Questions for generating decision trees.** Questions are used to cluster HMM states of similar phonetic contexts. Two ways of the question creation were examined: automatic (AUT, the default option in Kaldi) and manual expert-based (EXP, the only option in HTK), plus their combination (AUT+EXP).
- **Number of Gaussian mixtures.** The number of Gaussian mixtures to model output probability density function of each HMM state were examined both for *monophone* (M) and *triphone* (T) models: M4 (4 mixtures), M1k (1,000 mixtures), T1 (1 mixture), T10k (10,000 mixtures).

For non-pausal models, standard three-state left-to-right model topology (shown in Fig. 2a) often used for speech modeling was mostly employed. The various settings of the parameters are described in Table 1.

For both toolkits, the resulting boundaries were shifted by $\frac{l-s}{2}$ to the right to compensate for the shift imposed by the feature computation scheme [7] (l is the length of frames and s is the step used during the feature extraction).

3.2   Default Kaldi Setting vs. Tuned HTK Setting

Our first experiments concerned a comparison of the default Kaldi setting, set up according to available examples and recipes in the Kaldi repository, with the well-tuned HTK setting which was used to segment a speech corpus for unit-selection speech synthesis [6]. The motivation for this comparison was to find out what segmentation accuracy can be reached by a newcomer to Kaldi in comparison to the well-established HTK-based segmentation scheme.

The comparison is shown in Fig. 4 (the Kaldi-default and HTK-tuned bars). For the performance evaluation, percentage of boundaries deviating less than the given tolerance time region from the manually determined boundaries were taken into account. In our case, tolerance regions of 10 ms (corresponding to small segmentation errors), 20 ms and 30 ms (corresponding to grosser segmentation errors) were employed. As can be seen, the tuned HTK setting significantly outperforms the default Kaldi setting with respect to small segmentation errors. As for grosser segmentation errors, the differences between the toolkits tend to disappear.

3.3   Modifications to the Default Kaldi Setting

Pause models. In the default Kaldi setting, a single pause model SILK shown in Fig. 1b is used to model all pauses. On the other hand, HTK typically differentiate between two pause models—long pause (SILH), which is typically used to model longer sentence leading and trailing pauses, and short pause (SPH), which is used to model shorter sentence internal pauses (the exact topology of these pause models is shown in Fig. 1a). Therefore, we experimented with several pause models in Kaldi as well. The model topologies under consideration are shown in Fig. 1. Note that the SPH model cannot be used because Kaldi does not support model skipping. Instead, the SP model from Fig. 1c was used. The comparison of the pause models is shown in Fig. 3a. The models SIL+SP yield slightly better results than the others. The worst performance was obtained for SILK, the default pause model in Kaldi.

Questions for generating decision trees. Simply said, decision trees are used to cluster model states with similar phonetic contexts. Thus, decision trees make the modeling more robust. The decision trees are typically built using questions on the immediate phonetic context of each phone model. The two ways of question creation and their combination were described in Sect. 3.1. The results shown in Fig. 3b suggest that the combination of hand-crafted and automatically generated questions performs best, especially with respect to smaller segmentation errors.

Number of Gaussian mixtures. We also explored the impact of the number of Gaussian mixtures to model output probability density function of each HMM state on the segmentation accuracy. The configurations under examination were described in Sect. 3.1. The results in Fig. 3c show that the combination M4+T1 (which is actually the setting we currently use in HTK-based segmentation) performs best. The worst results were obtained for the combination M1k+T10k which is usually the default option in Kaldi.

HMM topology modification. Another modification (this time both to the Kaldi and HTK settings) concerned the topology of non-pausal HMMs. Whereas the original configuration consisted in standard three-state left-to-right topology (see Fig. 2a), the modified HMM topology duplicates lateral states but it also restricts the number of frames a lateral state can occupy. This is achieved by keeping the emission model shared between the lateral states, and replacing the self-transition loop by a transition to the next state (see Fig. 2b). Such five-state left-to-right topology was shown to be effective in duration controlling,

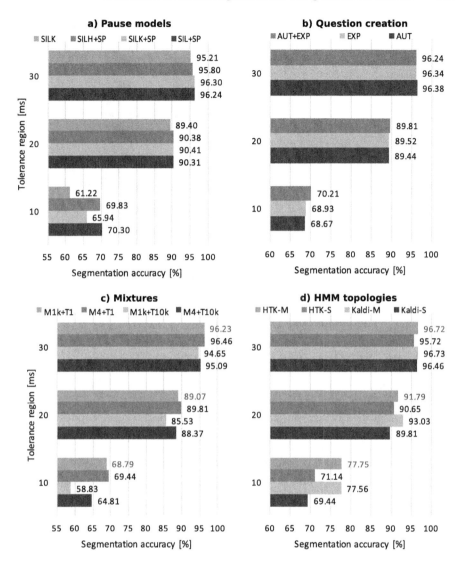

Fig. 3. Comparison of various modifications made to the default Kaldi setting: pause model topology modification (a), the way of decision-tree questions creation (b), the different number of Gaussian mixtures for monophone and triphone models (c), and non-pausal model topology modification both to the Kaldi and HTK systems (d).

reducing the dominance of one phone's lateral state over the other's [10,16]. Indeed, the results on our speech data confirm that the modified HMM topology (M) significantly outperforms the standard three-state left-to-right topology (S), especially with respect to small segmentation errors. This is true for both Kaldi- and HTK-based segmentation systems. Slightly better results were obtained for

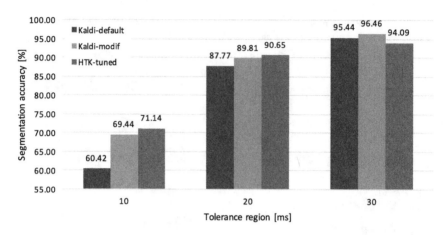

Fig. 4. Comparison of the default and modified Kaldi setting with the tuned HTK setting.

the Kaldi-based system (see Fig. 3d where S stands for the standard topology and M for the modified topology).

3.4   Modified Kaldi Setting vs. Tuned HTK Setting

In this experiment, we compared the modified Kaldi setting with the well-tuned HTK setting. The tuned HTK setting was the same as in the previous experiments. The modified Kaldi setting was based on the best results from the experiments described in Sect. 3.3, see Table 1. The standard three-state left-to-right topology of non-pausal models was used in this experiment.

As can be seen in Fig. 4b (the Kaldi-modif and HTK-tuned bars) the modified Kaldi system reaches almost the same segmentation accuracy as the tuned HTK system—slightly worse for small segmentation errors but better for gross segmentation errors. The largest increase in accuracy was achieved for small segmentation errors.

4   Conclusions

In this paper we explored the possibilities of automatic phonetic segmentation with the increasingly popular Kaldi toolkit. We showed that after some modifications the Kaldi toolkit reached segmentation results comparable to those obtained by the well-established HTK toolkit. We also confirmed that, for both toolkits, the modified five-state left-to-right topology significantly outperformed the standard three-state left-to-right model topology, especially with respect to small segmentation errors.

Our next steps will focus on incorporating more advanced speech-modeling and feature-extraction techniques available in Kaldi into the phonetic segmentation system. Since using some manually-aligned data as bootstrapping data for HMM initialization was reported to increase segmentation accuracy [2,6], we also plan to explore possibilities of the integration of bootstrap initialization into the Kaldi-based segmentation system.

References

1. Aylett, M.P., Dall, R., Ghoshal, A., Henter, G.E., Merritt, T.: A Flexible Front-End for HTS. In: INTERSPEECH, pp. 1283–1287. Singapore (2014)
2. Brognaux, S., Drugman, T.: HMM-based speech segmentation: improvements of fully automatic approaches. IEEE/ACM Trans. Audio Speech Lang. Process. **24**(1), 5–15 (2016)
3. Kala, J., Matoušek, J.: Very fast unit selection using Viterbi search with zero-concatenation-cost chains. In: IEEE International Conference on Acoustics Speech and Signal Processing, Florence, Italy, pp. 2569–2573 (2014)
4. Matoušek, J.: Automatic pitch-synchronous phonetic segmentation with context-independent HMMs. In: Matoušek, V., Mautner, P. (eds.) TSD 2009. LNCS, vol. 5729, pp. 178–185. Springer, Heidelberg (2009). doi:10.1007/978-3-642-04208-9_27
5. Matoušek, J., Romportl, J.: On building phonetically and prosodically rich speech corpus for text-to-speech synthesis. In: IASTED International Conference on Computational Intelligence, San Francisco, USA, pp. 442–447 (2006)
6. Matoušek, J., Romportl, J.: Automatic pitch-synchronous phonetic segmentation. In: INTERSPEECH, Brisbane, Australia, pp. 1626–1629 (2008)
7. Matoušek, J., Tihelka, D., Psutka, J.: Automatic segmentation for Czech concatenative speech synthesis using statistical approach with boundary-specific correction. In: INTERSPEECH, Geneva, Switzerland, pp. 301–304 (2003)
8. Matoušek, J., Tihelka, D., Psutka, J.: Experiments with automatic segmentation for Czech speech synthesis. In: Matoušek, V., Mautner, P. (eds.) TSD 2003. LNCS, vol. 2807, pp. 287–294. Springer, Heidelberg (2003). doi:10.1007/978-3-540-39398-6_41
9. Matoušek, J., Tihelka, D., Romportl, J.: Building of a speech corpus optimised for unit selection TTS synthesis. In: Language Resources and Evaluation Conference, Marrakech, Morocco, pp. 1296–1299 (2008)
10. Ogbureke, K.U., Carson-Berndsen, J.: Improving initial boundary estimation for HMM-based automatic phonetic segmentation. In: INTERSPEECH, Brighton, Great Britain, pp. 884–887 (2009)
11. Patc, Z., Mizera, P., Pollak, P.: Phonetic segmentation using KALDI and reduced pronunciation detection in causal Czech speech. In: Král, P., Matoušek, V. (eds.) TSD 2015. LNCS, vol. 9302, pp. 433–441. Springer, Cham (2015). doi:10.1007/978-3-319-24033-6_49
12. Plátek, O., Jurčíček, F.: Integration of an on-line Kaldi speech recogniser to the Alex dialogue systems framework. In: Text, Speech and Dialogue. LNCS, vol. 9302, pp. 433–441. Springer, Heidelberg (2015)
13. Potard, B., Aylett, M.P., Baude, D.A.: Idlak Tangle: an open source Kaldi based parametric speech synthesiser based on DNN. In: INTERSPEECH, San Francisco, USA, pp. 2293–2297 (2016)
14. Povey, D., Ghoshal, A., Boulianne, G., Burget, L., Glembek, O., Goel, N., Hannemann, M., Motlicek, P., Qian, Y., Schwarz, P., Silovsky, J., Stemmer, G., Vesely, K.: The Kaldi speech recognition toolkit. In: IEEE Workshop on Automatic Speech Recognition and Understanding, Hawaii, USA, pp. 1–4 (2011)

15. Psutka, J., Švec, J., Psutka, J.V., Vaněk, J., Pražák, A., Šmídl, L., Ircing, P.: System for fast lexical and phonetic spoken term detection in a Czech cultural heritage archive. EURASIP J. Audio Speech Music Process. **10** (2011)
16. Rendel, A., Sorin, A., Hoory, R., Breen, A.: Towards automatic phonetic segmentation for TTS. In: IEEE International Conference on Acoustics Speech and Signal Processing, Kyoto, Japan, pp. 4533–4536 (2012)
17. Švec, J., Šmídl, L.: On the use of phoneme lattices in spoken language understanding. In: Habernal, I., Matoušek, V. (eds.) TSD 2013. LNCS, vol. 8082, pp. 369–377. Springer, Heidelberg (2013). doi:10.1007/978-3-642-40585-3_47
18. Toledano, D., Gomez, L., Grande, L.: Automatic phonetic segmentation. IEEE Trans. Speech Audio Process. **11**(6), 617–625 (2003)
19. Young, S., Evermann, G., Gales, M.J.F., Hain, T., Kershaw, D., Liu, X., Moore, G., Odell, J., Ollason, D., Povey, D., Valtchev, V., Woodland, P.: HTK Book (for HTK Version 3.4). The Cambridge University, Cambridge, U.K. (2006)
20. Zen, H., Nose, T., Yamagishi, J., Sako, S., Masuko, T., Black, A.W., Tokuda, K.: The HMM-based speech synthesis system (HTS) version 2.0. In: Speech Synthesis Workshop, Bonn, Germany, pp. 294–299 (2007)

Language Independent Assessment of Motor Impairments of Patients with Parkinson's Disease Using i-Vectors

N. Garcia[1(✉)], J.C. Vásquez-Correa[1,2], J.R. Orozco-Arroyave[1,2], N. Dehak[3], and E. Nöth[2]

[1] Faculty of Engineering, Universidad de Antioquia UdeA, Medellín, Colombia
{nicanor.garcia,jcamilo.vasquez,rafael.orozco}@udea.edu.co
[2] Pattern Recognition Lab, Friedrich-Alexander-Universität Erlangen-Nürnberg, Erlangen, Germany
noeth@informatik.uni-erlangen.de
[3] Center for Language and Speech Processing, Jhons Hopkins University, Baltimore, USA
najim@jhu.edu

Abstract. Speech disorders are among the most common symptoms in patients with Parkinson's disease. In recent years, several studies have aimed to analyze speech signals to detect and to monitor the progression of the disease. Most studies have analyzed speakers of a single language, even in that scenario the problem remains open. In this study, a cross-language experiment is performed to evaluate the motor impairments of the patients in three different languages: Czech, German and Spanish. The i-vector approach is used for the evaluation due to its capability to model speaker traits. The cosine distance between the i-vector of a test speaker and a reference i-vector that represents either healthy controls or patients is computed. This distance is used to perform two analyses: classification between patients and healthy speakers, and the prediction of the neurological state of the patients according to the MDS-UPDRS score. Classification accuracies of up to 72% and Spearman's correlations of up to 0.41 are obtained between the cosine distance and the MDS-UPDRS score. This study is a step towards a language independent assessment of patients with neuro-degenerative disorders.

Keywords: Parkinson's disease · i-vectors · UPDRS score · Language independent assessment

1 Introduction

Parkinson's disease (PD) is a neuro-degenerative disorder which produces several motor and non-motor impairments. The motor symptoms include, among others tremor, rigidity, slowed movement, postural instability, lack of coordination and speech disorders [1]. Evaluating the condition of PD patients is difficult. Mobility

© Springer International Publishing AG 2017
K. Ekštein and V. Matoušek (Eds.): TSD 2017, LNAI 10415, pp. 147–155, 2017.
DOI: 10.1007/978-3-319-64206-2_17

problems make attending medical appointments burdensome, while speech disorders may hinder the communication with the medical experts [2]. Currently, the assessment of the disease in the motor capabilities is evaluated with the third section of the Movement Disorder Society, Unified Parkinson's Disease Rating Scale (MDS-UPDRS) [3]. This evaluation is subject to a clinical criterion and its intra- and inter-rater variability could be high. The diagnosis could be supported by computer aided systems, which could also improve the evaluation of the disease progression. On the other hand, only two of the 33 items of the MDS-UPDRS are related to the speech impairments of patients; however, speech disorders are among the most prevalent, and an early sign of further motor impairments [4]. In that way, speech signals could be used to assess the motor symptoms of PD patients.

There has been interest in the scientific community to develop computer aided tools to evaluate the condition of PD patients using information from speech. In the 2015 INTERSPEECH Computational Paralinguistics Challenge (ComPARE) the task of predicting the MDS-UPDRS score of PD patients from speech was addressed [5]. Speech recordings of 50 PD patients from the PC-GITA database [6] were considered for the train and development subsets. Recordings from eleven new patients were considered as the test set. All the speakers were native Spanish speakers. A Spearman's correlation coefficient of 0.39 was reported as baseline of the challenge. The winners of the challenge [7] grouped the speech tasks of each patient and used deep neural networks and Gaussian processes for the prediction, obtaining a correlation coefficient of up to 0.69. In [8] the authors classify the speech of PD patients vs. healthy controls (HC) speakers in different languages. Cross-language experiments were performed using data in three languages: Czech, German and Spanish. The reported accuracies range from 60% to 77%, depending on the languages used for train and test sets. In [9] a speaker model based on Gaussian mixture models-universal background models (GMM-UBM) was proposed to monitor the neurological state of PD patients. UBMs were trained with information from 61 PD and 50 HC speakers. Specific GMMs were adapted for seven PD patients recorded in three sessions. Then, the Bhattacharyya distance between the speaker models and the UBM was computed and correlated with the MDS-UPDRS score of the patients. A Pearson's correlation of up 0.60 was reported by the authors. In [10] the authors proposed a new regression method to track the progression of speech disorders. The method is based on a non-parametric learning strategy based on a probability distance measure between the speakers from the test and training sets. The authors consider data from 61 PD patients to predict the UPDRS score, obtaining a Pearson's correlation of up to 0.58.

Speaker models inspired by speaker verification and identification systems have shown promising results in evaluating PD from speech. The most recent breakthrough in speaker verification is the i-vector approach [11]. This strategy has also proven to be effective in many other speech tasks. Specially, it has shown excellent results in language identification [12,13]. Recently, i-vectors were used to identify the native language of a speaker from recordings in a second

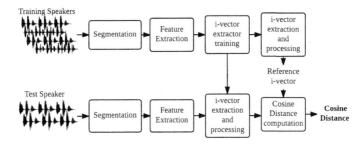

Fig. 1. General methodology followed in this study.

language [14]. According to the reviewed literature, a language independent strategy to evaluate the PD condition from speech has not been enough addressed. Developing a computer aided system that can evaluate PD from speech in different languages would be a major step towards an unified objective assessment of the disease. Additionally, the use of i-vectors has been successfully applied to model speaker traits in multiple languages, which indicates that it could also be used to evaluate the PD condition from speech in a cross-language approach. In this study, a strategy based on i-vectors is used to assess the PD condition from speech in three languages: Spanish, German and Czech. Cross-language experiments are performed, i.e., train the models with utterances from one language, and test with the speech recordings from the other ones. The proposed approach is tested in two scenarios: (1) classification of PD vs. HC subjects, and (2) the prediction of the MDS-UPDRS score of the patients. Different i-vector extractors are trained with features related to specific dimensions of speech, e.g., phonation, articulation and prosody with the aim of evaluating the information provided by each dimension to represent the PD condition of the patients.

2   Methods

The methodology proposed in this study comprises four steps: (1) several feature sets are computed to analyze different speech dimension from speech, (2) a subset of speakers are used to train an i-vector extractor, (3) the i-vectors of speech signals are extracted, and (4) the cosine distance between a reference i-vector and the speaker i-vector is computed. This process is summarized in Fig. 1.

2.1   Feature Extraction

Four feature sets were considered in this study to model the speech impairments of PD patients. The first set comprises the Mel-Frequency Cepstral Coefficients (MFCCs), which are the classical features used to train i-vectors. 19 MFCCs and the log-energy extracted from 30 ms windows with time-shift of 15 ms were used to form a 20-dimensional feature vector. Non-speech frames were discarded using an energy-threshold voice activity detector (VAD). The other feature sets are formed with descriptors to assess the articulation, phonation, and prosody dimensions of speech. To evaluate articulation, the energy content in 22 Bark bands (BBE) in the voiced/unvoiced and unvoiced/voiced transitions were considered, as in [15]. The features considered to evaluate phonation and articulation in voiced segments are: the log-energy, the fundamental frequency (F_0), first and second formants (F_1 and F_2) and their first and second derivatives. Additionally, perturbation features such as Jitter and Shimmer are also included. These descriptors form a 14-dimensional feature vector. These features were computed from voiced segments using 30 ms long analysis frames with a time-shift of 5 ms. To evaluate prosody we followed the approach introduced in [16]: The log-F_0 and the log-energy contours within analysis frames were approximated using Lagrange polynomials of order $P = 5$. Analysis frames of 200 ms with time-shift of 50 ms were used as in [13]. A 13-dimensional feature vector is formed concatenating the six coefficients computed from the log-F_0 and the log-energy contours, along with the number of voiced frames in the utterance.

2.2   i-Vectors

In this approach, factor analysis is used to define a new low-dimensional space known as the total variability space with the aim of modeling the speaker and the channel variability [11]. For applications related to pathological speech analysis, the speaker variability carries the information about the disorders in speech due to the disease. In the total variability space, an utterance is represented by a supervector M formed by concatenating the mean vectors of a GMM-UBM. The total variability space is expressed according to Eq. 1, where m is a speaker and channel independent supervector (the UBM), T is the total variability matrix and w corresponds to the i-vector.

$$\mathbf{M} = \mathbf{m} + \mathbf{Tw} \tag{1}$$

The i-vectors are processed in five steps: (1) i-vectors extracted from training speakers are normalized to zero mean and unit variance, i.e., Z-norm, (2) the normalized i-vectors computed from different speech tasks of a given speaker are averaged to obtain one i-vector per speaker, (3) the i-vectors of HC and PD

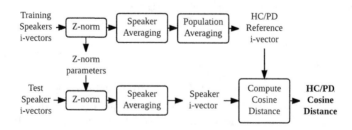

Fig. 2. i-vector processing.

speakers are averaged to obtain HC and PD reference i-vectors, respectively, (4) the i-vectors of a test speaker are normalized using the parameters from the training i-vectors, (5) the normalized i-vectors per utterace are averaged to obtain the speaker i-vector. Finally, the cosine distance between the HC/PD reference i-vectors and the speaker i-vectors is computed. The process is summarized in Fig. 2.

2.3   Cosine Distance

The cosine distance is used to compare two i-vectors w_1 and w_2. The distance is defined by Eq. 2. In this study, the i-vector of a single speaker is compared with a reference i-vector that represents the HC or PD population.

$$d_c(\mathbf{w}_1, \mathbf{w}_2) = 1 - \frac{\mathbf{w}_1 \cdot \mathbf{w}_2}{||\mathbf{w}_1|| ||\mathbf{w}_2||}. \tag{2}$$

2.4   Evaluation

The cosine distance between the test speaker i-vector and the reference i-vector is compared to a threshold to discriminate between PD patients and HC speakers. The development set is used to find the threshold that maximizes the accuracy. The prediction of the neurological state of a patient is evaluated using the Spearman's correlation coefficient between the real MDS-UPDRS score and the distance measure.

3   Data

Spanish- The PC-GITA database [6] is used in this study. It contains recordings of 50 PD patients and 50 healthy control (HC) speakers. All of them are native Colombian Spanish speakers. During the recordings, the participants were asked to perform different speech tasks including reading ten isolated sentences, and the repetition of /pa-ta-ka/, a diadochokinetic (DDK) exercise.

German- The German data contain recordings from 88 PD patients and 88 HC subjects. The speakers perform several speech tasks, including the repetition of /pa-ta-ka/, and reading five isolated sentences [17].

Czech- The Czech data are formed with recordings from 20 PD patients and 15 HC subjects. The patients were recorded at the time of diagnosis with PD, and none of them had been medicated before or during the recording session. The speech tasks performed by the speakers include the rapid repetition of /pa-ta-ka/, and several repetitions of a sentence [4].

4   Experiments and Results

Two speech tasks were analyzed independently in these experiments: the rapid repetition of /pa-ta-ka/, and read sentences. Data from the three languages are used in turn as training, development and test sets. All possible combinations are tested. The training data are used for several processes: (1) to train the UBM and the i-vector extractor, (2) to compute the normalization parameters, and (3) to obtain the HC and PD reference i-vectors. UBMs with different number of Gaussian components were trained in a range from $M = 2$ to $M = 2^9$ into powers of 2. The dimension of the i-vector $\dim_w$ was chosen following the relation $\dim_w = \log_2(M) \cdot \dim_f$, where M is the number of Gaussian components in the UBM and $\dim_f$ is the dimension of the feature vector.

Table 1. Accuracies (%) for the classification task.

Train Lang.	Test Lang.	HC reference				PD reference			
		MFCCs	Art.	Phon.	Pros.	MFCCs	Art.	Phon.	Pros.
DDK									
Czech	German	47.4	47.4	52.6	53.7	47.4	47.4	52.6	53.7
	Spanish	43.0	48.0	44.0	50.0	43.0	48.0	44.0	50.0
German	Czech	58.8	47.1	**67.6**	58.8	58.8	47.1	**67.6**	58.8
	Spanish	50.0	48.0	60.0	50.0	50.0	48.0	60.0	50.0
Spanish	Czech	**61.8**	47.1	58.8	**61.8**	**61.8**	47.1	58.8	**61.8**
	German	52.6	40.6	50.9	53.7	52.6	40.6	50.9	53.7
Sentences									
Czech	German	55.1	54.0	57.4	53.4	55.1	54.0	57.4	53.4
	Spanish	60.0	50.0	50.0	50.0	60.0	50.0	50.0	50.0
German	Czech	**72.2**	58.3	44.4	55.6	63.9	58.3	47.2	55.6
	Spanish	**68.0**	55.0	50.0	50.0	55.0	55.0	50.0	50.0
Spanish	Czech	50.0	**72.2**	50.0	**63.9**	50.0	**72.2**	50.0	**63.9**
	German	**60.2**	39.8	52.8	57.4	**60.2**	39.8	52.8	57.4

Table 1 shows the results for the classification of PD vs. HC speakers. For the DDK speech task, only the test in Czech language shows accuracies higher than 65%. This could be explained due to Czech patients being diagnosed at the time of the recording and being in an earlier state of the disease than the patients from the other two languages. Accuracies below 50% could be explained by the fact that ranges of the cosine distance are likely to be different in the development and test sets. On the other hand, for the sentences, the i-vectors extracted with MFCCs and articulation-based features show the best results in most of cases. The 50% results when Spanish is used for test can be explained by the mismatch of cosine distance ranges in the development and test sets. For this case, the threshold could be set so all speakers in the test dataset are classified either as PD or HC. The similar results found using the HC and PD reference i-vectors may be due to both vectors being antiparallel.

Table 2 shows the results for the prediction of the MDS-UPDRS. For this case, articulation-based features provide the best result when evaluating the DDK speech task. Phonation features show good results in some cases, specially when the Spanish language is used for test, but show poor results when testing on Czech. This maybe due to the fact that Spanish is a more voiced language than the other two languages, but further experimentation is required. Correlations with the MDS-UPDRS score of up to 0.4 were achieved. Slightly higher correlations were obtained using the DDK speech task due to the fact that such a task is language independent, i.e., the speakers in the corpora uttered the same sounds. Good results were also obtained when analyzing isolated sentences, which is a language dependent speech task. This is encouraging and indicates that other speech tasks could also be analyzed in a cross-language setting.

Table 2. Spearman's correlation for the prediction task.

Train Lang.	Test Lang.	HC reference				PD reference			
		MFCCs	Art.	Phon.	Pros.	MFCCs	Art.	Phon.	Pros.
DDK									
Czech	German	−0.14	−0.25	0.14	−0.06	0.14	0.25	−0.14	0.05
	Spanish	**0.32**	0.23	0.20	−0.13	**−0.32**	−0.29	−0.17	0.17
German	Czech	**0.26**	0.38	−0.09	0.11	**−0.30**	−0.38	0.16	−0.25
	Spanish	0.04	0.20	0.24	0.09	0.25	−0.21	**−0.39**	0.31
Spanish	Czech	−0.32	**0.41**	−0.19	−0.48	0.45	−0.25	0.12	0.16
	German	0.15	−0.14	0.11	−0.24	−0.17	0.14	−0.11	0.14
Sentences									
Czech	German	0.06	0.11	−0.14	0.15	−0.06	−0.11	0.14	−0.15
	Spanish	0.16	−0.15	−0.01	0.12	−0.15	0.15	0.01	−0.12
German	Czech	0.18	−0.11	−0.30	−0.02	−0.21	0.13	0.32	0.07
	Spanish	0.26	0.08	0.37	0.02	−0.27	−0.08	**−0.37**	−0.02
Spanish	Czech	−0.02	0.29	0.05	**0.36**	0.05	−0.29	−0.05	**−0.36**
	German	0.10	0.04	0.11	0.04	−0.10	−0.04	−0.19	−0.04

For comparison, language dependent results using the same Spanish database and the same i-vector methodology can be found in [18].

5   Conclusion

In this work we address the task of cross-language evaluation of Parkinson's Disease speech using the i-vector approach. Data in Czech, German and Spanish were used. One of the languages is used for train, while the other two were used for parameter selection and test. All possible combinations were considered. Two reference i-vectors were created. These reference i-vectors represent the population of HC speakers or PD patients. Then, the cosine distance between one of these reference i-vectors and the i-vector of a test speaker was computed. This distance was used in two experiments: to classify PD patients and HC speakers, and to assess the prediction of the neurological state of the patients. Results are promising, with classification rates around 70% when using MFCCs and articulation features. Similar classification results were obtained using both reference i-vectors. In many cases, a positive correlation between the labels and the cosine distance to the HC reference i-vector was found. This means that the more affected the speech, the larger the difference to healthy speakers. A similar reasoning can be followed for the negative correlations when comparing test speakers with respect to the PD reference i-vector, i.e., the more affected the speech, the lower the difference to the PD speakers. Future work includes evaluating the use of techniques that can eliminate the variability of language in the i-vector space with the aim of improving the results and obtain a language independent method to evaluate the condition of patients with neurodegenerative disorders.

Acknowledgments. Thanks to CODI from University of Antioquia by the grant Numbers 2015-7683 and PRV16-2-01 and to COLCIENCIAS by the grant Number 111556933858.

References

1. Ahmed, A.M., et al.: Motor symptoms in Parkinson's disease: a unified framework. Neurosci. Biobehav. Rev. **68**, 727–740 (2016)
2. Stamford, J.A., Schmidt, P.N., Friedl, K.E.: What engineering technology could do for quality of life in Parkinson's disease: a review of current needs and opportunities. IEEE J. Biomed. Health Inf. **19**(6), 1862–1872 (2015)
3. Goetz, C.G., et al.: Movement disorder society-sponsored revision of the unified Parkinson's disease rating scale (mds-updrs): scale presentation and clinimetric testing results. Mov. Disord. **23**(15), 2129–2170 (2008)
4. Rusz, J., et al.: Imprecise vowel articulation as a potential early marker of Parkinson's disease: effect of speaking task. J. Acoust. Soc. Am. **134**(3), 2171–2181 (2013)
5. Schuller, B., et al.: The INTERSPEECH 2015 computational paralinguistics challenge: nativeness, Parkinson's & eating condition. In: Proceedings of the 16th INTERSPEECH, pp. 478–482 (2015)
6. Orozco-Arroyave, J.R., Arias-Londoño, J.D., Vargas-Bonil, J.F., González-Rátiva, M.C., Nöth, E.: New Spanish speech corpus database for the analysis of people suffering from Parkinson's disease. In: Proceedings of the 9th LREC, pp. 342–347 (2014)
7. Grósz, T., Róbert, B.-F., Gábor, G., Tóth, L.: Assessing the degree of nativeness and Parkinson's condition using Gaussian processes and deep rectifier neural networks. In: Proceedings of the 16th INTERSPEECH, pp. 919–923 (2015)
8. Orozco-Arroyave, J.R., et al.: Automatic detection of Parkinson's disease from words uttered in three different languages. J. Acoust. Soc. Am. **139**(1), 481–500 (2016)

9. Arias-Vergara, T., Vasquez-Correa, J.C., Orozco-Arroyave, J.R., Vargas-Bonilla, J.F., Noth, E.: Parkinson's disease progression assessment from speech using GMM-UBM. In: Proceedings of the 17th INTERSPEECH, pp. 1933–1937 (2016)
10. Tu, M., Berisha, V., Liss, J.: Objective assessment of pathological speech using distribution regression. In: Proceedings of the IEEE International Conference on Acoustics, Speech, and Signal Processing (ICASSP) (2017)
11. Dehak, N., Kenny, P.J., Dehak, R., Dumouchel, P., Ouellet, P.: Front-end factor analysis for speaker verification. IEEE Trans. Audio Speech Lang. Process. 19(4), 788–798 (2011)
12. Dehak, N., Torres-Carrasquillo, P.A., Reynolds, D., Dehak, R.: Language recognition via i-vectors and dimensionality reduction. In: Proceedings of the 12th INTERSPEECH, pp. 857–860 (2011)
13. Martínez, D., Burget, L., Ferrer, L., Scheffer, N.: iVector-based prosodic system for language identification. In: Proceedings of the 37th ICASSP, pp. 4861–4864, March 2012
14. Senoussaoui, M., Cardinal, P., Dehak, N., Koerich, A.L.: Native language detection using the i-vector framework. In: Proceedings of the 17th INTERSPEECH, pp. 2398–2402 (2016)
15. Orozco-Arroyave, J.R., et al.: Towards an automatic monitoring of the neurological state of Parkinson's patients from speech. In: Proceedings of the 41st ICASSP, pp. 6490–6494 (2016)
16. Dehak, N., Dumouchel, P., Kenny, P.: Modeling prosodic features with joint factor analysis for speaker verification. IEEE Trans. Audio Speech Lang. Process. 15(7), 2095–2103 (2007)
17. Skodda, S., Visser, W., Schlegel, U.: Vowel articulation in Parkinson's disease. J. Voice 25(4), 467–472 (2012)
18. Garcia, N., Orozco-Arroyave, J.R., D'Haro, L.F., Dehak, N., Nöth, E.: Evaluation of the neurological state of people with Parkinson's disease using i-vectors. In: Proceedings of the 18th INTERSPEECH (2017, in Press)

ParCoLab: A Parallel Corpus for Serbian, French and English

Aleksandra Miletic[1]($\boxtimes$), Dejan Stosic[1], and Saša Marjanović[2]

[1] CLLE, CNRS & University of Toulouse, 5, Allées Antonio Machado,
31058 Toulouse, France
{aleksandra.miletic,dejan.stosic}@univ-tlse2.fr
[2] Faculty of Philology, University of Belgrade, Studentski Trg 3,
11000 Belgrade, Serbia
sasa.marjanovic@fil.bg.ac.rs

Abstract. ParCoLab is a trilingual parallel corpus containing texts in Serbian, French and English. It is developed at the CLLE-ERSS research unit (UMR 5263 CNRS) at the University of Toulouse, France, in collaboration with the Department of Romance Studies at the University of Belgrade, Serbia. Serbian being one of the less-resourced European languages, this is an important step towards the creation of freely accessible corpora and NLP tools for this language. Our main goal is to provide the scientific community with a high-quality resource that can be used in a wide range of applications, such as contrastive linguistic studies, NLP research, machine and computer assisted translation, translation studies, second language learning and teaching, and applied lexicography. The corpus currently contains 7.1M tokens mainly from literary works, but corpus extension and diversification efforts are ongoing. ParCoLab can be queried online and a part of it is available for download.

Keywords: Parallel corpus · Serbian · French · English · NLP resources

1 Introduction

ParCoLab is a Serbian-French-English corpus developed at the CLLE-ERSS research unit (UMR 5263 CNRS) at the University of Toulouse, France, in collaboration with the Department of Romance Studies at the University of Belgrade, Serbia. This work is an effort towards the creation of freely accessible resources for Serbian, which is still one of less-resourced European languages. At the time of writing this paper, the corpus contains 7.1M tokens. The content is predominantly literary, but diversification efforts are ongoing, especially towards including legal texts, subtitles and web content. It is conceived as a versatile resource, primarily intended for contrastive linguistic research and NLP, but as any parallel corpus, it can also find its uses in applied lexicography, second language learning and teaching and machine and computer-assisted translation. It can be queried online at the following address: http://parcolab.univ-tlse2.fr/sr/, and a part of its content is available for download.

K. Ekštein and V. Matoušek (Eds.): TSD 2017, LNAI 10415, pp. 156–164, 2017.
DOI: 10.1007/978-3-319-64206-2_18

There are several existing parallel corpora that incorporate Serbian. SrpFranKor [31] and SrpEnKor [13] are Serbian-French and Serbian-English parallel corpora containing respectively 1.7M and 4.4M words coming from literary texts, newspaper articles, scientific papers, movie subtitles, etc. SETimes [27] is a plurilingual corpus in English and a number of South-Eastern European languages including Serbian. It is based on newspaper articles and contains 9.2M words in the English-Serbian subcorpus. Another English-Serbian parallel corpus was built as part of the MULTEXT-East project [14]: it is based on Orwell's *1984* and contains 150K tokens. ParaSol [32] is comprised of belletristic texts in Slavic languages and French, German, English and Italian (1.3M tokens in Serbian). InterCorp [5] is part of the Czech National Corpus and boasts 30M tokens in Serbian, from fiction (10M) and movie subtitles (20M).

The content nature and the types of access available can vary greatly from one corpus to another, depending on their primary intended use. SETimes can be downloaded in TMX or text format and used for NLP or machine translation, but it does not have a query interface that would facilitate concordance extraction for linguistic research. Also, the originals and the translations are not distinguished within the corpus [28]. MULTEXT-East is similar: it can be downloaded in XML format and used in NLP, but has no dedicated search engine. In addition, it is based on a translation, and not an original text in Serbian. And whereas SrpFranKor and ParaSol do have web interfaces for searching the corpus, they contain only 2 original Serbian documents each. They are not available for download and therefore cannot be used in NLP or similar applications. InterCorp seems to be the only one that combines both approaches: it has a web interface for searching, and it is indicated as available on demand. However, it is unclear which portion of its content comes from original texts in Serbian.

The goal of ParCoLab is to combine these two aspects in providing freely available resources for Serbian: we aim to have high-quality, linguistically annotated content paired with a user-friendly interface allowing for easy applications in linguistics, while making a portion of the corpus suited for NLP research and available for download. In Sect. 2, we present the corpus content by language and by text type, Sect. 3 describes the ways of accessing and using the corpus, whereas Sect. 4 gives details on the efforts for the linguistic annotation of the corpus and creation of NLP ressources. Conclusions and perspectives for future work are given in Sect. 5.

2   ParCoLab: Content and Structure

At the time of writing this paper, ParCoLab contains 7.1M tokens, 2.4M of which are in Serbian, 3.1M in French, and 1.6M in English. The content is predominantly literary, but other types of documents, such as legal texts, subtitles and web content, are being included into the resource.

The documents are stored in a TEI–compliant XML format following the TEI-P5 Guidelines [6][1]. It comprises a <teiHeader> element with following metadata: title, subtitle, author, translator, year of creation, year of publication, publisher, place of publication, domain and genre, file language, original language (for translations) and derivation type (original text or translation). Some of these fields are available as search parameters (see Sect. 3). The corpus content is stored in a NoSQL database, which will allow for smooth integration of different annotation levels in the future.

2.1   Content Distribution by Language

ParCoLab incorporates original texts in Serbian, French, and English, aligned with their translations into one or both of the other languages of the corpus. In other words, there are three subcorpora, each having a different pivot language. Figure 1 indicates the number of tokens per subcorpus, both in the originals and in the translations.

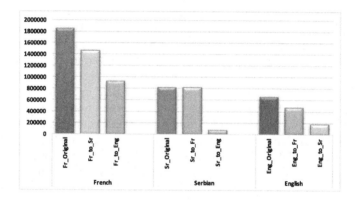

Fig. 1. Token distribution per language

Even though relatively small compared to large-scale web corpora, ParCoLab is growing steadily: in 2016, it went from 2M tokens to 7.1M, and is expected to cross the 10M threshold by the end of June 2017. The extensions are guided by two main principles: balancing out the languages in the corpus, and diversifying the genres. Up to now, the project has favoured the French-Serbian language pair, partly because there are more existing corpora for Serbian and English. However, one of our immediate goals is to extend the English part of the corpus, especially by including the English translations of Serbian texts already in the corpus. For genre diversification, see Sect. 2.2.

---

[1] The only two modifications we make is that we introduce an attribute @langOri used in the <teiHeader> in order to encode the language of the original text in the XML files containing translations, and the @id attribute used on the root <TEI> element, indicating the unique ID of the file inside the collection.

2.2   Corpus Content by Text Type

As mentioned above, in its current state, ParCoLab contains mainly literary texts. However, it also incorporates some legal documents, subtitles and different types of content coming from plurilingual web pages (see distribution in Table 1).

Table 1. Distribution of tokens per text type and language

Text type	English	French	Serbian	Total tokens
Literary texts	1 066 571	2 684 212	2 333 559	6 084 342
Web content	229 006	186 256	63 018	478 280
Legal texts	181 290	195 095	0	376 385
Subtitles	27 395	42 427	33 305	103 127

A part of the ParCoLab's **literary content** consists in copyright-free literary works acquired from internet databases of PDF files such as Gallica (http://gallica.bnf.fr) and Bibliothèque électronique du Québec (https://beq.ebooksgratuits.com/) for French, and Project Gutenberg (http://www.gutenberg.org/) for English. There is also an important number of more recent texts (still under copyright) for which we negotiated the rights to include them into the resource. In this case, the PDFs are generally obtained through manual scanning. They are then converted into text, manually cleaned, transformed into XML and imported into the database. The alignments are checked and corrected online.

The **web content** comes mostly from institutional websites such as embassies and cultural institutes. The bilingual sites are identified manually in order to guarantee the quality of the content and of the translations. This is also a constraint of the tool we use for web content extraction: Bitextor [7] performs automatic extraction of parallel content from web pages, but it needs a list of URLs as input.

The **legal texts** are the newest addition to the corpus. Given the fact that Serbia is not a member of the European Union, there are no Serbian texts in the Europarl Corpus [12] or the JRC-Acquis [25]. However, since Serbia has obtained the candidate status to join the EU, the translation of the JRC-Acquis into Serbian has been initiated. We have therefore started integrating the English and French texts, and are working on gaining access to the parts already translated into Serbian.

The work on the **subtitles** is done at the Department of Romance Studies at the University of Belgrade. The subtitles in the original language are downloaded from the Internet. The translations are either downloaded and submitted to a quality check, or created as part of the Translation Workshop by the French language students at the Department. There are two main types of subtitles: those from movies and TV shows ($\sim$30K tokens) and those from TEDtalks[2] ($\sim$60K tokens).

---

[2] TED is a platform for short talks on various subjects. See http://www.ted.com/.

3   Access and Use

Access to ParCoLab is free, but it is necessary to sign up and create a user account at http://parcolab.univ-tlse2.fr/sr/. With the user account, it is possible to make queries via the web interface. A sample of parallel texts from various sources containing 588K tokens can be downloaded, as well as a 150K token Serbian sub-corpus annotated with POS-tags and lemmas (see Sect. 4). Unfortunately, the whole corpus cannot be distributed because some of the works are under copyright and the rights we acquired are limited to online querying only. However, the remaining copyright-free texts, which represent approximately 4.2M tokens in total, will be progressively made available for download.

The search engine allows to define a search pattern composed of one or several tokens. It is possible to select the search language, limit the search to original texts or translations only, or choose an author. The search pattern can be defined in two or three languages, e.g. *kroz* 'through' in Serbian, and *par* 'through' in French. This would allow to search for sentence pairings in which the one in Serbian would contain *kroz*, and its French counterpart *par*. For now, only searches by inflected form are possible; searching by lemma is not yet available (see Sect. 4). However, wildcards replacing one or several characters can be used, and it is also possible to define the distance between different parts of the query pattern. All these parameters can be combined in order to formulate complex queries.

ParCoLab has already proven its usefulness for linguistic research. Up to now, it has been used in works on indefinite determiners in French and Serbian [24], on fictive motion [26], and on corpus-based phraseme selection for dictionary creation [17]. In order to optimize the corpus for these purposes, an annotation process is under way.

4   Towards a Linguistically Annotated Trilingual Corpus

One of the goals of the project is to add several layers of linguistic annotation to the content, including lemmatisation, POS-tagging and dependency parsing. This is not problematic for English and French, given the fact that there are a number of tools with state-of-the-art results in these tasks for both languages.[3] Unfortunately, this is not the case for Serbian. Despite recent advances in POS-tagging [8,30], parsing [10] and corpus building [14,16,30], Serbian remains one of the less-ressourced European languages when it comes to NLP. It is especially so regarding the gold-standard corpora for training: to the best of our knowledge, the only one for POS-tagging is the MULTEXT-East Serbian corpus [14], and there is still no readily available treebank. One of our goals is therefore to develop a suite of NLP ressources that would allow us to annotate the Serbian part of the corpus, including a general-purpose morphosyntactic lexicon, as well as

---

[3] See, e.g., [23] for POS-tagging and [4] for parsing of English; [21] for POS-tagging, [3] for parsing, and [22] for lemmatization of French.

training corpora and statistical models for lemmatisation, POS-tagging, fine-grained morphosyntactic analysis and dependency parsing. These resources will be freely available and shared with the NLP community.

The training corpora and statistical models for the 4 annotation levels mentioned above are being developed on a 150K token portion of the corpus for which we have obtained redistribution rights. The current state of the training corpus for each level of annotation and the performances of the models are given in Table 2.

Table 2. Training corpora and statistical models

Annotation level	Tagset size	Training corpus size	Tools tested	Average model accuracy
Lemmatisation	N/A	2.2M	CSTLemma [11]	95.52%
POS-tagging	47	100K	BTagger [8]	94.0%
Fine-grained morphosynt.	1045	80K	HunPOS [9]	85.0%
Dependency parsing	46	42K	Talismane [29]	76.3% LAS, 84.05% UAS

The POS-tagging training coprus from Table 2 is already available at the following address: http://parcolab.univ-tlse2.fr/en/about/resources/, and so is a lexicon: wikimorph-sr was mainly extracted from the Serbo-Croatian edition of the Wiktionary. It contains 1,2M inflected forms (117K lemmas, 3M unique triplets <*wordform, lemma, morphosyntactic description*>). For more detail, see [20].

For lemmatisation, CSTLemma was chosen based on its performance on Serbian and Croatian in [1] (96.3% and 97.78% accuracy respectively). It trains on a lexicon-like format. The 2.2M inflected form-lemma pairs we used were compiled from wikimorph-sr mentioned above and srLex, a wide coverage lexicon presented in [15].

The initial experiments in POS-tagging were done with BTagger for its bi-directional sequence classification algorithm, which makes it well suited for languages such as Serbian, with rich inflectional morphology and a flexible word order. It achieved an average accuracy of 94,17% on a coarse-grained set of 45 tags [19]. The experiments on fine-grained morphosyntactic annotation were done with HunPOS [9] based on its results (87.11% accuracy on Croatian, and 85.00% on Serbian using 600 tags) and on its reported speed in [1]. It showed to be very fast, and obtained solid results on a large tagset (>1000 detailed tags) even on small training corpora (78% on 20K words). It has also shown a stable learning curve, given in Table 3.

For parsing, we used Talismane, a transition-based parser achieving state-of-the-art results for French (cf. [29]). It is also a hybrid tool, allowing for the inclusion of hand-crafted rules, not as a pre- or post-processing method, but as a part of the decision-making process during parsing. However, the results reported in Table 2 are preliminary results obtained only using machine learning, on an

Table 3. HunPOS: learning curve

Training corpus size	20K	40K	60K	80K
Average accuracy	78.82%	82.37%	83.95%	85.00%
Gain		+3.55	+1.58	+1.05

initial 40K token training corpus, with a 4K test sample. It is worth noting that the scores given in Table 2 are somewhat better that those of MSTParser ([18]) on Croatian and Serbian [2]: 76.3% vs. 74.3% LAS, and 84.0 vs. 80.8% UAS. However, these results are based on a single evaluation run and are therefore indicative only.

Further tests are being conducted in order to optimize the performances on different annotation levels, and the work on expanding the training corpus for parsing is under way. All of the training corpora and the corresponding models will be made available by the end of June 2017.

5   Conclusions and Future Work

ParCoLab is a high-quality trilingual resource that has already proven its usefulness in contrastive lingustic studies, metalexicography and NLP resource-building for Serbian, an under-resourced language. It can also be used in a wide range of other applications, such as machine and computer-assisted translation, translation studies, second language learning and teaching, and applied lexicography. It has a growing user community, both in the academic and in the professional domain.

There are ongoing efforts for corpus extension and diversification. We are also actively working towards adding several layers of linguistic annotation to the content, thus optimizing the corpus for various types of theoretical and applied research. A number of NLP resources for Serbian are being developed as part of the project and will be made available in near future.

References

1. Agić, Ž., Ljubešić, N., Berović, D.: Lemmatization and morphosyntactic tagging of Croatian and Serbian. In: 4th Biennial International Workkshop on Balto-Slavic Natural Language Processing, BSNLP 2013 (2013)
2. Agić, Ž., Merkler, D., Berović, D.: Parsing Croatian and Serbian by using Croatian dependency treebanks. In: Proceedings of the Fourth Workshop on Statistical Parsing of Morphologically-Rich Languages (2013)
3. Candito, M., Nivre, J., Denis, P., Anguiano, E.H.: Benchmarking of statistical dependency parsers for French. In: Proceedings of the 23rd International Conference on Computational Linguistics: Posters, pp. 108–116. Association for Computational Linguistics (2010)
4. Carreras, X.: Experiments with a higher-order projective dependency parser. In: EMNLP-CoNLL, pp. 957–961 (2007)

5. Čermák, F., Rosen, A.: The case of InterCorp, a multilingual parallel corpus. Int. J. Corpus Linguist. **17**(3), 411–427 (2012)
6. Text Encoding Initiative Consortium (eds.): TEI P5: Guidelines for Electronic Text Encoding and Interchange. Text Encoding Initiative Consortium (2008)
7. Esplá-Gomis, M., Forcada, M.: Bitextor, a free/open-source software to harvest translation memories from multilingual websites. In: Proceedings of MT Summit XII, Ottawa, Canada. Association for Machine Translation in the Americas (2009)
8. Gesmundo, A., Samardžić, T.: Lemmatising Serbian as category tagging with bidirectional sequence classification. In: LREC, pp. 2103–2106 (2012)
9. Halácsy, P., Kornai, A., Oravecz, C.: Hunpos: an open source trigram tagger. In: Proceedings of the 45th Annual Meeting of the ACL on Interactive Poster and Demonstration Sessions, pp. 209–212. Association for Computational Linguistics (2007)
10. Jakovljević, B., Kovačević, A., Sečujski, M., Marković, M.: A dependency treebank for Serbian: initial experiments. In: Ronzhin, A., Potapova, R., Delic, V. (eds.) SPECOM 2014. LNCS, vol. 8773, pp. 42–49. Springer, Cham (2014). doi:10.1007/978-3-319-11581-8_5
11. Jongejan, B., Dalianis, H.: Automatic training of lemmatization rules that handle morphological changes in pre-, in- and suffixes alike. In: Proceedings of the Joint Conference of the 47th Annual Meeting of the ACL and the 4th International Joint Conference on Natural Language Processing of the AFNLP, pp. 145–153 (2009)
12. Koehn, P.: Europarl: a parallel corpus for statistical machine translation. In: MT Summit, vol. 5, pp. 79–86 (2005)
13. Krstev, C., Vitas, D.: An aligned English-Serbian corpus. In: ELLSIIR Proceedings (English Language and Literature Studies: Image, Identity, Reality), vol. 1, pp. 495–508 (2011)
14. Krstev, C., Vitas, D., Erjavec, T.: MULTEXT-East resources for Serbian. In: Zbornik 7. mednarodne multikonference Informacijska druzba IS 2004 Jezikovne tehnologije 9–15 Oktober 2004, Ljubljana, Slovenija, 2004. Erjavec, Tomaž and Zganec Gros, Jerneja (2004)
15. Ljubešić, N., Klubička, F., Agić, Ž., Jazbec, I.P.: New inflectional lexicons and training corpora for improved morphosyntactic annotation of Croatian and Serbian. In: Calzolari, N., Choukri, K., Declerck, T., Goggi, S., Grobelnik, M., Maegaard, B., Mariani, J., Mazo, H., Moreno, A., Odijk, J., Piperidis, S. (eds.) Proceedings of the Tenth International Conference on Language Resources and Evaluation, LREC 2016. European Language Resources Association (ELRA), Paris, May 2016
16. Ljubešić, N., Klubička, F.: {bs, hr, sr} WaC-web corpora of Bosnian, Croatian and Serbian. In: Proceedings of the 9th Web as Corpus Workshop (WaC-9), pp. 29–35 (2014)
17. Marjanović, S.: « Entrez, s'il vous plaît ! » : De la sélection lexicographique des phrasémes. In: Repenser le figement: enjeux et perspectives en phraséo-didactique des langues. Université Paris3 - Sorbonne Nouvelle (2016, forthcoming)
18. McDonald, R., Lerman, K., Pereira, F.: Multilingual dependency analysis with a two-stage discriminative parser. In: Proceedings of the Tenth Conference on Computational Natural Language Learning, pp. 216–220. Association for Computational Linguistics (2006)
19. Miletic, A.: Annotation morphosyntaxique semi-automatique d'un corpus litéraire serbe. Master's thesis, Université Charles de Gaulle - Lille 3 (2013)
20. Miletic, A.: Building a morphosyntactic lexicon for Serbian using Wiktionary. In: 6th Journées d'études Toulousaines, JéTou 2017 (2017, forthcoming)

21. Sagot, B.: Etiquetage multilingue en parties du discours avec MELT. In: Actes de la conférence conjointe JEP-TALN-RECITAL 2016 (2016)
22. Seddah, D., Chrupała, G., Çetinoğlu, Ö., Van Genabith, J., Candito, M.: Lemmatization and lexicalized statistical parsing of morphologically rich languages: the case of French. In: Proceedings of the NAACL HLT 2010 First Workshop on Statistical Parsing of Morphologically-Rich Languages, pp. 85–93. Association for Computational Linguistics (2010)
23. Shen, L., Satta, G., Joshi, A.: Guided learning for bidirectional sequence classification. ACL **7**, 760–767 (2007)
24. Stanojević, V., Durić, L.: Sur les indéfinis singuliers génériques en français et en serbe. Travaux de linguistique **1**, 121–133 (2016)
25. Steinberger, R., Pouliquen, B., Widiger, A., Ignat, C., Erjavec, T., Tufis, D., Varga, D.: The JRC-Acquis: A multilingual aligned parallel corpus with 20+ languages. In: 5th International Conference on Language Ressources and Evaluation, LREC2006 (2006)
26. Stosic, D., Fagard, B., Sarda, L., Colin, C.: Does the road go up the mountain? Fictive motion between linguistic conventions and cognitive motivations. Cogn. Process. **16**(1), 221–225 (2015)
27. Tiedemann, J.: News from Opus-a collection of multilingual parallel corpora with tools and interfaces. In: Nicolov, N., Bontchev, K., Angelova, G., Mitkov, R. (eds.) Recent Advances in Natural Language Processing, vol. 5, pp. 237–248 (2009)
28. Tyers, F.M., Alperen, M.S.: South-East European Times: a parallel corpus of Balkan languages. In: Proceedings of the LREC Workshop on Exploitation of Multilingual Resources and Tools for Central and (South-) Eastern European Languages, pp. 49–53 (2010)
29. Urieli, A.: Robust French syntax analysis: reconciling statistical methods and linguistic knowledge in the Talismane toolkit. Ph.D. thesis, Université Toulouse le Mirail-Toulouse II (2013)
30. Utvić, M.: Annotating the corpus of contemporary Serbian. In: Proceedings of the INFOtheca 2012 Conference (2011)
31. Vitas, D., Krstev, C.: Literature and aligned texts. Readings in Multilinguality, pp. 148–155 (2006)
32. von Waldenfels, R.: Compiling a parallel corpus of Slavic languages. Text strategies, tools and the question of lemmatization in alignment. In: Brehmer, B., Zdanova, V., Zimny, R. (eds.) Beiträge der Europäischen Slavistischen Linguistik, vol. 9, pp. 123–138 (2006)

Prosodic Phrase Boundary Classification Based on Czech Speech Corpora

Markéta Jůzová[✉]

New Technologies for the Information Society (NTIS)
and Department of Cybernetics, Faculty of Applied Sciences,
University of West Bohemia, Pilsen, Czech Republic
juzova@kky.zcu.cz

Abstract. The correct usage of phrase boundaries is an important issue for ensuring a natural sounding and easily intelligible speech. Therefore, it is not surprising that the boundary detection is also a part of text-to-speech systems. In the presented paper, large speech corpora are used for a classification based approach in order to improve the phrasing of synthesized sentences. The paper compares results of different classifiers to the deterministic approaches based on punctuation and conjunctions and shows that they are able to outperform the simple algorithms.

Keywords: Phrase boundary · Classification · Speech corpus · Speech synthesis

1 Introduction

There are two fundamental requirements on text-to-speech (TTS) systems – intelligibility and naturalness. Nowadays, TTS systems do not have problems with the first one and their developers are trying to increase the naturalness so that the synthesis is close to a real human voice. Leaving aside expressive speech, also the naturalness of "neutral" synthesis could be improved, namely by more natural prosody and segmentation (these two aspects are closely related). A sentence segmentation is considered to be a necessity, but the more natural prosody is achieved by appropriate sentence splitting into *prosodic phrases – phrasing* [12,16,19].

The term *phrasing* is used to describe the phenomenon that people group words within a sentence when speaking [8]. In speech, the phrase is mainly delimited by acoustic features of its boundaries and sometimes surrounded by pauses, it also usually contains an intonation peak and it could be distinguished by a special final intonation pattern. In the text form, it is sometimes connected to inter-sentence punctuation (commas, dashes etc.) and it is also closely related to the sentence syntactic structure.

This research was supported by Ministry of Education, Youth and Sports of the Czech Republic, project No. LO1506, and by the grant of the University of West Bohemia, project No. SGS-2016-039.

© Springer International Publishing AG 2017
K. Ekštein and V. Matoušek (Eds.): TSD 2017, LNAI 10415, pp. 165–173, 2017.
DOI: 10.1007/978-3-319-64206-2_19

In our current TTS system ARTIC [5], the phrasing issue is solved quite simply – the input sentences are split into phrases according to punctuation, especially commas. Although the usage of commas is very common in Czech – since besides listings, the comma is used to separate all subordinate and inserted clauses (contrary to English text) – it is not always a sufficient indicator for phrasing. This approach often causes creation of extremely long phrases, especially in the case of a long compound sentence – e.g. with a (EN: *and*) conjunction – where no comma is written. And these long phrases are, when synthesized, unnatural and also much more demanding on listeners' attention. In addition, there is a higher probability of a disturbing speech artefact appearance when a long sentence is synthesized. The author hopes that the improvement of input sentence phrasing should lead to more natural TTS system outputs.

The problem of insufficient sentence splitting by punctuation marks for inflected languages is also discussed e.g. in [7], where the authors used a special contextual grammar to find phrase breaks not inferred from the punctuation in Russian texts. However, it was decided to avoid designing of rules and to use a classification approach.

2   Our Previous Research in Prosodic Phrase Boundary Detection

The previous research in prosodic phrase detection on our department was, at first, focused on *a posteriori* phrasing – i.e. the phrasing of read, acoustically realized sentences (for more details please see [10,11,13]). This task seems to be easier since both acoustic (e.g. phone length, F_0 contour) and textual (e.g. part of speech tags, syntactical functions [20]) features can be used. However, the authors of [11] computed the overall agreement among human phrase breaks annotators which was not very high. This fact confirms that the phrase boundary problem is not perfectly clear, even when performed on speech data. For *a posteriori* phrasing, the authors used neural networks trained on a small number of sentences manually annotated with phrase breaks.

The *a priori* phrasing from the text was described in [14]. The authors used a template matching algorithm which utilize a speech corpus created for unit selection TTS system and assigned *a posteriori* phrasing as a "gold data". The syntactical functions [20] were assigned to words and used for the template matching.

3   New Approach

In the presented paper, the author decided to test a classification based, data-driven approach. Except for simple text information, like punctuation, position in the sentence etc., part-of-speech and morphological categories tags, which were widely used in many tasks and studies concerning text analysis and segmentation [2,15–17], are used as the features for training the classifiers.

3.1   Training Data

The classification problem needs a sufficient amount of data for training. To obtain enough data for phrasing, it would take a lot of time and manual work to prepare that – in ideal case, gained from more annotators since people differ in phrase splitting [11]. Therefore, the data for the algorithm, described in this paper, were prepared in a different way – the author utilizes speech corpora of semi-professional speakers' voices used in the TTS system [6], as did the authors of [14]. However, contrary to aforementioned study, no listening tests for phrase boundary annotations were used, which makes easier to extend the testing of proposed algorithm on different speech corpora. In addition, the corpora had been already manually annotated and automatically segmented on phones level [3,4] for the purposes of speech synthesis and so they also contain special units *breath* and *pause* which indicate the positions in the speech where the speaker has made a break. And since the speakers were professionals, the speech breaks (pauses/breaths) are expected to be in reasonable positions of read sentences. More detailed investigation of pauses in corpora showed that not all pauses were related to a comma, which means that the information about a pause (or a breath) in a speech can help to improve the phrasing of a sentence. Both corpora used in the presented paper contain about 10,000 sentences, the basic statistics are listed in Table 1. It follows from the table that the speaking style of the speakers differs. The speaker who has recorded *corpus2* made almost all pauses/breath at the comma punctuation and he also did not make many pauses elsewhere, while the speaker of *corpus1* has made more phrases and some of his speech breaks do not correspond to any punctuation mark.

Table 1. Basic statistics of speech corpora.

corpus	corpus1	corpus2
# words	91,003	79,888
# words **without** commas followed by a pause/breath	3,818	586
# words with commas **not** followed by a pause/breath	2,178	339
# words with commas followed by a pause/breath	9,419	11,383

Thus, as the training and testing data, the text sentences from the speech corpora supplemented by information about breath and pauses in the speech were used. In this way, a large amount of data has been easily and quickly gained, enough for training a classifier. This data were subsequently tagged with Czech morphological tags using our n-gram tagger – a combined unigram, bigram and trigram tagger with accuracy >90%.

3.2   Features and Classes

Now, let us describe the phrasing issue as a 2-class classification problem. Every sentence of N words $w_0, w_1, \ldots, w_{N-1}$, taken from the speech corpora, with morphological tags $t_0, t_1, \ldots, t_{N-1}$ can be represented as follows:

words+tags	(w_0,t_0)	(w_1,t_1)	(w_2,t_2)	$\ldots$		(w_{N-1},t_{N-1})
speech breaks	b_0	b_1	b_2	$\ldots$	b_{N-2}	
juncture	j_0	j_1	j_2	$\ldots$	j_{N-2}	

Note that the speech breaks b_i were set to 1 if the word w_i was followed by a breath or a pause in the speech corpus. The junctures j_i (the term was adopted from [16]) are the wanted phrase breaks ($j_i = 1$ for a *break*, $j_i = 0$ for *non-break*) determined from the speech corpus (see below). The classification task can be then described as a decision making whether there is or there is not a phrase break at every j_i for every i in the sentence.

The true "gold data" answers were gained from the speech corpus based on the following rules:

- $j_i = 1$ if w_i is followed by a comma,
- $j_i = 1$ if $b_i = 1$ (i.e. a word w_i is followed by a pause/breath in the corpus),
- $j_i = 0$ otherwise.

It was decided to consider both commas and speech pauses/breaths to be phrase breaks. First, making a phrase boundary at a punctuation seems to be reasonable and second, extra speech pauses b_i, which do not correspond to any comma, represent a "value added" and, hopefully, ensure more accurate phrasing compared to the phrasing just at commas.

The features for *a priori* phrasing must be designed so that they can be easily gained only from text (contrary to *a posteriori* phrasing in [10]) since this approach is designed to be employed on the text input of TTS system. Thus, similarly to other classifier approaches to phrasing (e.g. [2,15]), following features were used for every juncture j_i:

- word w_i,
- word w_i has or has not a comma,
- morphological tag t_i of the word w_i,
- morphological tag t_{i+1} of the word w_{i+1},
- bigram $t_i + t_{i+1}$,
- trigram $t_{i-1} + t_i + t_{i+1}$,
- sentence lenght N,
- position of the word w_i in the sentence $\frac{i}{N}$,
- distance from last comma $i - i_{LC}$,
- distance to next comma $i_{NC} - i$,

where i_{NC} and i_{LC} are the indexes of words followed by a comma, where $i_{NC} \geq i$ and $i_{LC} \leq i$ (or $i_{NC} = N - 1$ if none of words $w_{i+1} \ldots w_{N-1}$ is followed by a comma, and $i_{LC} = 0$ if none of words $w_0 \ldots w_{i-1}$ has a comma).

3.3 Classifiers and the Baseline Algorithms

For the training, the *scikit-learn* tool [9] was used for testing of several different classifiers with different parameters:

- *LogReg* – Logistic Regression,
  $C \in [2^{-4}, 2^{-3}, \ldots, 2^0, \ldots, 2^4]$,
- *SVC-lin* – Support Vector Machines with *linear* kernel,
  $C \in [2^{-4}, 2^{-3}, \ldots, 2^0, \ldots, 2^4]$,
- *SVC-rbf* – Support Vector Machines with *rbf* kernel,
  $C \in [2^{-4}, 2^{-3}, \ldots, 2^0, \ldots, 2^4]$ and $\gamma \in [2^{-4}, 2^{-3}, \ldots, 2^0, \ldots, 2^4]$,
- *Extrees* – Extremly Randomized Trees,
  $n \in [10, 20, \ldots 100]$.

As a baseline system, the author uses several deterministic algorithms which split the text to phrases in different ways:

- *Comma* – splits the text after every comma (it is performed in the current version of our TTS system ARTIC [5]),
- *CommaAnd* – splits the text after every comma and before every *a* (EN: *and*) conjunction (this one should, in theory, improve the phrasing of compound sentences compared to previous),
- *CommaConj* – splits the text after every comma and before every coordinating conjunction which can appear in Czech texts without a comma, e.g. *a*, *i*, *ani*, *nebo*, *či* (EN: *and, nor, or*).

Moreover, as a very simple baseline, an algorithm *NoBreaks*, which does not set any phrase break, can be defined.

Let us emphasize that the accuracy of *Comma* algorithm is, in general, quit high (compared e.g. to English), since commas are widely used in Czech texts which makes the phrase breaks detection task much easier. On the other hand, the word order is not fixed in Czech which could, on the contrary, make the phrasing more difficult.

4 Results

The data gained from speech corpora, as described in Sects. 3.1 and 3.2, were randomly split to training (80%) and testing (20%) data. The first mentioned were used for the 5-folds cross-validation to find the best parameters of the classifiers, using $F1$-score as a classifiers' tuning score. The results of the grid search are shown in Table 2.

The classifiers with the best parameters were then applied to the testing data. Table 3 summarizes all the results and compares the classifiers with the simple baseline algorithms, defined in Sect. 3.3, using 4 evaluation measures – *accuracy* (A), *precision* (P), *recall* (R) and *F1-score* $(F1)$.

Although the deterministic algorithms are quite simple, their *accuracy* and *F1-score* do not differs much from the scores of the classification methods.

Table 2. Results on training data – searching for the best parameters.

corpus	Classifier	Best $F1$-score	Best parameters
corpus1	*LogReg*	0.900	$C = 1.0$
	SVC-lin	0.905	$C = 0.125$
	SVC-rbf	0.902	$C = 2.0, \gamma = 0.0625$
	Extrees	0.899	$n = 80$
corpus2	*LogReg*	0.977	$C = 2.0$
	SVC-lin	0.977	$C = 2.0$
	SVC-rbf	0.976	$C = 4.0, \gamma = 0.125$
	Extrees	0.976	$n = 90$

Table 3. Classifiers comparison – results on testing data.

corpus	Classifier	A	P	R	$F1$
corpus1	*NoBreaks*	0.931	*nan*	*nan*	*nan*
	Comma	0.955	1.000	0.747	0.855
	CommaAnd	0.965	0.956	0.837	0.893
	CommaConj	0.961	0.918	0.853	0.884
	LogReg	0.970	0.948	0.868	**0.907**
	SVC-lin	0.970	0.948	0.865	0.905
	SVC-rbf	0.969	0.948	0.863	0.904
	Extrees	0.969	0.942	0.865	0.902
corpus2	*NoBreaks*	0.846	*nan*	*nan*	*nan*
	Comma	0.993	1.000	0.953	0.976
	CommaAnd	0.976	0.890	0.962	0.925
	CommaConj	0.968	0.852	0.963	0.904
	LogReg	0.994	0.997	0.962	**0.979**
	SVC-lin	0.994	0.996	0.963	0.979
	SVC-rbf	0.993	0.995	0.960	0.977
	Extrees	0.993	0.991	0.961	0.976

The detailed examination shows that the adding of conjunctions improves the *Comma* algorithm score only in *corpus1*, which is, probably, related to the difference in speakers' speaking styles and different pausing (see statistics in Table 1). For both corpora, the *precision* is decreasing which means that more "false alarms" are detected and the *recall* value is increasing (the algorithm detects more true phrase boundaries) – this confirms the original prediction in Sect. 3.3 that the use of conjunctions can improve the phrasing. In general, all score values are much higher in the case of *corpus2*, however, it is clear that these results closely correspond with the fact that almost all speech breaks b_i gained from this corpus match a comma.

The classifiers used in this experiments outperform the deterministic algorithms in *F1-score*, *LogReg* being the best for both corpora. However, it is worth emphasizing the fact that, contrary to majority of text analysis issues, the phrasing problem does not have fixed correct answers since the speech segmentation into phrases depends a lot on the speaking speed, style and a particular speaker, and also the text segmentation depends on annotator's feelings [16]. Thus, the false positive (*fp*) and false negative (*fn*) errors of classification were further investigated – the example of the confusion matrix for *LogReg* classifier on *corpus2* is shown in Table 4.

Table 4. Confusion matrix of *LogReg* classifier results on *corpus2*.

		Predicted values	
		phrase break $j_i = 1$	non-break $j_i = 0$
true values	phrase break $j_i = 1$	$tp = 2{,}405$	$fn = 94$
	non-break $j_i = 0$	$fp = 7$	$tn = 13{,}472$

It follows from the table that there were 7 "false alarms" (extra boundary detections) and 94 missed boundaries (not detected by a classifier). A human annotator was inspected the errors and find out that 6 of 7 detected "false alarms" are only another or extra phrase boundaries which have not been realized in the source speech corpus in any way but which are alternative possibilities of splitting the particular sentence into phrases and so they can be considered to be *true positives* instead of *false positives*. Similarly, the annotator checked 20 randomly selected missed boundaries and decided that 8 boundaries from the corpus (gained based on a pause/breath) were not much reasonable (it sometimes seems like a pause before a long difficult word or a break caused by a lack of breath before the sentence end) and so the classifier, in fact, made no mistake when missed them. These findings imply that the real *accuracy* and *F1-score* would be much higher if the training and testing data were more consistent.

Table 5 shows the results of *LogReg* classifiers (which obtained the best scores on testing data) trained on *corpus2* applied to *corpus1* and vice versa. These results are not so compelling, but, that is again caused by different speaking style of the speakers. And note that the miss-detections are usually not completely wrong, as described in the previous paragraph. Nevertheless, the *LogReg* classifier trained on *corpus2* outperformed all simple algorithms when used on *corpus1*. In any case, this table shows that the task of phrase boundary detection based on speech corpora data is slightly speaker-dependent. To eliminate the different speaking styles of different speakers, more data from more speakers could be used for training together.

Table 5. Results of boundary classification on the 2nd corpus.

corpus	Classifier	A	P	R	$F1$
corpus1	*LogReg-corpus2*	0.970	0.999	0.809	0.894
corpus2	*LogReg-corpus1*	0.977	0.901	0.972	0.935

5   Conclusion

The paper described a classification based approach to phrasing of text sentences, using data from large recorded speech corpora. The results presented in Sect. 4 show that this approach outperformed the deterministic approaches based on punctuation and conjunctions, which means that the idea to substitute the present algorithm for the trained classifier in the text processing module of TTS system was correct. In any case, the contribution to the overall quality of speech synthesis should be verified by listening tests [1,18] before releasing this modification into the publicly available version of our TTS system.

Planned future work includes the testing of hidden Markov models, conditional random fields or neural networks to phrasing, since these approaches take into account the position of other potential phrase boundaries in the particular sentence. One of the possible further work will also be the improvement of speech corpora segmentation focused on pause detection since not all phrases are correctly detected in the corpora, e.g. some short pauses are still considered to be a part of a plosive consonant signal.

References

1. Grůber, M., Matoušek, J.: Listening-test-based annotation of communicative functions for expressive speech synthesis. In: Sojka, P., Horák, A., Kopeček, I., Pala, K. (eds.) TSD 2010. LNCS, vol. 6231, pp. 283–290. Springer, Heidelberg (2010). doi:10.1007/978-3-642-15760-8_36
2. Hirschberg, J., Prieto, P.: Training intonational phrasing rules automatically for English and Spanish text-to-speech. Speech Commun. **18**(3), 281–290 (1996)
3. Legát, M., Matoušek, J., Tihelka, D.: A robust multi-phase pitch-mark detection algorithm. In: Proceedings of Interspeech 2007, vol. 1641–1644 (2007)
4. Matoušek, J., Romportl, J.: Automatic pitch-synchronous phonetic segmentation. In: Proceedings of Interspeech 2008, pp. 1626–1629. ISCA, Brisbane (2008)
5. Matoušek, J., Tihelka, D., Romportl, J.: Current state of Czech text-to-speech system ARTIC. In: Sojka, P., Kopeček, I., Pala, K. (eds.) TSD 2006. LNCS, vol. 4188, pp. 439–446. Springer, Heidelberg (2006). doi:10.1007/11846406_55
6. Matoušek, J., Romportl, J.: Recording and annotation of speech corpus for czech unit selection speech synthesis. In: Matoušek, V., Mautner, P. (eds.) TSD 2007. LNCS, vol. 4629, pp. 326–333. Springer, Heidelberg (2007). doi:10.1007/978-3-540-74628-7_43
7. Oparin, I.: Robust rule-based method for automatic break assignment in Russian texts. In: Matoušek, V., Mautner, P., Pavelka, T. (eds.) Text, Speech and Dialogue, pp. 356–363. Springer, Heidelberg (2005)

8. Palková, Z.: Rytmická výstavba prozaického textu. Studia ČSAV 13/1974, Academia (1974)
9. Pedregosa, F., Varoquaux, G., Gramfort, A., Michel, V., Thirion, B., Grisel, O., Blondel, M., Prettenhofer, P., Weiss, R., Dubourg, V., Vanderplas, J., Passos, A., Cournapeau, D., Brucher, M., Perrot, M., Duchesnay, E.: Scikit-learn: machine learning in Python. J. Mach. Learn. Res. **12**, 2825–2830 (2011)
10. Romportl, J.: Prosodic phrases and semantic accents in speech corpus for Czech TTS synthesis. In: Sojka, P., Horák, A., Kopeček, I., Pala, K. (eds.) TSD 2008. LNCS, vol. 5246, pp. 493–500. Springer, Heidelberg (2008). doi:10.1007/978-3-540-87391-4_63
11. Romportl, J.: Statistical evaluation of prosodic phrases in the Czech language. In: Proceedings of the Speech Prosody 2008, pp. 755–758. Editora RG/CNPq, Campinas, Brazil (2008)
12. Romportl, J.: Structural data-driven prosody model for TTS synthesis. In: Proceedings of the Speech Prosody 2006, pp. 549–552. TUDpress, Dresden (2006)
13. Romportl, J.: Automatic prosodic phrase annotation in a corpus for speech synthesis. In: Proceedings of Speech Prosody 2010. University of Illionois, Chicago, IL, USA (2010)
14. Romportl, J., Matoušek, J.: Several aspects of machine-driven phrasing in text-to-speech systems. Prague Bull. Math. Linguist. **95**, 51–61 (2011)
15. Sun, X., Applebaum, T.H.: Intonational phrase break prediction using decision tree and n-gram model. In: Proceedings of Eurospeech 2001, pp. 3–7 (2001)
16. Taylor, P.: Text-to-Speech Synthesis, 1st edn. Cambridge University Press, New York (2009)
17. Taylor, P., Black, A.W.: Assigning phrase breaks from part-of-speech sequences. Comput. Speech Lang. **12**(2), 99–117 (1998)
18. Tihelka, D., Grůber, M., Hanzlíček, Z.: Robust methodology for TTS enhancement evaluation. In: Habernal, I., Matoušek, V. (eds.) TSD 2013. LNCS, vol. 8082, pp. 442–449. Springer, Heidelberg (2013). doi:10.1007/978-3-642-40585-3_56
19. Tihelka, D., Matoušek, J.: Unit selection and its relation to symbolic prosody: a new approach. In: Proceedings of Interspeech 2006, vol. 1, pp. 2042–2045. ISCA, Bonn (2006)
20. Žabokrtský, Z., Ptáček, J., Pajas, P.: TectoMT: highly modular MT system with tectogrammatics used as transfer layer. In: Proceedings of StatMT 2008, pp. 167–170. Association for Computational Linguistics (2008)

Parliament Archives Used for Automatic Training of Multi-lingual Automatic Speech Recognition Systems

Jan Nouza[(⊠)] and Radek Safarik

Institute of Information Technology and Electronics,
Technical University of Liberec, Studentska 2, 461 17 Liberec, Czech Republic
{jan.nouza,radek.safarik}@tul.cz
https://www.ite.tul.cz/speechlab/

Abstract. In the paper we present a fully automated process capable of creating speech databases needed for training acoustic models for speech recognition systems. We show that archives of national parliaments are perfect sources of speech and text data suited for a lightly supervised training scheme, which does not require human intervention. We describe the process and its procedures in details and demonstrate its usage on three Slavic languages (Polish, Russian and Bulgarian). Practical evaluation is done on a broadcast news task and yields better results than those obtained on some established speech databases.

Keywords: Speech recognition · Cross-lingual bootstrapping · Parliament speech

1 Introduction

Since the early years of the automatic speech recognition (ASR) research, machine transcription of parliament speeches has been considered a challenging task and a nice application of the voice-to-text technology. ASR systems focused on parliaments have been developed for major languages [2,3,5,11] and later also for some other, including the Slavic ones [4,7,10,13]. The task's main challenge is a large variety of speaking styles, from read contributions, to less formal talks and even spontaneous utterances. Also the quality of the signal is not always high, namely because of reverberation in large halls. On the other side, the archives of these institutions contain thousands of spoken and written documents and hence it is not that difficult to fit the acoustic models (AM) and language models (LM) to the given task and achieve good results.

In this paper, we do not focus on developing an ASR system for a particular institution or a particular language. Instead, we want to show that parliament archives are very convenient data sources for fast and almost automatic development of acoustic models for ASR systems. The main advantage of these archives consists in the fact that they include both audio recordings of the sessions as well as their written (shorthand made) transcriptions. And even though the latter are

© Springer International Publishing AG 2017
K. Ekštein and V. Matoušek (Eds.): TSD 2017, LNAI 10415, pp. 174–182, 2017.
DOI: 10.1007/978-3-319-64206-2_20

not verbatim, it is still possible to utilize them for so called lightly-supervised AM training. They are particularly useful when we want to port an existing ASR system from one language to another without spending time (and money) for collecting special speech databases. In this paper, we show that we can built fairly good AMs for several languages within a short period, in a fully automated way and without an expert for each language.

2 Spoken and Written Documents in Public Parliament Archives

In many countries, national parliaments offer public access to the files that document their regular sessions. Usually, they have form of video files and written protocols. The latter contain lists of speakers and their talks. If the transcripts were verbatim, we could easily make a speech database suitable for AM training. We would just need to convert the video files to the audio ones, cut them into parts with speech of a single person, and attach a text transcript to each. Since many parliaments have several hundreds of members, we would obtain a collection whose content, size and speaker coverage would be comparable to specialized speech databases, e.g. GlobalPhone [9].

In reality, however, the transcripts are never verbatim. Even if they are produced by professional transcribers, they contain errors (e.g. typos, modified word order, synonyms instead of actually spoken words). Moreover, they represent an edited form of the actual talks, with grammatically corrected utterances, removed repeatedly spoken phrases, etc. (A real example is shown in Table 1). If we measured the matching score between actual utterances and their official transcripts in the same way as it is done in ASR evaluation, we would get values typically around 80 to 90 %.

Table 1. Difference between official text transcript of a speech fragment from Polish Sejm. (The example shows a typo in the first word and many omitted, but actually spoken, words.)

Text:	potrzebna			nie	zmian			na	wsi
Speech:	potrzeba	jest	naprawde	nie	zmian	tylko	jest	na	wsi

The collection of the recordings would be still useful for AM training, if somebody corrected the official transcripts so that they would agree with the spoken content. Yet, that would require a native speaker, a lot of manual work and rather long time.

Our approach is different and fully automated. We utilize our own ASR system (originally developed for another language) as a tool that transcribes the audio files and then searches for the fragments where ASR output fully agrees with the provided text. These fragments are cut and used (together with their automatically created phonetic annotations) for iterative retraining of the AM. The model is improving in time as more data from the target language is collected.

3   ASR System Used for Spoken Data Collection and Annotation

As stated above, we employ the ASR system to search for the fragments where the spoken and written content agree. This can be done even if we do not have the ASR system fitted to the target language.

We just need a modular large vocabulary continuous speech recognition (LVCSR) system, whose language specific components (a lexicon, a phonetic inventory, an acoustic model and a language model) can be separated from the LVCSR engine. Moreover, its decoder should produce not only recognized text, but also a precise time instant and pronunciation for each recognized word.

Then, for the target language we create a lexicon and an LM. This is not that difficult since the Internet offers large amounts of texts in many languages. The best sources are web pages of major newspapers and TV/radio stations. This data just needs to be downloaded and cleaned. The corpus is statistically analyzed to get a representative lexicon and LM. All that can be done automatically and within a short period [6,12]. We also add the texts from the parliament archive to the corpus. In this way we obtain a lexicon and an LM that is general enough to cover various topics and speaking styles present in common parliament talks, yet without restricting them to the formal language occurring in the official session protocols.

To build an ASR lexicon, we need to add pronunciation to each word. We have to define a phonetic inventory for the language and generate phonetic forms from the orthographic ones. In Slavic languages, this is not a difficult task as the relation between graphemes and phonemes is rather straight and the grapheme-to-phoneme (G2P) converter can be based on a limited number of rules, from which many are common for the whole language group [6,8].

The most critical component is the AM for the target language. To get it we need a large amount of annotated speech data. At the initial stage we have none, so we have to use any available AM from another language. In the process called bootstrapping we use that 'donor' AM to transcribe the spoken data from the archive and extract at least some fragments that could be used for training a better AM. This is repeated iteratively until we get enough data for training a good AM for the target language.

4   Automatic Extraction of Data for Acoustic Model Training

4.1   Written and Spoken Document Segmentation

Parliament session protocols have a firm structure with names of speakers followed by their talks. It is easy to parse the text and cut it into parts belonging to individual speakers. After that, we must cut also the audio records into segments that correspond to speaker's speech. This is the first task for the LVCSR system. It transcribes the whole session and its output is matched to the protocol as described in [1]. Time stamps attached to the words are used to split the audio into single speaker segments. These are further cut into short chunks whose length is restricted by setting minimum and maximum number of words. All the cuts are done during non-speech events (silence or noise) so that spoken phrases are not affected. The output of this procedure (we denote it DoSegmentation) is a list of signal chunks (ChunkList), each with a speaker label and a text fragment extracted from the original protocol.

4.2   Transcription Check

In procedure CheckTranscript, the LVCSR system is applied to all the files in the ChunkList. For each, the LVCSR text output is compared to the text assigned during the segmentation. If they are same, the transcription is considered correct and the file together with its orthographic and phonetic transcription (produced by the system) is moved from the ChunkList to the TrainList.

4.3   Iterative AM Retraining

After all the files from the ChunkList are processed, the data in the TrainList are used to train a new acoustic model. In the next iteration, the new AM is utilized for a repeated check and for making an updated TrainList. In several initial iterations, it is useful to repeat also the segmentation procedure and get a more appropriate assignment of the text to the chunks. The iterative process is stopped when the size of the TrainList exhibits only minor changes.

4.4   Cross-Lingual Bootstrapping

When we start to work on a new target language (TL), we have to borrow an AM from another 'donor' language (DL). This is possible but we have to modify the pronunciation lexicon so that it uses only the DL phonemes. It is just a temporary modification and it must be designed so that it is reversible. In practice, we make a table that maps TL phonemes to the closest DL ones (or to their combinations), see e.g. [8]. In the procedure named MapPhonemes we apply the table to the lexicon. After that we can start the above described procedures.

During initial iterations, however, the number of successfully extracted speech chunks is too small for training an AM. Therefore, we initialize the TrainList by a certain amount (e.g. 10 h) of DL training data. Hence, in the bootstrapping phase we train a model that mixes DL and TL data. When the size of the latter exceeds that of the former, we finish this stage by running RemapPhonemes procedure. It (a) returns back to the original lexicon, and (b) remaps the phonetic transcriptions of the training data to the TL phonetic set (using the lexicon as a reverse look-up table). After that we remove the DL data from the TrainList and train the first genuine AM for the target language.

4.5   Formal Description of Complete Process

The key part of the process is the iterative retraining procedure. It can be formally described by the following scheme:

```
IterativeRetraining:
```

1. For each file in ArchiveFileList
       DoSegmentation
2. For each Chunk in ChunkList
       CheckTranscription
3. Retrain AM on data from TrainList
4. Repeat from step 1 or 2

In the bootstrapping phase, the above procedure is combined with routines that ensure compatibility with the data and models from the donor language.

```
Boostrapping:
```

1. MapPhonemes
2. Add DL data to Trainlist
3. IterativeRetraining
4. RemapPhonemes
5. Remove DL data from Trainlist
6. Retrain AM on data from TrainList

The complete process consists of the bootstrapping stage followed by the standard iterative retraining procedure. During the latter, we monitor the size of the training set and also the amount of speech from individual speakers in order to keep the training set balanced. The process is stopped when the required amount of data is collected.

5   Practical Evaluation in Three Slavic Languages

We have already used the described method of the rapid AM development for several languages. Here, we will document it on three Slavic ones, Polish (PL), Russian (RU) and Bulgarian (BG). For each, one can access public web pages of their national parliaments: Polish *Sejm*[1], Russian *Duma*[2] and Bulgarian *Narodno sabranie*[3]. We selected 10 to 20 sessions from each institution and downloaded their videos and written protocols. The number of sessions was chosen so that we got about 50 h of audio per language. For each language we have collected a large corpus of (mainly) newspaper texts, mixed it with the parliament protocols, created a lexicon with G2P generated pronunciations, and computed a bigram LM. After that we could start the data extraction and AM training process

---

[1] http://www.sejm.gov.pl/.

[2] http://www.duma.gov.ru/.

[3] http://www.parliament.bg/tv/.

described in Sect. 4. Our LVCSR system works with triphones and can use both GMM-HMM and DNN-HMM acoustic models. The former (based on Gaussian mixture models) are more convenient for initial iterations since their training is faster, while the latter (with deep neural nets) are beneficial in later stages when more training data is available. For bootstrapping, we used Czech AM (with 40 phonemes) and 10 h of Czech training data. Some basic facts on the developed language specific modules are in Table 2.

Table 2. Basic facts on the developed components for three Slavic languages

Language	Polish (PL)	Russian (RU)	Bulgarian (BG)
Phonemes	42	53	40
Text corpus size [GB]	2.3	2.8	0.9
Lexicon size [#words]	303k	310k	284k
Data extracted for AM training [hours]	21.6	19.9	15.2
Number of speakers in train set	68	59	47

We wanted to compare the quality of the obtained AMs with that of the AMs trained on some established databases, like e.g. GlobalPhone (GP). The Polish part of the GP contains recordings read by 100 speakers, each with 100 utterances. The first 10 speakers make the test set (1.5 h of speech with 14894 words), the rest is used for training (23.3 h). In Table 3, one can compare the WER (word error rate) values achieved on the GP test set with different AMs.

Table 3. Experiments with Polish GlobalPhone test set to compare various AMs

Acoustic model	Model type	WER [%]
Czech AM used for bootstrapping	GMM	46.8
Polish AM based on Sejm data	GMM	33.1
Polish AM based on GP train set	GMM	25.7
Polish AM based on Sejm data	DNN	28.9
Polish AM based on GP train set	DNN	22.2

We can see that the Czech AM is (obviously) not good for practical transcription but proved to work in bootstrapping. We admit that the AM trained on Sejm data is significantly worse than that trained on the GlobalPhone and it is true for both the GMM and DNN models. However, we should realize that the test set used in these experiments is a part of the GlobalPhone and hence the Sejm based AM is largely disqualified by train/test mismatch (different acoustic conditions and speaking styles) when compared to the GP one.

Therefore, in Table 4 we compare the same two AMs on some other test data. One is a set of 10 Sejm talks (from speakers not present in the train set), and the other is made of 4 complete TV and radio (TVR) 30-minute-long broadcast news. We can see that in these two cases the AM trained on the Sejm data performs better than that of the GlobalPhone. It is evident especially in the TVR experiment where none of the two AMs can be considered as task fitting.

Table 4. WER achieved with two Polish AMs on 3 different test sets: GlobalPhone (14894 words), Sejm talks (5997 words), and TV/radio news (14731 words)

	WER		
	GlobalPhone	Sejm talks	TVR news
Polish AM based on Sejm data (DNN)	28.9	33.2	28.9
Polish AM based on GP train set (DNN)	22.2	49.7	41.2

Now, let us have a look at all the three Slavic languages. We run very similar experiments in each of them and tested the acoustic models created with the public data from national parliaments. For our project, the most relevant results were those achieved on the independent broadcast data. We utilized standardized test sets that cover complete TV/Radio news shows in selected Slavic languages[4,5]. Their sizes (measured in the number of spoken words) were: PL (14731), RU (12277) and BG (15197). Our results are presented in Table 5.

Table 5. WER achieved on TVR data (broadcast news) with 2 types of AMs

Acoustic models	WER [%]
Polish AM trained on Sejm (21.6 h, DNN)	28.9
Polish AM trained on all data (126 h, DNN)	20.1
Russian AM trained on Duma (19.9 h, DNN)	27.4
Russian AM trained on all data (58 h, DNN)	22.1
Bulgarian AM trained on NS (15.2 h, DNN)	28.6
Bulgarian AM trained on all data (41 h, DNN)	20.8

In Table 5 we compare the performance of two types of acoustic models for each language. The first one is that created from national parliament archives in a fully automated way presented in this paper. The second one is the currently best available model for that language. For its training we utilized additional available data, mainly the transcriptions of many broadcast programs and also

---

[4] https://gitlab.ite.tul.cz/SpeechLab/EastSlavicTestData.
[5] https://gitlab.ite.tul.cz/SpeechLab/SouthSlavicTestData.

some read speech data sets. (The amounts of training data is shown.) We can see that the difference in performance is significant. Yet, the results achieved with the AMs trained solely on the automatically collected data can be considered as usable. In fact, we used these AMs to process and collect speech data from the mentioned additional sources in a way similar to that presented here, though not that straightforward like in case of parliament archives.

In all the experiments, we worked with just a small part of the parliament archives (10–20 sessions as mentioned earlier). We could use much more audio and text data but the goal of this research was not to create AMs perfectly fitting talks in each of the parliaments. Our aim was to get similar amount of training data as in the GlobalPhone set so that we could perform fair comparisons.

6   Conclusion

We present a fully automated process for creating annotated speech databases that are applicable for training acoustic models needed in automatic speech recognition systems. Our approach utilizes data from publically available archives of national parliaments. Their main advantage consists in the fact that they contain large amounts of spoken data as well as their transcriptions. Even though the latter are not verbatim, our method is capable of extracting the exactly matching fragments and utilize them for data collection and iterative acoustic model training. The process can be fully automated and it is possible to develop all modules for a new language (i.e. a lexicon, a language model and namely the acoustic model) within a short period of one month. We demonstrate it on 3 Slavic languages (each from a different subgroup) and obtain results that are generally better than those than could be achieved with a dedicated, commercially available speech database.

The presented method is applicable also to some other sources that contain both speech and text, like e.g. TV programs equipped with subtitles, though some minor modifications are needed.

Acknowledgements. The research was supported by the Technology Agency of the Czech Republic (project TA04010199) and by the Student Grant Scheme at the Technical University of Liberec.

References

1. Boháč, M., Blavka, K.: Text-to-speech alignment for imperfect transcriptions. In: Habernal, I., Matoušek, V. (eds.) TSD 2013. LNCS, vol. 8082, pp. 536–543. Springer, Heidelberg (2013). doi:10.1007/978-3-642-40585-3_67
2. Kawahara, T.: Transcription system using automatic speech recognition for the Japanese parliament (diet). In: Proceedings of IAAI, pp. 2224–2228 (2012)
3. Makhoul, J., Kubala, F., Leek, T., Liu, D., Nguyen, L., Schwartz, R., Srivastava, A.: Speech and language technologies for audio indexing and retrieval. Proc. IEEE **88**(8), 1338–1353 (2000)

4. Marasek, K., Koržinek, D., Brocki, Ł.: System for automatic transcription of sessions of the Polish senate. Arch. Acoust. **39**(4), 501–509 (2014)
5. Neves, L., Martins, C., Meinedo, H., Neto, J.: Domain adaptation of a broadcast news transcription system for the Portuguese parliament. In: Teixeira, A., Lima, V.L.S., Oliveira, L.C., Quaresma, P. (eds.) PROPOR 2008. LNCS, vol. 5190, pp. 163–171. Springer, Heidelberg (2008). doi:10.1007/978-3-540-85980-2_17
6. Nouza, J., Safarik, R., Cerva, P.: ASR for south Slavic languages developed in almost automated way. In: Proceedings of Interspeech, pp. 3868–3872 (2016)
7. Pražák, A., Psutka, J.V., Hoidekr, J., Kanis, J., Müller, L., Psutka, J.: Automatic online subtitling of the Czech parliament meetings. In: Sojka, P., Kopeček, I., Pala, K. (eds.) TSD 2006. LNCS, vol. 4188, pp. 501–508. Springer, Heidelberg (2006). doi:10.1007/11846406_63
8. Safarik, R., Nouza, J.: Methods for rapid development of automatic speech recognition system for Russian. In: Proceedings of the IEEE Workshop ECMSM, pp. 1–6 (2015)
9. Schultz, T.: Globalphone: a multilingual speech and text database developed at Karlsruhe university. In: Proceedings of Interspeech, pp. 345–348 (2002)
10. Staš, J., Hládek, D., Juhár, J.: Language model speaker adaptation for transcription of Slovak parliament proceedings. In: Ronzhin, A., Potapova, R., Fakotakis, N. (eds.) SPECOM 2015. LNCS, vol. 9319, pp. 259–267. Springer, Cham (2015). doi:10.1007/978-3-319-23132-7_32
11. Stüker, S., Fügen, C., Kraft, F., Wölfel, M.: The ISL 2007 English speech transcription system for European parliament speeches. In: Proceedings of the Interspeech, pp. 2609–2612 (2007)
12. Vu, N.T., Schlippe, T., Kraus, F., Schultz, T.: Rapid bootstrapping of five Eastern European languages using the rapid language adaptation toolkit. In: Proceedings of the Interspeech, pp. 865–868 (2010)
13. Zgank, A., Rotovnik, T., Grasic, M., Kos, M., Vlaj, D., Kacic, Z.: Sloparl - Slovenian parliamentary speech and text corpus for large vocabulary continuous speech recognition. In: Proceedings of the Interspeech, pp. 197–200 (2006)

Recent Results in Speech Recognition for the Tatar Language

Aidar Khusainov[✉]

Institute of Applied Semiotics of the Tatarstan Academy of Sciences,
Kazan Federal University, Kazan, Russia
khusainov.aidar@gmail.com

Abstract. This paper presents a comparative study of several different systems for speech recognition for the Tatar language, including systems for very large and unlimited vocabularies. All the compared systems use a corpus based approach, so recent results in speech and text corpora creation are also shown. The recognition systems differ in acoustic modelling algorithms, basic acoustic units, and language modelling techniques. The DNN based system with the sub-word based language model shows the best recognition result obtained on the test part of speech corpus.

Keywords: Speech recognition · Acoustic modelling · Language modelling · Tatar language

1 Introduction

The conventional way of building speech recognition systems is to obtain required acoustic models, a pronunciation dictionary, a language model, and use some of the decoders. The situation can be worse whenever you have to recognize the speech of an under-resourced language. In that case some (or all) of the required resources and algorithms may not exist.

In this article we present our recent results in creating continuous speech recognition systems for the Tatar language. We used the word based approach to create a very large vocabulary speech recognition system and the sub-word based approach for the case of unlimited vocabulary recognition.

Tatar is spoken by several million people and is the second spoken language in Russia. There are 4.2 million of speakers in Russia and near 5.2 million of speakers in the world [8]. The Cyrillic alphabet (unified in 1939) consists of 39 characters. There are 12 vowel and 28 consonant sounds. Different dialects of Tatar can be identified: Western, Kazan (Middle) and Eastern. Based on the existing language classification [7], in 2013 it was assigned to the under-resourced language class [6]. However, recent results in machine translation [1], speech analysis and synthesis [3] fields can change this situation. The main feature of Tatar in case of creating speech recognition systems is the agglutinative nature of the language, so one word can have tens of surface forms. It leads to a very high OOV rate problem.

© Springer International Publishing AG 2017
K. Ekštein and V. Matoušek (Eds.): TSD 2017, LNAI 10415, pp. 183–191, 2017.
DOI: 10.1007/978-3-319-64206-2_21

Therefore, we have to use approaches for very large or unlimited vocabulary recognition systems. We need to determine an acoustic alphabet, to record and annotate speech corpora, to build models with different existing approaches and to evaluate the recognition quality of combinations of the systems parts.

The paper is organized in the following way. Section 2 gives an overview of speech and text corpora of the Tatar language that are used to train and test acoustic and language models. Section 3 shows different types of developed Tatar speech recognition systems. Section 4 discusses the experiment results of proposed recognition systems. Last section concludes the paper.

2   Data

The size and the quality of training data play an essential role in modern recognition approaches. To train speech recognizers for the Tatar language we have created a continuous speech data set representing different types of speakers [2]. The data set consists of read speech mostly spoken by native speakers with a common Kazan dialect. As for training language models we have used the preprocessed Tatar National Corpus [5].

2.1   Speech Corpus

The most modern systems use speech corpora with a total duration of hundreds and thousands hours to create robust acoustic models. This amount of training data gives a possibility to create robust recognizers. The robustness in that case means relatively equal recognition accuracy for male and female speakers, speakers of different sex and age, noise conditions, etc. Building and annotating the multi-speaker speech corpus for the Tatar language is currently in progress. Now it consists of two main parts. The first part—"Core part"—has been created to cover all the Tatar phonemes pronounced by the large number of speakers. Due to this goal, each speaker has been asked to utter approximately 2 min: 11 sentences from literature texts, 13 separate words, and 7 sentences from news. Each of the resulting set of 31 text fragments has been adjusted to contain all the Tatar phonemes and the maximum number of phonemes contexts (left and right phonemes) based on 2- and 3-grams number. The "Core part" is now used in several algorithms in their first stages to create initial monophone models (see Sect. 3 for details).

The second part of the corpus is "Read part". We asked people to read randomly selected texts for 30 min. The source of used texts is the Tatar National Corpus described in Sect. 2.2. The only text adaptation was transcribing of all the abbreviations and numbers.

Both "Core" and "Read" parts have been manually annotated, for now the speech corpus contains speech files, corresponding text and phonetic transcriptions. The corpus also contains additional meta-information about speakers (sex, age, mother tongue) and an expert's score of speakers' proficiency in Tatar. In

Table 1. The characteristics of multi-speaker speech corpus for the Tatar language

Parameter	Value
Number of speakers	377
Male speakers	109
Female speakers	268
Duration	57:55:09
Number of speakers in training subcorpus	361
Duration in training subcorpus	52:50:15
Number of speakers in test subcorpus	16
Duration in test subcorpus	5:04:54
Number of speakers in Core part	251
Duration of Core part	8:12:16
Average duration per speaker in Core part	1:58
Number of speakers in Reading part	126
Duration of Reading part	49:42:53
Average duration per speaker in Reading part	23:40

addition, we plan to continue recording and annotating the "Spontaneous" and start collecting/transcribing "Radio and TV" parts of this corpus.

The main characteristics of the speech corpus are presented in Table 1.

2.2   Text Corpus

Texts that have been used to create language models are from the Tatar National Corpus. Some preprocessing steps have been implemented to prepare texts for language modelling:

1. Duplicate fragments have been removed;
2. All the texts have been lower-cased;
3. Abbreviations, numbers, dates have been transcribed;
4. The texts have been split into separate sentences.

The main characteristics of the text corpus after the processing steps are presented in Table 2.

3   Systems Description

3.1   General Overview

We have built and evaluated several recognition systems that differ in an acoustic modelling unit, the size of used training data, and in the algorithms used for

Table 2. The characteristics of the text corpus

Parameter	Value
Number of files	217 294
Number of words	69 810 033
Number of words in learning part	64 629 794
Number of words in test part	5 180 239
Number of syllables	186 014 478 (2,66 per word)
Number of morphemes	110 280 448 (1,58 per word)
Number of letters	434 636 548 (6,23 per word)

models' creation and decoding phases. All the training and evaluation have been done using the Kaldi toolkit [4].

We experimented with two different types of acoustic units: monophones and triphones. As we have already mentioned in Sect. 2.1, we used short utterances from the "Core part" of the speech corpus to create acoustic models in the initial training stages (see Training audio data column in Table 3).

The basic speech feature we used is mel-frequency cepstral coefficients (MFCCs), but also delta and delta-delta coefficients have been used to form 39-dimension feature vector. In more advanced systems (Tri2, Tri3, Tri4, NN) we have used LDA/MLLT feature transformation algorithm, SAT and fMLLR speaker adaptation techniques [10].

LDA-MLLT stands for Linear Discriminant Analysis – Maximum Likelihood Linear Transform. LDA reduces feature space for all the data, whilst MLLT takes this reduced feature space from LDA and calculates a transformation for each speaker to implement speaker normalization.

SAT stands for Speaker Adaptive Training and performs speaker and noise normalization. After SAT training, the acoustic models are trained on speaker-normalized features. fMLLR stands for Feature Space Maximum Likelihood Linear Regression. The inverse of the fMLLR matrix is used to remove the speaker identity from the original features.

Two types of language models have been created using word and sub-word based approaches. To create the sub-word based language model we split low-frequency words into syllables, that allows to reduce the OOV rate. For both word and sub-word language models pruned and full 3-gram and 4-gram models were built.

We have used step by step training process so each next system should improve recognition quality by using more training data, advanced adaptation techniques and larger language models. The resulting set of speech recognition systems is presented in Table 3, where, for example, MonoSW has the same components as Mono system except of using the sub-word language model.

The latter NN system from listed in Table 3 is the DNN-based recognizer. This neural network uses a p-norm activation function and predicts the posterior

Table 3. The overview of built Tatar speech recognition systems

System	Acoustic unit	Training audio data	Features	Language models
Mono, MonoSW	Monophone	Separate words	MFCCs	Small 3-gram
Tri1, Tri1SW	Triphone	Separate words	+delta, delta-delta	+3-gram full
Tri2, Tri2SW	Triphone	Core part	+LDA/MLLT	As above
Tri3, Tri3SW	Triphone	Core part	+fMLLR	As above
Tri4, Tri4SW	Triphone	Full training corpus	As above	+4-gram
NN, NNSW	Triphone	Full training corpus	As above	As above

probabilities of context-dependent states [14]. It has been trained on the training corpus data aligned by Tri4 recognition system.

All the Mono, Tri1-Tri4 and NN systems use the same 200k words vocabulary consisting of most frequent words. Sub-word (SW) based systems use the vocabulary of 50k most frequent words combined with 10k most frequent syllables.

3.2 Acoustic Models

In this work we have created acoustic models for rather good quality recordings: 16 bits, 16 kHz. We could use them to recognize speech in offices, in front of home PC, to analyze speech in not very noisy conditions. In future, we plan to create separate acoustic models for speech transmitted over telephone, TV and radio channels.

The simplest monophone and triphone acoustic models are created for 32 non-silence and 2 silence phones. Context dependency is introduced with left and right adjacent phonemes. Therefore, each context-dependent (CD) phoneme is presented with a triple of context-independent phonemes designated as a–b+c where b is a central phoneme name, a and c are names for left and right context phonemes respectively.

Phonemes names are taken from the basic phoneme alphabet for Tatar (a, ae, b, ch, d, dzh, e, f, g, h, i, j, k, kh, l, m, n, ng, o, oe, p, r, s, sh, t, ts, u, ue, v, y, z, zh) accomplished with a silence sil and short-pause sp, which makes total 34 items. Grapheme-to-phoneme conversion have been done using the tool described in [13].

Acoustic models for sub-word based systems trained using preprocessed text transcriptions: words (except of 50k most frequent words) were divided into syllables. It can help the system to be more robust in recognition of sub-word units.

3.3 Language Models

A language model creation task arises in many applications from spellchecking to machine translation systems. In all the cases, a language model has to describe language grammar rules and has the ability to estimate the probabilities of a word sequence in a specified language. We have built the language model for the Tatar language using the SRILM toolkit (Speech Technology and Research (STAR) Laboratory) [12]. This tool has the functionality to create n-gram models, can interpolate different models and estimate the quality of the built models. The common way to use SRILM is as follows:

1. Executing the "ngram-count" function to calculate the count of n-grams.
2. Executing the "ngram-count" function to build the language model based on the results of the first step. A smoothing algorithm has to be specified.
   (a) Executing the "ngram" function with a prune option with a threshold as a value.
3. Model quality estimation using "ngram" function with a ppl parameter.

In addition to conventional 3- and 4-gram models, we have built a pruned 3-gram model. Language model pruning can help in dealing with the limited amount of memory in computing device. Used algorithm prunes n-gram probabilities if their removal causes the perplexity of the model to increase by less than the threshold value [11].

According to the limit of the corpus size, the developed word based language models cannot be complete. Thus, there will be unseen n-grams with a zero probability. As the probability of the entire speech utterance is calculated as the multiplication of separate n-grams, this can lead to the situation, in which even one unseen n-gram zeroes out the total utterance probability. To overcome this drawback, we used the Kneser-Ney smoothing algorithm [9].

The Tatar language belongs to the agglutinative language family. Thus, its main characteristic is rich morphology. In case of word-based models, the only approach to cover the entire lexicon better is to use a larger vocabulary. Our experiments have shown that 20k words vocabulary gives 17% OOV rate, 50k – 10%, even 200k vocabulary shows 4.4% OOV on test data set. In case of sub-word models, we achieved 0% OOV rate using 60 K mixed words and syllables vocabulary. The quality of the built models have been evaluated on the following parameters: memory usage and perplexity (model the confidence level in the analysis of the test data set), Table 4.

Table 4. Language models features

Language models	Memory	Perplexity	OOV
Word based models			
Pruned 3-gram	31 MB	1041.9	4.4%
3-gram	170 MB	315.9	4.4%
4-gram	204 MB	278.9	4.4%
Sub-word based models			
Pruned 3-gram	29 MB	242.8	0%
3-gram	218 MB	65.7	0%
4-gram	360 MB	45.0	0%

3.4   Evaluation Method

The most common performance metric in speech recognition is word error rate (WER), that is computed as follows:

$$WER = \frac{I + D + S}{N} \times 100\%,$$

where I is the number of insertion errors, D – deletion errors, S – substitution errors, N – the total number of words in an uttered text. The agglutinative nature of the Tatar language can lead to such a situation where one incorrectly recognized affix will be counted as a whole word error in WER. For example, in the third record from the test subcorpus word "kaltyradym" have been recognized as "kaltyrady" and the WER statistics can't show the real quality of this recognition result since it observes only the substitution error. To give a different source of evaluation information we have computed an additional metric: syllable error rate (SER).

One of the applications of speech recognition systems is the dictation system. In this type of programs users estimate the recognition quality mostly by the number of corrections they have to make in recognized texts. Therefore, such characteristic as character error rate (CER) can be representative.

4   Results

Table 5 shows the performance of the recognizers on the 5-hours test subcorpus.

Table 5. Evaluation results for Tatar speech recognition systems

System	Language models	WER		SER		CER	
		Word based	Sub-word	Word based	Sub-word	Word based	Sub-word
Mono	Pruned 3-gram	52,06	-	39,65	-	28,70	-
Tri1	Pruned 3-gram	28,80	24,08	18,32	13,98	12,54	8,18
Tri1	3-gram	22,59	18,42	14,09	10,84	9,78	6,44
Tri2	Pruned 3-gram	24,14	21,20	13,95	11,46	8,69	6,40
Tri2	3-gram	19,08	16,17	10,86	9,11	6,91	5,82
Tri3	Pruned 3-gram	21,16	18,67	11,35	9,74	6,67	5,33
Tri3	3-gram	17,21	14,90	9,04	7,81	5,37	4.91
Tri4	Pruned 3-gram	18,57	19,70	9,29	10,08	5,24	5,54
Tri4	3-gram	15,19	16,09	7,46	8,29	4,18	4,59
Tri4	4-gram	15,10	15,71	7,41	8,05	4,15	4,44
NN	Pruned 3-gram	16,47	17,17	8,27	8,13	4,94	4,29
NN	3-gram	12,99	13.25	6,44	6,37	3,86	3,41
NN	4-gram	12,89	**12,79**	6,38	**6,14**	3,83	**3,29**

The analysis of the WER values for the word based systems shows the main component of the errors: the substitution errors. For the conducted experiments the number of substitution errors is from 5 to 10 times bigger than the insertion or deletion errors. For example, for the best NN 4-gram LM system error numbers are as follows: 496 insertions, 395 deletions, 2362 substitutions. We have found two possible reasons for these results. The first one is the OOV rate, and the second one is the rich morphological structure of the Tatar words. The number of OOV words in the speech test corpus is near 1% (213 from 25240 words). The influence of rich morphological structure can be seen in SER and WER difference: syllable based error rates are nearly two times less than word based.

The sub-word based systems show the recognition accuracy comparable with the word based systems. However, there are two main differences between these approaches. First, the sub-word based systems allow to cover all the Tatar lexicon, so no OOV words have been detected. But at the same time it's harder to acoustically recognize small parts of words.

5   Conclusions

In this paper we present some recent results in creating large and unlimited vocabulary speech recognition systems for the Tatar language. First multi-speaker speech corpus has been created and used to model Tatar acoustic units (monophones and triphones). The Tatar National Corpus has been used to build the word and sub-word based language models. The best result obtained on the test part of the speech corpus – more than 87% word recognition accuracy – has been shown by the DNN-based system with the sub-word based language model. The sub-word based speech recognition system showed 0% OOV rate, so it can be used in the applications where this factor can play an essential role, for example, in dialogue systems.

References

1. Yandex translate (2017). https://translate.yandex.com/
2. Khusainov, A.: Design and creation of speech corpora for the Tatar speech recognition and synthesis tasks. In: Proceedings of Third International Conference on Turkic Languages Processing TurkLang-2015, Kazan, Russia, pp. 475–484 (2015)
3. Khusainov, A.: Speech human-machine interface for the Tatar language, FRUCT Oy, Helsinki, pp. 60–65 (2016)
4. Povey, D., Ghoshal, A., Boulianne, G., Burget, L., Glembek, O., Goel, N., Hannemann, M., Motlicek, P., Qian, Y., Schwarz, P., et al.: The Kaldi speech recognition toolkit. In: Proceedings of ASRU, pp. 1–4 (2011)
5. Suleymanov, D., Nevzorova, O.A., Khakimov, B.: National corpus of the Tatar language Tugan tel: structure and features of grammatical annotation. In: Proceedings International Conference Georgian Language and Modern Technology, Tbilisi, pp. 107–108 (2013)
6. Khusainov, A.: Tekhnologiya avtomatizatsii sozdaniya I otsenki kachestva programmnikh sredstv analiza rechi c uchetom osobennostey maloresursnykh yazikov. Ph.D. thesis, Kazan (2014)
7. Krauwer, S.: The basic language resource kit (BLARK) as the first milestone for the language resources roadmap. In: Proceedings of International Workshop Speech and Computer SPEECOM, Moscow, Russia, pp. 8–15 (2003)

8. Lewis, M., Paul Simons, G.F., Fennig, C.D. (eds.): Ethnologue: Languages of the World, 9th edn. (2016). http://www.ethnologue.com. Accessed 15 Jan 2017
9. Kneser, R., Ney, H.: Improved backing off for m-gram language modeling. In: Proceedings of the IEEE International Conference on Acoustics, Speech and Signal Processing, vol. 1 (1995)
10. Rath, S.P., Povey, D., Vesely, K., Cernocky, J.H.: Improved feature processing for deep neural networks. In. Proceedings of InterSpeech (2013)
11. Stolcke, A.: Entropy-based pruning of backoff language models. In: Proceedings DARPA Broadcast News Transcription and Understanding Workshop, Lansdowne, pp. 270–274 (1998)
12. Stolcke, A.: SRILM an extensible language modeling toolkit. In: Proceedings of International Conference on Spoken Language Processing, Denver, vol. 2, pp. 901–904 (2002)
13. Robeiko, V., Sazhok, M.: Bidirectional text-to-pronunciation conversion with word stress prediction for Ukranian. In: Proceedings UkrObraz 2012, Kyiv, pp. 43–46 (2025)
14. Zhang, X., Trmal, J., Povey, D., Khudanpur, S.: Improving deep neural network acoustic models using generalized maxout networks. In: 2014 IEEE International Conference on Acoustics, Speech and Signal Processing (ICASSP), pp. 215–219 (2014)

Data-Driven Identification of German Phrasal Compounds

Adrien Barbaresi[1,3]([✉]) and Katrin Hein[2]

[1] Zentrum Sprache, Berlin-Brandenburg Academy of Sciences,
Jägerstraße 22/23, 10117 Berlin, Germany
barbaresi@bbaw.de
[2] Lexical Department, Institute for the German Language,
R5, 6-13, 68161 Mannheim, Germany
hein@ids-mannheim.de
[3] Academy Corpora, Austrian Academy of Sciences,
Sonnenfelsgasse 19, 1010 Vienna, Austria
adrien.barbaresi@oeaw.ac.at
http://www.ids-mannheim.de/lexik/personal/hein.html

Abstract. We present a method to identify and document a phenomenon on which there is very little empirical data: German phrasal compounds occurring in the form of as a single token (without punctuation between their components). Relying on linguistic criteria, our approach implies to have an operational notion of compounds which can
be systematically applied as well as (web) corpora which are large and
diverse enough to contain rarely seen phenomena. The method is based
on word segmentation and morphological analysis, it takes advantage of a
data-driven learning process. Our results show that coarse-grained identification of phrasal compounds is best performed with empirical data,
whereas fine-grained detection could be improved with a combination of
rule-based and frequency-based word lists. Along with the characteristics of web texts, the orthographic realizations seem to be linked to the
degree of expressivity.

Keywords: Corpus linguistics · Word segmentation · Morphological
analysis · Web corpora

1 Introduction

Composition, that is "the combination of two or more lexemes (roots, stems or
freely occuring words) in the formation of a new, complex word", is a productive process of German word formation [27, p. 364]. German is indeed considered as a language which makes extensive use of compounds [31], for example
Biowahn, *Freigeist*, or *zitronengelb*. Compounding does not always operate by
a simple string concatenation, it can involve linking elements (e.g. the *ens* in
Schmerzensschrei) as well as the elision of word-final characters in the non-head
constituent of a compound [18]. In the non-head position of such determinative

© Springer International Publishing AG 2017
K. Ekštein and V. Matoušek (Eds.): TSD 2017, LNAI 10415, pp. 192–200, 2017.
DOI: 10.1007/978-3-319-64206-2_22

compounds, not only lexical categories, but also syntactic units can be inserted – a phenomenon called phrasal compounding.

This paper discusses the automatic detection of german phrasal compounds (PCs) like *"Man-muss-doch-über-alles-reden-können"-Credo* or *Habdichliebpolitik*[1] in web corpora. PCs display a specific type of determinative compounds and can be defined as "complex words with phrases in modifier position" [24, p. 153]; their study is worthwhile in theoretical terms alone and sheds light on the general process of composition [16,17,22,24,25,34]. Phrasal compounding is not restricted to German, but can be found in other languages as well [35], for example English: "cut-and-run meal" [34].

While [16] (cf. also [29]) presented the first elaborated, large corpus-based investigation of German PCs whose immediate constituents are separated by hyphens (e.g. *Second-Hand-Liebe*, cf. [15, pp. 349–353] for orthographic variants), until now no systematic study has ever put into focus PCs which are written as one word, i.e. without hyphens or blanks between their component parts (e.g. *Heileweltsache*). We want to outline the methodological challenges of their automatic detection in this paper, however the investigation of this PC variant in itself can also be seen as a desideratum.

The absence of linguistical or computational approximations to PCs can notably be explained by the lack of attested data. Within the inventory of nominal compounds, PCs account for an amount of 3.2% [28]. As it is assumed that PCs written as one word are even less prominent [16], possible hits are to be found at the lower end of the frequency spectrum. Thus, the annotated samples which we put together and use can themselves be considered as a precious and unprecedented source of linguistic evidence.

The automatic detection is indeed particularly challenging, especially the distinction between complex prototypical determinative compounds and PCs which are written in one word, or the distinction between PCs and types of phrasal derivations like *Wasser-in-Wein-Verwandler* [22]. It implies to have an operational notion of compounds which can be systematically applied as well as corpora which are large and diverse enough to contain rarely seen phenomena. Web corpora built for linguistic research relying on scientific criteria [2] seem particularly suitable for this endeavor, as they may comprehend a significant amount of texts from a large gamut of sources.

2   Method

The detection method grounds on a morphological compound analysis operating on token level. Since German is considered to be a morphologically rich language, state-of-the-art approaches do not always perform well on words absent from the dictionary, which is typically the case for phrasal compounds. Thus, in order to get information on potential segmentation patterns, we use the morphological

---

[1] "One-should-be-able-to-talk-about-everything motto", "I-like-you-policy". All examples appear in their original graphic realization, as found in previous studies [19] and billion-token web corpora [5,6].

analyzer SMOR [32] in combination with a data-driven morphological segmentation based on affix and component trees. This combination follows from two different goals: to design a robust detection process and to be able to estimate the degree of lexicalization of complex compounds.

2.1 Previous Work

The combination is deemed to be necessary as SMOR can be subject to coverage issues. In previous work on non-standard text in an under-resourced variety of written German, namely retro-digitized newspaper texts from former East Germany, we showed that a data-driven method could overcome data sparsity and trump SMOR's full-fledged morphological analysis to predict whether a given token is to be considered as part of the language or as an OCR error [4].

A similar approach has also been used to build an unsupervised morphological model for a number of different languages and language varieties for the *Discriminating between Similar Languages* shared task [3,23]. Criteria resulting from the segmentation analysis are statistically relevant and can be used as a sparse feature in a model to discriminate similar languages. A reasonable hypothesis is that they add new linguistically motivated information, dealing with the morpho-lexical logic of the languages to be classified while also yielding insights on linguistic typology.

2.2 Data-Driven Segmentation Process

The method is based on segmentation and affix analysis. The original idea behind this simple yet efficient principle appears to go back to Harris' letter successor variety which grounds on transitional probabilities to detect morpheme boundaries [14]. The principle has proven valuable to construct stem dictionaries for document classification [13], it has been used in the past by spell-checkers [20,30], as it is both linguistically relevant and computationally efficient. Relevant information is stored in a trie [11], a data structure allowing for prefix search and its reverse opposite in order to look for sublexicons, which greatly extends lexical coverage. Forward (prefix) and backward (suffix) tries are used in a similar fashion, albeit with different constraints. This approach does not necessarily perform evenly across languages; it has for example led to considerable progress in morphologically-rich languages such as Arabic [7] or Basque [1]. Similar approaches have been used successfully to segment words into morphemes in an unsupervised way and for several languages. A more recent implementation has been the *RePortS* algorithm which gained attention in the context of the *PASCAL challenge* [8,9,21] by outperforming most of the other systems. The present approach makes similar assumptions as the work cited and adapts the base algorithm to the task at hand. In this regard, this experiment also tests if the data-driven morphological analysis of surface forms can be useful in the context of phrasal compounds.

2.3   Implementation

In order to build the corresponding model, a dictionary is composed by observing unigrams in the training data, then prefix and suffix trees are constructed using this dictionary. Additionally, an affix candidate list is constituted by decomposing the tokens present in the training data. The identification algorithm aims at the decomposition into possibly known parts. It consists of two main phases: first a prefix/suffix search over respective trees in order to look for the longest possible known subwords, and secondly sanity checks including a series of known composition rules to see if the rest could itself be an affix or a word out of the dictionary. If $\alpha\beta$ is a concatenation absent from the dictionary and if α and β are both in training data, then $\alpha\beta$ is considered to be a valid token. The segmentation can be repeated if necessary, in order to identify all necessary components of long words. It is only performed backward here since the nominal basis is a discriminative criterion in this study.

Once the word has been decomposed into potentially meaningful parts, the morphological analysis tool SMOR [32] is used to determine the grammatical category of the identified root, i.e. the last matched subword on the right. If it is considered to be a valid noun, the rest of the subwords is analyzed in the same way. Adjectival and adverbial combinations on a noun base are used as a cue that the token is a phrasal compound.

For example, the token *Allerweltsfragen* is not necessarily in the dictionary, but it can be decomposed into *Allerwelt+s+fragen* and ultimately into *aller+welt+s+fragen*. *Fragen* and *Welt* are identified as in-dictionary nouns, *aller* is a valid component, and *s* is among a fixed number of composition rules. Thus, this token is classified as PC.

3   Evaluation

The evaluation is performed on a gold standard of manually annotated samples: lists of PCs (coarse-grained and fine-grained) and a list of other similar compounds (noise). There are 123 *coarse*, 103 *fine*, and 504 *noise* tokens. The samples mainly come from experiments with billion-token web corpora [5,6], completed by results from previous studies [19]. They are annotated manually following expert criteria defined in [16].[2]

Apart from morphological criteria such as binarity or constraints for the realization of linking elements, the question whether the non-head can be considered as a phrasal element is crucial for the identification as PC. We assume a gradual understanding of the category "phrase" in this context, with congruency between the elements of the non-head being an important criterion, cf. *Harte-Jungs-Gerede* or *1000-Stunden-Jahr*, but not *Dreibettzimmerzuschlag*. In addition to

---

[2] One PC-type defined is not captured by our automatic detection: Word formations whose non-head consists of not explicitly coordinated NPs, e.g. *Frage-Antwort-Stunde*, cf. p. 194 f.

these classical cases, entities whose status as a phrase is less clear are also considered here, e.g. *Coca-Cola-trink-Unterhaltungs-Freundschaft* (contains only a verb stem as verbal element). Both syntactically complete structures (e.g. *"Der-Reporter-macht-sich-langsam-auf-den-Weg-in-die-Redaktion"-Stunde*) as well as sentence-like elliptical structures (e.g. *"Jetzt geht's los"-Motto*) in non-head position are considered to be within the category "sentence".

Entities which do not correspond to our criteria are gathered in the *noise* list, whose purpose resides in emulating larger datasets by entailing long tokens such as proper names, complex compounds, and compounds which share a formal similarity with phrasal compounds, notably complex nominal compounds (*Waschsalontristesse*).

In order to do justice to the numerous entities which fulfill certain, but not all of the criteria linked to the PC-status, we make use of a *coarse* inter-category. Constructs which have something to do with PCs from a coarse-grained perspective, particularly constructs with a phrasal component, are collected in this list: *Immernacktschlafende*, product of phrasal derivation [22]; *Grünkohlinderbadewannewaschens*, product of phrasal conversion [22]; *Afterwork[ichraffmichgradsonochauf]FeierabendSportler*, phrasal element in the middle of a complex construction [10]; *Einpersonenhaushalt*, lack of congruency within the non-head, potential phrase because of the A+N-non-head; *Mehr-Aufmerksamkeitsheischerei*, realization of a non paradigmatic linking element in combination with a non-lexicalized non-head [16]. Because they share certain properties with fine-grained PCs, the entities from the coarse list can be useful for the automatic detection of PCs that are fully compatible with the theoretical model.

After empirical testing, the smallest possible token length for learning and searching is fixed to 4 characters, the upper bound on token and subword length during learning phase is 16 characters. We test several lists which are expected to contain valuable information on variation at morphological level. On one hand, morphologically motivated word lists which have been made available for training and/or experiments on German words by the MarMoT [26] and GermaNet [18] systems are tested in this particular context. On the other hand, an empirical dataset is used for comparison, it stems from a combination of common tokens from a german reference corpus [12] and from newspaper corpora [2] (as described in [4]).

Table 1 summarizes the results. The training data from the MarMoT morphological toolkit lead to the best general results (*coarse+fine*) as well as the best compromise between precision and recall (F1-measure), but the empirical data from a selection of corpora reaches the best accuracies in all three settings. Therefore, it appears that coarse-grained detection of phrasal compounds is best performed with empirical learning data, whereas fine-grained detection can benefit from a combination of rule-based and frequency-based word lists.

Table 1. Results of evaluation on manually annotated samples (coarse and fine-grained). MMT = MarMoT, GNT = Germanet, CPS = corpus data. Higher is better.

Measure	Coarse+Fine			Coarse			Fine		
	MMT	GNT	CPS	MMT	GNT	CPS	MMT	GNT	CPS
Precision	.452	.405	.546	.250	.198	.368	.330	.310	.383
Recall	.527	.199	.394	.390	.138	.350	.689	.301	.447
F1	**.487**	.267	.458	.305	.163	**.358**	**.447**	.305	.413
Accuracy	.656	.662	**.711**	.651	.721	**.754**	.710	.768	**.784**

4 Discussion

More seldom seen combinations can be problematic and lead to a decrease in recall, mostly because of segmentation and component identification issues. First, the token *Halsringreitjungfrau* is correctly segmented into *Halsring+reit+jungfrau*, but since neither the "reit" modifier (corresponding to the verb *reiten*) nor the potential noun "Reitjungfrau" are present in the training data, the case is left undecided at the current stage of implementation. The greediness of the search algorithm could also be fine-tuned: the (non-PC) token *Ichhaballesimgrifflern* is decomposed into *ichhaballesim* – an unknown string resulting from the algorithm not being greedy enough, whereas the decomposition into *lern* and *griff* is too greedy and misses the nominal formation in dative form.

Second, the dictionary coverage obtained from reference and newspaper corpora affects precision. *Grünpanzerschildkröte* (a rare species of tortoise) is wrongly considered to be a PC (lack of congruency) since the token is decomposed correctly, *grün* is an adjective, and *Grünpanzer* does not appear to be lexicalized. Component parts coming from other languages are also problematic, such as in *Mainstream+medien+superfrau*, where the lexical class of *Mainstream* is hard to assess automatically due to sparse data. Additionally, a systematic notion of congruence as well as a refined analysis of combining elements seem to be necessary to improve the detection process: so are *Grün+gemüse+n+spendeaktion* and *Nicht+eintreten+s+entscheid* both analyzed as PCs, although this is no clear-cut case and only the first one has been annotated as a potential/coarse-grained one.

Finally, the model does not presently yield information about frequency effects. As it is restricted to concatenative morphology, the fact that a stem has to be in the dictionary is a strong limitation impeding recall in particular [9]. However, an overall conservative setting has been kept so far as it prevents the model from overgenerating.

5   Conclusions

We have presented a study to identify and document a rare phenomenon on which there is very little empirical evidence, phrasal compounds occurring in the form of as a single token without punctuation between their component parts. Our method implies a systematic approach as well as corpora which are large and diverse enough. It operates at the crossroads of qualitative and quantitative research, in such a way that both approaches benefit from each other. On one hand, we need empirical data to draw conclusions on this rarely observed phenomenon. On the other hand, trying to replicate fine-grained decisions also makes for more stringent and thorough criteria. Our method is based on word segmentation and morphological analysis, the first takes advantage of a data-driven learning process while the latter uses existing software. As documented examples are quite scarce, machine-based scans through large web corpora are one possible way to look at these compounds in all their (so far unsuspected) variety.

Since one specific communicative function of – at least one sort of – PCs can be seen in producing expressivity effects [16,25], this word formation type seems to be predestined for a productive use in computer-based communication, which web corpora entail. Our results show that coarse-grained detection of phrasal compounds is best performed with empirical data, whereas fine-grained detection could be improved with a combination of rule-based and frequency-based word lists. Additionally, the investigation of PCs in web corpora displays a fruitful supplement to the newspaper-based investigations. If we compare results from both sources, there seem to be parallels inasmuch as certain lexemes are more predestined than others to appear as a head word in PCs. For example, there are many PCs whose head word is a denomination of a person (e.g. *Frau*) in the present findings. Such heads are often combined with non heads which express a stereotypical property of a person or a type of person [33], such as *Wäschewaschaufhängbüglezusammenlegfrau* or *buchcoverfotosmitsmartphonecamerabildermachfrau.*

Looking at the PCs which we have automatically extracted from web corpora, one can conclude that their orthographic realization, i.e. the missing use of punctuation between the component parts, seems to increase their degree of expressivity. This holds especially for very complex/long PCs like *AfterworkichraffmichgradsonochaufFeierabendSportler.* Catching attention or being creative seems to be more important in this case than facilitating the reception for the reader by the use of punctuation. Moreover, creatively formed PCs such as *Oberflächlichallesmöglicheabernichtsrichtiglerner* contribute to reject the claim [22] that predominantly lexicalized PCs (e.g. *Armeleuteessen*) are written as one word. Future work includes giving answers to linguistically relevant open questions with respect to the proportion of PCs written together in comparison to PCs with hyphens, the potential existence of a systematic difference between both PC types, and further properties of compounds which are suitable both for qualitative analysis and automatic identification.

References

1. Agirre, E., Alegria, I., Arregi, X., Artola, X., de Ilarraza, A.D., Maritxalar, M., Sarasola, K., Urkia, M.: XUXEN: a spelling checker/corrector for Basque based on two-level morphology. In: Proceedings of the 3rd Conference on Applied Natural Language Processing, pp. 119–125. Association for Computational Linguistics (1992)
2. Barbaresi, A.: Ad hoc and general-purpose corpus construction from web sources. Ph.D. thesis, École Normale Supérieure de Lyon, France (2015)
3. Barbaresi, A.: An unsupervised morphological criterion for discriminating similar languages. In: Malmasi, S., Zampieri, M., Ljubešić, N., Nakov, P., Ali, A., Tiedemann, J. (eds.) Proceedings of the 3rd VarDial Workshop, pp. 212–220 (2016)
4. Barbaresi, A.: Bootstrapped OCR error detection for a less-resourced language variant. In: Dipper, S., Neubarth, F., Zinsmeister, H. (eds.) Proceedings of the 13th Conference on Natural Language Processing (KONVENS 2016), pp. 21–26. University of Bochum (2016)
5. Barbaresi, A.: Efficient construction of metadata-enhanced web corpora. In: Cook, P., Evert, S., Schäfer, R., Stemle, E. (eds.) Proceedings of the 10th Web as Corpus Workshop, pp. 7–16. Association for Computational Linguistics (2016)
6. Barbaresi, A., Würzner, K.M.: For a fistful of blogs: discovery and comparative benchmarking of republishable German content. In: Beißwenger, M., Zesch, T. (eds.) KONVENS 2014, NLP4CMC Workshop Proceedings, pp. 2–10. Hildesheim University Press (2014)
7. Ben Hamadou, A.: A compression technique for Arabic dictionaries: the affix analysis. In: Proceedings of the 11th Conference on Computational Linguistics, pp. 286–288. Association for Computational Linguistics (1986)
8. Dasgupta, S., Ng, V.: High-performance, language-independent morphological segmentation. In: HLT-NAACL, pp. 155–163 (2007)
9. Demberg, V.: A language-independent unsupervised model for morphological segmentation. In: Annual Meeting of the Association for Computational Linguistics, vol. 45, pp. 920–927 (2007)
10. Finkbeiner, R., Meibauer, J.: Boris "Ich bin drin" Becker ("Boris I am in Becker"). Syntax, semantics and pragmatics of a special naming construction. Lingua **181**, 36–57 (2016)
11. Fredkin, E.: Trie memory. Commun. ACM **3**(9), 490–499 (1960)
12. Geyken, A.: The DWDS corpus: a reference corpus for the German language of the 20th century. In: Fellbaum, C. (ed.) Collocations and Idioms: Linguistic, Lexicographic, and Computational Aspects, pp. 23–41. Continuum Press (2007)
13. Hafer, M.A., Weiss, S.F.: Word segmentation by letter successor varieties. Inf. Storage Retrieval **10**, 371–385 (1974)
14. Harris, Z.S.: From phoneme to morphemes. Language **31**(2), 190–222 (1955)
15. Hein, K.: Phrasenkomposita - ein wortbildungsfremdes Randphänomen zwischen Morphologie und Syntax? Deutsche Sprache **39**, 331–361 (2011)
16. Hein, K.: Phrasenkomposita im Deutschen. Empirische Untersuchung und konstruktionsgrammatische Modellierung. Narr (2015)
17. Hein, K.: Modeling the properties of German phrasal compounds within a usage-based constructional approach. In: Trips, C., Kornflit, J. (eds.) Further Investigations into the Nature of Phrasal Compounding. Language Science Press, Berlin (2017, to appear)

18. Henrich, V., Hinrichs, E.W.: Determining immediate constituents of compounds in GermaNet. In: Proceedings of Recent Advances in Natural Language Processing, pp. 420–426 (2011)
19. IDS: Deutsches Referenzkorpus/Archiv der Korpora geschriebener Gegenwartssprache 2011-I. Technical report, Institut für Deutsche Sprache Mannheim (2011). www.ids-mannheim.de/dereko
20. Jones, M.A., Silverman, A.: A spelling checker based on affix classes. In: Agrawal, J.C., Zunde, P. (eds.) Empirical Foundations of Information and Software Science, pp. 373–379. Springer, Boston (1985)
21. Keshava, S., Pitler, E.: A simpler, intuitive approach to morpheme induction. In: Proceedings of 2nd Pascal Challenges Workshop, pp. 31–35 (2006)
22. Lawrenz, B.: Moderne deutsche Wortbildung. Phrasale Wortbildung im Deutschen: Linguistische Untersuchung und sprachdidaktische Behandlung. Dr. Kovaĉ (2006)
23. Malmasi, S., Zampieri, M., Ljubešić, N., Nakov, P., Ali, A., Tiedemann, J.: Discriminating between similar languages and Arabic dialect identification: a report on the third DSL shared task. In: Proceedings of the 3rd VarDial Workshop (2016)
24. Meibauer, J.: Phrasenkomposita zwischen Wortsyntax und Lexikon. Zeitschrift für Sprachwissenschaft **22**, 153–188 (2003)
25. Meibauer, J.: How marginal are phrasal compounds? Generalized insertion, expressivity, and I/Q-interaction. Morphology **17**, 233–259 (2007)
26. Müller, T.: General methods for fine-grained morphological and syntactic disambiguation. Ph.D. thesis, LMU Munich (2015)
27. Olsen, S.: Composition. In: Müller, P.O., Ohnheiser, I., Olsen, S., Rainer, F. (eds.) Word-formation. An International Handbook of the Languages of Europe, II: Units and Processes in Word-formation I: General Aspects, vol. 1, pp. 364–386. De Gruyter Mouton, Berlin/Boston (2015)
28. Ortner, L., Müller-Bollhagen, E.: Substantivkomposita. Deutsche Wortbildung: Typen und Tendenzen in der Gegenwartssprache, Schwann (1991)
29. Particke, H.J.: Phrasenkomposita: eine morphosyntaktische Beschreibung und Korpusstudie am Beispiel des Deutschen. Diplomica-Verlag, Hamburg (2015)
30. Peterson, J.L.: Computer programs for detecting and correcting spelling errors. Commun. ACM **23**(12), 676–687 (1980)
31. Schlücker, B.: Die deutsche Kompositionsfreudigkeit. Übersicht und Einführung. In: Gaeta, L., Schlücker, B. (eds.) Deutsche als kompositionsfreudige Sprache. Strukturelle Eigenschaften und systembezogene Aspekte, pp. 1–25. de Gruyter (2012)
32. Schmid, H., Fitschen, A., Heid, U.: SMOR: a German computational morphology covering derivation, composition and inflection. In: Proceedings of LREC, pp. 233–259 (2004)
33. Steyer, K., Hein, K.: Satzwertige usuelle Wortverbindungen und gebrauchsbasierte Muster. In: Engelberg, S., Lobin, H., Steyer, K., Wolfer, S. (eds.) Wortschätze: Dynamik, Muster, Komplexität, Jahrbuch des Instituts für Deutsche Sprache 2017. de Gruyter (2018, to appear)
34. Trips, C.: The relevance of phrasal compounds for the architecture of grammar. In: ten Hacken, P. (ed.) The Semantics of Compounding, pp. 153–177. Oxford University Press (2016)
35. Trips, C., Kornfilt, J. (eds.): Phrasal compounds from a typological and theoretical perspective. Special issue of STUF. Language Typology and Universals (2015)

Automatic Preparation of Standard Arabic Phonetically Rich Written Corpora with Different Linguistic Units

Fadi Sindran[1(✉)], Firas Mualla[1], Tino Haderlein[1], Khaled Daqrouq[2], and Elmar Nöth[1]

[1] Lehrstuhl für Informatik 5 (Mustererkennung), Friedrich-Alexander-Universität Erlangen-Nürnberg (FAU), MartensstraßE 3, 91058 Erlangen, Germany
fadi.sindran@faui51.informatik.uni-erlangen.de,
{firas.mualla,tino.haderlein,noeth}@cs.fau.de
[2] Department of Electrical and Computer Engineering, King Abdulaziz University, Jeddah 22254, Saudi Arabia
haleddaq@yahoo.com
http://www5.cs.fau.de
http://ee.kau.edu.sa/Default.aspx?Site_ID=135005&lng=EN

Abstract. Phonetically rich and balanced speech corpora are essential components in state-of-the-art automatic speech recognition (ASR) and text-to-speech (TTS) systems. The written form of speech corpora must be prepared carefully to represent the richness and balance in the linguistic content. There is a lack of this type of spoken and written corpora for Standard Arabic (SA), and the only one available was prepared manually by expert linguists and phoneticians. In this work, we address the task of automatic preparation of written corpora with rich linguistic units. Our work depends on a comprehensive statistical linguistic study of SA based on automatic phonetic transcription of texts with more than 5 million words. We prepared two written corpora: the first corpus contains all allophones in SA with at least 3 occurrences of each allophone and 17 occurences of each phoneme. The second corpus contains, in addition to all allophones, 90.72% of diphones in SA.

Keywords: Phonetically rich · SA written corpora · Linguistic content · Allophones · Diphones

1 Introduction

Phonetically rich and balanced sentences are sentences that contain all allophones in a language with respect to their frequency in natural speech [10]. Naturally balanced sentences (phonetically balanced sentences), however, require the availability of very large data sets in order to accurately estimate the phones' frequency distribution as in natural speech. The training of a TTS or ASR system in this case may become complicated and time-consuming. For this reason, the so-called uniformly balanced sentences, which contain all phonetic events as uniformly distributed as possible, are utilized [9,10]. Many ASR systems are based on phonetically rich and balanced corpora. In [6], it was shown that the adoption of this

© Springer International Publishing AG 2017
K. Ekštein and V. Matoušek (Eds.): TSD 2017, LNAI 10415, pp. 201–209, 2017.
DOI: 10.1007/978-3-319-64206-2_23

type of speech corpus in an Arabic ASR system can achieve a high performance. Building this corpus depends primarily on the good preparation and design of a phonetically rich and balanced sentences and phrases. The same holds for developing Arabic TTS systems with high intelligibility. For SA, there is a lack of these corpora. In the literature, the only freely available written Arabic corpus, based on phonetically rich and balanced words, is presented in [7]. It contains 367 sentences covering all phonemes according to their phonetic distribution in SA. Experts created this corpus manually. In this paper, we automatically prepared written corpora based on different linguistic units: phonemes, allophones, and diphones. The novelty of this work lies in the following contributions:

- Preparing the corpora is based on our previous works [11,12], where we performed a rule-based phonetization of SA and a comprehensive statistical linguistic study on four levels: phoneme, allophone, syllable, and allosyllable. In [11], the numbers of the mentioned linguistic units in SA were obtained, and the equations that best fit the ranked frequency distribution at the level of each unit were verified.
- Preparing the corpora was fully automatic using a software package we developed and tested for doing the automatic phonetic transcription of SA texts with very high accuracy (>99%).
- In addition to preparing a phonetically rich corpus covering all allophones in SA, we prepared another written corpus. It covers more than 90% of all diphones in SA with the least possible number of words. This corpus can be used efficiently in Arabic TTS, and Computer-Aided Pronunciation Learning (CAPL) systems.

This paper is organized as follows: Sect. 2 discusses the works related to the preparation of SA corpora. In Sect. 3, we describe the steps of automatic preparation of SA text corpora, and the corpora from which the desired written corpora regarding different linguistic units will be automatically extracted. Henceforth, these corpora will be called "**input corpora**" and the desired extracted corpora "**output corpora**". In Sect. 4, we present the algorithm we adopted for extracting the desired output corpora, while the evaluation and comparison to related works will be presented and discussed in Sect. 5.

2   Related Works

Regarding the manual preparation and design of phonetically rich and balanced SA speech and text corpora, the work done by King Abdulaziz City of Science and Technology (KACST) in Saudi Arabia and presented in [7] is the most considerable work here. The method they used to prepare a phonetically rich and balanced written corpus can be described as follows:

- Preparing a minimal number of phonetically rich words: A list of 663 different words was prepared.
- Putting the words into phrases and sentences, taking into account the following restrictions:

- The corpus must have a minimal number of repeated words.
- The number of words in each sentence or phrase should be between two and nine.
- The pronunciation of the words should be easy.
- The desired corpus must have the least possible number of sentences and phrases with the largest possible number of phonetically rich words in each one.

The result of the work performed by Alghamdi et al. [7] was a corpus of 367 sentences containing 1812 words, 1333 of them were unique. One advantage of this corpus is that it was done with respect to SA as a whole taking into account the different positions of every phoneme: the beginning, middle, and end of a sentence. However, it was done manually and some allophones are not included in it. One of our resulting corpora will be compared with this set in terms of number of words and content of different linguistic units. Regarding the automatic preparation of phonetically rich and balanced SA corpora, there is only one work presented by Yuwan et al. in [14]. A set of 180 phonetically rich and balanced verses from the Holy Quran was automatically prepared. Each verse in this set has between four and seven words. The work in [14] depends on a certain style of Quran reciting resulting in 42 quranic sounds. However, these sounds do not include only phonemes but also special symbols for quranic transliteration, and the prepared corpus is suitable to represent a special style of Quran reciting, not the SA as a whole. Additionally, many phonetic variations were not considered. On the other side, the corpus is phonetically balanced (with respect to only 29 sounds) and automatically prepared. The verses prepared in [14], however, are not publically available for a direct comparison.

3 Automatic Preparation of SA Text Corpora

3.1 Processing Steps

Our work can be divided into three steps as follows:

(1) Preparing and pre-processing of input corpora
(2) Automatic phonetic transcription of the resulting texts from the first step
(3) Applying an appropriate algorithm to extract the output corpora with desired properties depending on the field where these corpora will be utilized.

In the first step, different text corpora from both Classical Arabic (CA) and Modern Standard Arabic (MSA) will be prepared. Every corpus will be segmented into words, phrases, and sentences so that a single line has only one sentence, phrase, or word. The correct full vocalization must be verified, and all dates, numbers, acronyms, abbreviations, and special symbols must be transcribed to a proper word or sequence of words manually. When preparing text corpora for use in TTS or ASR systems, the number of words in each line must be reasonably low. Two to ten words in each line is a good choice for easy reading and recording. The second step includes the automatic transcription and processing of input corpora in order to get the unique phonemes, allophones, syllables, allosyllables, and diphones according to their frequencies for each sentence or phrase. In the last step, the appropriate algorithm, which is a greedy algorithm [13], in different scenarios related to the desired output corpus, will be applied to get the desired corpus containing a minimal number of words and sentences.

3.2   Input Corpora

Six corpora have been used in our work:

(1) **Quran_4_Corpus:** the Holy Quran [4], rewritten in four words per line
(2) **Nahj_6_Corpus:** the text of the book [Nahj al-Balagha] (in Arabic: "نهج البلاغة") [5], which is a book written by Sharif Razi containing sermons, letters, and sayings of Imam Ali ibn Abi Talib, rewritten in six words per line
(3) **MSA_2-9_Corpus:** a corpus containing 2350 sentences and phrases collected and processed from newspapers, stories, and different texts on the Al Jazeera website for Arabic learning [1]
(4) **Bibel_3-14_Corpus:** a set of 655 sentences and phrases selected from the Van Dyck Version of the Holy Bible [3]
(5) **Bobzin_Corpus:** A set of 306 sentences and phrases prepared by Katharina Bobzin from her book [Arabic Basic Course] (in German: "Arabisch Grundkurs") [8]
(6) **Poetry_Corpus:** contains 1538 verses of ancient and modern rhymed Arabic poetry collected and processed from the website [2].

Table 1 gives more information about these corpora.

Table 1. Information about input corpora (BEC: Beginning and End of a sentence Consideration, which means all phonemes appearing at the beginning or end of a sentence must be considered as allophones when calculating the number of allophones)

Input corpus	Quran_4_Corpus	Nahj_6_Corpus	MSA_2-9_Corpus	Bibel_3-14_Corpus	Bobzin_Corpus	Poetry_Corpus
Number of words	78296	63722	14763	4828	1342	12873
Vocabulary words	17578	27854	8677	2471	809	8445
Words per line	4	6	2–9	3–14	1–11	2–14
Unique phonemes	36	36	36	36	36	36
Unique allophones with BEC	155	155	154	134	127	142
Unique allophones without BEC	92	92	93	89	83	90
Unique syllables	2448	2782	1759	987	629	1603
Unique diphones	969	1006	873	738	541	931

4   Output Corpora Extraction Algorithm

After performing the automatic transcription of the pre-processed input corpora at different linguistic levels, an appropriate algorithm must be utilized to extract the best sentences and phrases with the desired content of linguistic units and with the least possible number of words and sentences. A greedy algorithm fits well for this task. We implemented it as follows:

(1) The user can specify the desired percentage of phonemes, allophones, syllables, allosyllables, and diphones in the output corpus, regarding either the input corpus or the SA as a whole. Figure 1 is a screenshot of the program that we developed in Matlab for this purpose. It is also possible to specify the number of sentences wanted. If one or more of the levels are not relevant, the user can put '0' in the respective fields. Additionally, the number of output corpora can be given. Each of the corpora fulfills the above requirements, if possible. In every case, statistics about the percentage of phonemes, allophones, syllables, allosyllables, and diphones in the output corpora will be provided.

(2) A score representing the uniqueness of the linguistic content of every line (a sentence, a phrase, or a word) in the input corpus is calculated, and the line with the highest score is selected und put in a separated set called "**output_set**".

(3) The same score is calculated again for the content of **output_set** with every line in the input corpus. The line corresponding to the highest score will be added to **output_set**.

(4) Step 3 is repeated until the desired percentage of linguistic units in **output_set** is achieved.

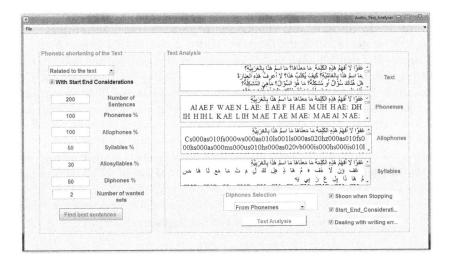

Fig. 1. A screenshot of our program for the automatic preparation of written output corpora

(5) In case of more than one output corpus being required, the sentences or phrases in **output_set** are removed from the input corpus, and the steps 2, 3, and 4 are performed for each output corpus.

(6) The algorithm terminates if the linguistic content in **output_set** did not change after applying step 3.

5   Evaluation and Comparison

From the first three input corpora (Sect. 3), we extracted three output sub-corpora. Each of these sub-corpora contains all allophones in the input corpus from which it is automatically extracted, and with the least possible number of sentences and words. These sub-corpora are:

(1) **Quran_4_All_Alloph_Corpus**: extracted from **Quran_4_Corpus**
(2) **Nahj_6_All_Alloph_Corpus**: extracted from **Nahj_6_Corpus**
(3) **MSA_2-9_All_Alloph_Corpus**: extracted from **MSA_2-9_Corpus**

Table 2 gives more information about these sub-corpora.

Table 2. Information about output sub-corpora (CIC: Corresponding Input Corpus)

Output corpus	Percentage of allophones related to the CIC (%)	Percentage of allophones related to SA (%)	Unique allophones	Number of words	Percentage of words related to the CIC (%)	Number of sentences and phrases
Quran_4_All_Alloph_Corpus	100	98.73	155	200	0.26	50
Nahj_6_All_Alloph_Corpus	100	98.73	155	275	0.43	46
MSA_2-9_All_Alloph_Corpus	100	98.09	154	320	2.17	50

In order to include all allophones of SA in each sub-corpus (with BEC), one sentence or more must be added to each sub-corpus manually. There are only two allophones of SA not included in the first two corpora **Quran_4_All_Alloph_Corpus** and **Nahj_6_All_Alloph_Corpus**. In the third sub-corpus **MSA_2-9_All_Alloph_Corpus**, there are three allophones missing. We added 6 sentences in total in order to complete the missing allophones in these sub-corpora. These three corpora were then merged together to get the corpus named **SA_All_Allophones_Corpus**. It contains all allophones in SA with every allophone occurring at least three times in order to take into account different phone contexts. The resulting corpus consists of 152 sentences and phrases. Table 3 shows some statistics about this corpus.

Table 3. Statistics about **SA_All_Allophones_Corpus**

Output corpus	Minimum occurrence of each phoneme	Unique allophones with BEC	Unique syllables	Unique diphones	Number of words	Vocabulary words
SA_All_Allophones_Corpus	17	157	588	563	821	690

The next step is to get a corpus which has all allophones in SA in addition to the largest possible number of diphones. This corpus must be extracted from all input corpora together in order to cover most diphones of SA. In this case, the computational complexity is very high. Since only a standard computer with 8 GB RAM was available, we applied a compromise solution by dividing the input corpora into smaller subsets. Each subset must contain less than 3000 sentences and phrases and the algorithm is applied on each subset separately. After that, several output subcorpora with less than overall 3000 sentences and phrases are merged, and the algorithm is re-applied. This step was repeated until we obtained the final desired corpus called "**SA_Alloph_Diph_Corpus**". Table 4 provides a comparison between this corpus and the KACST corpus prepared manually in [7].

Table 4. Comparison between **SA_Alloph_Diph_Corpus** (automatically computed) and KACST corpus (manually determined)

Corpus name	SA_Alloph_Diph_Corpus	KACST corpus
Number of sentences and phrases	306	367
Number of words	2051	1812
Vocabulary words	1704	1425
Words per line	2–13	2–9
Unique phonemes	36	36
Unique allophones with BEC	157	143
Unique allophones without BEC	93	90
Unique syllables	1023	1070
Unique diphones	1046	1008

Based on Table 4, we conclude the following:

- Regarding the allophones, the **SA_Alloph_Diph_Corpus** contains 100% of the allophones in SA, while the KACST corpus has about 91.1%.
- Regarding the diphones, about 90.7% of all diphones in SA are included in the **SA_Alloph_Diph_Corpus**, and about 87.4% in the KACST corpus.
- The KACST corpus contains more syllables, about 21.7% of all syllables in SA, while the **SA_Alloph_Diph_Corpus** includes about 20.8%. This is because we did not take the number of syllables into account when preparing our corpus.
- The **SA_Alloph_Diph_Corpus** has about 83.1% vocabulary words, while about 78.6% of the words in the KACST corpus are unique, i. e. the number of repeated words is smaller in our corpus.

6   Conclusion

We accomplished the task of automatic preparation of SA text corpora with respect to different linguistic units. This work is based on a comprehensive statistical linguistic study of SA. Moreover, the transcription of input corpora was fully automatic using a software package that we developed and where we achieved a very high accuracy (>99%) at both phonemic and phonetic level. A greedy algorithm has been utilized to extract the desired output corpora with the least possible number of sentences and phrases (their length was limited to a maximum of 14 words). Comparing the results with a state-of-the-art manual method, the automatically prepared corpora outperform the manually prepared ones and save a lot of time and effort required by linguists and phoneticians. The output corpora can be utilized in SA ASR, diphones-based TTS, and CAPL systems. In order to build a sufficient corpus for unit selection speech synthesis systems, various prosodic features, in addition to the rich context of phones and diphones, will be considered in our future work.

References

1. Al Jazeera Website For Learning Arabic, in Arabic: "موقع الجزيرة لتعليم العربية", March 2017. http://learning.aljazeera.net/Arabic
2. Diwan of Standard Arabic Poetry, in Arabic "ديوان الشعر الفصيح", March 2017. http://www.aldiwan.net/poem.html?Word=%C7%E1%DF%C7%E3%E1& Find=meaning
3. Holy Bible, in Arabic: "الكتاب المقدس", March 2017. http://ar.arabicbible.com/ arabic-bible/word.html
4. Holy Quran, in Arabic: "القرآن الكريم", March 2017. http://www.holyquran.net/ quran/index.html
5. Nahj al-Balagha, in Arabic: "نهج البلاغة", March 2017. http://ia600306.us.archive. org/7/items/98472389432/nhj-blagh-ali.pdf
6. Abushariah, M., Ainon, R., Zainuddin, R., Khalifa, O., Elshafei, M.: Phonetically rich and balanced arabic speech corpus: an overview. In: International Conference on Computer and Communication Engineering, pp. 1–6. IEEE, Kuala Lumpur (2010)
7. Alghamdi, M., Alhamid, A.H., Aldasuqi, M.M.: Database of Arabic sounds: sentences, in Arabic: "قاعدة بيانات الأصوات العربية: جمل". Technical report, King Abdulaziz City of Science and Technology (KACST), Riyadh, Saudi Arabia (2003)
8. Bobzin, K.: Arabic Basic Course, in German: "Arabisch Grundkurs". Harrassowitz Verlag, Wiesbaden (2009)
9. Gibbon, D., Moore, R., Winski, R.: Handbook of Standards and Resources for Spoken Language Systems. Mouton De Gruyter, Berlin (1997)
10. Matoušek, J., Romportl, J.: On building phonetically and prosodically rich speech corpus for text-to-speech synthesis. In: Proceedings of the Second IASTED International Conference on Computational Intelligence, pp. 442–447. ACTA Press, San Francisco (2006)
11. Sindran, F., Mualla, F., Haderlein, T., Daqrouq, K., Nöth, E.: Automatic phonetization-based statistical linguistic study of standard Arabic. Int. J. Comput. Linguist. (IJCL) **7**, 38–53 (2016)
12. Sindran, F., Mualla, F., Haderlein, T., Daqrouq, K., Nöth, E.: Rule-based standard arabic phonetization at phoneme, allophone, and syllable level. Int. J. Comput. Linguist. (IJCL) **7**, 23–37 (2016)

13. Cormen, Thomas H., Leiserson, Charles E., Rivest, Ronald L., Stein, Clifford: Introduction to Algorithms. The MIT Press, Massachusetts (2009)
14. Yuwan, R., Lestari, D.P.: Automatic extraction phonetically rich and balanced verses for speaker-dependent quranic speech recognition system. In: 14th International Conference of the Pacific Association for Computational Linguistics, pp. 65–75. Springer, Bali (2015)

Adding Thesaurus Information into Probabilistic Topic Models

Natalia Loukachevitch[1(✉)] and Michael Nokel[2]

[1] Lomonosov Moscow State University, Moscow, Russia
louk_nat@mail.ru
[2] Yandex, Moscow, Russia
mnokel@gmail.com

Abstract. In this paper we present an approach of introducing thesaurus information into probabilistic topic models. The main idea of the approach is based on the assumption that the frequencies of semantically related words and phrases, which met in the same texts, should be enhanced and this action leads to their larger contribution into topics found in these texts. The experiments demonstrate that the direct implementation of this idea using WordNet synonyms or direct relations leads to great degradation of the initial model. But the correction of the initial assumption improves the model and makes it better than the initial model in several measures. Adding ngrams in similar manner further improves the model.

Keywords: Thesaurus · Multiword expression · Probabilistic topic models

1 Introduction

Currently, the probabilistic topic models become important tools for improving automatic text processing including information retrieval, text categorization, summarization, etc. Besides, they have the prominent role in supporting expert analysis of document collections [1–3]. To facilitate this analysis, such approaches as automatic topic labeling and various visualization techniques are proposed [2,5].

Boyd-Graber et al. [4] indicate that to be understandable by humans, topics should be specific, coherent, and informative. Relationships between the topic components can be inferred. In [2] four topic visualization approaches are compared. The authors of the experiment the authors conclude that manual topic labels include considerable number of phrases; users prefer shorter labels with more general words. Blei and Lafferty [5] visualize topics with ngrams consisting of words mentioned in these topics. These works confirm that phrases and knowledge about word relations are important for topic representation.

In this paper we describe an approach to integrate large manual lexical resources such as WordNet or EuroVoc into probabilistic topic models, as well

© Springer International Publishing AG 2017
K. Ekštein and V. Matoušek (Eds.): TSD 2017, LNAI 10415, pp. 210–218, 2017.
DOI: 10.1007/978-3-319-64206-2_24

as automatically extracted n-grams to improve coherence and informativeness of generated topics. The structure of the paper is as follows. In Sect. 2 we consider related works. Section 3 describes the proposed approach and text collections. Section 4 enumerates used quality measures. Section 5 presents the experiments and obtained results.

2   Related Work

The approaches of utilizing prior knowledge in topic models are limited to single words. Andrzejewski et al. [6] incorporated knowledge by Must-Link and Cannot-Link primitives represented by a Dirichlet Forest prior. These primitives were then used in [7], where similar words are encouraged to have similar topic distributions. However, all such methods incorporate knowledge in a hard and topic-independent way, which is a simplification since two words that are similar in one topic are not necessarily of equal importance for another topic.

Xie et al. [8] proposed a Markov Random Field regularized LDA model (MRF-LDA), which utilizes the external knowledge to improve the coherence of topic modeling. Within a document, if two words are labeled as similar according to the external knowledge, their latent topic nodes are connected by an undirected edge and a binary potential function is defined to encourage them to share the same topic label. Distributional similarity of words is calculated beforehand on a large text corpus.

Approaches of using ngrams in topic models can be subdivided into two groups. Most studies belong to the first kind of methods and are limited to bigrams: i.e. the Bigram Topic Model [9] and LDA Collocation Model [10]. In [11] the Topical N-Gram Model was proposed to allow the generation of ngrams based on the context. However, all these models are enough complex and hard to compute on real datasets.

The second group of methods is based on preliminary extraction of ngrams and then use part of them in topics generation. Initial studies of this approach used only bigrams [12,13]. Nokel and Loukachevitch [14] proposed the PLSA-SIM algorithm, which integrates top-ranked ngrams and terms of information-retrieval thesauri into topic models. They create similarity sets of expressions having the same word components and sum up frequencies of similarity set members if they co-occur in the same text.

The approaches to integrate whole manual thesauri into topic models together with generated ngrams were not studied before.

3   Approach to Integration Whole Thesauri into Topic Models

In our approach we develop the idea of [14] that proposed to construct similarity sets between ngram phrases between each other and single words. Phrases and words are included in the same similarity set if they have the same component word, for example, *weapon – nuclear weapon – weapon of mass destruction; discrimination – racial discrimination*. It was supposed that if expressions from the same similarity set co-occur in the same document then their contribution into the document topics is really more than it is presented with their frequencies, therefore their frequencies should be enhanced. In such an approach, the algorithm can "see" similarities between different multiword expressions with the same component word.

In our approach, at first, we include related single words and phrases from a thesaurus such as WordNet or EuroVoc in these similarity sets. Then, we add preliminarily extracted ngrams into these sets and, this way, we use two different sources of external knowledge.

We use the same LDA-SIM algorithm as described in [14] but study what types of different content can be introduced into such similarity sets. The pseudocode of LDA-SIM algorithm is presented in Algorithm 1, where $S = \{S_w\}$ is a similarity set. Expressions in similarity sets can comprise single words, thesaurus phrases or generated noun compounds.[1]

Algorithm 1. LDA-SIM algorithm

Input: collection D, vocabulary W, number of topics $|T|$, initial $\{p(w|t)\}$ and $\{p(t|d)\}$, sets of similar expressions S, hyperparameters $\{\alpha_t\}$ and $\{\beta_w\}$

Output: distributions $\{p(w|t)\}$ and $\{p(t|d)\}$

1   **while** *not meet the stop criterion* **do**

2   **for** $d \in D, w \in W, t \in T$ **do**

3       $p(t|d,w) = \dfrac{p(w|t)p(t|d)}{\sum\limits_{u \in T} p(w|u)p(u|d)}$

4   **for** $d \in D, w \in W, t \in T$ **do**

5       $n'_{dw} = n_{dw} + \sum\limits_{s \in S_w} n_{ds}$

6       $p(w|t) = \dfrac{\sum\limits_{d \in D} n'_{dw} p(t|d,w) + \beta_w}{\sum\limits_{d \in D} \sum\limits_{w \in W} n'_{dw} p(t|d,w) + \sum\limits_{w \in W} \beta_w}$

7       $p(t|d) = \dfrac{\sum\limits_{w \in d} n'_{dw} p(t|d,w) + \alpha_t}{\sum\limits_{w \in W} \sum\limits_{t \in T} n'_{dw} p(t|d,w) + \sum\limits_{t \in T} \alpha_t}$

For experiments we use several English text collections and one Russian collection (Table 1). We experiment with three thesauri: WordNet, information-retrieval thesaurus of the European Union EuroVoc (comprises 15161 terms)[2], and Russian thesaurus RuThes (115 thousand words and expressions)[3].

---

[1] n_{dw} is the frequency of w in the document d.

[2] http://eurovoc.europa.eu/drupal/.

[3] http://www.labinform.ru/pub/ruthes/index_eng.htm.

Table 1. Text collections for experiments

Text collection	Number of texts	Number of words
English part of Europarl corpus	9672	≈56 mln
English part of JRC-Acquiz corpus	23545	≈53 mln
ACL anthology reference corpus	10921	≈48 mln
NIPS conference papers (2000–2012)	17400	≈5 mln
Russian banking texts	10422	≈32 mln

4   Measures to Estimate the Quality of Topic Models

To estimate the quality of topic models, we use two main automatic measures: topic coherence and kernel uniqueness. For human content analysis, measures of topic coherence and kernel uniqueness are both important and complement each other. Topics can be coherent but have a lot of repetitions. On the other hand, generated topics can be very diverse, but incoherent within each topic.

Topic coherence is an automatic metric of interpretability. It was shown that the coherence measure has the high correlation with the expert estimates of topics interpretability [15, 16]. We calculate automatic topic coherence in two variants. The coherence of a topic is the median PMI (NPMI) of word pairs representing the topic, usually it is calculated for n ($n = 10$) most probable elements in the topic. The coherence of the model is the median of the topic coherence. To make this measure more objective, it should be calculated on an external corpus [16]. In our case, we use Wikipedia dumps.

$$PMI(w_i, w_j) = log\frac{p(w_i, w_j)}{p(w_i)p(w_j)} \qquad NPMI(w_i, w_j) = \frac{PMI(w_i, w_j)}{-log(p(w_i, w_j))} \qquad (1)$$

Human-constructed topics usually have unique main words. The measure of kernel uniqueness shows to what extent topics are different from each other and is calculated as number of unique elements among most probable elements of topics (kernels) in relation to the whole number of elements in kernels.

$$U(\Phi) = \frac{|\cup_t kernel(T_i)|}{\sum_{t \in T} |kernel(T_i)|} \qquad (2)$$

If uniqueness of the topic kernels is closer to zero then many topics are similar to each other, contain the same words in their kernels. In this paper the kernel of a topic means the ten most probable words in the topic. We also calculated perplexity as the measure of language models. We use it for additional check of the model quality.

5   Experiments

At the preprocessing step, documents were processed by morphological analyzers. Also, we extracted noun groups in the form of the regular expression

$((Adj|Noun)^+|(Adj|Noun)^* (Noun\ Prep)^? (Adj|Noun)^*)^* Noun$ [17]. We take into account only such ngrams since topics are mainly identified by noun groups.

As baselines, we use the unigram LDA topic model and LDA topic model with added 1000 ngrams with maximal NC-value [17] extracted from the collection under analysis. As it was described before [12,14], simple addition of ngrams considerably worsens the perplexity because of the vocabulary growth (for perplexity the less is the better) and practically does not change other quality measures (Table 2).

Table 2. Integration of WordNet into topic models

Collection	Method	TC-PMI	TC-NPMI	Kernel uniq	Perplex.
Europarl	LDA unigram	1.20	0.24	0.33	1466
	LDA+1000 ngram	1.19	0.23	0.35	2497
	LDA-Sim+WNsyn	1.05	**0.26**	0.16	1715
	LDA-Sim+WNsynrel	1.20	**0.25**	0.18	4984
	LDA-Sim+WNsr/hyp	1.47	0.24	0.33	1502
	LDA-Sim+WNsr/hyp+Ngrams	2.08	0.23	**0.42**	1929
	LDA-Sim+WNsr/hyp+Ngrams/l	**2.46**	**0.25**	**0.43**	1880
JRC	LDA unigram	1.42	0.24	0.53	807
	LDA+1000 ngrams	1.46	0.22	0.56	1140
	LDA-Sim+WNsyn	1.32	0.25	0.44	854
	LDA-Sim+WNsynrel	1.26	**0.27**	0.28	1367
	LDA-Sim+WNsynrel/hyp	**1.57**	0.24	0.54	823
	LDA-Sim+WNsr/hyp+Ngrams	1.54	0.19	0.64	1093
	LDA-Sim+WNsr/hyp+Ngrams/l	**1.58**	0.18	**0.68**	1064
ACL	LDA unigram	1.63	0.24	0.51	1779
	LDA+1000 ngrams	1.55	0.23	0.51	2277
	LDA-Sim+WNsyn	1.42	0.26	0.47	1853
	LDA-Sim+WNsynrel	1.26	0.27	0.35	2554
	LDA-Sim+WNsynrel/hyp	1.56	0.24	0.51	1785
	LDA-Sim+WNsr/hyp+Ngrams	2.72	**0.28**	0.69	2164
	LDA-Sim+WNsr/hyp+Ngrams/l	**3.04**	**0.28**	**0.76**	2160
NIPS	LDA unigram	1.60	0.24	0.41	1284
	LDA+1000 ngrams	1.54	0.24	0.41	1969
	LDA-Sim+WNsyn	1.34	0.26	0.39	1346
	LDA-Sim+WNsynrel	1.20	0.27	0.29	2594
	LDA-Sim+WNsynrel/hyp	1.78	0.25	0.43	1331
	LDA-Sim+WNsr/hyp+Ngrams	3.18	**0.31**	0.62	1740
	LDA-Sim+WNsr/hyp+Ngrams/l	**3.27**	**0.30**	**0.67**	1741

We add the Wordnet data in the following steps. At the first step, we include WordNet synonyms (including multiword expressions) into the proposed similarity sets (LDA-Sim+WNsyn). At this step, frequencies of synonyms found in the same document are summed up in process LDA topic learning as described in Algorithm 1. We can see that the kernel uniqueness becomes very low, topics are very close to each other in content (Table 2: LDA-Sim+WNsyn). Then, we add word direct relatives (hyponyms, hypernyms, etc.) to similarity sets. Now the frequencies of semantically related words are added up enhancing the contribution into all topics of the current document.

Table 3. Integration of EuroVoc into topic models

Collection	Method	TC-PMI	TC-NPMI	Kernel uniq	Perplex.
Europarl	LDA unigram	1.20	0.24	0.33	1466
	LDA+1000 ngram	1.19	0.23	0.35	2497
	LDA-Sim+EVsyn	1.57	0.24	0.43	1655
	LDA-Sim+EVsynrel	1.39	0.24	0.35	1473
	LDA-Sim+EVsr/hyp+Ngrams	**2.51**	**0.26**	**0.50**	1957
	LDA-Sim+EVsr/hyp+Ngrams/l	**2.5**	**0.25**	0.45	1882
JRC	LDA unigram	1.42	0.24	0.53	807
	LDA+1000 ngrams	1.46	0.22	0.56	1140
	LDA-Sim+EVsyn	1.65	**0.25**	0.57	857
	LDA-Sim+EVsynrel	1.71	**0.24**	0.57	844
	LDA-Sim+EVsr/hyp+Ngrams	**1.91**	0.21	**0.68**	1094
	LDA-Sim+EVsr/hyp+Ngrams/l	1.5	0.18	**0.67**	1061

The Table 2 shows that these two steps lead to great degradation of the topic model in most measures in comparison with the initial unigram model: uniqueness of kernels abruptly decreases, perplexity at the second step grows by several times (Table 2: LDA-Sim+WNsynrel). It is evident that at this step the model is very bad. If we look at the topics, the cause of the problem seems to be clear. We can see the overgeneralization of the obtained topics. The topics are built around very general words such as "person", "organization", "year", etc. These words were initially frequent in the collection and then received additional frequencies from their frequent synonyms and related words.

Then we suppose that these general words were used in texts to discuss specific events and objects, therefore we change the constructions of the similarity sets in the following way: we do not add word hyponyms to its similarity set. Thus, hyponyms that are usually more specific and concrete should obtain additional frequencies from upper synsets and increase their contributions into the document topics. But the frequencies and contribution of hypernyms into the topic of the document are not changed. And we see the miracle: the kernel uniqueness considerably improves, perplexity decreases to levels comparable with the unigram model, topic coherence characteristics also improve for most collections (Table 2: LDA-Sim+WNsynrel/hyp).

We further add ngrams having the same component words into the similarity sets as described in [14]. All measures significantly improve for all collections

(Table 2: LDA-Sim+WNsr/hyp+Ngrams). At the last step we try to apply the same approach to ngrams that were previously utilized to hyponym-hypernym relations: frequencies of shorter ngrams and words are summed to frequencies of longer ngrams but not vice versa. In this case we try to increase the contribution of more specific longer ngrams into topics. It can be seen (Table 2) that the kernel uniqueness grows significantly, at this step it is more than for the baseline models in 1.3–1.6 times achieving 0.76 on the ACL collection (Table 2: LDA-Sim+WNsr/hyp+Ngrams/1).

At the second series of the experiments, we applied EuroVoc information retrieval thesaurus to two European Union collections: Europarl and JRC. In content, the EuroVoc thesaurus is much smaller than WordNet, it contains terms from economic and political domains and does not include general abstract words. The results are shown in Table 3. It can be seen that inclusion of EuroVoc synsets improves the topic coherence and increases kernel uniqueness (in contrast to results with WordNet). Adding ngrams further improves the topic coherence and kernel uniqueness.

At last we experimented with the Russian banking collection and utilized RuThes thesaurus. In this case we obtained improvement already on RuThes synsets and again adding ngrams further improved topic coherence and kernel uniqueness (Table 4). In Table 5 one of topics generated for the banking collection together with found thesaurus relations (similarity sets) is shown. The high coherence of this topic is clearly seen.

Table 4. The results obtained for Russian banking collection

Collection	Processing	TC-PMI	TC-NPMI	Kernel uniq	Perplex.
Banking collection	LDA unigram	1.81	0.29	0.54	1654
	LDA+1000 ngrams	2.01	0.30	0.60	2497
	LDA-Sim+RTsyn	2.03	0.29	0.63	2189
	LDA-Sim+RTsr/hyp+Ngrams	2.72	**0.33**	**0.70**	2396
	LDA-SIM+RTsr/hyp+Ngrams/1	**3.02**	0.31	0.68	2311

It is worth noting that adding ngrams sometimes worsens the TC-NPMI measure, especially on the JRC collection. This is due to the fact that in these frameworks, topic beginnings contain a lot of multiword expressions, which rarely occur in Wikipedia, used for the coherence calculation, therefore the TC-NPMI measure in fact does not have enough evidence for correct estimates.

6   Conclusion

In this paper we presented the approach for introducing thesaurus information into topic models. The main idea of the approach is based on the assumption that if related words or phrases are met in the same text, that their frequencies

Table 5. Example of the information technology topic from Banking collection with indication of thesaurus relations in form of similarity set numbers.

Topic Elements	Translation	96	125	300	334	476	518	712	891	988
программное обеспечение	program software				X	X			X	
информационная технология	information technol.		X	X						
информационная система	information system	X					X	X		
автоматизированная система	automated system							X		X
защита информации	information protect.								X	
информационная безопасность	information security								X	
программный продукт	program product	X				X	X			
новая технология	new technology		X	X						

should be enhanced and this action leads to their mutual larger contribution into topics found in this text.

In the experiments on four English collections it was shown that the direct implementation of this idea using WordNet synonyms and/or direct relations leads to great degradation of an initial model. But the correction of initial assumptions and excluding of hypernyms from frequencies adding magically improves the model and make it much better than the initial model in several measures. Adding ngrams in the similar manner further improves the model.

Introducing information from domain-specific thesaurus EuroVoc led to improving the initial model without the additional assumption which can be explained with the absence of general abstract words in such information-retrieval thesauri.

Acknowledgments. This work was partially supported by Russian National Foundation, grant N16-18-02074.

References

1. Blei, D.: Probabilistic topic models. Commun. ACM **55**(4), 77–84 (2012)
2. Smith, A., Lee, T.Y., Poursabzi-Sangdeh, F., Boyd-Graber, J., Elmqvist, N., Findlater, L.: Evaluating visual representations for topic understanding and their effects on manually generated labels. Trans. Assoc. Comput. Linguist. **5**, 1–15 (2016)
3. Chang, J., Boyd-Graber, J., Wang, C.H., Gerrich S., Blei, D.: Reading tea leaves: how humans interpret topic models. In: Proceedings of the 24th Annual Conference on Neural Information Processing Systems, pp. 288–296 (2009)
4. Boyd-Graber, J., Mimno, D., Newman, D.: Care and Feeding of Topic Models: Problems, Diagnostics, and Improvements. CRC Handbooks of Modern Statistical Methods. CRC Press, Boca Raton (2014)
5. Blei, D., Lafferty, J.: Visualizing topics with multi-word expressions (2009). https://arxiv.org/pdf/0907.1013.pdf
6. Andrzejewski, D., Zhu, X., Craven, M.: Incorporating domain knowledge into topic modeling via Dirichlet forest priors. In: Proceedings of the 26th Annual International Conference on Machine Learning, pp. 25–32 (2011)

7. Newman, D., Bonilla, E., Buntine, W.: Improving topic coherence with regularized topic models. In: Advances in Neural Information Processing Systems, pp. 496–504 (2011)

8. Xie, P., Yang, D., Xing, E.: Incorporating word correlation knowledge into topic modeling. In: Proceedings of NAACL 2015, pp. 725–734 (2015)

9. Wallach, H. Topic modeling: beyond bag-of-words. In: Proceedings of the 23rd International Conference on Machine Learning, pp. 977–984 (2006)

10. Griffiths, T., Steyvers, M., Tenenbaum, J.: Topics in semantic representation. Psychol. Rev. **114**(2), 211–244 (2007)

11. Wang, X., McCallum, A., Wei, X.: Topical N-Grams: phrase and topic discovery, with an application to information retrieval. In: Proceedings of the 2007 Seventh IEEE International Conference on Data Mining, pp. 697–702 (2007)

12. Lau, J., Baldwin, T., Newman, D.: On collocations and topic models. ACM Trans. Speech Lang. Process. **10**(3), 1–14 (2013)

13. Nokel, M., Loukachevitch, N.: A method of accounting bigrams in topic models. In: Proceedings of the 11th Workshop on Multiword Expressions (2015)

14. Nokel, M., Loukachevitch, N.: Accounting ngrams and multi-word terms can improve topic models. In: Proceedings of the 11th Workshop on Multiword Expressions (2016)

15. Mimno, D., Wallach, H., Talley, E., Leenders, M., McCallum, A.: Optimizing semantic coherence in topic models. In: Proceedings of EMNLP 2011, pp. 262–272 (2011)

16. Lau, J., Newman, D., Baldwin, T.: Machine reading tea leaves: automatically evaluating topic coherence and topic model quality. In: Proceedings of the European Chapter of the Association for Computational Linguistics (2014)

17. Frantzi, K., Ananiadou, S.: The C-value/NC-value domain-independent method for multi-word term extraction. J. Nat. Lang. Process. **6**(3), 145–179 (1999)

Dialogue Modelling in Multi-party Social Media Conversation

Subhabrata Dutta and Dipankar Das[(⊠)]

Jadavpur University, Kolkata 700032, West Bengal, India
subha0009@gmail.com, dipankar.dipnil2005@gmail.com

Abstract. Social Media is a rich source of human-human interactions on exhausting number of topics. Although dialogue modeling from human-human interactions is not new, but there is no previous work as far as our knowledge attempting to model dialogues from social media data. This paper implements and compares multiple supervised and unsupervised approaches for dialogue modelling from social media conversation; each approach exploiting and unfolding special properties of informal conversations in social media. A new frequency measure is proposed especially for text classification problem in these type of data.

Keywords: Dialogue modelling · Facebook comment · Frequency weight score · Sentiment polarity

1 Introduction

1.1 Dialogue Processing and Social Media

A *dialogue* is a written or spoken conversational exchange between two or more people. Automated understanding of human-human informal interaction and discourse processing in dialogues are one of the most challenging issues in present day computational linguistics. In the field of dialogue processing, much work has been done in speaker identification [1], topic identification [2], designing intelligent conversational agents [3] etc. Identifying discourse structure in multi-party dialogues is still in its infancy [4].

A problem while working on informal human-human interaction is non-availability of sufficient data, both speech and textual, and mostly due to privacy issues. Thankfully after the emergence of social media, a certain variation of human-human interaction in textual form is available as public data, and this data is huge and ever growing. One important such platform is Facebook. As of December 2016, on average there are 1.86 billion monthly active Facebook users worldwide[1]. On 2012, a statistical survey[2] revealed that every 60 seconds on Facebook, there are 510,000 posted comments and 293,000 status updates on average.

---

[1] http://newsroom.fb.com/company-info/.

[2] http://thesocialskinny.com/100-social-media-statistics-for-2012/.

© Springer International Publishing AG 2017
K. Ekštein and V. Matoušek (Eds.): TSD 2017, LNAI 10415, pp. 219–227, 2017.
DOI: 10.1007/978-3-319-64206-2_25

1.2   Conversation on Facebook: Posts, Threads and Comments

Users on Facebook interact with *comments* put under posts. Under a certain post from an individual user or a page, people posts their comment to express their opinion regarding that topic. Under each such comment there is a 'reply' option using which people can put comment around that specific comment. Multiple comments under a single comment constitutes a *thread*. As shown in the left part of Fig. 1, Individual comments and the number of replies under them (which is, the total number of comments in a thread) are shown.

On the right part of Fig. 1, a single thread is shown with arrows showing who have replied to whom, and how a *Multi-party Dialogue* is constituted by these comments.

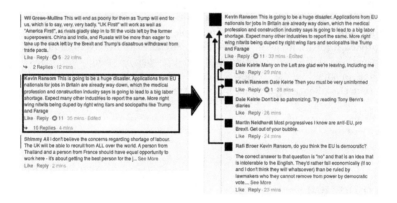

Fig. 1. Multiple threads under a post (left) and individual thread (right)

1.3   Dialogue Modelling Problem

Dialogue Modelling from an unstructured utterance corpus is the process of assigning order to each utterance from the corpus to form a meaningful set of dialogues. Although Facebook offers a huge collection of conversations, to use them for various dialogue processing frameworks, we need to model those comments into well formed dialogues. For dialogue modelling from human-human interactions there are already some works available [5]. But these works aim to model real or imitated 'personal' dialogues, i.e., conversation between two or more persons in real world. As for our knowledge there is no previous work on dialogue modelling from Facebook comment threads. This paper aims to approach this problem: how can we design a system which will decide, given a pair of comments from thread, whether these two comments are consecutive utterance of a dialogue or not.

2   Dataset

The dataset prepared till date contains Facebook page comments of posts regarding (1) the Jallikattu debate from The Hindu, Times of India and The Economic Times, (2) the US presidential election 2017 from The New York Times, (3) BREXIT debate from The BBC, (4) Bhopal encounter from Al-Jazeera, (5) immigration and terror attacks in Europe and (6) Feminism in media. A single post in a Facebook page contains multiple comments. To each comment, there are multiple replies which construct a conversation thread. Our motivation was to extract threads containing more replies and comments. Thus, we considered the topics based on controversial issues. Threads from each of the posts are collected and stored in Json format using Facebook Graph API[3]. This is a continuous in-house effort to prepare a dataset based on social dialogues. The details of the dataset collected are mentioned in Table 1.

Table 1. Topic wise statistics for the dataset

Topic	Threads	Comments	Sentences	Words
Jallikattu	4	128	323	2107
US Election	4	109	298	1804
Immigration	5	239	727	5781
Brexit	2	119	342	2604
Bhopal encounter	2	95	262	1589
Feminism	3	107	264	1413

Initially, it was observed that most of the posts on the Jallikattu debate contain code-mixed contents. Thus, in order to avoid the dealing with code mixed contents, comments in non-English languages have been removed manually. Also, to get rid of irrelevant comments, any comment containing word frequency less than six has been removed. However, the nested comments of having length two or less have been discarded to avoid the trivial thread instances.

Abrevial forms (e.g., 'u' for 'you', 'wht' for 'what', 'tht' for 'that' etc.) are replaced by originals using a dictionary, so that a comment e.g., "*Wht u call ur culture is nthng bt cruelty*" will be read as "*What you call your culture is nothing but cruelty*". After all the pre-processings, we get a dataset with specifications as given in Table 1. In order to prepare a gold standard data for training purpose, each comment is tagged manually to which previous comment it is replied to.

3   Methodology

We have developed two separate approaches for our problem. We take pairs of comments $[C_i, C_j]$ and implemented a supervised classification framework which

---

[3] developers.facebook.com/tools/explorer?method=GET&path=&version=v2.8.

is composed of three different types of classification strategies by using Support Vector Machine and employing specific sets of the following features. We also implemented a purely unsupervised rule-based frame work, where each comment is matched against all the previous comments in that thread, with aggregate scores assigned using the following features.

3.1 User Name Mention Matching

User Name Mentions are specific style of posting a comment where, the comment usually starts or ends with the name of the user to whose comment this is a reply. For example take the following pair of comments:

1. **User1:** *So you're going to complain about not having enough time to hang out with your friends? Don't go to college then.*
2. **User2:** *User1 I didn't complain. I just explained why you cannot put 'every-thing' into a 'limited' syllabus.*

Second comment mentions the name of the user of the first comment. In our dataset, 35.96% of the comments mentions a previous user name. To find whether a comment mentions some user, we tag Named Entities in every comment and match them with *User Name list*, which contains 502 unique users commenting. We define this binary valued feature as *UNMM*.

3.2 Reply to Starting Comment

In Facebook comment threads, it is usually the first comment around which most of the conversations revolve. In our dataset, 46.55% of the comments are reply to the first comment of the thread. We define this binary valued feature as *RSC*.

3.3 Longest Common Continuous Word Sequence

Common Continuous Word Sequence between two comments is the longest segment of a sentence which is present in both the comments. While commenting against a specific comment, users tend to quote a certain portion of that comment; or even express opinion against a certain clause.

But again components like 'phrasal' verbs (which are short and frequent) can be used by a user without noticing whether a previous comment contains the same component. Thus longer the common word sequence, higher is the probability of two comments being part of a dialogue.

Let S_1 and S_0 be two set of comment-pairs both containing named entities defined as,

$$[C_i, C_j] \in S_1 \quad \text{if } C_j \text{ is a reply to } C_i \tag{1}$$

$$[C_i, C_j] \in S_0 \quad \text{if } C_j \text{ is not a reply to } C_i \tag{2}$$

In our dataset, average length of longest common word sequence over S_1 is 1.3774; in case of S_0, this value drops to 1.0534. We define this continuous valued feature as *LCCWS*.

3.4   Named Entity Overlap

Named Entity Overlap is the score equal to the frequency of common Named Entities among two comments in a comment-pair. Usually Named Entities are more important than other words present in a sentence in terms of identifying the topic that sentence is related to. More the number of common Named Entities between two comments, more are chances of them being close in terms of topic.

In our dataset, for $[C_i, C_j] \in S_1$ containing named entities, average *Named Entity Overlap* over S_1 is 0.2424; for $[C_i, C_j] \in S_0$ containing named entities, it is 0.1489; S_1 and S_0 being as defined in (1) and (2). We define this continuous valued feature as *NEO*.

3.5   Frequency Weight Measures for Common Terms

For text classification techniques, counting *tf-idf* for words is a vividly used approach. The motivation is to assign a weight to individual terms which will count for their 'relevance' in a document. *Simple tf* is defined as the frequency of a term t in a single document d. Higher the frequency, higher the chances of that word being related to the topic of that document. *Simple idf* or Inverse Document Frequency is defined as $\log(1 + \frac{N}{df})$ where N is the total number of documents and df is the number of documents where t is present.

In our problem, each comment can be defined as a document, thereby measuring the relevance of terms in individual comments using $tf - idf$ scores. But there is one statistical problem with it. Facebook comments are very short, and come as informal texts from different users of varying linguistic styles. Which means, If we take 20 different comments for example, it is likely that even simple verbs like '*do*', '*go*' etc. occur rarely, and a high *idf* is assigned to them. Worst is the case when terms which are related to the topic of a thread occur multiple times within that thread and acquire a very low *idf* (For example words like '*Muslim*' or '*economy*' in threads related to Immigration and Terrorism in Europe or '*tradition*', '*cruelty*' etc. in Jallikattu). Taking into account these features, we use a modified version of *idf* score in our problem.

An *ITF* or Inverse Thread Frequency score is defined for a term t over the dataset D as,

$$ITF(t) = \log(1 + \frac{N_D}{f_T}) \tag{3}$$

where N_D is no. the threads in D and f_T is the no. of threads where t is present. Another score *ICF* or Inverse Comment Frequency is defined for a term t over a thread T as,

$$ICF(t) = \log(1 + \frac{N_T}{f_C}) \tag{4}$$

where N_T is the no. of comments in T and f_C is the no. of comments where t is present.

The combined *tf-itf-icf* score is used for assigning weights to terms from each comment. Also, we have selected only the nouns, verbs, adverbs and adjectives to construct the term set. While matching any two terms from two comments, their base forms are matched. For example, *'nationalism'* is matched with *'nationalist'*. This approach is to take into account the short length of comments as documents, where a single comment mostly contains a single form of a word, and different users use different sentence structure (and use different word forms).

While using *tf-itf-itc* scores as weights assigned to terms, we used common terms between a comment pair $[C_i, C_j]$ instead of representing each comment as a vector. Reasoning behind this is the small size of the featureset (<4000) compared to the size of each feature vector resulting in (>6000). *"Curse of dimensionality"* [6] thereby barred us from using vector representation of individual comments.

For the unsupervised rule-based system, and also for the next feature using sentiment polarity measure, we defined two threshold values for itf and icf scores of individual terms, ITF_{lim} and ICF_{lim}. They are defined as,

$$ITF_{lim} = \log(1 + \frac{N_D}{m}) \tag{5}$$

where N_D and f_T are as defined in (3) and m is maximum no. of threads in a single topic.

$$ICF_{lim} = \log(1 + \frac{N_T}{(\frac{\Sigma f_C}{N_2})}) \tag{6}$$

where N_T and f_C are as defined in (4) and N_2 is no. of terms in that thread.

ITF_{lim} is actually the lower bound of itf score of a term which occurs in topic-specific threads. For example, the term *'tradition'* occurs only in the four threads of Jallikattu and thus have itf greater than ITF_{lim}. Likewise, ICF_{lim} is the upper bound on the icf score of a term in a thread around which the discussion of that thread has proceeded.

For comment-pairs $[C_i, C_j]$, let

$$\tau = (t_0, \ldots, t_n) \tag{7}$$

be the set of common terms between C_i and C_j. Then the feature *Frequency Weight Measures* for $[C_i, C_j]$ is defined as

$$FWM(C_i, C_j) = \sum_{t \in \tau} TF(t) * ITF(t) * ICF(t) \tag{8}$$

For S_1 and S_0 defined in (1) and (2), average FWM defined over S_1 is 7.7457, and over S_0 it is 2.8191.

3.6   Sentiment Polarity of Comments

The motivation of using *Sentiment Polarity of Comments* as a feature for our problem comes from a simple observation in our dataset. Given a single comment, 92.4% of the replies to it show tendency to oppose the statement done by it. That is, given a comment-pair $[C_i, C_j]$, if we can decide C_j opposes the statement of C_i, then C_j is likely to be a reply to C_i.

But bluntly calculating the sentiment polarity of two comments do not serve our interest. For example, these two comments *"Jallikattu is not about cruelty. It's not PETA who have the right to decide!"* and *"This fight to protect our tradition must and will go on"* show opposite compound polarity. But as we can see, the second comment is less likely to be a reply to the first one. To use sentiment polarity measure properly, we choose only those comments with common terms having itf score greater than ITF_{lim} and icf score less than ICF_{lim}. Such terms, as discussed in Sect. 3.5 are the terms related to the topic of that thread. We measure compound sentiment polarity of the sentences containing these 'high itf low icf' terms and check whether they bear opposite polarity or not. Simply, we decide whether the two given comments express a positive or negative sentiment polarity against a common term which is under discussion of that thread. We define this continuous valued feature as *SPC*.

4   Results

We build following three different classifiers using Support Vector Machine with *rbf* kernel, train them on half of the dataset using different feature combinations:

- **Classifier A:** uses features *UNMM, RSC, NEO, LCCWS* and basic $tf - idf$ scores
- **Classifier B:** uses features *UNMM, RSC, NEO, LCCWS* and *FWM*
- **Classifier C:** uses features *UNMM, RSC, NEO, LCCWS, FWM* and *SPC*

Results of testing A, B and C on the other half of our dataset yields results given in Table 2. Performances of the rule-based system D is also shown.

Table 2. Scores of the supervised classifiers on different topic-threads

Topic	A	B	C	D
Jallikattu	69.23	70.12	73.13	52.04
Immigration	82.57	91.88	85.42	55.17
US election	91.67	97.22	94.66	50.01
Brexit	67.86	75.00	76.92	48.22
Bhopal	74.23	76.54	76.24	47.13
Feminism	76.79	82.14	79.55	38.98

The rule-based system, which used sum of all the features with equal weight, clearly lags way behind the supervised systems. This proves that not all the features are of equal importance in our classification problem. Also, comparing the results of classifier A (using *simple tf-idf*) with those of classifier B (using *tf-itf-icf*) we can clearly conclude the efficiency of our proposed *tf-itf-icf* measures in these kinds of data.

An interesting phenomenon can be observed from the comparison of performance by classifier B and C. C, which was added with an extra feature *Sentiment Polarity* generally yields less accurate scores than B. But it outperforms B in threads related to Jallikattu and Brexit. Also in threads related to US Election or Bhopal, their performances are nearly equal. If we take a closer look into the threads where C performs better than B, we can see that comments in these threads contain very sharp opinion polarity; the number of 'high *itf* low *icf*' terms is small in these threads, and they usually have lower *icf* than their counterparts in other threads. To put simply, in threads related to Jallikattu or Brexit, users who put comments used to have a sharp opinion around some of these terms like *'culture'*, *'cruelty'*, *'rights'*, *'EU'* etc. Such characteristics have been captured well by the *Sentiment Polarity* feature. In these threads, whether two comments are 'close' to each other in terms of words used seldom means they are connected in a dialogue, because the majority of the comments have used the same set of words to express their opinion. Thus *Frequency Weight Measure for Common Terms* fell to be as successful as other threads.

5   Further Works

For the supervised classification approach we implemented, all of the features except the *Sentiment Polarity* exploited either word level or thread structure level properties of the data. It is already discussed that Facebook (and other social media platforms) comments are very short and put by a large variety of users using different linguistic style. This limits the effectivity of word-level features to perform accurately to a great extent. The *Sentiment Polarity* metric we used is also based on a naive approximation that, simple sentiment score reflects the opinion of a comment on a topic.

As discussed in Sect. 3.5 we tried matching words by their base forms. This surely increases performance by a great degree. But still there is a scope of further development if we can match synonyms or mostly similar words with a certain similarity score. For example, in phrases like *'stop animal slaughter'* and *'ban illegal slaughter house'*, the verb *stop* carries similar meaning as *ban*. The framework we proposed is not provisioned identify these similarities.

A lot of work is left in part of identifying if the modelling done is accurate in discourse-level. To state simply, classifying two comments using measures of topic-level similarity does not necessarily ensure that, they are actually two consecutive utterances of a dialogue. As we can see in real world human-human interaction that often an utterance put in reply to another utterance does not necessarily contain similar words or opposite opinions. A deep study of discourse-connectives in dialogue is needed for this improvement.

References

1. Furui, S.: 50 years of progress in speech and speaker recognition research. ECTI Trans. Comput. Inf. Technol. (ECTI-CIT) **1**(2), 64–74 (2005)
2. Purver, M., Griffiths, T.L., Körding, K.P., Tenenbaum, J.B.: Unsupervised topic modelling for multi-party spoken discourse. In: 21st International Conference on Computational Linguistics and the 44th Annual Meeting of the Association for Computational Linguistics, Association for Computational Linguistics, pp. 17–24 (2006)
3. Bickmore, T., Cassell, J.: Social dialongue with embodied conversational agents. In: van Kuppevelt, J.C.J., Dybkær, L., Bernsen, N.O. (eds.) Advances in Natural Multimodal Dialogue Systems, pp. 23–54. Springer, Netherlands (2005). doi:10.1007/1-4020-3933-6_2
4. Perret, J., Afantenos, S., Asher, N., Morey, M.: Integer linear programming for discourse parsing. In: NAACL-HLT, pp. 99–109 (2016)
5. Gandhe, S., Traum, D.: First steps toward dialogue modelling from an un-annotated human-human corpus. In: IJCAI Workshop on Knowledge and Reasoning in Practical Dialogue Systems, Hyderabad, India (2007)
6. Hughes, G.: On the mean accuracy of statistical pattern recognizers. IEEE Trans. Inf. Theor. **14**(1), 55–63 (1968)

Markov Text Generator for Basque Poetry

Aitzol Astigarraga, José María Martínez-Otzeta[✉], Igor Rodriguez,
Basilio Sierra, and Elena Lazkano

Department of Computer Sciences and Artificial Intelligence, University of the
Basque Country UPV/EHU, 20018 Donostia-San Sebastian, Spain
{aitzol.astigarraga,josemaria.martinezo,igor.rodriguez,b.sierra,
e.lazkano}@ehu.eus
http://www.sc.ehu.es/ccwrobot

Abstract. Poetry generation is a challenging field in the area of natural
language processing. A poem is a text structured according to predefined
formal rules and whose parts are semantically related. In this work we
present a novel automated system to generate poetry in Basque language
conditioned by non-local constraints. From a given corpus two Markov
chains representing forward and backward 2-grams are built. From these
Markov chains and a semantic model, a system able to generate poems
conforming a given metric and following semantic cues has been designed.
The user is prompted to input a theme for the poem and also a seed word
to start the generating process. The system produces several poems in
less than a minute, enough for using it in live events.

Keywords: Poetry generation · Basque language · N-grams

1 Introduction

Poetry is one of the most expressive -and challenging- ways to use language. It
is commonly accepted that quality of good poetry arises from an equilibrium
between content and form, both content and form contributing to its aesthetic
value. But to what extent each one affects the overall result is still a matter of
debate. Oral poetry is considered poetry constructed without the aid of writing
[17]. Oral meant that a work was composed and performed at the moment,
with no prior preparation. Poets and story-tellers from many different cultures
have historically used such oral improvisation in performances. Nowadays many
improvisational oral practices exists around the world, such as Serbo-Croatian
guslars, freestyle rap and Basque bertsolaritza.

In this work we present a novel system that is able to generate Basque poetry
under the constraints of *bertsolaritza*, a form of oral and improvised poetry.
Those constraints are local (the metric of the verses and the rhyming pattern)
and non-local (the semantic similarity with a given theme and the time allotted
to produce the finished poem). The mechanism for poetry generation relies in
N-gram models [18] created from a corpus of Basque poetry previously collected.
Semantic similarity between lines is measured.

© Springer International Publishing AG 2017
K. Ekštein and V. Matoušek (Eds.): TSD 2017, LNAI 10415, pp. 228–236, 2017.
DOI: 10.1007/978-3-319-64206-2_26

An experiment where poems have been generated with a given theme and metric has been performed, and the result has been subject to evaluation by four individuals familiar with Basque poetry. We present the summary of their evaluations, along with some poems.

The rest of the paper is organized as follows: in Sect. 2 a brief account of Basque poetry is presented. In Sect. 3 related work is surveyed, while Sect. 4 is devoted to present our approach. Experimental setup and results are shown in Sect. 5, and finally the conclusions are summarized in Sect. 6.

2   Some Words About Basque Language and *bertsolaritza*

Basque is the language of the inhabitants of the Basque Country. And *bertsolaritza*, an improvised contest poetry, is one of the manifestations of traditional Basque culture. Events and competitions in which improvised verses, *bertso-s*, are composed are very common. In such performances, one or more verse-makers, named *bertsolari-s*, produce impromptu compositions about topics or prompts which are given to them by an emcee (theme-prompter). Then, the verse-maker takes a few seconds, usually less than a minute, to compose and sing a poem along the pattern of a prescribed verse-form that also involves a rhyme scheme.

Xabier Amuriza, a famous verse-maker who modernized the *bertsolari* movement, defined *bertsolaritza* in a verse as:

Neurriz eta errimaz	*Through meter and rhyme*
kantatzea hitza	*to sing the word*
horra hor zer kirol mota	*that is what kind of sport*
den bertsolaritza.	*bertsolaritza is.*

When constructing an improvised verse a number of formal requirements must be taken into account. Rhyme and meter are inseparable elements in improvised verse singing (in the above example, odd lines, which must rhyme with each other, have seven syllables and even lines six). A person able to construct and sing a *bertso* with the chosen meter and rhyme is considered as having the minimum skills required to be a *bertsolari*. But the true quality of the *bertso* does not only rely on those demanding technical requirements. The real value of the *bertso* resides on its dialectical, rhetorical and poetical value [9]. Thus, a *bertsolari* must be able to express a variety of ideas and thoughts in an original way while dealing with the mentioned technical constraints. In this balance lies the magic of a *bertso*.

3   Related Work

Computational modeling for poetry generation has become a topic in the artificial intelligence community in the last years. People with a background closer to humanities made early efforts in systematic generation of poetry. We could mention works related to generating variations over a predetermined set of verses [21], or to select a template to produce poems from it [20].

According to [12], two main strategies can be outlined in the field of computer generation of poetry:

- **Corpus-based approach:** computer is used to harvest and reuse text already formated into poem-like structure of lines. This approach can be formulated as an information retrieval task, where the objective is to extract and select existing lines to compose new poems. The reuse and ordering of written text was introduced by Queneau [21]. Many computer-based systems rely nowadays in this method. Most relevant are: [19,22,23]. This procedure is adopted in [2,3] where two methods to ensure internal coherence of poems were presented.
- **Composition from scratch:** alternative methods rely on building a text from scratch, character by character or word by word, and establishing a distribution of the resulting text into poem lines by some additional procedure. A popular -and rather simple- method to generate text is the N-gram model, which is the simplest Markov model. N-grams assign probabilities to sequences of words and the generated model can be used to stochastically generate sequences of words based on the generated distributions [14,18]. An N-gram probability is the conditional probability of a word given the previous N-1 words. Markov chains have been widely used as the basis of poetry generation systems as they provide a clear ans simple way to model some syntactic and semantic characteristics of language [16]. Popular and recent examples of N-gram poem generators are [4,6,12]. But text poetry hold non-local properties such as rhyme and metric that cannot be modeled by an ordinary N-gram model. Therefore, the above mentioned methods need additional procedures for distributing the resulting text into poem lines with metrical and rhyming constraints.

For a more thorough review of systems related to automatic generation of poetry, we point the reader to [10,15].

4 Proposed Approach

The formal constraints of *bertsolaritza* (and poetry as well) can be viewed as intra-line constraints and inter-line constraints. Intra-line constraints include the number of syllables allowed on each line, that is, the metrical scheme. Inter-line constraints consist of rules that state rhyming constraints between the lines and the semantical relatedness of the text respect to the proposed theme. The *bertsolari* has a basic strategy that is used in a systematic way [8]: think up the end first. On hearing a proposed theme, the *bertsolari* turns on their mental machinery and starts thinking about what is going to say and the order in which (s)he is going to say it, keeping the main idea for the end.

Our approach implements the same strategy used by *bertsolaris* for the creation of impromptu verses, and in a few seconds - less than minute - assembles a new poem along the prescribed verse-form. Although our work focuses on *bertsolaritza*, it can be generalized as the automatic poem generation.

The main characteristics of our word-based N-gram approach are the following:

- Live generation of verses.
- Poem generation in the style of an existing author.
- Satisfy structural constraints, both, inter-line metrical constraints and between-line rhyming ones.
- Semantic similarity respect to the theme given to construct a verse and between the lines that compose the verse (internal coherence).

4.1 Training the Language Model

As the text generation module relies on an N-gram model of language to produce sequences of text that are word to word coherent, the overall style of the resulting poems is strongly determined by the sources used to train the content generators. Therefore, different sources of text were used in the experimental setup in order to experiment with them and extract conclusions.

- Mixed: 18913 lines and 94314 words, of which them 21411 are unique. This corpus is a compilation of sentences mined from Basque newspaper Egunkaria[1] (85%) alongside poetry sung in *bertsolari* contests by different performers (15%).
- Txirrita: 2127 lines and 24277 words, of which them 6998 are unique. This corpus is a compilation of poetry by a famous *bertsolari* called *Txirrita*[2].

4.2 Text Applications

Several linguistic tools have been developed and used in the verse-maker project.

- **Rhyme search:** finding words that rhyme with a given word is an essential task that the verse-generation system must perform. Basque rhyme schemes are mainly consonant. The widely consulted rhyming dictionary *Hiztegi Errimatua* [1] contains a number of documented phonological alternations that are acceptable as off-rhymes. These alternations have been implemented using regular expressions.
- **Syllable counter:** counts the number of syllables present in the given text. For the syllabification itself, the approach describing the principal elements of Basque language structure [13] has been implemented.

---

[1] https://en.wikipedia.org/wiki/Egunkaria.
[2] https://en.wikipedia.org/wiki/Txirrita.

– **Similarity measure:** the main purpose of this module is to measure the semantic relationship between pairs of words and sentences. The module computes for each pair of words/sentences a score that evaluates how similar they are. Latent Semantic Analysis (LSA) method [7,24] has been implemented to capture the semantic relatedness. This semantic model has been generated with news mined from the Egunkaria newspaper.

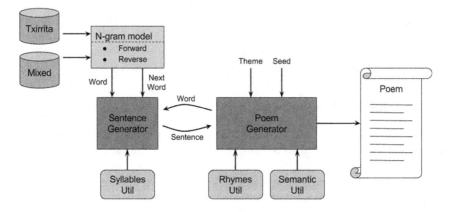

Fig. 1. System architecture.

The aforementioned modules have been integrated into a verse-maker system for automatic poem generation, as shown in Fig. 1.

5   Experimental Setup and Results

In the basic scenario, a topic is given by the user and the proposed method then aims to give as output a novel poem following *bertsolaris'* strategy for poem creation (think up the last line, find rhyming words and generate the poem line by line). The output satisfies the formal constraints and also shows coherent content related to the given topic.

The procedure to create a poem involves prompting the user for a *theme*, a *first word* to start the last line, a *metric* and a *corpus* from which extract the 2-grams that will make the poem. The poem will consist of four lines, each 13 syllables long, and all of them sharing a common rhyme. This is the metric of *Zortziko-txikia*, commonly used in *bertsolaritza* contests.

With these parameters the system follows these steps:

– The last line of the poem is generated using the forward-looking chain created for *corpus*, taking as starting point the *first word* given by the user and then chaining 2-grams until the *metric* is exactly fulfilled. A number of candidates (60) are generated in this way.

- The candidate lines are then ordered according to their semantic similarity with respect to the *theme*.
- Starting from the most similar candidate, the system tries to complete a poem. The feasibility of this task is dependent on the number of possible rhyming words existing in the corpus as well as in the existence of a chain of 2-grams finishing in that rhyme with the given *metric*.
- To create the remaining lines, the possible rhyming words are ordered according to their semantic similarity with respect to the *theme*.
- The N-gram generation system generates poem lines in the following manner: at each step the sequence of words are extended with new words that have a non-zero probability of appearing after the last word according to the N-gram model. At each step, word choice is made randomly.
- The first five poems built in this way are returned by the system.

By means of this procedure, the poems with the last line and rhymes more similar to the theme are returned, given the constraints faced by the system. In less than a minute the created poems are obtained, making it possible to use this approach in a live event of improvised poetry.

In order to test the capabilities of the above described approach, an experiment with two different corpora has been carried out. We have tested 2-gram and 3-gram models and the preliminary results have shown that the 3-gram model tended to replicate the corpus almost verbatim. Therefore we have used 2-grams, because we wanted our system to produce tentative solutions somewhat different to the original lines, and this is more likely with low order N-grams. Table 1 shows the number of total 2-grams for the two corpora along with the number of them appearing 1, 2, 3 and 4 times is shown. As the corpora are not big, no minimum threshold has been imposed over the frequency of 2-gram terms.

In Table 2 we show two poems composed by the automated system, in its original form and in an approximate English translation.

Objective evaluation of poetry is difficult, if not impossible, to assess in an automatic way, because as the saying goes, beauty is in the eye of the beholder. As Gervás [11] and Cardoso [5] stated, human evaluators are needed to assess the degree of creativity of a computational creation. We have presented the generated poems to four people familiar with *bertsolaritza*, explicitly telling them that

Table 1. 2-gram frequency table. The number of 2-grams appearing with a frequency of 1, 2, 3 and 4 is shown.

Corpus	Txirrita	Mixed
#2-grams	21665	85720
Unique	19500	81321
2-times	1320	3144
3-times	402	603
4-times	176	242

Table 2. Two poems created by the system. On the left with the theme 'man' and from the Txirrita corpus, and on the right with the theme 'war' and from Mixed corpus.

Basque	Nun dezu bada nere baserritar ona zabaldu eta orain zeruan dagona biartzen bada kotxe bakoitzak komona bertso birekin egin nai zuen gizona	Basque	Orain globo bat jarri dezagun betiko nik ahal izan baino lehen itzuliko gizarteak bizkarra eman eta kito eta biolentzia ez da eroriko
English	Where you have so my good farmer that has spread and is now in the sky if each car is forced a service wanted to create a man with two verses	English	Let's have a balloon forever I will be back as soon as possible society turns its back and ready and violence will not decay

such poems were the product of an automated system. Each of them analyzed twenty poems, ten from each corpus. They have been asked to give their overall impression about the quality, similarity with the theme, internal coherence and style. Their impression has been positive, stating that they were well-formed poems, although not of human-produced quality. They found a general sense of semantic relationship with the given theme, even diluted, but they also found that the internal coherence of the whole poem was pretty poor. It was also stated that the poems created from the Txirrita corpus sounded more natural and closer to the style of the *bertsolaritza* than the poems from the Mixed corpus. This corroborates, as some authors suggest, that the N-gram model imposes a certain overall style on the texts that can be produced.

6   Conclusions and Further Work

In this paper we have presented an automated system to generate poetry in Basque language. As a preliminary step, two Markov chains are generated from a given corpus, with their nodes representing the words and the directed edges the existence of a 2-gram with the words in the given order. The proposed approach uses one forward and one backward Markov chain because we construct some lines from the start and other from the end. From a theme, a first word and a metric, poems are generated.

The result of a experiment with two different corpora has been shown, with the system been able to generate five poems in less than a minute, fast enough to be used in a live event. The evaluation of the poems has been made by several people familiar to *bertsolaritza* and their opinions have been reflected. The overall impression has been positive, although the quality is still far from a human composer.

As further work, we will try to speed up the generating process, in order to be able to explore more candidate lines and build higher quality poems. Other bigger or more diverse corpus will be used to generate the Markov chains. The semantic model could be learned from a corpus with a higher percentage of fiction documents, that could convey semantic relationships closer to those in poetry[3].

Acknowledgments. This paper has been supported by the Spanish Ministerio de Economía y Competitividad, contract TIN2015-64395-R (MINECO/FEDER, UE), as well as by the Basque Government, contract IT900-16. The authors gratefully acknowledge Bertsozale Elkartea (Association of the Friends of Bertsolaritza), whose verse corpora has been used to test and develop the proposed method.

References

1. Amuriza, X.: Hiztegi Errimatua. Lanku (1981)
2. Astigarraga, A., Jauregi, E., Lazkano, E., Agirrezabal, M.: Textual coherence in a verse-maker robot. In: Hippe, Z.S., Kulikowski, J.L., Mroczek, T., Wtorek, J. (eds.) Human-Computer Systems Interaction: Backgrounds and Applications 3. AISC, vol. 300, pp. 275–287. Springer, Cham (2014). doi:10.1007/978-3-319-08491-6_23
3. Astigarraga, A., Agirrezabal, M., Lazkano, E., Jauregi, E., Sierra, B.: Bertsobot: the first minstrel robot. In: 2013 The 6th International Conference on Human System Interaction (HSI), pp. 129–136. IEEE (2013)
4. Barbieri, G., Pachet, F., Roy, P., Esposti, M.D.: Markov constraints for generating lyrics with style. In: Proceedings of the 20th European Conference on Artificial Intelligence, pp. 115–120. IOS Press (2012)
5. Cardoso, A., Veale, T., Wiggins, G.A.: Converging on the divergent: the history (and future) of the international joint workshops in computational creativity. AI Mag. **30**(3), 15 (2009)
6. Das, A., Gambäck, B.: Poetic machine: computational creativity for automatic poetry generation in bengali. In: 5th International Conference on Computational Creativity, ICCC (2014)
7. Deerwester, S., Dumais, S.T., Furnas, G.W., Landauer, T.K., Harshman, R.: Indexing by latent semantic analysis. J. Am. Soc. Inf. Sci. **41**(6), 391 (1990)
8. Egaña, A.: The process of creating improvised bertsos. Oral Tradit. **22**(2), 117–142 (2007)
9. Garzia, J., Sarasua, J., Egaña, A.: The art of bertsolaritza: improvised Basque verse singing. Bertsolari liburuak (2001)
10. Gervás, P.: Computational modelling of poetry generation. In: Artificial Intelligence and Poetry Symposium, AISB Convention 2013. The Society for the Study of Artificial Intelligence and the Simulation of Behaviour, University of Exeter, United Kingdom (2013)
11. Gervás, P.: Deconstructing computer poets: making selected processes available as services. Comput. Intell. **33**(1), 3–31 (2015)
12. Gervás, P.: Constrained creation of poetic forms during theme-driven exploration of a domain defined by an N-gram model. Connection Sci. **28**(2), 111–130 (2016)
13. Jauregi, O.: Euskal testuetako silaba egituren maiztasuna diakronikoki. Anuario del Seminario de Filología Vasca "Julio de Urquijo" **37**(1), 393–410 (2013)
14. Jurafsky, D., James, H.: Speech and Language Processing: An Introduction to Natural Language Processing, Computational Linguistics, and Speech Recognition. Prentice Hall, Upper Saddle River (2000)

---

[3] http://www.bertsozale.com/en.

15. Lamb, C., Brown, D.G., Clarke, C.L.: A taxonomy of generative poetry techniques. In: Bridges Finland Conference Proceedings, pp. 195–202 (2016)

16. Langkilde, I., Knight, K.: Generation that exploits corpus-based statistical knowledge. In: Proceedings of the 36th Annual Meeting of the Association for Computational Linguistics and 17th International Conference on Computational Linguistics, vol. 1, pp. 704–710. Association for Computational Linguistics (1998)

17. Lord, A.B., Mitchell, S.A., Nagy, G.: The Singer of Tales, vol. 24. Harvard University Press, Cambridge (2000)

18. Manning, C.D., Schütze, H., et al.: Foundations of Statistical Natural Language Processing, vol. 999. MIT Press, Cambridge (1999)

19. Gonçalo Oliveira, H., Cardoso, A.: Poetry generation with PoeTryMe. In: Besold, T.R., Schorlemmer, M., Smaill, A. (eds.) Computational Creativity Research: Towards Creative Machines. ATM, vol. 7, pp. 243–266. Atlantis Press, Paris (2015). doi:10.2991/978-94-6239-085-0_12

20. Oulipo, A.: Atlas de litterature potentielle. Gallimard, Collection Idees (1981)

21. Queneau, R.: 100.000.000.000.000 de poemes. Gallimard Series, Schoenhofs Foreign Books (1961)

22. Toivanen, J., Toivonen, H., Valitutti, A., Gross, O., et al.: Corpus-based generation of content and form in poetry. In: Proceedings of the Third International Conference on Computational Creativity, Dublin, Ireland, pp. 175–179 (2012)

23. Toivanen, J.M., Järvisalo, M., Toivonen, H., et al.: Harnessing constraint programming for poetry composition. In: Proceedings of the Fourth International Conference on Computational Creativity, Sydney, Australia, pp. 160–167 (2013)

24. Zelaia, A., Arregi, O., Sierra, B.: Combining singular value decomposition and a multi-classifier: A new approach to support coreference resolution. Eng. Appl. AI **46**, 279–286 (2015)

Neural Machine Translation for Morphologically Rich Languages with Improved Sub-word Units and Synthetic Data

Mārcis Pinnis$^{(\boxtimes)}$, Rihards Krišlauks, Daiga Deksne, and Toms Miks

Tilde, Vienibas gatve 75A, Riga, Latvia
{marcis.pinnis,rihards.krislauks}@tilde.lv

Abstract. This paper analyses issues of rare and unknown word splitting with byte pair encoding for neural machine translation and proposes two methods that allow improving the quality of word splitting. The first method linguistically guides byte pair encoding and the second method limits splitting of unknown words. We also evaluate corpus re-translation for a new language pair – English-Latvian. We show a significant improvement in translation quality over baseline systems in all reported experiments. We envision that the proposed methods will allow improving the translation of named entities and technical texts in production systems that often receive data not represented in the training corpus.

Keywords: Neural machine translation · Morphologically rich languages · Sub-word units

1 Introduction

In 2016 neural machine translation (NMT) systems (e.g., [1,4,8,12], etc.) showed to achieve state of the art for multiple language pairs including English-German, English-Czech, and English-Romanian [2,16] continuing a recent paradigm shift from phrase-based machine translation (SMT) technologies (e.g., [9,10]). Recent research in NMT involves analysis of factored input support [15], character level NMT (in order to remove the necessity of data pre- and post-processing; e.g., [11]), methods for improved attention mechanisms [13], multi-language NMT [6], and multi-modal NMT [3].

For languages with a rich morphology and limited resources, data sparseness is a notable issue, which arises due to a greater number of word surface forms compared to morphologically simpler languages. The use of sub-word units as introduced by Sennrich et al. [17] greatly alleviates the problem gaining much of the benefits of character-level NMT systems while retaining the speed and robustness of word-level systems. The byte-pair encoding (BPE) algorithm's word splitting model used in sub-word NMT, however, acts solely on the character statistics found in the training corpus. The word parts produced by the model do not necessarily align with the morphological structure of the words. We feel

© Springer International Publishing AG 2017
K. Ekštein and V. Matoušek (Eds.): TSD 2017, LNAI 10415, pp. 237–245, 2017.
DOI: 10.1007/978-3-319-64206-2_27

that a more accurate word splitting strategy could provide further improvements in NMT system training for highly inflected languages such as Latvian.

Although BPE ensures that words composed of characters belonging to the training set will be covered by the system's vocabulary, there might be situations in which out-of-vocabulary word parts are encountered. If after training we translate a sentence containing, e.g., foreign characters then these will be replaced with the $<UNK>$ placeholder. Since the NMT system will have had little to no experience in dealing with such tokens during training, they might get mistranslated. To solve this issue, we propose to mark the whole word as $<UNK>$ instead of individual characters and to synthesize additional training data to artificially increase the number of $<UNK>$ tokens by randomly replacing known words.

Another avenue we explore is synthetic data augmentation. Sennrich et al. [16] showed that monolingual data in the target language can be re-translated to increase the amount of parallel data. We re-produce the experiment for a new language pair – English-Latvian.

The paper is further structured as follows: Sect. 2 describes the data and experiment set-up, Sect. 3 describes the experiments on morphology-driven word splitting (MWS), Sect. 4 describes the experiments on improved handling of unknown words, in Sect. 5 we describe the experiments with monolingual data re-translation, Sect. 6 discusses evaluation results, and finally in Sect. 7 we conclude the paper.

2   Experiment Set-Up

For NMT system training we use the Nematus toolkit[1] which implements an attention-based encoder-decoder model with gated recurrent units. We set the hyper-parameters to the values used by Sennrich et al. [16] in their WMT 2016 submission[2], except for the vocabulary size which was set to 100,000.

To train the systems we used a large English-Latvian parallel corpus covering texts from a broad domain, e.g., legal texts, news, information technology, medicine, mechanical engineering, tourism and other sources.

We pre-processed the training data using the standard data pre-processing workflow of the Tilde MT [20] platform. At first, the data were cleaned (e.g., by normalising punctuation, whitespace, removing control symbols, decoding XML and HTML entities, etc.) and filtered (e.g., by deleting duplicates, sentences with word count differences and alphanumeric symbol and other symbol ratio differences higher than a threshold). Then non-translatable tokens were identified (e.g., e-mail addresses, file addresses, codes, etc.) and replaced with specific placeholders (e.g., $\beta EMAIL\beta$, $\beta URL\beta$, etc.) to allow creating more generalised models (and reduce data sparsity). Finally, the data were tokenised and truecased.

---

[1] https://github.com/rsennrich/nematus.

[2] https://github.com/rsennrich/wmt16-scripts/blob/master/sample/config.py.

Two different baseline NMT systems were created – one for the MWS experiments, and the other for the synthetic data experiments. The systems differ only in the data-sets used – for the MWS experiments, the data were further filtered so that no single source sentence exceeds 50 word parts as produced by the BPE algorithm for the baseline NMT system, or 75 word parts as produced by the MWS algorithm (further described in Sect. 3). The number 75 was selected, because the average length of sentences in English increased from 20 to 30 parts.

The final sentence counts of the data sets are as follows: 7,300,666 for the synthetic data experiment baseline, 13,886,764 for the unknown word experiment, 14,601,332 and 21,901,998 for the re-translated data experiments, and 6,684,461 for the MWS experiments.

3 Morphology-Driven Word Splitting

As explained above, BPE allows creating open vocabulary NMT systems. Because BPE is a language agnostic process, it does not take the morphological structure of words into account. Therefore, BPE can split different surface forms of a single word inconsistently. Table 1 shows an example where BPE inconsistently separated the root from the different affixes for English and Latvian words. In the examples, "@@" identifies the position where the words are split using the two different word splitting methods.

Table 1. Differences between BPE and MWS

Word	BPE	MWS
English		
legalization	leg@@ alization	legal@@ ization
legalize	leg@@ alize	legal@@ ize
legalized	leg@@ alized	legal@@ iz@@ ed
legalizes	legaliz@@ es	legal@@ izes
legalizing	legaliz@@ ing	legal@@ iz@@ ing
Latvian ("atbalss" is translated as "echo")		
atbalss	at@@ balss	atbals@@ s
atbalsis	atbal@@ sis	atbals@@ is
atbalsi	atbal@@ si	atbals@@ i
atbalsīs	at@@ bals@@ īs	atbals@@ īs
atbalsīm	at@@ balsīm	atbals@@ īm

To address the inconsistent BPE behaviour, we propose to split words using a morphological analyser prior to BPE. For each word, we perform: (1) compound splitting, (2) separation of common prefixes, e.g., for English we separate *above-*, *some-*, *every-*, *any-*, *bio-*, *post-*, *vice-*, *multi-*, etc., and for Latvian we separate pie-, uz-, ne-, pa-, pār-, vis-, etc., 3) separation of suffixes. For Latvian, we separate the ending for non-reflexive nouns and adjectives and the complete suffix after the root for other words. For English, we separate plural endings and common suffixes, e.g., -logy, -logic, -ly, -ist, -less, -dom, -ment, -ful, -down, -al, -ality, -ship, -ness, -able, -out, -over, -around, -ing, -ism.

After applying the morphologically aware splitting, the data were processed with BPE, thereby ensuring support for an open vocabulary. The third column in Table 1 shows that the method increases consistency in splitting of words in different surface forms and words that share the same roots.

4   Unknown Words as Placeholders

Although BPE allows handling unknown words, it does not completely eliminate the unknown word problem. Words (e.g., person or organisation names) or character sequences (e.g., unicode emoticons), which are written using symbols that appear neither in the source, nor target sides of the training data, will be treated as unknown words. Also, because the BPE model is shared between source and target languages, but the dictionaries are kept separate, the BPE model can potentially create parts for source language words that are unknown in the source language. Finally, if the size of the source language's vocabulary is the same as the unique count of BPE parts in the training corpus, the training data will not have any unknown word parts present for the model to learn how to handle them and the surrounding context around the unknown word parts. This means that the translation of sentences that have the above-described phenomena can be unpredictable (a common consequence of this issue is mistranslation of the whole sentence).

To address the unknown word problem, we propose to:

1. Modify the splitting algorithm such that if a word contains unknown parts, it should be kept together as a single unknown word. Note that words consisting of unknown word parts will most likely be named entities written in original alphabets, Unicode symbol sequences not present in the training data or phrases written in the target language, therefore, we believe that keeping such words together is natural.
2. Treat consecutive unknown words as a single unknown word. This processing step will help, for instance, to translate multi-word named entities as single units.
3. Supplement the training data with sentence pairs with artificially introduced unknown words. The intuition of using a synthetic unknown word corpus (and the unknown word handling method) is to make the system robust when encountering unknown symbols or word parts and to make the system learn how to translate the context around the unknown word tokens.

To create the synthetic corpus, first, we performed word alignment of the training data using *fast-align* [5]. Then we identified unambiguous (one-to-one) word alignments in each segment and randomly selected one to three content words that have unambiguous alignments to be replaced with the $<UNK>$ tokens. As content words we consider tokens that (1) consist of only letter sequences, and (2) are not to be found in function word lists from the English and Latvian morphological analyser databases. An alignment example showing potential content words (underlined) for replacement is given in Fig. 1. The process produced additional 6,586,098 sentence pairs that were added to the training data.

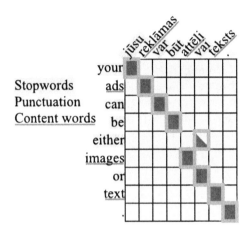

Fig. 1. Alignment example of potential content words for substitution as $<UNK>$ placeholders (image created using WAV [7])

Table 2. Example of translations with and without unknown word merging

Source sentence	Māris Kučinskis is the Prime Minister of Latvia
After BPE	Mār@@ is Ku@@ č@@ in@@ skis is the Prime Minister of Latvia
Unknown words (no UNK merging)	UNK is Ku@@ UNK in@@ skis is the Prime Minister of Latvia
Unknown words (UNK merging)	UNK is the Prime Minister of Latvia
Baseline transl.	Due ir Kuveita, kas ir Latvijas Ministru prezidents
Improved transl.	Māris Kučinskis ir Latvijas Ministru prezidents

When the data were prepared, we trained a new NMT system using the new training data and the same training set-up as for the baseline system. Table 2 shows an example sentence translated with the baseline system (without any explicit unknown word support) and the improved system (with unknown word support). The example shows that the improved system is able to transfer the unknown word sequence to the target side, whereas the baseline system performs unreliably by producing a wrong translation. As further supported by the evaluation results, our method ensures that unknown words are handled reliably in different contexts, thereby making the model more robust.

5 Supplementing Training Data with Re-translated Monolingual Data

Following the successful experiments of Sennrich et al. [16], we trained a Latvian-English NMT system and re-translated 7.3 and 14.6 million randomly selected sentences from a large monolingual corpus. This amounts for 100% and 200% of the size of the parallel training data.

Then, we trained two new English-Latvian NMT systems using the two new synthetic data sets, and compared them to the baseline NMT system trained on the initial parallel data set. These experiments validate the findings of Sennrich et al. [16] and show (see Sect. 6 for more details) that the additional corpus can increase translation quality. However, we also show that the larger corpus allows achieving a lower quality increase due to the increase of noise in the training data.

6 Evaluation

We evaluated all systems automatically using three evaluation metrics: BLEU [14], CharacTER [21] and BEER 2.0 [19]. For evaluation, we used the ACCURAT balanced evaluation set (512 sentences) [18] and the newsdev2017 data set from the WMT shared task[3] (2000 sentences). In order to better simulate unknown word presence in the evaluation data, we prepared a third evaluation set (1781 sentences) using the method from Sect. 4. Each sentence in this data set contains from one to three unknown word placeholders. This data set will show how reliably the different NMT systems handle unknown words. The results are given in Table 3.

The results for the MWS experiments (first two rows) show that MWS allows improving the overall translation quality. However, the quality increase is relatively low. This can be explained by the fact that the improvements in word splitting affect mostly rarely occurring words, but the out-of-vocabulary (OOV) rate of English words is just 0.2% and 1.5% in the ACCURAT and *newsdev2017* data sets. In the future, we plan to carry out human comparative evaluation on a more diverse data set with a higher OOV rate to see whether the improvements become more evident.

The results for the unknown word handling and re-translated data experiments show better improvements over the baseline system. Also here the limited improvement of better unknown word handling is due to the limited number of unknown words in the evaluation data. However, the results for the synthetic

---

[3] http://statmt.org/wmt17/translation-task.html.

Table 3. Evaluation results (the results are significant with $p = 0.01$†, $p = 0.05$‡, and $p = 0.1'$; B2 stands for BEER 2.0 and CT stands for CharacTER)

Training scenario	ACCURAT			newsdev2017			Synth. newsdev2017		
	BLEU	CT	B2	BLEU	CT	B2	BLEU	CT	B2
Baseline (BPE only)	24.10	0.5746	0.5833	18.43	0.5711	0.5497	-	-	-
Improved (Morph. + BPE)	24.81′	0.5565	0.5891	18.89	0.5816	0.5497	-	-	-
Baseline	23.95	0.5769	0.5835	18.73	0.5843	0.5481	17.61	0.6035	0.5441
Baseline + 14.6 million synthetic segments	25.15‡	0.5553	0.5885	**24.73**†	0.5136	0.5866	**20.85**†	0.5602	0.5658
Baseline + 7.3 million synthetic segments	**25.97**†	0.5485	0.5914	**24.34**†	0.5156	0.5861	**21.76**†	0.5600	0.5702
Baseline + synthetic UNK placeholder segments	24.80′	0.5644	0.5862	18.95	0.5805	0.5483	**21.40**†	0.5793	0.5719

unknown word evaluation set clearly show that the system that is trained on the training data set with artificially introduced unknown words translates sentences that contain unknown words significantly better than the baseline system. It is also evident that there is a decrease in translation quality for the systems trained on the re-translated training data due to the lack of unknown words in the training data of these systems.

The unknown word handling method allows to train NMT systems that are more stable when translating unknown words. The results also show that there is a great potential for combining all three methods to achieve even better results. However, this is an area for future experiments.

7   Conclusion

In the paper, we analysed issues of rare word and unknown word splitting with BPE for NMT and we proposed two methods that allow improving the word splitting quality. The first method, MWS, linguistically guides the BPE algorithm by pre-processing the data with a morphological analyser. The second method limits splitting of unknown words in BPE and supplements the parallel training data with a synthetic corpus. Both methods show promising evaluation results. For all evaluation metrics, we observe improvements over the baseline systems. However, we acknowledge that further investigation and evaluation is necessary to analyse the benefits of MWS over BPE due to the limited number of rare and unknown words in the evaluation sets.

It is evident from the evaluation results that the second method allows to train NMT systems that are significantly more stable than the baseline system when translating sentences with unknown words. We envision that the unknown word handling method with the synthetic corpus will allow improving the translation of named entities and technical texts in production systems that often receive data not represented in the training corpus of NMT systems.

We also analysed the effect on translation quality when training NMT systems with training data supplemented with a significant amount of re-translated segments. We showed that a larger re-translated corpus does not allow achieving the highest translation quality due to the increased noise. However, in future experiments we plan to perform more fine-grained analysis to identify the threshold of noise where the quality starts to be affected negatively.

Acknowledgments. The research has been supported by the European Regional Development Fund within the research project "Neural Network Modelling for Inflected Natural Languages" No. 1.1.1.1/16/A/215.

References

1. Bahdanau, D., Cho, K., Bengio, Y.: Neural machine translation by jointly learning to align and translate. In: Proceedings of the International Conference on Learning Representations (ICLR) (2015)
2. Bojar, O., Chatterjee, R., Federmann, C., Graham, Y., Haddow, B., Huck, M., Jimeno Yepes, A., Koehn, P., Logacheva, V., Monz, C., Negri, M., Neveol, A., Neves, M., Popel, M., Post, M., Rubino, R., Scarton, C., Specia, L., Turchi, M., Verspoor, K., Zampieri, M.: Findings of the 2016 conference on machine translation. In: Proceedings of the First Conference on Machine Translation (WMT 2016), vol. 2, pp. 131–198 (2016). Shared Task Papers
3. Caglayan, O., Barrault, L., Bougares, F.: Multimodal Attention for Neural Machine Translation (2016). http://arxiv.org/abs/1609.03976
4. Devlin, J., Zbib, R., Huang, Z., Lamar, T., Schwartz, R.M., Makhoul, J.: Fast and robust neural network joint models for statistical machine translation. In: ACL (1), pp. 1370–1380. Citeseer (2014)
5. Dyer, C., Chahuneau, V., Smith, N.A.: A Simple, fast, and effective reparameterization of IBM model 2. In: Proceedings of NAACL HLT 2013, pp. 644–648, Atlanta, June 2013
6. Firat, O., Cho, K., Bengio, Y.: Multi-way, multilingual neural machine translation with a shared attention mechanism. In: NAACL-HLT 2016, pp. 866–875 (2016)
7. Girgždis, V., Kāle, M., Vaicekauskis, M., Zariņa, I., Skadiņa, I.: Tracing mistakes and finding gaps in automatic word alignments for Latvian-English translation. In: Proceedings of Baltic HLT 2014, pp. 87–94. IOS Press (2014)
8. Jean, S., Firat, O., Cho, K., Memisevic, R., Bengio, Y.: Montreal neural machine translation systems for WMT15. In: Proceedings of WMT 2015, pp. 134–140 (2015)
9. Koehn, P., Hoang, H., Birch, A., Callison-Burch, C., Federico, M., Bertoldi, N., Cowan, B., Shen, W., Moran, C., Zens, R., Dyer, C., Bojar, O., Constantin, A., Herbst, E.: Moses: open source toolkit for statistical machine translation. In: Proceedings of the 45th Annual Meeting of the ACL on Interactive Poster and Demonstration Sessions, ACL 2007, pp. 177–180. Association for Computational Linguistics, Stroudsburg (2007)
10. Koehn, P., Och, F.J., Marcu, D.: Statistical phrase-based translation. In: Proceedings of NAACL HLT 2013, pp. 48–54. Association for Computational Linguistics (2003)
11. Lee, J., Cho, K., Hofmann, T.: Fully Character-Level Neural Machine Translation without Explicit Segmentation (2016)

12. Luong, M.T., Pham, H., Manning, C.D.: Effective approaches to attention-based neural machine translation. In: Proceedings of EMNLP 2015, pp. 1412–1421. Association for Computational Linguistics, Lisbon (2015)
13. Meng, F., Lu, Z., Li, H., Liu, Q.: Interactive attention for neural machine translation. In: Proceedings of the 26th International Conference on Computational Linguistics, Osaka, Japan, pp. 2174–2185 (2016)
14. Papineni, K., Roukos, S., Ward, T., Zhu, W.J.: BLEU: a method for automatic evaluation of machine translation. In: Proceedings of the 40th Annual Meeting on Association for Computational Linguistics, pp. 311–318. Association for Computational Linguistics (2002)
15. Sennrich, R., Haddow, B.: Linguistic input features improve neural machine translation. In: Proceedings of the First Conference on Machine Translation (WMT 2016), vol. 1, pp. 83–91 (2016). Research Papers
16. Sennrich, R., Haddow, B., Birch, A.: Edinburgh neural machine translation systems for WMT 16. In: Proceedings of the First Conference on Machine Translation (WMT 2016), vol. 2 (2016). Shared Task Papers
17. Sennrich, R., Haddow, B., Birch, A.: Neural machine translation of rare words with subword units. In: Proceedings of the 54th Annual Meeting of the Association for Computational Linguistics (ACL 2016), Berlin, Germany (2016)
18. Skadiņš, R., Goba, K., Šics, V.: Improving SMT for baltic languages with factored models. In: Proceedings of the Fourth International Conference on Human Language Technologies: The Baltic Perspective, Baltic HLT 2010, vol. 219, pp. 125–132. IOS Press (2010)
19. Stanojevic, M., Sima'an, K.: BEER: BEtter evaluation as ranking. In: Proceedings of the Ninth Workshop on Statistical Machine Translation, pp. 414–419 (2014)
20. Vasiļjevs, A., Skadiņš, R., Tiedemann, J.: LetsMT!: a cloud-based platform for do-it-yourself machine translation. In: Proceedings of the ACL 2012 System Demonstrations, pp. 43–48. Association for Computational Linguistics, Jeju Island (2012)
21. Wang, W., Peter, J.T., Rosendahl, H., Ney, H.: CharacTER: translation edit rate on character level. In: Proceedings of the First Conference on Machine Translation (WMT 2016), Berlin, Germany, vol. 2, pp. 505–510 (2016). Shared Task Papers

Evaluation of Dictionary Creating Methods for Under-Resourced Languages

Eszter Simon[✉] and Iván Mittelholcz

Research Institute for Linguistics, Hungarian Academy of Sciences,
Benczúr u. 33, Budapest 1068, Hungary
{simon.eszter,mittelholcz.ivan}@nytud.mta.hu

Abstract. In this paper, we present several bilingual dictionary building methods applied for Northern Saami–{English, Finnish, Hungarian,
Russian} language pairs. Since Northern Saami is an under-resourced
language and standard dictionary building methods require a large
amount of pre-processed data, we had to find alternative methods. In
a thorough evaluation, we compared the results for each method, which
proved our expectations that the precision of standard lexicon building
methods is quite low. The most precise method is utilizing Wikipedia title
pairs extracted via inter-language links, but Wiktionary-based methods
also provided useful result.

Keywords: Bilingual dictionaries · Evaluation · Under-resourced languages · Dictionary building methods

1 Introduction

Bilingual dictionaries play a critical role not only in machine translation [5] and
cross-language information retrieval [8], but also in other NLP applications such
as computational semantics and several tasks requiring reliable lexical semantic
information [16]. Since manual dictionary building is time-consuming and takes
a significant amount of skilled work, it is not affordable in the case of lesser
used languages. However, completely automatic generation of clean bilingual
resources is not possible according to the state of the art, but it is possible to
create certain lexical resources, termed proto-dictionaries, that can support lexicographic and NLP work. Proto-dictionaries contain candidate translation pairs
produced by bilingual dictionary building methods. Depending on the method
used, they either comprise more incorrect translation candidates and provide
greater coverage, or provide precise word pairs at the expense of some decrease
in recall; their right size depends on the specific needs.

The standard dictionary building methods are based on parallel corpora.
However, such corpora are still available only for the best-resourced language
pairs – this is the reason of the increased interest in compiling comparable corpora. The standard approach of bilingual lexicon extraction from comparable
corpora is based on context similarity methods (e.g. [7,11]). Recently, source

© Springer International Publishing AG 2017
K. Ekštein and V. Matoušek (Eds.): TSD 2017, LNAI 10415, pp. 246–254, 2017.
DOI: 10.1007/978-3-319-64206-2_28

and target vectors are learned as word embeddings in neural networks based on gigaword corpora (e.g. [15]). These methods need a large amount of (pre-processed) data and a seed lexicon which is then used to acquire additional translations of the context words. One of the shortcomings of this approach is that it is sensitive to the choice of parameters such as the size of the context, the size of the corpus, the size of the seed lexicon, and the choice of the association and similarity measures.

The research demonstrated in this paper is part of a project whose general objective is to provide linguistically based support for several small Finno-Ugric (FU) digital communities in generating online content and help revitalize the digital functions of some endangered FU languages. The practical objective of the project is to create bilingual dictionaries for six small FU languages (Udmurt, Komi-Permyak, Komi-Zyrian, Hill Mari, Meadow Mari and Northern Sami) paired with four major languages which are important for these small communities (English, Finnish, Hungarian, Russian).

The status of each language of the world is usually described using the Expanded Graded Intergenerational Disruption Scale (EGIDS) [9], which gives an estimate of the overall development versus endangerment of the language. In this scale – quite counterintuitively – the highest level is 0, where languages are world-wide used *koinés*, while languages on level 10 are already extinct. Northern Saami is on the highest level among the aforementioned FU languages: its level is 2 (provincial), thus it is used in education, work, mass media, and government within some officially bilingual region of Norway, Sweden and Finland. In the case of the Meadow Mari language, the EGIDS level is 4 (educational), which means that it is in vigorous use, with standardization and literature being sustained through a widespread system of institutionally supported education. The EGIDS level of the other FU languages (Komi-Zyrian, Komi-Permyak, Hill Mari, Udmurt) is 5, i.e. they are developing, which means that there is literature which is available in a standardized form, though it is not yet widespread or sustainable.

Consequently, all these languages are under-resourced, therefore we could not collect enough data for building parallel and comparable corpora. Even if we found some text material in these languages, we could not automatically pre-process them, since – with only rare exceptions – standard text processing tools for these languages are lacking. For these reasons, the aforementioned standard dictionary building methods cannot be used for these languages. Therefore, conducting experiments with alternative methods was needed. We made experiments with several lexicon building methods utilizing crowd-sourced language resources, such as Wikipedia and Wiktionary [3,12].

Having the proto-dictionaries, they were merged for each language pair, and repeated lines were filtered out. These files were then the object of manual validation by native speakers and experts of the languages. In the last phase of the project, we will deploy the enriched lexical material on the web in the framework of the collaborative dictionary project Wiktionary.

The rest of the article is as follows. In Sect. 2, the methods used for creating the proto-dictionaries are presented. We conducted thorough evaluation of the resulted dictionaries for language pairs where the source language is Northern Saami and the target language is English, Finnish, Hungarian or Russian. In Sect. 3, the results of the evaluation is presented: Sect. 3.1 contains the description of the process of the manual validation of the merged proto-dictionaries, while in Sect. 3.2, we detail the performance of each dictionary creating method applied here. The article ends with some conclusions and future directions in Sect. 4.

2   Creating the Proto-dictionaries

2.1   Wikipedia Title Pairs

Wikipedia is not only the largest publicly available database of comparable documents, but it also can be used for bilingual lexicon extraction in several ways. For example, Erdmann et al. [6] used pairs of article titles for creating bilingual dictionaries, which were later expanded with translation pairs extracted from the article texts. Mohammadi and Ghasem-Aghaee [10] extracted parallel sentences from the English and Persian Wikipedia using a bilingual dictionary generated from Wikipedia titles as a seed lexicon. We followed the approach which is common in both articles, thus we created bilingual dictionaries from Wikipedia title pairs using the interwiki links, which resulted in a few hundred candidates for each language pair.

Opinions differ in the literature on how the set of the resulting title pairs is viewed. Some (such as [6]) consider it as a dictionary on its own with a significant amount of multi-word expressions, while others (such as [4]) regard as a parallel corpus and proceed with further steps to extract word translations using methods based on word co-occurrences. In our work, entries where both the source and target language words are one-word units are considered as entries of a bilingual dictionary. The remaining pairs were handled as a parallel corpus, and additional word translations were extracted from it using a procedure based on word co-occurrences (for details, see [3]), but the proto-dictionaries coming from this method are not part of the evaluation presented in this paper.

2.2   Wiktionary-Based Methods

Besides Wikipedia, Wiktionary is also considered as a crowd-sourced language resource which can serve as a source of bilingual dictionary extraction. Although Wiktionary is primarily for human audience, the extraction of underlying data can be automated to a certain degree. Ács et al. [2] extracted translations from the so-called translation tables. Since their tool Wikt2dict is freely available[1], we could apply it for our language pairs. We parsed the English, Finnish, Russian and Hungarian editions of Wiktionary looking for translations in the small FU

---

[1] https://github.com/juditacs/wikt2dict.

languages we deal with. With this method, we gathered several translation candidates for almost all language pairs.

Ács [1] expanded the collection of translation pairs, discovering previously non-existent links between translations with a triangulation method. It is based on the assumption that two expressions are likely to be translations, if they are translations of the same word in a third language. With the triangulation mode of Wikt2dict, we could create proto-dictionaries with a few hundred candidates for each language pair.

3   Evaluation

The proto-dictionaries for each language pair were merged, and repeated lines were filtered out. These merged files were then manually validated by a linguist expert of Northern Saami. The instructions for the validator were as follows. The source and the target word must be a valid word in the language concerned, they must be dictionary forms, and they must be translations of each other. If the source word is not a valid Northern Saami word, the word pair is treated as wrong. If the source word is a valid word but not a dictionary form, the correct dictionary form should be manually added. If the target word is a good translation of the source word but is not a dictionary form, similarly to the former case, the correct dictionary form should be added. If the target word is not a good translation, a new translation should be given.

The following categories come from these instructions:

- ok-ok: The source and the target word are valid words, they are dictionary forms, and they are translations of each other.
- ok-nd: The source and the target word are valid words, they are translations of each other, but the target word is not a dictionary form.
- nd-ok: The source and the target word are valid words, they are translations of each other, but the source word is not a dictionary form.
- nd-nd: The source and the target word are valid words, they are translations of each other, but none of them are dictionary forms.
- ok-wr: The source word is a valid word, it is a dictionary form, but the target word is not a valid word or it is not a correct translation of the source word.
- nd-wr: The source word is a valid word but not a dictionary form, and the target word is not a valid word or it is not a correct translation of the source word.
- wr-xx: The source word is not a valid word.

3.1   Evaluation of the Merged Dictionaries

We made experiments with several lexicon building methods, as detailed above. Applying each method resulted in bilingual resources containing translation candidates for all language pairs. These dictionary files will then be used as the starting point to create the final dictionaries.

Besides the aforementioned proto-dictionaries, the large merged file also contains a proto-dictionary which was not created by us but was downloaded from the Opus corpus [13]. For the Northern Saami–{English, Finnish, Hungarian} language pairs, there are available dictionaries which are lists of "reliable" alphabetic token links extracted from the automatic word alignment created with GIZA++ and the Moses toolkit. First, word pairs where the source and target words were character-level equivalents of each other were removed, since they are probably incorrect word pairs and remaining parts after (or in the lack of) boilerplate removal. The remaining part of the dictionary was also merged into the large dictionary, serving as an interesting example of applying standard lexicon extraction tools for an under-resourced language. The text material from which the Opus proto-dictionaries come is a parallel corpus of KDE4 localization files, where the Northern Saami–English parallel data contain 0.9M tokens, the Northern Saami–Finnish data contain 0.6M tokens, and the Northern Saami–Hungarian data contain 0.8M tokens. At the time of creating the proto-dictionaries, there was no available dic file for Northern Saami–Russian in the Opus corpus.

The large merged dictionaries were evaluated for each category described above; the results can be seen in Table 1. The first impression is that the ok-ok category is much better for sme–rus[2] than for the other language pairs, whose reason is that the sme–rus merged dictionary does not contain translation candidates from the automatically generated Opus dictionary (KDE4). As expected, the standard dictionary creation methods based on parallel texts do not have good performance for under-resourced languages, as pointed out in Sect. 3.2. This is also proved by the fact that the total number of wrong word pairs (ok-wr + nd-wr + wr-xx) is more than 10% lower for sme–rus than for the other language pairs. Similarly, the total number of word pairs from whose words at least one is not a dictionary form (ok-nd + nd-ok + nd-nd) is also significantly lower in the case of sme–rus. It may be because the KDE4 dictionaries were generated from running text containing suffixed word forms as well, while Wikipedia titles and Wiktionary entries usually are lemmas.

As mentioned in Sect. 1, the manually validated word pairs will be used as the source material of newly created Wiktionary entries, which contain several obligatory elements. These elements containing morphological, etymological and lexico-semantic information will be generated as automatically as possible. For instance, in the case of the Northern Saami–English language pair, the title of the entry will be the Northern Saami word, while its English definition will be its English translation equivalent.

For this purpose, we need to extract all useful word pairs from the merged dictionary for each language pair. Table 1 contains the number of all word pairs for each language pair and the ratio of the number of useful word pairs and the number of all word pairs. In this case, useful word pairs comprise all word pairs minus the wr-xx category, since correct dictionary forms and translation

---

[2] We use ISO 639-3 language codes in the article: sme: Northern Saami, eng: English, fin: Finnish, hun: Hungarian, rus: Russian.

equivalents were manually added by the human validator. Repeated lines were filtered out; so that the number of lines in the remaining part is the number of useful word pairs.

Table 1. Results for the merged dictionaries

Lang pair	All (#)	Useful (%)	ok-ok (%)	ok-nd (%)	nd-ok (%)	nd-nd (%)	ok-wr (%)	nd-wr (%)	wr-xx (%)
sme–eng	6,042	92.29	53.26	0.43	9.17	4.10	20.94	4.39	7.71
sme–fin	7,100	91.44	42.28	3.59	6.17	12.48	19.31	7.59	8.56
sme–hun	4,969	90.72	49.57	1.99	6.72	6.36	16.28	9.80	9.28
sme–rus	4,373	95.95	71.74	0.57	3.27	0.14	19.48	0.75	4.05

3.2   Evaluation of the Methods

Category tags given to word pairs in the merged dictionaries were projected onto the corresponding word pairs in the proto-dictionaries. Results for each method were then summed up across all language pairs, as can be seen in Table 2. Abbreviations of the name of the methods are as follows: WikiTitle: Wikipedia title pairs, W2D ext: Wikt2dict extraction mode, W2D tri: Wikt2dict triangulation mode, KDE4: dic files generated from KDE4 parallel files. Besides category tags, the total number of dictionary entries of proto-dictionaries is presented in the first column.

Table 2. Results for the methods

Method	All (#)	ok-ok (%)	ok-nd (%)	nd-ok (%)	nd-nd (%)	ok-wr (%)	nd-wr (%)	wr-xx (%)
WikiTitle	2,989	94.58	0.33	1.20	0.70	1.97	0.33	0.67
W2D ext	921	91.75	0.00	3.69	0.00	3.04	0.33	1.09
W2D tri	11,714	60.94	0.79	4.23	0.20	26.26	1.05	6.49
KDE4	8,401	29.23	3.61	11.25	16.83	13.81	14.13	10.97

Methods are presented in a descending order based on their performance in the ok-ok category. This score is the *precision* of a method, i.e. the ratio of the number of the correct word pairs and the total number of word pairs. Depending on the research purpose, word pairs containing non-dictionary forms can also be treated as correct translations, thus precision metrics may vary among approaches. Here we use it in a strict sense, thus a word pair is correct if it is in the ok-ok category.

Some precision-like metrics are generally used for the evaluation of automatically generated bilingual dictionaries. For example, Vulic et al. [14] use Precision@1 score, which is the percentage of words where the first word from the list of translations is the correct one, and mean reciprocal rank (MRR), where for a source word w, $rank_w$ denotes the rank of its correct translation within the retrieved list of potential translations. All these metrics are based on the assumption that the method used produces a list of translation candidates along with some confidence or probability measures. Even though it is not the case in our work, we can treat figures in the ok-ok column in Table 2 as Precision@1 scores calculated for a one-unit list of translation candidates.

Not surprisingly, using Wikipedia title pairs as a dictionary is proved to be the most precise method. This resource has very valuable translation texts since these translations were manually made by Wikipedia editors. The second most precise method is using Wikt2dict in extraction mode thus extracting translation equivalents from Wiktionary translation tables. Similarly to that of in the case of Wikipedia, word pairs coming from this method are quite reliable, since Wiktionary entries are manually created. The third method is using Wikt2dict in triangulation method, but there is a 30% decrease in the performance of this method compared to that of the first two ones. As this method does not directly use manually created links, its output may contain incorrect translations. The ok-wr figure for this method is the highest, mainly due to polysemy. The worst result was produced by the method used in the Opus corpus, which is a standard dictionary building method based on parallel text material, using standard alignment and word pair extraction tools developed for well-resourced languages.

Figures of the last method are more flat, i.e. word pairs more uniformly spread among the categories compared to the other methods. It may have several reasons. First, the KDE4 dictionaries were generated from running text containing inflected and derived word forms and lemmas as well. Therefore, the number of non-dictionary forms and wrong translations is higher. (Inflected word forms were treated as valid words in non-dictionary form, while derived forms were categorized as wrong by the validator.) Second, the tools used within the Opus corpus project are not really feasible for under-resourced languages therefore produced more non-dictionary forms and wrong word pairs.

If the number of created dictionary entries can be treated as a kind of *coverage*, it can be said that the Wikt2dict triangulation method has the best coverage, since it produced the largest number of translation candidates. As expected, the method with the worst precision has a quite good coverage. Reversing this logic, the method with the best precision should have the worst coverage, but this is not the case. That is a sign of that evaluating the coverage of a dictionary is greatly challenging. We could gather much more word pairs from Wikipedia titles than from Wiktionary translation tables, which is likely due to the fact that Wikipedia contains more articles compared to the number of translations in Wiktionary's translation tables. Moreover, the number of articles and entries highly depends on the activity of editors knowing the Northern Saami language and willing to create new articles and entries. Coverage of a dictionary can also be measured by comparing the number of its entries to that of another – ideally hand-crafted – dictionary, such as in [4]. For this purpose, we plan to use Wiktionary, which is not an expert-built lexicon but manually edited by thousands of contributors.

4 Conclusions and Future Work

We presented several bilingual dictionary building methods applied for the Northern Saami–{English, Finnish, Hungarian, Russian} language pairs. Since Northern Saami is an under-resourced language and standard dictionary building methods require a large amount of pre-processed data, we had to find alternative methods. In a thorough evaluation, we compared the results for each method, which proved our expectations that the precision of standard lexicon building methods is quite low. The most precise method is using Wikipedia title pairs extracted via inter-language links, but Wiktionary-based methods also provided useful result.

Wiktionary is not only used for extracting data from it, but we want to give our results back to the community, thus translation pairs enriched with obligatory pieces of linguistic information will be uploaded as new entries into Wiktionary. Before uploading new entries, it must be checked whether an entry with the same word already exists in Wiktionary. From this, the number of brand new entries created by us can be easily counted, along with a kind of coverage, if we compare the number of the word pairs in the merged dictionaries to the number of the Northern Saami words in the version of Wiktionary in the language concerned. This, however, remains for future work.

Acknowledgements. The research reported in the paper was conducted with the support of the Hungarian Scientific Research Fund (OTKA) grant #107885.

References

1. Ács, J.: Pivot-based multilingual dictionary building using Wiktionary. In: 9th Language Resources and Evaluation Conference. ELRA, Reykjavik (2014)
2. Ács, J., Pajkossy, K., Kornai, A.: Building basic vocabulary across 40 languages. In: 6th Workshop on Building and Using Comparable Corpora, pp. 52–58. ACL, Sofia (2013)
3. Benyeda, I., Koczka, P., Váradi, T.: Creating seed lexicons for under-resourced languages. In: GLOBALEX 2016 workshop. ELRA, Portorož (2016)
4. Bharadwaj, G.R., Tandon, N., Varma, V.: An iterative approach to extract dictionaries from Wikipedia for under-resourced languages. In: 8th International Conference on Natural Language Processing. Macmillan Publishers, India (2010)
5. Brown, R.D.: Automated dictionary extraction for "knowledge-free" example-based translation. In: 7th International Conference on Theoretical and Methodological Issues in Machine Translation, pp. 111–118 (1997)
6. Erdmann, M., Nakayama, K., Hara, T., Nishio, S.: An approach for extracting bilingual terminology from Wikipedia. ACM Trans. Multimed. Comput. Commun. Appl. **5**(4), 1–17 (2009)
7. Fung, P., Yee, L.Y.: An IR approach for translating new words from nonparallel, comparable texts. In: 17th International Conference on Computational Linguistics, pp. 414–420. ACL, Stroudsburg (1998)
8. Grefenstette, G.: The problem of cross-language information retrieval. In: Grefenstette, G. (ed.) Cross-Language Information Retrieval, pp. 1–9. Kluwer Academic Publishers, Boston (1998)

9. Lewis, M.P., Simons, G.F.: Assessing endangerment: expanding Fishman's GIDS. Revue Roumaine de Linguistique **55**(2), 103–120 (2010)
10. Mohammadi, M., Ghasem-Aghaee, N.: Building bilingual parallel corpora based on Wikipedia. In: 2nd International Conference on Computer Engineering and Applications, pp. 264–268 (2010)
11. Rapp, R.: Identifying word translations in non-parallel texts. In: 33rd Annual Meeting of the Association for Computational Linguistics, pp. 320–322. ACL, Stroudsburg (1995)
12. Simon, E., Benyeda, I., Koczka, P., Ludányi, Zs.: Automatic creation of bilingual dictionaries for Finno-Ugric languages. In: 1st International Workshop on Computational Linguistics for Uralic Languages, Tromsø (2015)
13. Tiedemann, J.: News from OPUS - a collection of multilingual parallel corpora with tools and interfaces. In: Nicolov, N., Angelova, G., Mitkov, R. (eds.) Recent Advances in Natural Language Processing V: Selected Papers from RANLP 2007, pp. 237–248. John Benjamins, Borovets (2009)
14. Vulić, I., De Smet, W., Moens, M.F.: Identifying word translations from comparable corpora using latent topic models. In: 49th Annual Meeting of the Association for Computational Linguistics, pp. 479–484. ACL, Stroudsburg (2011)
15. Vulić, I., Moens, M.F.: Bilingual word embeddings from non-parallel document-aligned data applied to bilingual lexicon induction. In: 53rd Annual Meeting of the Association for Computational Linguistics, pp. 719–725. ACL, Stroudsburg (2015)
16. Zesch, T., Müller, C., Gurevych, I.: Extracting lexical semantic knowledge from Wikipedia and Wiktionary. In: 6th Language Resources and Evaluation Conference. ELRA, Marrakech (2008)

Comparative Evaluation and Integration of Collocation Extraction Metrics

Victor Zakharov[(✉)]

Saint-Petersburg State University, Universitetskaya emb., 7-9, 199034 Saint-Petersburg, Russia
v.zakharov@spbu.ru

Abstract. The paper deals with collocation extraction from corpus data. A whole number of formulae have been created to integrate different factors that determine the association between the collocation components. The experiments are described which objective was to study the method of collocation extraction based on the statistical association measures. The work is focused on bigram collocations. The obtained data on the measure precision allow to establish to some degree that some measures are more precise than others. No measure is ideal, which is why various options of their integration are desirable and useful. We propose a number of parameters that allow to rank collocates in an combined list, namely, an average rank, a normalized rank and an optimized rank.

Keywords: Collocation extraction · Association measures · Evaluation · Ranking · Average rank · Normalized rank · Optimized rank

1 Introduction

Let's speak about the notion of collocation. There are different approaches to this term. Sometimes a collocation is meant as a synonym of a word combination, sometimes it is a special type of a set phrase. S. Evert suggests the following definition: "A collocation is a word combination whose semantic and syntactic properties can't be fully predicted on the basis of information about its constituents and which therefore should be added to the dictionary (lexicon)" [1: 17]. But there are many set phrases whose meaning is equal to the sum of the meanings of their constituents, despite the fact that such phrases function as a single unit, with the stability rather than idiomatic nature being the main feature. A threshold of stability should be chosen to range them, above which a word combination can be called a set phrase. This approach assumes a probabilistic nature of collocations. Many modern authors and most of corpus linguists understand collocations as statistically determined set phrases. In this case, not only idioms but also multiword terms, named entities (real-world objects, such as persons, locations, organisations, products, etc.,) and other types of free combinations could be regarded as set phrases.

The above approach is the basic point of our paper which is aimed at evaluation of various statistical methods of automatic collocation extraction.

© Springer International Publishing AG 2017
K. Ekštein and V. Matoušek (Eds.): TSD 2017, LNAI 10415, pp. 255–262, 2017.
DOI: 10.1007/978-3-319-64206-2_29

2 State of the Art

Nowadays, there are several ways to calculate the degree of coherence of parts of a collocation. A whole number of formulae have been created to integrate different factors that determine the association between the collocation components. Usually, such formulae are called association measures. P. Pecina provides 82 measures, and describes their mathematical foundations including their formulae and key references [2: 44–45, 48]. The most popular measures seem to be *MI, t-score,* and *log-likelihood.*

One should not forget also that words which tend to collocate with each other cannot be found in a random order in any case, as there exist grammar rules which imply that "the language system is a probabilistic one and it is a grammatical probability that word frequency shows in a text" [3: 31]. There are methods that take into account the syntactic nature of collocations. B. Daille claims that the linguistic knowledge drastically improves the quality of stochastic systems [4: 192]. One of the methods to take syntax in account are so-called word sketches, which are lists of statistical collocations, each one for each syntactic relation [5]. These syntax-based collocations are described in detail by V. Seretan [6: 59–101]. But in this paper, the grammatical probability is not taken into consideration, only the statistical one.

Lexical association measures being applied to a key word (node) occurrence and context statistics extracted from the corpus for all collocation candidates result in their association scores. But proper formulae are different, which is why collocation ranks obtained by different measures do not coincide. It is known, too, that some measures bring similar results and others are significantly different [7: 246–247].

The research on and evaluation of various association measures has been done for quite a long time and has been quite intensive. It is known that *t-score* extracts most frequent collocations. *Log-likelihood* was eventually preferred for its good behaviour on all corpus sizes and also for promoting less frequent candidates. On the contrary, the *MI* measure allows to reveal low-frequency multiword terms and proper names.

Besides, association score depends on the type of the units (lemmas or word forms) whose statistics are used for the calculations. The analysis described in [8: 340] has shown that in some cases word form collocations overwhelmingly have significantly bigger value.

The very number of the calculated collocates and the association scores are also dependent on the "window" between the node and the collocate that has been chosen for the calculations. When the window size is increased, besides meaningful syntagmas, words from a general lexico-semantic field are found as collocation candidates.

3 Collocation Extraction: An Experiment

The experiments were conducted on the basis of the Araneum corpora of Russian (http://unesco.uniba.sk), with the access provided through the NoSketch Engine [9]. We used 2 corpora, Russicum Minus (120 mln tokens) and Russicum Russicum Maius (1,20 bln). These corpora belong to the family of web corpora being created by the wacky technology [10].

Our objective was to study the method of collocation extraction based on the statistical association measures. We extracted collocations for the word *вода* (water) by means of the tool Collocations of the NoSketch Engine system using 7 association measures: *T-score, MI, MI3, log likelihood (LL), minimum sensitivity (MS), logDice and MI.log_f* [11].

The result for the query *вода* (water) was represented by a list of collocates (collocations) organized for each of the 7 above association measures ranged according to the association score in the form of a table (see an example in Table 1).

Table 1. List of collocates for *вода* (water) (a fragment)

Collocates	Co-occurrence count	Candidate count	MI.log_f score
Сточный (sewer)	12479	13791	100,505
Питьевой (drinkable)	11288	14006	97,878
Грунтовый (ground)	8672	11598	94,132
Кипяченый (boiled)	3635	4502	86,016
Горячий (hot)	20665	102240	84,393
...	...	...	...

A rank has been assigned to every collocate (i.e. collocation) according to the score of the each measure. The number of ranked collocates for each measure was 100.

4   Evaluation of the Effectiveness of Association Measures

Usually, comparison to some "gold standard" or expert evaluation are used to evaluate the results of automated systems. When methods of collocation extraction are evaluated both options appear to be problematic. There is no "gold standard" that would fully or significantly cover the set phrases. We could try to build it *ad hoc* for selected key words based on various dictionaries, but, due to the incomplete nature of dictionaries, the quality would be doubtful. As to expert evaluation, it is very expensive, taking into account time and human resources. Unfortunately, the quality of automated methods is often evaluated based on the examples taken from the top units of ranked lists, and from a small number of the resulting collocates [6: 70].

In this work, we have used expert evaluation on rather big amount of collocations obtained, namely, 100 for each measure. Further, we calculated the number of "true" collocations for each measure individually (Table 2).

The sum in each column can be interpreted as the precision indicator (in percentage) for the upper part of the ranked list.

However, it is not only the number of the true collocations extracted using each measure that is important: the rank of the relevant collocations is significant, too. This is why it would be prudent to introduce a weight of true collocations for each measure taking into account the place of the collocates in a sorted table. In order to evaluate the efficiency of each of the association measures the Kharin-Ashmanov method, which evaluates the relevance of the information retrieval results, was used [12].

Table 2. Distribution of the number of "true" collocations for each measure

Ranks	T-score	MI	MI3	LL	MS	log-Dice	MI.log_f
1–10	0	5	4	2	5	6	8
11–20	4	3	4	4	2	2	3
21–30	2	2	3	3	4	1	4
31–40	1	7	4	3	0	4	6
41–50	1	1	3	2	2	2	2
51–60	2	5	2	4	1	2	2
61–70	0	5	1	4	0	2	1
71–80	3	4	7	0	2	1	3
81–90	1	4	4	2	2	2	1
91–100	3	3	1	2	3	0	1
Total	17	39	33	26	21	22	31

Based on the expert evaluation of the extracted collocates and their place in the ranked list with regard to each association measure, a characteristic set was formed. A characteristic set for each measure means the number of the true collocations from the ranked list (precision value) obtained with this measure. According [12], we select characteristic sets that contain 5 elements – that is the precision values for the first 10, 30, 50, 70 and 100 collocates from the top of the list.

A weight is assigned to each element of the characteristic set (5, 4, 3, 2, and 1, respectively). Each element is "weighed": each of 5 precision values is multiplied by the its weight and divided by 15 (the sum of all weights). The sum of the weighed elements is the resulting precision of the characteristic set, i.e. the precision for appropriate measure.

Here is an example for the *MI* measure that has 5 true collocates in the top ten candidates (precision is 0.5), 10 true collocates $(5 + 3 + 2$, see Table 2) in the top thirty (precision is 0.33), $18(5 + 3 + 2 + 7 + 1)$ in the top fifty (0.36), 28 in the top seventy (0.40), and 39 in the top hundred (0.39). Then, each element was normalized (weighed) and the resulting precision will be equal to $0.5 * 5/15 + 0.33 * 4/15 + 0.36 * 3/15 + 0.4 * 2/15 + 0.39 * 1/15 = 0.167 + 0.088 + 0.072 + 0.053 + 0.026 = 0.406$.

The values of the precision calculated like that for all seven measures are given below (Table 3).

Table 3. Precision values for association measures

	t-score	MI	MI3	LL	MS	log-Dice	MI.log_f
Precision	0.115	**0.406**	0.366	0.262	0.357	**0.391**	**0.562**
Place	7	2	4	6	5	3	1

So, in this case the best measures seem to be *MI.log_f, MI* and *log-Dice*. Of course, this result based on a single keyword is not enough for a safe generalization. Nevertheless, experiments with other keywords mostly confirm above list adding to it *min. sensitivity* (the sequence of measures can differ).

5   Integration of Different Association Measures

The next part of the study is aimed at developing methods for the integrated use of different measures of association. We used 7 collocation lists obtained in the first experiment. The lists of collocates were processed in the following manner. Meaningless collocations with punctuation marks were removed. Due to errors of lemmatization, some collocates were presented in several different word forms. For such cases, non-lemmatized word forms of the same word were united into a single unit, with the highest association value being chosen. "Clean" ranged lists of collocates were obtained as a result. Then, 7 tables (with 100 collocates in each) were merged into a new one in such a way so as to the collocates that were obtained through several measures were merged into a single line of the combined table, with their rank for each measure being provided. When a collocate was not available among the first hundred collocates for some measure it had no rank (see Table 4). The combined table counted 247 collocations. By the way, according to expert evaluation 86 of them were marked as true.

Table 4. Combined table of collocates for *вода (water)* with integrated ranks (a fragment)

Collocates	T-score	MI	MI3	LL	MS	log-Dice	MI.log_f	n	R_{av}	R_{norm}
Сточный (sewer)	5	25	1	2	5	4	1	7	6.14	6.14
Питьевой (drinkable)	7	39	2	4	7	6	2	7	9.57	9.57
Грунтовый (ground)	13	53	4	7	13	10	3	7	14.71	14.71
...	...	...	...	...	...	...	...	...	...	...
Родниковый (spring)	-	78	70	-	-	-	30	3	59.33	103.23
Туалетный (cologne)	73	-	37	45	75	57	31	6	53.00	59.36

It is clear that the same collocations with the word *вода* in the ranked lists of different measures have different rank, i.e. different measures estimate the syntagmatic association strength (collocability) between the components of a collocation in a different way. So, there is an idea that the collocation lists obtained through different measures should be merged. Then a question arises: what is the rank of a certain collocation in such merged list, or, in other words, what unique single rank should be assigned for each collocation.

The following hypotheses were made:

(1)   the more the number of the measures that identified a relevant collocate, the stronger the collocability of a given collocation;

(2)   the less the sum of the ranks or the average rank for a relevant collocate, the stronger the collocability;

(3)   if both above conditions are observed then the "value" of a given collocation is higher, which is why we introduce the notion of a normalized rank.

As a result, the following indicators (parameters) have been added to combined table (Table 4):

(1)   the number of association measures (n) that have "calculated" a given collocate (within first 100 lines);

(2) the *average rank* of the collocate (R_{av}): the sum of all non-empty ranks divided by their number;

(3) the *normalized rank* of the collocate.

The normalized rank (R_{norm}) is calculated as follows:

$$R_{norm} = k \cdot R_{av},$$

where k is the coefficient calculated by the following formula:

$$k = log_2(1 + 7/n),$$

where n is the number of the successful measures for this collocate.

It is safe to say that the average and the normalized ranks "objectify" (integrate) the functionality of various association measures.

6 Optimized Rank

However, ranks are based on a association measure score, which is why it is our task (including within this article) to correlate the ranks, i.e. the association strength, with some truth criterion concerning the effectiveness of appropriate measures.

The data on precision values for association measures (Sect. 4) allows to establish to some degree that such measures as *MI.log_f, log-Dice, MI*, and *MS* are more preferable. Having obtained "objective" evaluation of the efficiency of individual measures, we suggest introducing an indicator that is calculated taking into account the preference of the measures. We will call it the *optimized average rank*.

It is calculated as follows: all products of non-zero ranks multiplied by the coefficient of the measure significance are summed up and are divided into the number of measures used for a given collocate. In our case the measure significance coefficients are set, with their precision taken into account (Table 3): *MI.log_f* – 0.4, MI – 0.5, *logDice* – 0.6, *MI3* – 0.7, *min. sensitivity* – 0.8, *log-likelihood* – 0.9, *T-score* – 1.0. As a result, the rank of the collocations extracted by more efficient measures is reduced, and the relevant collocate in the combined table goes up. See the example in Table 5.

Table 5. Optimized rank for individual collocations

No.	Collocate	Average rank	Optimized rank
1.	Поверхностный (surface)	**81.5**	59.8
2.	Крещенский (baptismal)	82.0	**36.0**
3.	Обычный (usual)	**61.0**	34.1
4.	Газированный (sparkling)	63.0	**27.9**

Let's compare the collocates with even and odd numbers by pairs (*поверхностный vs. крещенский, обычный vs. газированный*). We can see that the latter, having collocations with *water* as the node, still have a bit higher average rank than the former. However, as per our suggestion, following the optimisation, the latter will have a lower rank and go up in the ranked list (once again: the higher the rank in this list the higher the collocability degree). It seems the "odd" collocations really are stronger.

Naturally, it is so far only the idea. For real practice measure significance coefficients have to be chosen more reasonably, on a bigger experimental basis.

7   Conclusion and Further Work

To sum it up, the experiments have produced important results that characterise the efficiency of individual association measures. We also offer a method of assessing the effectiveness of statistical association measures.

Merging several lists of collocates obtained by different measures into one could improve the efficiency of statistical tools in total. We offer several options that allow to assess "the quality" of collocations in the combined list.

It is important to stress that the experiments were conducted using representative corpora, with large amount of the resulting collocations being under study. This was also confirmed in experiments with other words.

The evaluation procedure needs also special attention. Available lexical resources are both impure and incomplete. Therefore, the expert assessment remains one of the main methods but it needs thorough elaborated preprocessing and enrichment with terminological information.

Further research will be as follows:

1. Develop the programming tool that allows to make a single list of collocates with all the necessary parameters and to calculate integrated ranks.
2. Study how the efficiency of the association measures is associated with the width of the window (to the left and to the right of the key word) within which collocates are selected, and estimate the degree of such efficiency.
3. Identify the inter-relation between "syntagmatic" and "paradigmatic" collocates on the one hand and "idiomatic" and "statistical" on the other hand within the same search results, and identify the dependence of such inter-relation on the width of the window.

Acknowledgments. This work was partly supported by the grant of the Russian Foundation for Humanities (research project No. 16-04-12019).

References

1. Evert, S.: The statistics of word cooccurences word pairs and collocations. Ph.D. thesis, Institut für Maschinelle Sprachverarbeitung (IMS), Stuttgart (2004)
2. Pecina, P.: Lexical association measures and collocation extraction. Lang. Resour. Eval. **44**(1–2), 137–158 (2009). Prague
3. Halliday, M.: Current Ideas in Systemic Practice and Theory. Pinter, London (1991)
4. Daille, B.: Mixed approach for the automatic extraction of terminology: lexical statistics and linguistic filters [Approche mixte pour l'extraction automatique de terminologie: statistiques lexicales et filtres linguistiques]. Ph.D. thesis, Université Paris 7 (1994)

5. Kilgarriff, A., Tugwell, D.: Sketching words. In: Correard, M.H. (ed.) Lexicography and Natural Language Processing: A Festschrift in Honour of B.T.S. Atkins, pp. 125–137. Euralex, Goteborg (2002)
6. Seretan, V.: Syntax-Based Collocation Extraction. Text, Speech and Language. Springer, Dordrecht (2011)
7. Křen, M.: Collocation Measures and the Czech Language: Comparison on the Czech National Corpus data [Kolokační míry a čeština: srovnání na datech Českého národního korpusu], pp. 223–248. Kolokace, Praha (2006)
8. Zakharov, V., Khokhlova, M.: Syntagmatic relations in Russian corpora and dictionaries. In: Schoepe, K., et al. (eds.) Pragmantax II. The Present State of Linguistics and its Sub-Disciplines, pp. 333–344. Peter Lang, Frankfurt a.M. (2014)
9. Rychlý, P.: Manatee/Bonito – a modular corpus manager. In: 1st Workshop on Recent Advances in Slavonic Natural Language Processing, pp. 65–70. Masaryk University, Brno (2007)
10. Benko, V.: Aranea: yet another family of (comparable) web corpora. In: Sojka, P., Horák, A., Kopeček, I., Pala, K. (eds.) TSD 2014. LNCS, vol. 8655, pp. 247–256. Springer, Cham (2014). doi:10.1007/978-3-319-10816-2_31
11. Statistics Used in Sketch Engine. https://www.sketchengine.co.uk/documentation/statistics-used-in-sketch-engine/. Accessed 3 Feb 2017
12. Ashmanov, I., Grigoryev, S., Gusev, V., Kharin, N., Shabanov, V.: Using statistical method for intelligent computer-based text processing [Primenenie statisticheskih metodov dlja intellektual'noj komp'juternoj obrabotki tekstov]. In: The Proceedings of the Dialog 1997 International Seminar on Computational Linguistics and Its Applications, pp. 33–37 (1997)

Errors in Inflection in Czech as a Second Language and Their Automatic Classification

Tomáš Jelínek[(✉)]

Faculty of Arts, Institute of Theoretical and Computational Linguistics,
Charles University, Prague, Czech Republic
tomas.jelinek@ff.cuni.cz

Abstract. When analyzing language acquisition of inflective languages like Czech, it is necessary to distinguish between errors in word stems and errors in inflection. We use the data of the learner corpus CzeSL, but we propose a simpler error classification based on levels of language description (orthography, morphonology, morphology, syntax, lexicon), which takes into account the uncertainty about the causes of the error. We present a rule-based automatic annotation tool, which can assist both the task of manual error classification and stochastic automatic error annotation with preliminary results of types of errors related to the language proficiency of the text authors.

Keywords: Morphology · Language acquisition · Automatic error classification

1 Introduction

The analysis of texts of non-native speakers has become an important tool for understanding the process of learning a second language and the development of adequate teaching methodologies. When dealing with a highly inflective language like Czech, it is necessary to distinguish between errors in word stems and errors in inflection. In this article, we use the learner corpus CzeSL ([3,5]), composed of short texts written by students of Czech as a second language with various levels of proficiency. We want to show a concept of error annotation of learner language that takes into account the uncertainty about the causes of observed deviations from the standard language, and propose a method of automatically classifying errors in inflection. We will present some results of automatic error classification using the data of the manually annotated part of the CzeSL corpus.

2 Czech as a Highly Flective Language

In Czech, as an inflective language, the syntactic functions of words are mostly expressed by their form, whereas word order is mostly free. Czech nouns have seven cases, with distinct forms for singular and plural, which means that any

© Springer International Publishing AG 2017
K. Ekštein and V. Matoušek (Eds.): TSD 2017, LNAI 10415, pp. 263–271, 2017.
DOI: 10.1007/978-3-319-64206-2_30

noun may have up to 14 different forms (some forms are identical in every declension paradigm). There are 14 basic paradigms for nominal declension, and a larger number of paradigm subtypes. For example the paradigm *žen|a* (the most frequent for feminine nouns) has 10 forms, e.g. *žen|ě* is the form for dative and locative singular. Similarly, adjectives, pronouns, numerals and verbs have a rich inflection with many paradigms and inflective forms. It is nonetheless not necessary to master the entire Czech inflectional system in order to successfully communicate in Czech. It is enough to know how to use properly most of the Czech cases and the most used verbal forms for the more frequent paradigms. For example, there is an error in the nominal declension in (1):

(1) Oslavil jsem Vánoce se svými příbuznými v jejich domě v *Úvalách.
Oslavil jsem Vánoce se svými příbuznými v jejich domě v Úvalech.
'I celebrated Christmas with my relatives at their home in Úvaly.'

The understanding of the sentence is not disrupted by the error *Úvalách/Úvalech*, an error in the form of the locative plural of the name of a Czech town *Úvaly*: the case ending *-ách* used incorrectly instead of *-ech* is an existing ending, used to express the same morphosyntactic properties (locative plural) of nouns of another paradigm. As the incorrect form denotes the same nominal case of the noun, it may be noticed as unexpected by a native speaker, but it will not hinder the understanding of the whole sentence. Another type of inflection error, found in (2), can be a bit more of a problem for understanding:

(2) V životě dávám přednost *rodinu.
V životě dávám přednost rodině.
In life give precedence family.
'In my life, I prefer family.'

The form *rodinu* 'family' is a form for accusative singular of the noun *rodina*, in a construction where the dative form *rodině* is expected. The sentence is still understandable, as it is composed of only a few words, but the use of incorrect case is more challenging for understanding by a native speaker. When a completely random ending is used unrelated to paradigm or case form, the understanding is even more disrupted.

3   Classification of Errors in Inflection

We use the term error in the analysis of learner language to designate a word form which differs from the form that would be used by a native speaker in the same context. The corpus CzeSL-man offers manual emendations of word forms. The aim of the error annotation is to specify the reason for these deviations from the norm. As we do not know the learner and his background, we can only base the error annotation on the sentence context. Sometimes, many interpretations of the error are possible, as in sentence (3):

(3) Tam žije 200 *lidi.
Tam žije 200 lidí.
There live 200 people.
'200 people live there.'

The form *lidi* 'people' is incorrect, the correct form is *lidí*. The form *lidi* is a form of nominative or accusative plural, the correct form *lidí* is a form of genitive plural required in Czech for quantified nouns following cardinal numbers (5 and more). The diacritic (acute) over the vowel *i* denotes a longer vowel. The causes of the deviation from the correct form can be several:

- orthography: the student forgot to mark the appropriate diacritic over the character *i* (a common error for native speakers)
- phonology: the student does not register (hear) the phonological difference between a short (*i*) and a long vowel (*í*)
- morphology: the student considers the ending -*i* as the correct one for the genitive plural case of the word *lidé*
- syntax: the student assumes that the correct case in this context is nominative plural (as the nominal phrase *200 lidí* is the subject of the sentence).

Such ambiguous errors do occur in learner's texts, but usually the uncertainty about the cause of the error is limited to one or two levels of language description. It is possible to create an error annotation ordered by levels of language description (orthography, phonology, morphology, syntax, lexicon), taking into account the uncertainty about the cause of the error. This approach can contribute to a better understanding of the process of language acquisition. Existing error classification systems in projects like Merlin (see [1,8]) or CzeSL (see [5,6]) usually disregard this uncertainty about the origin of the error. Sentence (4) can further illustrate the need to distinguish between various levels of language description in error classification, and sometimes combine some of them.

(4) Během *dovoleny *šla s *kamaradku na *pláži každý den.
Během dovolené chodila s kamarádkou na pláž každý den.
During holidays went with friend on beach every day.
During holidays, she went to the beach with her friend every day.

There are four incorrectly used words in the sentence (4): *dovoleny/dovolené*, *šla/chodila*, *kamaradku/kamarádkou* and *pláži/pláž*. One of these words, *kamaradku* 'friend', has two independent errors (*a/á* and *u/ou*). The form *dovoleny* 'holidays', where only the ending -*y* differs from the appropriate form *dovolené*, is apparently formed from a correct word base and an incorrect ending -*y*, which is a correct ending for the appropriate case (feminine singular) for other paradigms. The error in *dovoleny/dovolené* is therefore undoubtedly an error in morphology, as the author of the text apparently does not know the correct ending for the word he uses, but the case seems to be correct. The form *šla*, 'went' used instead of the correct form *chodila*, 'used to go', is a correct Czech word, with a correct form. It's use is inappropriate in the given context, it cannot be used with the expression *každý den* 'every day'. We classify this error as an error in lexicon/usage. The omission of the diacritic on the vowel *a* in the word *kamaradku/kamarádkou* 'friend' can be classified as an error in either orthography, or phonology (sometimes, non-native speakers of Czech do not distinguish between 'short' and 'long" vowels). The second error in the word *kamaradku* 'friend', i.e. the inappropriate use of the ending *-u* (which is a correct ending for the accusative singular) instead of *-ou* (the ending for instrumental singular, required here after the preposition *s* 'with'), is either an error in morphology (the author of the text does not know what ending to choose to form the instrumental case), or an error in syntax (the author does not know which case should be used with the preposition *s* 'with'). The last erroneous form *pláži* 'beach', used instead of *pláž* is a correct form of locative singular, a form that can be used after the preposition *na* 'on' (so that both *na pláži* and *na pláž* can be correct, depending on the context), but it is inappropriate with a verb of movement, such as *jít/chodit* 'go', so the error can be interpreted as an error in syntax. However, the ending *-i* is used in other feminine paradigms to form the accusative case, so we cannot exclude the possibility of an error in morphology (the student formed incorrectly the appropriate accusative case).

4   The CzeSL-Man Corpus

The corpus CzeSL-man [7] is a manually annotated part of the corpus CzeSL. There are two target hypothesis sentences (emendations) for every original sentence, the first target hypothesis corrects individual forms, disregarding the context; the second target hypothesis corrects the words in context, with a correct Czech sentence as a result. Every correction on both levels has an error label, with some 30 error labels assigned manually, completed by some 50 automatically assigned error labels, see [5]. Figure 1 shows an example of the two-level error annotation in the CzeSL corpus.

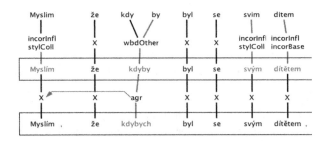

Fig. 1. Example of the two-level annotation of CzeSL corpus

The annotation covers a wide range of errors in various language phenomena. By using two levels of target hypothesis, it separates formal errors from syntactic and lexical errors. The annotation of the CzeSL-man corpus may be too complex for some uses, both for NLP and actual human use. It is also hard to reproduce automatically.

5   Error Classification by Levels of Language Description

In order to simplify NLP tasks and allow access to learner corpora for users without extensive linguistic background, we propose a different, (semi)automatic error classification of Czech learner texts. Only one, final target hypothesis will be used. Five general categories of errors with a short list of optional subcategories for each category will be used: orthography, morphonology, morphology, syntax and lexicon. The domain of orthography covers errors caused by ignoring the conventions of Czech writing, such as capitalization (*praha/Praha* 'Prague'), conventions of transcription of some combinations of phonemes, e.g. *ě* representing the phonemes *j* and *e* in *vjec/věc* 'thing', the use of accents (*děti/děti* 'children') etc. Many of these errors are common for native speakers. The domain of morphonology includes errors in phonology, e.g. the transcription of voiced/voiceless consonants (*sůstala/zůstala* 'stayed'), or the distincion between *r* and *l* consonants (sometimes ignored by Czech non-native speakers of Asian origin, e.g. *na kluku/na krku* 'on a boy/on the neck'), and incorrect forms of morphemes unrelated to inflection, e.g. *učiteka/učitelka* 'teacher'). As errors in morphology we classify only errors related to nominal declension and verbal conjugation, including both non-words (*na Erasmuse/Erasmu* 'on the Erasmus'; *studovám/studuju* 'I study') and existing forms of the given word, inappropriate in the given context (in this case, the error can be either morphological or syntactic). The domain of syntax covers errors caused by the incorrect use of word forms and function words (e.g. prepositions) in a given context, it includes errors in valency, agreement, quantification, word order etc. The errors in the lexicon domain concern cases when the original word is replaced in the correction by a different word with a different meaning and it is not a result of a random morphonological error. If necessary, two or more error domains can be used for the classification of any error.

6   Automatic Identification of Errors in Inflection

The process of error classification, both manual and automatic, can be simplified by a preliminary automatic rule-based analysis of the errors, which can add useful information about each error. It can determine whether an error meets the criteria for an orthographic error or distinguish between errors in the bases of words and errors in inflection. For the errors in inflection, it cat further analyze the properties of the incorrect ending. The input for the automatic identification of errors is the original text and the final target hypothesis (TH), aligned word-to-word. We use the manual emendation from CzeSL-man for now, but we expect to use automatically corrected word forms in the future.

In the first phase, the words from the TH are analyzed, and in case of inflectional words, they are split into a word-base and inflectional suffixes (and sometimes prefixes), using the morphological tag, lemma and a dictionary of paradigms and their properties.

Then the original word is analyzed, comparing the original word with the TH word character by character, allowing for a list of minor errors: changes in diacritics; changes of characters such as r/l, g/h, j/y typical for Czech learners; character alternations such as palatalization h/z, k/c; metathesis (*jesm/jsem*) etc. If the ratio of character changes is not too high, i.e. the stem can be seen as identical, the most likely ending of the original word is deduced.

For example, if the original, incorrect form is *stromom* 'tree' and the TH is *strom* with empty inflectional ending, the system should not compare only the two last characters of both words, but compare the whole stems and determine that the ending of the original word is *-om* (*strom|om*). Using the stems and endings for both the original and TH word, a two-dimensional comparison of the stems and of the inflectional affixes is then performed. If the stems (original and TH) differ, two facts are checked: whether there are any minor (orthographic, phonologic) errors in the stem, and whether the original stem is an allomorph of the stem from the TH word as in *v Prahe/Praze* 'in Prague', where the original stem *Prah* (incompatible with the ending -e) is used to form other (correct) cases of the same lemma, e.g. *Prah|a, Prah|y* etc.

If the endings differ, they are also checked for minor changes (orthography, e.g. diacritics), and independently, whether the incorrect ending is used within the given paradigm for other morphosynt. properties or whether the ending is used with other paradigms to express the same morphosynt. properties. The observed differences correspond roughly to the proposed error classification scheme: all errors in orthography and most of the errors in morphonology can be identified automatically. Incorrect endings indicate an error in morphology; if the incorrect ending is an existing one, expressing the same morphosyntactic properties, it may be an error only in morphology, otherwise it has to be seen as a possible error in syntax as well. If the original word is correct and has the same morphosyntactic properties as the TH word, but the lemma is different, the error may belongs to the domain of lexicon (except for functional words). The relationship between the automatic classification and the classification into error domains is not straightforward. A manual test on a sample of 500 learner's errors shows that the approach is reliable with more than 90% of categories determined correctly. As the system is rule-based, it can be fine-tuned by modifying the rules.

7   Results of Automatic Classification of Errors in the Inflection of Nouns

We tested the rule-based system on the data of the CZESL-man corpus, concentrating only on Czech nouns (tokens disambiguated as nouns on the level of the target hypothesis), as there is a higher percentage of errors and more error variety. Using text metadata, we divided the data by language proficiency of the authors (CEFR, see [2]). The levels are not evenly distributed, as shown in Table 1.

We performed two analyses of nouns in the CZESL-man corpus: one more general, determining the proportion of incorrect nouns in the corpus, one detailed, focused only on errors in inflection endings of nouns.

Table 1. Data distribution by language proficiency

	A1	A2	B1	B2	C1	Total
Number of tokens	6 961	42 252	39 987	28 182	5 522	122 904
Percentage of the data	5.66	34.38	32.54	22.93	4.49	100.00

Table 2 shows the proportion of correct nouns, nouns with an incorrect ending (*jeskyne/jeskyně* 'cave'), with an incorrect stem and a correct ending (*Prahe/Praze* 'Prague') and with both stem and ending incorrect (*delki/délky* 'length') or impossible to split automatically (*těmy/tématu* 'theme').

Table 2. Proportion of correct and incorrect nouns by proficiency levels

	A1	A2	B1	B2	C1	Total
Correct	61.32	68.15	77.45	78.01	95.09	74.25
Incorr. ending	9.10	10.68	6.30	6.30	0.97	7.72
Incorr. stem	18.91	14.20	11.18	11.23	3.17	12.31
Incorr. whole	19.77	17.65	11.37	10.76	1.74	13.44
Total	100.00	100.00	100.00	100.00	100.00	100.00

The proportion of correct nouns increases with the proficiency level, but there is little change between B1 and B2. There is an unexpectedly big difference between B2 and C1 in the proportion of correct nouns. The highest proportion of incorrect endings is in the texts of the level A2.

We analyzed in more detail the errors in nominal endings: all nouns with either a correct word base, or with minor changes compared with the TH were examined. Two parameters were observed: whether the error in the ending can be an error in orthography, and whether the ending is an existing Czech ending used either to express the same case, number and gender in other paradigms, or is used in the same paradigm to express other morphosynt. properties. Table 3 shows the analysis of errors in nouns in CzeSL. Six subtypes of nouns with incorrect ending were registered:

– Other paradigms: an ending that is used for other paradigms (*Úvalách/Úvalech* 'Úvaly, town name'), likely syntactically correct
– Other paradigms & orthography: nouns with an ending that is used for other paradigms and is an error in orthography in the same time (*Práže/Praze* 'Prague')
– Paradigm: nouns with an ending used inside the paradigm for other morphosynt. properties (*na procházka/procházku* 'on/for a walk'); likely an error in syntax
– Paradigm & orthography: as above, the previous error with an error in orthography (*lidi/lidí* 'people')

Table 3. Proportion of types of errors in endings of nouns

	A1	A2	B1	B2	C1	Total
Other paradigm	12.19	14.90	14.16	18.97	16.20	15.23
Other p. & orthography	8.87	4.51	7.08	6.91	7.82	6.83
Paradigm	19.38	30.77	22.83	24.03	24.02	25.34
Paradigm & orthography	4.03	3.24	5.25	5.14	11.73	4.40
Orthography	7.65	2.94	3.65	7.00	7.82	4.06
Other	47.88	43.64	47.03	37.94	32.40	44.14
Total	100.00	100.00	100.00	100.00	100.00	100.00

- Orthography: only an error in orthography, none of the above (*pracé/práce* 'work')
- Other: all other instances

We observe a steady decrease of "Other" errors, and an increase in the proportion of orthographic errors with language proficiency levels (the authors with a higher proficiency make less errors in general, but keep omitting diacritics). The system allows also for the analysis of individual endings: we observed, for example, that endings with high ambiguity such as *-e, -i, -í* are more prone to errors (already noted in [4], p. 220).

8   Conclusion

When analyzing learner's texts, we cannot be always sure about the causes of observed deviations from the standard language. We propose an error annotation scheme where the errors are annotated using the levels of language description: orthography, morphonology, morphology, syntax, lexicon. Two or more language domains can be used to describe an error, if the cause is not clear. A rule-based automatic system analyzing errors has been developed to assist the task of error annotation of learner's texts, both by NLP tools and manually. This rule-based system could be adapted for example to provide feedback in e-learning software for non-native speakers of Czech.

The rule-based system is too interconnected with Czech morphology to easily adapt to another language, but the approach to error annotation of learner language according to the levels of language description with the possibility of assigning multiple potential error labels to one error is, in our opinion, useful for error annotation of learner's texts in any language.

Acknowledgments. This research was supported by the Grant Agency of the Czech Republic through the grant 16-10185S (Non-native Czech from the Theoretical and Computational Perspective).

References

1. Boyd, A., Hana, J., Nicolas, L., Meurers, D., Wisniewski, K., Abel, A., Schöne, K., Štindlová, B., Vettori, C.: The MERLIN corpus: learner language and the CEFR. In: Calzolari, N., et al. (eds.) Proceedings of LREC 2014. ELRA, Reykjavik (2014)
2. Council of Europe: Common European Framework of Reference for Languages: Learning, Teaching, Assessment. Press Syndicate of the University of Cambridge, Cambridge (2001)
3. Šebesta, K., Škodová, S.: Žákovský korpus a jeho využití pro češtinu jako druhý jazyk. In: 20 let vývoje didaktiky cizích jazyků. TUL, Liberec (2011)
4. Hudousková, A.: Jmenné koncovky v češtině pro cizince - distribuce, frekvence a fonetika. prvnísonda. In: Petkevič, V., et al. (eds.) Radost z jazyků. Sborník k 75. narozeninám prof. Františka Čermáka, Studie z korpusové lingvistiky, vol. 20, pp. 215–230. NLN, Prague (2014)
5. Jelínek, T., Štindlová, B., Rosen, A., Hana, J.: Combining manual and automatic annotation of a learner corpus. In: Sojka, P., Horák, A., Kopeček, I., Pala, K. (eds.) TSD 2012. LNCS, vol. 7499, pp. 127–134. Springer, Heidelberg (2012). doi:10.1007/978-3-642-32790-2_15
6. Petkevič, V., Rosen, A., Štindlová, B., Jelínek, T., Hnátková, M., Jäger, P.: Anotace chybových textů v českém žákovském korpusu. In: Šebesta, K., Škodová, S. (eds.) Čeština - cílový jazyk a korpusy, p. 61–88. TUL, Liberec (2012)
7. Rosen, A.: Building and using corpora of non-native Czech. In: Brejová, B. (ed.) Proceedings of the ITAT/SloNLP 2016. CEUR Workshop Proceedings, vol. 1649, pp. 80–87. CreateSpace, Bratislava (2016)
8. Wisniewski, K., Woldt, C., Schöne, K., Abel, A., Blaschitz, V., Štindlová, B., Vodičková, K.: The MERLIN annotation scheme for the annotation of German, Italian, and Czech learner language (2014). www.merlin-platform.eu

Speaker Model to Monitor the Neurological State and the Dysarthria Level of Patients with Parkinson's Disease

J.C. Vásquez-Correa[1,3](✉), R. Castrillón[2], T. Arias-Vergara[1],
J.R. Orozco-Arroyave[1,3], and E. Nöth[3]

[1] Faculty of Engineering, Universidad de Antioquia UdeA,
Calle 70 No. 52-21, Medellín, Colombia
jcamilo.vasquez@udea.edu.co
[2] Universidad Católica de Oriente, Rionegro, Colombia
[3] Pattern Recognition Lab, Friedrich-Alexander-Universität Erlangen-Nürnberg,
Erlangen, Germany

Abstract. The progression of the disease in Parkinson's patients is commonly evaluated with the unified Parkinson's disease rating scale (UPDRS), which contains several items to assess motor and non–motor impairments. The patients develop speech impairments that can be assessed with a scale to evaluate dysarthria. Continuous monitoring of the patients is suitable to update the medication or the therapy. In this study, a robust speaker model based on the GMM–UBM approach is proposed for the continuous monitoring of the state of Parkinson's patients. The model is trained with phonation, articulation, and prosody features with the aim of evaluating deficits on each speech dimension. The performance of the model is evaluated in two scenarios: the monitoring of the UPDRS score and the prediction of the dysarthria level of the speakers. The results indicate that the speaker models are suitable to track the disease progression, specially in terms of the evaluation of the dysarthia level of the speakers.

Keywords: Parkinson's disease · UPDRS · Dysarthria · Phonation · Articulation · Prosody · Speaker model

1   Introduction

Parkinson's disease (PD) is a neurological disorder characterized by the progressive loss of dopaminergic neurons in the midbrain producing several motor and non–motor impairments [1]. The progression of the disease is currently evaluated with the third section of the movement disorder society unified Parkinson's disease rating scale (UPDRS) [2]. Only one of 33 items of the scale are related to the speech impairments of the patients; however, speech disorders are among the most prevalent, and an early sign of further motor impairments [3]. As general speech impairments developed by PD patients are described as *hypokinetic dysarthria*, a scale to assess dysarthria such as the Frenchay dysarthria assessment (FDA) [4] could be used to assess speech of PD patients.

© Springer International Publishing AG 2017
K. Ekštein and V. Matoušek (Eds.): TSD 2017, LNAI 10415, pp. 272–280, 2017.
DOI: 10.1007/978-3-319-64206-2_31

There are several studies focused on monitoring the disease progression of PD. In [5] the authors aims to predict the PD progression from speech using information from sustained vowels. Speech of 42 PD patients was recorded once per week during six months, and neurologist experts evaluated the patients three times during the study. The best result reported by the authors corresponds to a mean absolute error (MAE) of 7.5 points in the predictions of the UPDRS score; however the MAE is not a good performance measure for the addressed problem, besides the authors performed a linear interpolation to obtain most of the labels. In [6] 80 PD patients were recorded in two sessions separated between 12 and 88 months. In both sessions the patients were evaluated by neurologist experts according to the UPDRS score. The speech deficits of the patients were also perceptually rated in terms of voice, articulation, prosody, and fluency. The authors found significant differences for acoustic features such as shimmer, pause ratio, and vowel articulation index when features from the first recording session were compared with the same features calculated upon the second session. In [7] the authors recorded four male PD patients every week during one month to monitor the disease progression. Several features were computed to describe phonation deficits. The features were correlated with the Hoehn and Yahr (H&Y) scale in each recording session. The authors found that tremor and biomechanical features evolve differently with the treatment; however, the authors claimed that different time intervals between evaluations should be considered to obtain more conclusive results. Recently, in [8] a speaker model based on Gaussian mixture models–universal background models (GMM–UBM) was proposed to evaluate the neurological state of PD from speech. UBMs were trained with information from 61 PD patients and 50 healthy control (HC) speakers. Specific GMMs were adapted for seven PD patients recorded in three sessions. The Bhattacharyya distance was computed between the UBMs and the speaker models. Then, the distance was correlated with the UPDRS score of the patients. A Pearson's correlation of up to 0.60 was reported between the UPDRS scores and the distance measures. In this study, a more robust approach of the presented in [8] is performed. GMM–UBMs speaker models are trained using features based on phonation, articulation, and prosody with the aim to analyze the deviation of each speech dimension relative to the UBM. GMMs for phonation, articulation and prosody are combined following two strategies described in Sect. 3.3, and they are correlated with the UPDRS score and with a modified version of the FDA score (m–FDA) introduced previously [9]. The original version of FDA requires the patient to be with the examiner, which is not possible in most of the cases with PD patients due to their reduced mobility. The m–FDA can be applied even when only speech recordings of the patient are available. According to results, it is possible to monitor the disease progression of PD patients following the speaker models.

2   Data

The PD and HC speakers were labeled by three phoniatricians according to the m–FDA [4]. The original version of the FDA needs the patient to be with the examiner. The proposed version considers only the speech recordings and evaluates 13 items including among others the movements of the lips, larynx, palate and tongue, the respiration, and the intelligibility. The evaluation of each item ranges from 0 to 4, for a total range from 0 to 52 (0 normal, and 52 completely dysarthric). The phoniatricians agreed in the first ten evaluations, and then performed the evaluation of the other recordings. The inter–rater reliability among the labelers is 0.75. The median among the labels of the evaluators was considered as the label of the speaker. Previous studies have considered the m–FDA score to assess the speech disorders of the patients [10].

An extended version of the PC–GITA database [11] is considered. The data contain speech utterances from 68 PD patients and 50 HC subjects. All of them are Colombian Spanish native speakers. 33 of the patients were recorded in several sessions between 2012 and 2016. Seven of them were recorded in four sessions. This set of seven patients comprises our test set, i.e., they were used to adapt and test the proposed approach. The remaining 61 patients and 50 HC subjects were used to train the UBMs. Table 1 details the information of the seven patients used for test. The speech tasks performed by the subjects contain ten isolated sentences, a read text, a monologue, and six diadochokinetic (DDK) exercises such as the rapid repetition of the syllables /pa-ta-ka/, /pa-ka-ta/, /pe-ta-ka/, /pa/, /ta/, and /ka/.

Table 1. List of patients used to test the models. M–FDA and UPDRS scores in each session (S1, S2, S3, S4) are included.

	P1 male 64 years old		P2 female 57 years old		P3 female 51 years old		P4 female 55 years old		P5 (male 59 years old		P6 (male 68 years old		P7 (female 55 years old	
	UPDRS	m–FDA	UPDRS	m–FDA	UPDRS	m–FDA	UPDRS	m–FDA	UPDRS	m–FDA	UPDRS	m–FDA	UPDRS	m–FDA
S1	31	28	29	41	15	38	13	43	23	6	29	14	23	29
S2	15	19	36	35	21	49	12	10	37	8	23	25	28	26
S3	19	13	26	33	14	44	24	19	19	24	24	7	23	26
S4	–	–	34	33	20	45	19	–	22	21	28	–	21	30

3   Methods

The proposed method consists of 4 stages: (1) phonation, articulation, and prosody features are extracted, (2) UBMs are trained with the different feature sets and GMMs are adapted from the UBMs, (3) a distance measure is computed between the UBMs and the GMMs for phonation, articulation, and prosody, and (4), the distances for each speech dimension are combined using two proposed strategies (similarity and area). Pearson's correlation (r) is used to evaluate the predicted scores.

3.1   Feature Extraction

Several studies have described the speech impairments of PD patients in terms of different dimensions such as phonation, articulation, and prosody [3,12].

Phonation is related to the capability of the speaker to make the vocal folds vibrate [5,13]. Articulation is related with the modification of the position, stress, and shape of several limbs and muscles to produce speech [3,12,14]. Finally, prosody reflects variation of loudness, pitch, and timing to produce natural speech [14].

Phonation– The analysis comprises nine descriptors: the first and second derivative of the fundamental frequency, jitter, shimmer, the log–energy and their first two derivatives, the amplitude perturbation quotient (APQ), and the pitch perturbation quotient (PPQ). The descriptors are computed for frames of 40 ms length with a time-shift of 10 ms. PPQ and APQ are computed every eleventh and fifth frame, respectively due to those descriptors model the long term variability of the pitch and amplitude contours. Six statistical functionals are calculated per descritor (mean, standard deviation, skewness, kurtosis, maximum, and minimum), forming a 54–dimensional feature vector per utterance [13].

Articulation– The features include twelve MFCCs and their derivatives, the first two formant frequencies (F1 and F2) with their derivatives, and the energy distributed into 22 frequency bands according to the Bark scale computed in the transitions from unvoiced to voiced segments (onset), and from voiced to unvoiced (offset). The same six statistical functionals are computed, forming a 560–dimensional feature vector [12].

Prosody– The features are based on the duration of voiced segments, the fundamental frequency, and the energy. The same six statistical functionals are computed on the contours of the fundamental frequency, the energy, and the duration of voiced segments. The feature vector is completed with the voiced rate and the pause rate, forming a 20–dimensional feature vector per utterance.

3.2   Speaker Models

The speaker models are based on the GMM–UBM approach introduced previously to evaluate the progress of PD [8]. Individual UBMs are trained using phonation, articulation, and prosody–based features. The number of Gaussian components is estimated using the Akaike information criterion (AIC) [15]. After training the UBMs, individual GMMs are adapted for the seven speakers of the test set using the maximum a posteriori (MAP) strategy. Finally, a distance measure is computed between the UBMs and the adapted models for each speaker. Such distances are correlated with the severity of the disease of the patients.

3.3   Distance Computation and Model Combination

The Bhattacharyya distance (δ) is used to compare the UBMs and the speaker models. Such a distance measures the change between two multivariate Gaussian probability density functions considering both the mean vectors and the covariance matrices [16]. The distance is computed for the models trained with phonation, articulation, and prosody features. Those distances are combined using two strategies. The first strategy is a non–linear transformation of the distances with the aim of obtaining a similarity score S between the UBMs and the speaker models. The transformation is performed using Eq. 1, where δ_{phon}, δ_{art}, and δ_{pros} correspond to the distance between the speaker models and the UBMs for phonation, articulation, and prosody, respectively. If the distances are close to zero, S will be close to one, indicating that the speaker model is close to the UBM; conversely, if the distance measures are higher, S will tend to zero, indicating that the difference between the speaker model and the UBM is larger.

$$S = \frac{1}{1 + (\delta_{phon} + \delta_{art} + \delta_{pros})} \tag{1}$$

The second combination strategy is based on computing the area of a triangle, where each vertex corresponds to the distance obtained on each speech dimension: phonation, articulation, and prosody. Larger areas of the triangles indicate larger differences between the speaker model and the UBM. Figure 1 shows an example of the triangle. The most deviated speech dimension can be observed in the triangle as the largest distance from the centroid to the vertex.

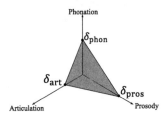

Fig. 1. Triangle created with the distance between the UBM and the speaker models for phonation, articulation, and prosody.

4   Experiments and Results

The speaker models are adapted from UBMs trained with features extracted from three groups of speaker: PD patients, HC speakers, and the combination of PD and HC speakers. The distances between the speaker models and the UBMs are correlated with the UPDRS and m–FDA scores. Results for individual models based on phonation, articulation, and prosody are reported. The combination of the models using the two proposed strategies is also tested. The results of the correlation between the speaker models and the UPDRS scores are shown in Table 2. Note that the correlations obtained using the area of the triangle and the similarity measure are always opposite. This fact is because the area is proportional to the level of impairment, i.e., larger areas indicate more deficits in phonation, articulation, and/or prosody, while the similarity is inverse, i.e., larger similarity indicates less level of impairment. In general the combination strategies provides the highest correlation, specially for patients P1, P2, and P4. Note that for P5 the results are inverse to the obtained for the remaining patients

Table 2. Pearson's correlation between the Bhattacharyya distance and the UPDRS score of the seven patients (P1, ..., P7).

UBM	Analysis	P1	P2	P3	P4	P5	P6	P7	Avg.
HC + PD	Area	**0.91**	0.50	**0.67**	**0.95**	−0.99	−0.08	0.30	**0.32**
	Similitude	**−0.90**	**−0.63**	−0.48	**−0.96**	0.99	0.06	**−0.46**	**−0.34**
	Phonation	−0.55	0.24	−0.71	−0.82	0.82	0.11	**0.76**	−0.02
	Articulation	−0.51	0.59	−0.44	0.23	0.20	−0.42	−0.63	−0.14
	Prosody	−0.99	−0.92	−0.11	−0.99	**0.93**	0.02	−0.42	−0.36
HC	Area	**0.91**	**0.95**	−0.78	**0.97**	−0.84	−0.04	**0.22**	0.20
	Similitude	**−0.91**	**−0.93**	0.75	**−0.98**	0.90	−0.29	−0.07	−0.22
	Phonation	−0.58	−0.54	0.45	−0.99	0.95	0.14	0.04	−0.08
	Articulation	−0.90	−0.84	0.12	−0.56	**0.99**	0.87	−0.87	−0.17
	Prosody	−0.99	−0.94	0.75	−0.99	0.77	−0.63	0.25	−0.26
PD	Area	−0.72	−0.66	−0.15	−0.66	0.72	**0.61**	0.10	−0.11
	Similitude	**−0.91**	**−0.80**	**−0.40**	**−0.97**	0.87	0.34	−0.43	−0.33
	Phonation	−0.67	−0.59	0.07	−0.84	0.82	−0.03	**0.35**	−0.13
	Articulation	−0.20	0.31	−0.74	0.64	0.06	0.73	−0.47	0.05
	Prosody	−0.99	−0.90	−0.23	−0.99	**0.84**	0.47	−0.23	−0.29

although a high correlation is obtained with the prosody model ($r = 0.93$). The average correlation indicates that the best results are obtained with the UBMs trained with PD and HC speaker, which confirms previous observations [8].

Table 3 shows the results obtained predicting the m–FDA score of the seven patients. High correlations are obtained, specially for the patients P1, P3 and P6. In general the combination of speech dimensions using the area or the similarity strategies improves the results relative to the separate prediction; however, note that for patient P4 a high correlation is obtained only with the articulation model. Figure 2 shows the triangles obtained for the seven patients across the recording sessions. For most of patients (P3, P4, P5, P6, and P7) the highest distance is observed for the prosody models, indicating that such a speech dimension could be the most impaired one for those patients. Note also that the areas of the triangle are changing (mainly compressing) across the sessions. For instance, a triangle with a certain area is observed for patient P1 in session 1, but the area becomes smaller across the sessions. Finally, in order to assess the progression of the disease of the patients, Fig. 3 illustrates the change in

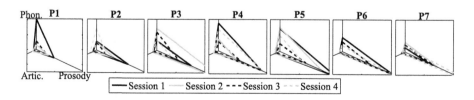

Fig. 2. Triangles for the seven patients in several recording sessions.

Table 3. Pearson's correlation between the Bhattacharyya distance and the m–FDA score of the seven patients (P1, ..., P7).

UBM	Analysis	P1	P2	P3	P4	P5	P6	P7	Avg.
HC + PD	Area	**0.99**	**0.57**	**0.87**	−0.53	**0.66**	**0.99**	−0.40	**0.45**
	Similitude	**−0.99**	**0.52**	**−0.76**	0.50	**−0.68**	**−0.99**	0.56	**−0.41**
	Phonation	−0.86	−0.64	−0.38	0.72	−0.69	−0.14	−0.12	−0.30
	Articulation	−0.07	−0.81	−0.76	**0.85**	−0.32	0.29	−0.47	−0.18
	Prosody	−0.94	−0.09	−0.58	0.16	−0.59	−0.93	**0.58**	−0.34
HC	Area	**0.99**	−0.15	−0.11	−0.49	**0.25**	**0.99**	0.38	0.27
	Similitude	**−0.99**	0.07	0.07	0.45	**−0.40**	**−0.89**	**−0.51**	−0.31
	Phonation	−0.88	−0.30	0.03	0.15	−0.50	−0.96	0.29	−0.31
	Articulation	−0.60	−0.33	−0.39	**0.58**	−0.70	−0.65	−0.12	−0.32
	Prosody	−0.83	**0.39**	0.04	0.28	−0.45	−0.67	−0.72	−0.28
PD	Area	−0.96	0.01	−0.47	**0.89**	−0.60	−0.89	**0.71**	−0.19
	Similitude	**−0.90**	**−0.63**	**−0.48**	**−0.96**	0.99	0.06	−0.46	−0.34
	Phonation	−0.93	0.11	−0.13	0.70	−0.63	−0.41	0.70	−0.08
	Articulation	−0.62	−0.29	−0.89	0.55	−0.43	0.58	−0.64	−0.25
	Prosody	−0.91	−0.32	−0.66	0.14	−0.53	−0.92	0.60	−0.37

the values of the area of the triangles for the seven patients across the recoding sessions. The area of the triangles is compared to the UPDRS and the m–FDA scores. Note that the speaker model is highly accurate for patients P1, P3, and P7. Note the behavior of patient P5, where the speaker model is not able to trace the UPDRS score, but it is highly accurate to trace the progression of the m–FDA score.

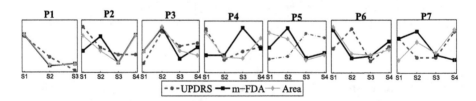

Fig. 3. Comparison of the area of each triangle with the progression of the disease both for the UPDRS and the m–FDA scores

5 Conclusion

A speaker model based on GMM–UBMs trained with phonation, articulation, and prosody features is proposed to track the disease progression in PD patients. A novel approach is introduced to combine phonation, articulation, and prosody models, which allows to evaluate which dimension of speech is more impaired.

The performance of the speaker models is tested by predicting the UPDRS score and a dysarthria level according to a modified version of the FDA scale of seven PD patients recorded in four sessions between 2012 and 2016. The speaker models are able to trace the disease progression, specially in terms of the dysarthria level of the speakers. This result confirms previous observations reported in [9], where the researchers conclude that, although the speech production is a motor process, it is maybe not convenient to pretend the prediction of a neurological scale (like the UPDRS) based only on speech signals. The combination of phonation, articulation, and prosody models improves the performance of the method. The proposed approach should be validated with information from more patients with the aim of obtaining more conclusive results. The proposed model will also be implemented in a mobile platform to perform an unobtrusive monitoring of the patients.

Acknowledgments. The work reported here was started at JSALT 2016, and was supported by JHU via grants from DARPA (LORELEI), Microsoft, Amazon, Google and Facebook. Thanks also to CODI from University of Antioquia by the grant Numbers 2015–7683 and PRV16-2-01.

References

1. Hornykiewicz, O.: Biochemical aspects of Parkinson's disease. Neurology **51**(2 Suppl 2), S2–S9 (1998)
2. Goetz, C.G., et al.: Movement Disorder Society-sponsored revision of the Unified Parkinson's Disease Rating Scale (MDS-UPDRS): scale presentation and clinimetric testing results. Mov. Disord. **23**(15), 2129–2170 (2008)
3. Rusz, J., Cmejla, R., Tykalova, T., Ruzickova, H., Klempir, J., Majerova, V., Picmausova, J., Roth, J., Ruzicka, E.: Imprecise vowel articulation as a potential early marker of Parkinson's disease: effect of speaking task. J. Acoust. Soc. Am. **134**(3), 2171–2181 (2013)
4. Enderby, P.M., Palmer, R.: FDA-2: Frenchay Dysarthria Assessment: Examiner's Manual. Pro-Ed, Texas (2008)
5. Tsanas, A., Little, M., McSharry, P.E., Ramig, L.: Accurate telemonitoring of Parkinson's disease progression by noninvasive speech tests. IEEE Trans. Biomed. Eng. **57**(4), 884–893 (2010)
6. Skodda, S., Grönheit, W., Mancinelli, N., Schlegel, U.: Progression of voice and speech impairment in the course of Parkinson's disease: a longitudinal study. Parkinson's Dis. **2013**, 1–8 (2013). Article ID 389195
7. Gómez-Vilda, P., Vicente-Torcal, M.C., Ferrández-Vicente, J.M., Álvarez-Marquina, A., Rodellar-Biarge, V., Nieto-Lluis, V., Martínez-Olalla, R.: Parkinson's disease monitoring from phonation biomechanics. In: Ferrández Vicente, J.M., Álvarez-Sánchez, J.R., de la Paz López, F., Toledo-Moreo, F.J., Adeli, H. (eds.) IWINAC 2015. LNCS, vol. 9107, pp. 238–248. Springer, Cham (2015). doi:10.1007/978-3-319-18914-7_25
8. Arias-Vergara, T., Vásquez-Correa, J.C., Orozco-Arroyave, J.R., Vargas-Bonilla, J.F., Nöth, E.: Parkinson disease progression assessment from speech using GMM-UBM. In: Annual Conference of the International Speech Communication Association (INTERSPEECH), pp. 1933–1937 (2016)

9. Nöth, E., et al.: Remote monitoring of neurodegeneration through speech. In: Final Presentation of the Third Frederick Jelinek Memorial Summer Workshop (JSALT), August 2016

10. Vásquez-Correa, J.C., Orozco-Arroyave, J.R., Arora, R., Nöth, E., Dehak, N., Christensen, H., Rudzicz, F., Bocklet, T., Cernak, M., Chinaei, H., et al.: Multiview representation learning via GCCA for multimodal analysis of Parkinson's disease. In: 2017 IEEE International Conference on Acoustics, Speech, and Signal Processing (ICASSP 2017) (2017)

11. Orozco-Arroyave, J.R., Arias-Londoño, J.D., Vargas-Bonilla, J.F., Gonzalez-Rátiva, M.C., Nöth, E.: New Spanish speech corpus database for the analysis of people suffering from Parkinson's disease. In: Language Resources and Evaluation Conference, (LREC), pp. 342–347 (2014)

12. Orozco-Arroyave, J.R., Vásquez-Correa, J.C., Hönig, F., Arias-Londoño, J.D., Vargas-Bonilla, J.F., Skodda, S., Rusz, J., Nöth, E.: Towards an automatic monitoring of the neurological state of the Parkinson's patients from speech. In: 41st International Conference on Acoustic, Speech, and Signal Processing (ICASSP), pp. 6490–6494 (2016)

13. Orozco-Arroyave, J.R., Belalcazar-Bolaños, E.A., et al.: Characterization methods for the detection of multiple voice disorders: neurological, functional, and laryngeal diseases. IEEE J. Biomed. Health Inf. 19(6), 1820–1828 (2015)

14. Skodda, S., Visser, W., Schlegel, U.: Vowel articulation in parkinson's disease. J. Voice 25(4), 467–472 (2011)

15. Akaike, H.: A new look at the statistical model identification. IEEE Trans. Autom. Control 19(6), 716–723 (1974)

16. You, C.H., Lee, K.A., Li, H.: GMM-SVM kernel with a Bhattacharyya-based distance for speaker recognition. IEEE Trans. Audio Speech Lang. Process. 18(6), 1300–1312 (2010)

A Lightweight Regression Method to Infer Psycholinguistic Properties for Brazilian Portuguese

Leandro Borges dos Santos[1(✉)], Magali Sanches Duran[1],
Nathan Siegle Hartmann[1], Arnaldo Candido Jr.[2], Gustavo Henrique Paetzold[3],
and Sandra Maria Aluisio[1]

[1] Institute of Mathematics and Computer Sciences,
University of São Paulo, São Paulo, Brazil
`leandrobs@usp.br`, `magali.duran@uol.com.br`, {`nathansh,sandra`}`@icmc.usp.br`
[2] Federal Technological University of Paraná (UTFPR),
Medianeira, Medianeira, Brazil
`arnaldoc@utfpr.edu.br`
[3] Department of Computer Science, University of Sheffield, Sheffield, England
`g.h.paetzold@sheffield.ac.uk`

Abstract. Psycholinguistic properties of words have been used in various approaches to Natural Language Processing tasks, such as text simplification and readability assessment. Most of these properties are subjective, involving costly and time-consuming surveys to be gathered. Recent approaches use the limited datasets of psycholinguistic properties to extend them automatically to large lexicons. However, some of the resources used by such approaches are not available to most languages. This study presents a method to infer psycholinguistic properties for Brazilian Portuguese (BP) using regressors built with a light set of features usually available for less resourced languages: word length, frequency lists, lexical databases composed of school dictionaries and word embedding models. The correlations between the properties inferred are close to those obtained by related works. The resulting resource contains 26,874 words in BP annotated with concreteness, age of acquisition, imageability and subjective frequency.

Keywords: Psycholinguistic properties · Brazilian Portuguese · Lexical resources

1 Introduction

Besides frequency, form, and meaning, words have several other less known properties, such as imageability, concreteness, familiarity, subjective frequency, and age of acquisition (AoA). These are subjective psycholinguistic properties, which depend on the experiences individuals had using the words. According to [14], word imageability is the ease and speed with which a word evokes a mental

© Springer International Publishing AG 2017
K. Ekštein and V. Matoušek (Eds.): TSD 2017, LNAI 10415, pp. 281–289, 2017.
DOI: 10.1007/978-3-319-64206-2_32

image; concreteness is the degree to which words refer to objects, people, places, or things that can be experienced by the senses; experiential familiarity is the degree to which individuals know and use words in their everyday life; subjective frequency is the estimation of the number of times a word is encountered by individuals in its written or spoken form, and AoA is the estimation of the age at which a word was learned. Psycholinguistic properties have been used in various approaches: for Lexical Simplification [11]; for Text Simplification at the sentence level, aiming to reduce the difficulty of informative text for language learners [17]; to predict the reading times (RTs) of each word in a sentence to assess sentence complexity [13] and also to create robust text level readability models [16].

Because of its inherent costs, the measurement of subjective psycholinguistic properties is usually used in the creation of datasets of limited size [2,6,7,14]. For the English language, the most well known database of this kind is the MRC Psycholinguistic Database[1], which contains 27 subjective psycholinguistic properties for 150,837 words. For BP, there is a psycholinguistic database[2] containing 21 columns of information for 215,175 words, but no subjective psycholinguistic properties. We aim to overcome this gap by automatically inferring the psycholinguistic properties of imageability, concreteness, AoA and subjective frequency for a large database of 26,874 BP words, using a resource-light regression approach. This work relies heavily on the results of [11] who proposed an automatic bootstrapping method for regression to populate the MRC Database. We explore here three research questions: (1) is it possible to achieve high Pearson and Spearman correlations values and low MSE values with a regression method using only word embedding features to infer the psycholinguistic properties? (2) which size a database with psycholinguistic properties should have to be used in regression models? Does merging databases from different sources yield better correlation and lower MSE scores? (3) can the inferred values help in creating features that result in more reliable readability prediction models? In addition, we assessed interrater reliability (Cronbachs alpha) between ratings generated by our method and the imageability and concreteness produced for 237 nouns by [8]. We also analyzed the relations between the inferred ratings and other psycholinguistic variables.

2   Related Works

To the best of our knowledge there are only two studies that propose regression methods to automatically estimate missing psycholinguistic properties in the MRC Database [4,11]. [4] propose a computational model to predict word concreteness, by using linear regression with word attributes from WordNet [3], Latent Semantic Analysis (LSA) and the CELEX Database[3] and use these attributes to simulate human ratings in the MRC database. The lexical features

---

[1] websites.psychology.uwa.edu.au/school/MRCDatabase/mrc2.html.

[2] www.lexicodoportugues.com.

[3] celex.mpi.nl/.

used were 19 lexical types from WordNet, 17 LSA dimensions, hypernymy information from WordNet, word frequencies from the CELEX Database, and word length (i.e., number of letters), totalling 39 attributes. The Pearson correlation between the estimated concreteness score and the concreteness score in the test set was 0.82.

[11] automatically estimate missing psycholinguistic properties in the MRC Database through a bootstrapping algorithm for regression. Their method exploits word embedding models and 15 lexical features, including the number of senses, synonyms, hypernyms and hyponyms for word in WordNet and also minimum, maximum and average distance between the words senses in Word-Net and the thesaurus root sense. The Pearson correlation between the estimated score and the inferred score for familiarity was 0.846; 0.862 for AoA; 0.823 for imagenery and 0.869 for concretness, which is better than the results of [4].

3   A Lightweight Regression Method to Infer Psycholinguistic Properties of Words

The fact that the methods developed by [4,11] are based on a large, scarce lexical resource as WordNet, led us to raise the question "Could we have a similar performance with a simpler set of features which are easily obtainable for most languages?". Therefore we decided to build our regressors using only word length, frequency lists, lexical databases composed of school dictionaries and word embeddings models. One critical difference between the strategy of [11] and ours is that they concatenate all features to train a regressor, while we take a different approach. Although simply combining all features is straightforward, it can lead to noise insertion, given that the features greatly contrast among them (e.g. word embeddings and word length). Instead, we adopted a more elegant solution, called Multi-View Learning [18]. In a Multi-View Learning, multiple regressors/classifiers are trained over different feature spaces and then combined to produce a single result. Here, the fusion stage is made by averaging the values predicted by the regressors [18].

3.1   Adaptation of Databases with Psychological Norms for Portuguese Words

We present in Table 1 surveys involving the subjective psycholinguistic properties of words focused in this study (concreteness, age of acquisition, imageability and subjective frequency), both for European Portuguese (EP) and BP.

If, in one hand, manually produced resources fulfill the needs for which they were collected, on the other hand, they are very limited for Natural Language Processing purposes, given their limited size. There are ways, however, to automatically infer subjective psycholinguistic properties for several words, using the existing ones. To achieve this goal, however, we need, first of all, to rely on a set of words with values for each aimed property.

Table 1. Norms for Portuguese on the focused psycholinguistic properties.

Study	Participants	Words	Property	Variant	Scale
[14]	2,357	3,789	Concreteness, imageability, subjective frequency	EP	1–7
[2]	685	1,748	AoA	EP	1–9
[6]	719	909	Concreteness	BP	1–7
[7]	110	834	AoA	EP	1–7
[8]	103	249	Imageability, concreteness	EP	1–7

In BP, we only have 909 words with concreteness values [6]. Therefore, we decided to incorporate EP resources to our set, as well as to combine different resources containing values for the same property. In order to turn EP resources usable for our study in BP, we executed adjustments in the word lists. Most of them were in orthography, as for example: acção/acão (action), adopção/adoção (adoption), amnistia/anistia (amnesty). Other adjustments pertained to concepts that the two variants of Portuguese lexicalize in different ways, such as: ficheiro/arquivo (file), assassínio/assassinato (murder), apuramento/apuração (calculation). Finally, some words have been discarded, as they lexicalize concepts related to fauna, flora, and culinary traits native to Portugal, such as: faneca (pout), faia (beech) and rebuçado (candy). In theory, there should not be a problem in concatenating two or more lists of words with the same psycholinguistic property. We did this for concreteness, merging the list of [14], once adapted for BP, with the one of [6], which was created for BP. As both lists rated concreteness using a Likert scale of 7 points, the values were comparable. However, in what concerns AoA, the two lists available, [2,7], rated concreteness using Likert scales of 7 and 9 points, respectively. It is worth mentioning that both lists contain AoA ratings produced by adults (AoA could be, alternatively, gathered using the proficiency of children of different ages in object naming tasks). Therefore, to turn them comparable, we had to convert the scale of 9 points into a scale of 7 points. After concluding the lexical adjustments, converting the scale of 9 points into 7 points for AoA lists, merging the lists and eliminating duplicated words, we obtained sizeable datasets for all word properties addressed in this study. Table 2 shows the number of entries obtained for each property, between the parentheses.

3.2   Features

Our regressors use 10 features from several sources, grouped in: (i) lexical (1–8); (ii) Skip-Gram word embeddings (9) [9]; and (iii) GloVe word embeddings (10) [12]:

1. Log of Frequency in SUBTLEX-pt-BR [15], which is a database of BP word frequencies based on more than 50 million words from film and television subtitles;

2. Log of Contextual diversity in SUBTLEX-pt-BR, which is the number of subtitles that contain the word;
3. Log of Frequency in SubIMDb-PT [10]: this corpus was extracted from subtitles of family, comedy and children movies and series;
4. Log of Frequency in the Written Language part of Corpus Brasileiro, a corpus with about 1 billion words of Contemporary BP;
5. Log of Frequency in the Spoken Language part of Corpus Brasileiro;
6. Log of Frequency in a corpus of 1.4 billion tokens of Mixed Text Genres in BP;
7. Word Length;
8. Lexical databases from 6 school dictionaries for specific grade-levels;
9. Word's raw embedding values of word embeddings models created using the Skip-Gram algorithm [9], with word vector sizes of 300, 600 and 1,000;
10. Word's raw embedding values of word embeddings models created using the GloVe algorithm [12], with word vector sizes of 300, 600 and 1,000.

Reading time studies provide evidence that more processing time is allocated to rare words than high-frequency words. Besides that, the logarithm of word frequency was used here because reading times are linearly related to the logarithm of word frequency, not to raw word frequencies [1]. We trained our embedding models using Skip-Gram word2vec and GloVe over a corpus of 1.4 billion tokens, and 3.827.725 types composed by mixed text genres, including subtitles to cover spoken language besides written texts.

3.3   Using Regression in a Multi-View Learning Approach

We used a linear least squares regressor with L2 regularization, which is also known as Ridge Regression or Tikhonov regularization. We choose this regression method due to the promising results reported by [11]. We trained three regressors in different feature spaces: lexical features, Skip-Gram embeddings, and GloVe embeddings.

4   Evaluation

We experimented with several dimensions of word embeddings, but for space reasons, here we include only the best results: Skip-Gram and GloVe embeddings with 300 word vector dimensions. We used 20×5-fold cross-validation in order to perform our experiments. As evaluation metrics, we used Mean Square Error (MSE), Spearman's (ρ), and Pearson's (r) correlation. For the MSE metric, a repeated measures ANOVA with Dunnet post-test was used to compare the best regressors with the others to significance level of 0.05. Table 2 shows the evaluation results of our method. For subjective frequency, the best result was given by the combination of Lexical, Skip-gram and GloVe embeddings. For AoA, the best result was given by the combination of Lexical and GloVe embeddings. For AoA, the three better lexical features responsible for such results were grade-level lexical databases, the log frequency in Sub-IMDb-PT and word length. The regressors in bold presented statistically significant differences when compared with the others. However, for AoA, we did not find a statistically significant difference between the Lexical + GloVe regressor and the Lexical + Skip-gram + GloVe regressor.

Table 2. MSE and Pearson and Spearman correlation scores of the regression models.

Regressors	Concreteness (4088)			Subjective frequency (3735)			Imageability (3735)			AoA merging (2368)		
	MSE	r	ρ	MSE	r	ρ	MSE	r	ρ	MSE	r	ρ
Lexical	1.24	0.54	0.56	0.55	0.72	0.73	0.74	0.58	0.59	0.67	0.73	0.73
Skip-gram	0.52	0.84	0.84	0.58	0.70	0.71	0.46	0.77	0.77	0.81	0.66	0.66
GloVe	0.62	0.80	0.81	0.40	0.81	0.81	0.49	0.75	0.75	0.63	0.75	0.75
Lexical + Skip-gram	0.64	0.82	0.82	0.44	0.79	0.79	0.47	0.77	0.78	0.59	0.77	0.77
Lexical + GloVe	0.70	0.80	0.80	0.39	0.81	0.81	0.50	0.75	0.76	**0.54**	**0.79**	**0.79**
Skip-gram + GloVe	**0.49**	**0.85**	**0.85**	0.41	0.80	0.80	**0.42**	**0.79**	**0.79**	0.62	0.75	0.75
Lexical + Skip-gram + GloVe	0.55	0.85	0.84	**0.38**	**0.82**	**0.82**	0.43	0.79	0.78	0.54	0.79	0.79

Portuguese databases with AoA properties are small in size, therefore we evaluated three different databases for this property. The first has 765 words [7], the second has 1717 words [2], and the third is composed by a merging of the first and the second converted into a 7-scale; the merging resulted in a database with 2368 different words, which is still small compared to the other 3 properties evaluated here. The resulting correlations and MSE for AoA (see Table 3) show that merged datasets yield better results. There was a drop of 0.26 in MSE scores and an increase of 0.07 and 0.08 of Pearson and Spearman values.

Table 3. MSE, Pearson, and Spearman correlations of the regression models.

Regressors	AoA (765)			AoA (1717)			AoA merge (2368)		
	MSE	r	ρ	MSE	r	ρ	MSE	r	ρ
Lexical	0.91	0.67	0.66	1.04	0.76	0.75	0.67	0.73	0.72
Skip-gram	1.30	0.56	0.58	1.36	0.68	0.65	0.81	0.66	0.66
GloVe	1.18	0.62	0.63	0.93	0.79	0.75	0.63	0.75	0.75
Lexical + GloVe	**0.80**	**0.72**	**0.71**	**0.79**	**0.83**	**0.80**	**0.54**	**0.79**	**0.79**

We also compared the four properties' interdependency among themselves by using Pearson correlation. Table 4 presents the results obtained for our comparisons, as well as the results obtained on similar comparisons from related contributions. In Table 4, dashes represent evaluations which were not performed on a given study. Our results are close to those reported in the literature, except for the correlation between age of aquisition and concreteness, which is stronger in

Table 4. Pearson correlations among properties.

Properties compared	OURS	[7]	[2]	[14]	[7] vs. [14]	[2] vs. [14]
AoA vs. Concreteness	−0.37	−0.52	−0.61	-	−0.49	−0.54
AoA vs. Imageability	−0.73	−0.69	−0.66	-	−0.66	−0.62
AoA vs. Sub. Freq	−0.52	-	-	-	−0.65	−0.60
Imageability vs. Sub. Freq	0.11	-	-	0.04	−0.10	-
Concreteness vs. Sub. Freq	−0.05	-	-	−0.09	-	-
Imageability vs. Concreteness	0.92	-	-	0.88	0.82	-

other studies. This may be related to the fact that a full dictionary has a larger proportion of rare words than observed in the lists of those studies.

To validate the reliability of our automatically inferred psycholinguistic properties, we conducted internal consistency analyses. We calculated alpha scores between our automatically produced imageability and concreteness properties and those present in the psycholinguistic dataset of [8]. In total, 237 words were considered. The alpha scores for imageability and concreteness are 0.921 and 0.820, which are similar to the values achieved by [14], and suggest that our features do, in fact, accurately capture the psycholinguistic properties being targeted.

We built a database of plain words which was populated with the inferred values for the four psycholinguistic properties targeted in this study. For this, we exploited Minidicionário Caldas Aulete's entries [5] and their respective first grammatical category. We selected only nouns, verbs, adjectives and adverbs. All loanwords (foreign words used in BP) were discarded. Then we searched the frequency of each word in the large corpus of 1.4 billion words we used to train our word embeddings models, and after a manual analysis, we decided to disregard words with less than 8 occurrences, as they are very uncommon. The final lexicon is available[4] and contains 26,874 words, being 15,204 nouns, 4,305 verbs, 7,293 adjectives and 72 adverbs with the information of the four inferred psycholinguistic properties using the better results with less features (shown in bold in Table 2).

5 Evaluating Psycholinguistic Features in Readability Prediction

In order to evaluate the use of psycholinguistic properties in predicting the readability level of BP informative texts from newspapers and magazines for early school years, we trained a classifier using a corpus of 1,413 texts which were classified as easy to read for 3rd to 6th graders. These texts were annotated by two linguists (0,914 weighted kappa) and the corpus distribution by grade levels

---

[4] http://nilc.icmc.usp.br/portlex/index.php/en/psycholinguistic.

is 183 texts for 3rd grade, 361 texts for 4th grade, 537 texts for 5th grade and 332 texts for 6th-grade year. We are still in the annotation process of a dataset similar in size to the one of [16]. We compared the use of our four psycholinguistic features and six traditional readability formulas: Flesch Reading-Ease adapted to BP Brunét, Honoré, Dale-Chall, Gunning Fox and Moving Average Type-Token Ratio (MATTR). We used the mean and standard deviation of our psycholinguistic properties as features to train SVM classifiers with RBF kernel for each psycholinguistic information and a single-view classifier for all four properties. All results presented here were obtained by a 10-fold cross-validation process. Table 5 shows that subjective frequency provided better results than the other psycholinguistic features in classifying grade levels and achieved the 3rd best result for individual features. The single-view classifier of psycholinguistic features overcame all traditional formulas but MATTR and Brunét Index for grade-level classification in F1-measure. Both MATTR and Brunét Index measure the lexical diversity of a text and are independent of text length. Their high performance in this evaluation suggests that lexical diversity is a strong proxy to distinguish grade levels in primary school years.

Table 5. Evaluating Psycholinguistic and Classic readability formulas for readability prediction.

Features	Flesch	Honoré	Concreteness	Familiarity	AoA	Dale-Chall	Gunning Fox	Subjective frequency	Psycholin-guistics	MATTR	Brunét
F1	0.26	0.29	0.27	0.23	0.25	0.36	0.37	0.32	0.45	0.48	**0.54**

6   Conclusion and Future Work

In this work, we set our aims at finding a light set of features available for most languages to build regressors that infer psycholinguistic properties for BP words. We have made publicly available a large database of 26,874 BP words annotated with psycholinguistic properties. With respect to our research questions (1) and (2), we have shown we can infer psycholinguistic properties for BP using word embeddings as features. Nonetheless, our regressors need a reasonably large number of training instances (at least, more than two thousand examples), as well as complementary lexical resources to yield top performance for AoA and subjective frequency. As for research question (3), our results show that psycholinguistic properties can potentially aid readability prediction. These results ratify the claims of [13], which state that (i) words with higher concreteness are easier to imagine, comprehend, and memorize and therefore increase the readability of texts, and (ii) age of acquisition is helpful in predicting reading difficulty. As future work, we propose to extend our extrinsic evaluation to other tasks, to use new modelling techniques for our psycholinguistic features (besides the average and standard deviation of the inferred values) and to use a more robust approach to perform the fusion of regressors, e.g. stacking regression.

Acknowledgments. This work was supported by CNPq and FAPESP.

References

1. Graesser, A.C., McNamara, D.S.: Computational analyses of multilevel discourse comprehension. Top. Cogn. Sci. **3**(2), 371–98 (2011)
2. Cameirao, M.L., Vicente, S.G.: Age-of-acquisition norms for a set of 1,749 Portuguese words. Behav. Res. Meth. **42**(2), 474–480 (2010)
3. Fellbaum, C.: Wordnet: An Electronic Lexical Database. MIT Press, Cambridge (1998)
4. Feng, S., Cai, Z., Crossley, S.A., McNamara, D.S.: Simulating human ratings on word concreteness. In: Proceedings of 24th FLAIRS Conference, pp. 1–6 (2011)
5. Geiger, P.: Minidicionário contemporâneo da língua portuguesa (2011)
6. Janczura, G., Castilho, G., Rocha, N., van Erven, T., Huang, T.: Normas de concretude para 909 palavras da língua portuguesa. Psicologia: Teoria e Pesquisa, pp. 195–204 (2007)
7. Marques, J.F., Fonseca, F.L., Morais, S., Pinto, I.A.: Estimated age of acquisition norms for 834 Portuguese nouns and their relation with other psycholinguistic variables. Behav. Res. Meth. **39**(3), 439–444 (2007)
8. Marques, J.F.: Normas de imagética e concreteza para substantivos comuns. Laboratório de Psicologia **3**, 65–75 (2005)
9. Mikolov, T., Chen, K., Corrado, G., Dean, J.: Efficient estimation of word representations in vector space. arXiv preprint arXiv:1301.3781 (2013)
10. Paetzold, G., Specia, L.: Collecting and exploring everyday language for predicting psycholinguistic properties of words. In: Proceedings of COLING 2016, pp. 1669–1679 (2016)
11. Paetzold, G.H., Specia, L.: Inferring psycholinguistic properties of words. In: Proceedings of NAACL-HLT 2016, pp. 435–440 (2016)
12. Pennington, J., Socher, R., Manning, C.D.: Glove: global vectors for word representation. In: Proceedings of EMNLP 2014, pp. 1532–1543 (2014)
13. Singh, A.D., Mehta, P., Husain, S., Rajkumar, R.: Quantifying sentence complexity based on eye-tracking measures (2016)
14. Soares, A.P., Costa, A.S., Machado, J., Comesana, M., Oliveira, H.M.: The Minho Word Pool: norms for imageability, concreteness, and subjective frequency for 3,800 portuguese words. Behav. Res. Meth. **49**(3), 1065–1081 (2017)
15. Tang, K.: A 61 million word corpus of Brazilian Portuguese film subtitles as a resource for linguistic research. UCL Work Pap. Linguist. **24**, 208–214 (2012)
16. Vajjala, S., Meurers, D.: Readability assessment for text simplification from analysing documents to identifying sentential simplifications. Recent Adv. Autom. Readability Assess. Text Simplification **165**(2), 194–222 (2014)
17. Vajjala, S., Meurers, D.: Readability-based sentence ranking for evaluating text simplification. arXiv preprint arXiv:1603.06009 (2016)
18. Xu, C., Tao, D., Xu, C.: A survey on multi-view learning. arXiv preprint: arXiv:1304.5634 (2013)

Open-Domain Non-factoid Question Answering

Maria Khvalchik$^{(\boxtimes)}$ and Anagha Kulkarni

Computer Science Department, San Francisco State University,
1600 Holloway Ave, San Francisco, CA 94132, USA
{mkhvalch,ak}@sfsu.edu

Abstract. We present an end-to-end system for open-domain non-factoid question answering. We leverage the information on the ever-growing World Wide Web, and the capabilities of modern search engines to find the relevant information. Our QA system is composed of three components: (i) query formulation module (QFM) (ii) candidate answer generation module (CAGM) and (iii) answer selection module (ASM). A thorough empirical evaluation using two datasets demonstrates that the proposed approach is highly competitive.

Keywords: Question answering · Learning to rank · Neural network · BLSTM

1 Introduction

The popularity of QA websites such as Quora, and Yahoo! Answers highlights users' preference to express information needs as natural language questions, rather than keyword based queries. Currently, the person who posts the question has to wait for the answer until someone responds with the correct answer. Often the answer to the posted question is already out there on the World Wide Web (WWW), as an answer to a similar question, or embedded in the content of a web page. This observation was inspired evaluation forums such as the TREC LiveQA Track[1], and QALD Challenge[2] that are facilitating the research on the automated QA problem.

Developing an automated QA system that is capable of answering factoid or non-factoid questions from any domain (e.g. health, sports, cooking, etc.) is a challenging problem. In this paper we present our take on this problem. The end-to-end system that we have developed consists of three modules: (i) QFM converts the free-text question into a boolean query that can be processed by a commercial search engine; (ii) CAGM extracts all the promising candidate answers from the top ranked web pages returned by the search engine and (iii) ASM employs different ranking and classification approaches to select the best answer from the set of candidates. In the subsequent sections we describe our

---

[1] https://sites.google.com/site/trecliveqa2015/.
[2] https://project-hobbit.eu/challenges/qald2017/.

© Springer International Publishing AG 2017
K. Ekštein and V. Matoušek (Eds.): TSD 2017, LNAI 10415, pp. 290–298, 2017.
DOI: 10.1007/978-3-319-64206-2_33

novel approach which combines several machine learning models together and is trained on a large sanitized dataset.

The subproblem of answering factoid questions using a static collection of documents has been researched since late 60's [6, 12, 20]. One of the recent examples is work by Bian et al. [4] where they built a framework which allows to extract facts from the data source and rank them. Their work defines a set of textual features some of which we use in our work. Suryanto et al. [18] proposed a similar method using the reputation of the question asker and the answerer to determine the relevance of the answer. Both papers are focused on factoid QA using Yahoo! Answers data. Our method is focused on non-factoid questions with using the entire Web as the data source.

Soricut and Brill [16] published one of the first papers on non-factoid question answering, and many others have followed [13, 14, 17]. As a training set they used a corpus of 1M question-answer pairs from FAQ collected on the Web. To search for the answer candidates they used MSNSearch and Google. Our work uses different algorithm for QFM, is trained using Yahoo! Answers dataset and uses learning to rank techniques which started to advance in mid-00s. In recent years the advancements in NLP/ML techniques and availability of large QA datasets have propelled research and contests on answering open-domain non-factoid questions [3]. Wang et al. [21, 22] works were the winner of two subsequent TREC LiveQA competitions. In the first paper they trained an answer prediction model using BLSTM Neural Network. In the second - Neural Machine Translation techniques to train the model which generates the answer itself given only a question. We use their method as a baseline comparing our work against to.

2   Open Domain Factoid/Non-factoid QA System

We use a typical architecture for our QA system. The natural language question is transformed to a keyword based boolean query by the QFM. A commercial search engine, Bing, is then used to obtain the relevant web pages to the boolean query. CAGM mines the downloaded web pages for candidate answers to the original question. Finally, ASM identifies the best answer from all the candidates, and presents it to the user.

Query Formulation Module (QFM): The QFM transforms the natural language question to a well-formed boolean conjunctive query that can be evaluated by a search engine. This is a challenging problem as questions are often verbose. They contain information that is useful for a human but is superfluous, or even misleading, if included in the query. We address this verbosity problem at multiple levels. First, not every sentence in the question contributes to the query. Only sentences that start with WH-words (e.g. Who, When, Where, Why) and end with a question mark do [19]. Second, within a sentence only certain parts of the question are included in the query. For example, transforming the following question *Why's juice from orange peel supposed to be good for eyes?* into a boolean query: *(orange) AND (peel) AND (juice) AND (good) AND (eyes)* is not effective because most of the retrieved web pages are about *orange juice* and not about *orange peel juice*. In order to achieve this, QFM performs grammatical analysis of the question. Specifically, we use the Stanford Dependency Parser [8] to obtain the grammatical structure of the sentence, and then apply a recursive logic to identify various phrases (noun, verb, preposition, and adjective phrases). This allows us to identify important phrases, rather than just individual words. For the above question this approach selects a noun phrase *orange peel*, an adjective phrase *good for eyes*, and a single word *juice*. The final boolean conjunctive query is constructed as follows: *(juice) AND (orange peel) AND (good for eyes)*. This query is successful at retrieving web pages about *orange peel juice* rather than about *orange juice* even though the latter has the more dominant presence on the web.

The English *closed class* terms (pronouns, determiners, prepositions) in the question are often ignored since they do not capture the topic of the question. However, in certain situations the prepositions should be included in the query. In case of the following question *How much should I pay for a round trip direct flight from NYC to Chicago in early November?*, if the preposition words, *from* and *to*, are ignored then the information about the travel direction is lost. To address this issue, the grammatical tree structure of the sentence is leveraged (using Stanford parser) to identify the preposition phrases, such as, *from NYC* and *to Chicago*, and these are included as-is in the boolean query.

The verb phrases are also important because the verb alone is too broad to be a standalone keyword in the query. For question *What should I have in my disaster emergency kit stored outside my house?*, without the verb phrase detection, the system generates *(disaster emergency kit) AND (outside house) AND (store)*. Some of the web pages retrieved by this query are about stores that sell disaster emergency kit. Whereas with verb phrase detection logic, a better query is generated: *(disaster emergency kit) AND (store outside house)*.

Candidate Answer Generation Module (CAGM): The boolean query created by QFM is executed against the commercial search engine, Bing. The top 20 web pages returned for the query are downloaded, and each page is passed through the following text processing pipeline. The first step extracts ASCII text from the web page using an html2text library[3]. We refer to the extracted text as a document. This document is next split into *passages*, where each passage consists of four consecutive sentences, the most popular answer length in Yahoo! Answers dataset. A sliding span of four consecutive sentences is used to generate the passages. Thus a document containing 5 sentences would generate two passages. This approach generates many passages, specifically, $1 + (n - 4)$, where n is the total number of sentences in the document. Passages that do not contain any of the query terms, or that contain more than 2 line breaks, or more than 10 punctuation marks, or non-printable symbols are eliminated. Also, passages that are not in English are filtered out. The langdetect library[4] is employed for language identification. All the passages that survive the filtering step are considered as *candidate answers*.

---

[3] https://pypi.python.org/pypi/html2text.
[4] https://pypi.python.org/pypi/langdetect.

Answer Selection Module (ASM): In this final step of the QA pipeline, the best answer from all the candidate answers is chosen. We experiment with three algorithms for this task: 1. Learning To Rank based LambdaMart algorithm [7], 2. Neural Network based BLSTM algorithm [11], and 3. A combination approach that employs both, LambdaMart and BLSTM.

There is a rich history of LeToR approaches being applied to automated QA [1,5,17]. Following on this tradition, for the baseline approach, we employ the LambdaMart algorithm to learn a ranking model for scoring the candidate answers, and the highest scored answer is selected as the final answer. We refer to this answer selection approach as LLTR. A subset of the *Webscope Yahoo! Answers L6* dataset[5] is used for training the LLTR model. For many questions in this dataset one of the answers for the question is identified as the best answer. For training LLTR the best answer is assigned the highest rank label, and the remaining answers are assigned a rank label proportional to their BM25 score with the best answer. The following feature set is computed for each <question, answer>pair: Okapi BM25 score, cosine similarity, number of overlapping terms, number of punctuation marks in the passage, number of words in the answer, number of characters in the answer, query likelihood probability, largest distance between two query terms in the answer, average distance between two terms, number of terms in longest continuous span, maximum number of terms matched in a single sentence, maximum number of terms in order. Before computing each of these features, all terms from query and candidate answer were stemmed using Porter.

Recurrent Neural Network (RNN) based approaches have received a lot of attention from the QA community recently [10,15,21,22]. Since carefully feature engineering is completely unnecessary for NNs these networks lend themselves very well to the QA problem where it is difficult to defining features that generalize well. In fact, the best performing system (Encoder-Decoder) at the TREC 2016 LiveQA track employed a recurrent neural network based approach. In our work we have employed the Bidirectional Long Short Term Memory (BLSTM) neural network because it adapts well to data with varying dependency spans length. The bidirectional property of this network allows for tracking of both, forward and backward relations in the text. We use a modification of network architecture implemented in [21]. The network consists of several layers: the word embedding layer followed by BLSTM layer, dropout layer to reduce overfitting, mean pooling, and dense layer for the output. The output for the network is a number from 0 to 1 identifying how likely the answer matches the question. It was trained with *ADAM* optimizer, with binary cross-entropy as a target loss function. To train the network a subset containing 384K <question, answer>pairs from the *Webscope Yahoo! Answers L6* dataset was used.

The third answer selection approach that we investigate simply combines the above two approaches. The score assigned by BLSTM to each <question, answer>pair is used as an additional feature in the feature set used by the LLTR ranking algorithm.

---

[5] http://webscope.sandbox.yahoo.com.

3   Experimental Setup and Evaluation Data

In order to compile the subsets of L6 dataset that are used to train the LLTR and BLSTM models, we used two steps to filter out low-quality question-answers. We were discarding: questions with less than 3 answers, questions (or answers) that were too small (less than two sentences) or too long (greater than 1000 characters). A subset of 48,000 question-answers from the L6 dataset was used to train the LLTR ranking model[6].

For training BLSTM, the answers voted as the best one for the questions were assigned the positive label, and answers for another random question were assigned the negative label. For the embedding layer we used the pretrained Google News word2vec model[7]. It was found that the most efficient to use the word2vec vector size of 200 with 15000 most popular words (i.e. all words except these are discarded). We couldn't use more words or word2vec dimensionality because of the overfitting. The input size for BLSTM was 128 words and the dropout was set to 0.5.

To evaluate the ASM we employed the LiveQA track data from TREC 2015 and TREC 2016, which both contain the answers from all participant systems for approximately 1000 questions. Each answer is rated by human judges on the scale from 1 (poor) to 4 (excellent). The effectiveness of the three answer selection approaches was evaluated with the above two datasets. Standard evaluation metrics were used for this task: NDCG (Normalized Discounted Cumulative Gain), MAP (Mean average precision) at rank X, and MRR (Mean Reciprocal Rank) at rank X. As a point of reference we also define a baseline QA approach: the original question is used as-is for the query (stopwords excluded), the top web page retrieved by Bing is downloaded, and the passage with highest BM25 score with respect to the question, is selected as the final answer.

4   Results and Analysis

Table 1 provides the results for the evaluation of the ASM. For the TREC 2015 LiveQA evaluation set, the results for the best performing system for this task are available [22], and are included in the table (Encoder-Decoder). LLTR is less effective than the state-of-the-art approach, Encoder-Decoder, across all the metrics. However, the neural network based approach, BLSTM performs substantially better than Encoder-Decoder and LLTR for both datasets. The results for LLTR+BLSTM illustrate that two approaches have complementary strengths that can be combined to obtain the best results for the task. The difference between LLTR and LLTR+BLSTM is statistically significant.

We believe that quality of the model can be improved by sanitizing the training dataset. Currently, two main problems are: (i) presence of words with misspellings which make computations of statistical features imprecise; (ii) quality of the best answers manually selected by voters. There exists a few approaches

---

[6] The datasets will be shared after publication.

[7] https://code.google.com/archive/p/word2vec/.

Table 1. Results of answer ranking

	NDCG	MAP@			MRR@		
		2	3	4	2	3	4
TREC 2015							
Encoder-Decoder	0.6346	0.5124	0.3390	0.1657	0.5645	0.3672	0.1779
LLTR	0.6222	0.4843	0.3162	0.1551	0.5490	0.3522	0.1562
BLSTM	0.6562	0.5462	0.3470	0.1744	0.5874	0.3790	0.2046
LLTR+BLSTM	0.6602	0.5498	0.3487	0.1763	0.5901	0.3810	0.2059
TREC 2016							
LLTR	0.6484	0.5124	0.3463	0.2165	0.6211	0.3806	0.2410
BLSTM	0.6712	0.5591	0.3788	0.2541	0.6478	0.4033	0.2879
LLTR+BLSTM	0.6754	0.5674	0.3835	0.2567	0.6504	0.3990	0.2928

to diminish impact of both issues such as [9] for misspellings and [2] for keeping only high-quality answers.

Evaluating the end-to-end QA system is tricky because the generated answer might change if the search engine results change, and thus manual assessment of answer relevance cannot be a one-time activity. As a compromise, we attempt to provide quantitative evaluation by computing similarity between the answer generated by our system and the best answer assessed by TREC annotators. The intuition being that higher the similarity score the more effective the system is.

Table 2. Overall system quality results

Metric name	Jaccard coefficient			Cosine similarity			KLD similarity		
Question length (in tokens)	[0;10]	[11;32]	[33;)	[0;10]	[11;32]	[33;)	[0;10]	[11;32]	[33;)
Number of questions	134	152	148	134	152	148	134	152	148
BM25	0.05 ±0.01	0.05 ±0.01	0.06 ±0.01	0.07 ±0.01	0.08 ±0.02	0.07 ±0.01	0.43 ±0.05	0.48 ±0.05	0.44 ±0.06
LLTR	0.12 ±0.02	0.11 ±0.01	0.11 ±0.02	0.23 ±0.03	0.25 ±0.02	0.25 ±0.03	0.64 ±0.04	0.65 ±0.05	0.67 ±0.06
BLSTM	0.13 +0.02	0.12 ±0.02	0.12 ±0.02	0.23 ±0.02	0.26 ±0.03	0.26 ±0.04	0.65 ±0.05	0.65 ±0.06	0.68 ±0.06
LLTR+ BLSTM	0.13 ±0.02	0.13 ±0.02	0.13 ±0.02	0.24 ±0.03	0.26 ±0.03	0.27 ±0.04	0.67 ±0.04	0.68 ±0.05	0.70 ±0.05

Table 3. Answers returned by the system compared to best human answers

---

Question: How do you fix a computer programme that run for itself instead of serving you?

System's Answer: Run AdwCleaner. Click the "Scan" button in AdwCleaner to scan your computer for infections. This may take 20 min or so to complete. Once the "Clean" to remove any infections that AdwCleaner found.

Best Human Answer: More information would be useful.

Question: What's the best tablet to get?

System's Answer: 7-inch Apple iPad Pro is the best tablet yet. We love the sleek design and stunning True Tone display, which makes the most of the 1 million tablet-specific apps available in the iOS App Store. Not only can this slate's 12-megapixel camera shoot in 4K, the A9X processor inside can also edit multiple 4 K streams without a hiccup. Plus, this more portable Pro tablet lasts a remarkable 10 h and 53 min on a charge, and it supports the $99 Apple Pencil.

Best Human Answer: Transformer prime if your looking for customization and uniqueness. iPad if you like it simple.

Question: Should the WWE hire AJ Styles? What do you think?

System's Answer: WWE would hire Sting before they hire AJ Styles.not only that.they'd make him cover his tattoo of "AJ" and then change his name and make out of some NXT rookie and give him some generic name like, "Kyle Evans". AJ isnt a rookie and shouldnt be treated as such.

Best Human Answer: WWE has tried several times to sign AJ Styles. AJ Styles has turned WWE down every time. AJ Styles can make more money working for New Japan Pro Wrestling, Ring of Honor and Independent Professional Wrestling Promotions all over the world. AJ Styles works where he wants, when he wants and as often as he wants.

Question: Can midline shift be corrected by elastics? I got upper braces and I'm getting the bottom ones soon.

System's Answer: You should let your dentist know so he can correct his own work (that's what you paid him for, right?) or at least ask for more rubber bands. I wanted to prevent my midline from shifting due to my braces but my dentist waited to correct mine, so I certainly understand your concern. I began asking for extra rubber bands of the kind I needed at every dentist visit so I could start correcting it on my own.

Best Human Answer: What you need to realize is that what "a lot of people" have had has nothing to do with what you need. Every case is different. Patient "A" can have a midline shift and will be corrected with elastics. Patient "B" can have a midline shift that may need surgery. There are many other factors, including the formation of the jaw, that come into play. Mention using elastics to your orthodontist and s/he will be able to tell you, and show you in x-rays, why surgery is being recommended for your particular situation.

Question: Which is worse? Gambling or cheating with PEDs on the game of baseball? Why?

System's Answer: That gambling creates the moral hazard of being potentially so in hock to bookies that they require you to throw a game – something that apparently didn't happen to Rose but quite easily could have – must be noted as well. No one takes PEDs to lose or only takes them before important games. Results are more important than statistics. This would seem self-evident, but it's a claim Rose makes explicit.

Best Human Answer: PEDs. using a illegal substance and not working hard to improve your craft is worse. Gambling is just stupid, as there's still a chance you could gamble wrong

The absolute similarity scores are not the focus, but the relative difference in system scores is worth analyzing.

Table 2 reports the similarity score using Jaccard, Cosine, and Negative & Symmetric KL Divergence metrics for the TREC 2015 LiveQA evaluation dataset. The presented results are also binned based on question-length, so as to analyze the effects of question-length on the system's performance. Compared to the baseline (BM25-based QA system), all the versions of our system perform better, with the combination approach (LLTR+BLSTM) being the best one. This trend is seen across the three metrics, and the different query-lengths. In future we plan to employ more powerful metrics, such as, METEOR[8] and doc2vec[9] that can capture synonyms and paraphrasing. In Table 3 we present examples of the answers given by our system for some of the dataset questions. In some cases, such as a question about computer programs the QA system produces more comprehensive answer than humans. Also, it is worth noting that the quality of the questions varies dramatically which is one of the big challenges for this problem.

5   Conclusions and Future Work

In this paper we presented our attempt at tackling the challenging problem of answering open-domain non-factoid questions. The empirical evaluation illustrates that the simple approach of combining LLTR and BLSTM outperforms the state-of-the-art system. The qualitative evaluation shows that the system is capable of producing high-quality answers. Possible directions for future research include use of recurrent neural networks for summarizing the question and answer to generate better queries and more concise answers[10]. The quality of the training dataset could also be improved to increase the performance of answer selection models.

References

1. Agarwal A., et al.: Learning to rank for Robust question answering. In: Proceedings of CIKM (2012)
2. Agichtein E., et al.: Finding high-quality content in social media. In: Proceedings of WSDM (2008)
3. Agichtein E., et al.: Overview of the TREC 2015 LiveQA track. In: Proceedings of TREC (2015)
4. Bian J., et al.: Finding the right facts in the crowd: factoid question answering over social media. In: Proceedings of WWW (2008)
5. Bilotti M.W., et al.: Rank learning for factoid question answering with linguistic and semantic constraints. In: Proceedings of CIKM (2010)

---

[8] http://www.cs.cmu.edu/~alavie/METEOR/.
[9] https://radimrehurek.com/gensim/models/doc2vec.html.
[10] https://research.googleblog.com/2016/08/text-summarization-with-tensorflow.html.

6. Bobrow D.G.: A question-answering system for high school algebra word problems. In: Proceedings of FJCC (1964)
7. Burges C.: From ranknet to lambdarank to lambdamart: an overview. Learning **11**, 81 (2010)
8. Chen D., Manning, C.: A fast and accurate dependency parser using neural networks. In: Proceedings of EMNLP (2014)
9. Chen, Q., Li, M., Zhou, M.: Improving query spelling correction using web search results. In: Proceedings of EMNLP-CoNLL (2007)
10. Cohen, D., Croft, B.: End to end long short term memory networks for non-factoid question answering. In: Proceedings of ICTIR (2016)
11. Graves, A., Schmidhuber, J.: Framewise phoneme classification with bidirectional LSTM and other neural network architectures. Neural Netw. **18**(5), 602–610 (2005)
12. Green, C.: Theorem proving by resolution as a basis for question-answering systems. Mach. Intell. **4**, 183–205 (1969)
13. Higashinaka, R., Isozaki, H.: Corpus-based question answering for why-questions. In: Proceedings of IJCNLP (2008)
14. Oh, J.H., et al.: Why question answering using sentiment analysis and word classes. In: Proceedings of EMNLP-CoNLL (2012)
15. Severyn A., Moschitti A.: Learning to rank short text pairs with convolutional deep neural networks. In: Proceedings of SIGIR (2015)
16. Soricut, R., Brill, E.: Automatic question answering using the web: beyond the factoid. Inf. Retrieval. **9**, 191–206 (2006)
17. Surdeanu M., et al.: Learning to rank answers to non-factoid questions from web collections. Comput. Linguist. **37**, 351–383 (2011)
18. Suryanto, M.A., et al.: Quality-aware collaborative question answering: methods and evaluation. In: Proceedings of WSDM (2009)
19. Varanasi, S., Neumann, G.: Question/answer matching for Yahoo! Answers using a corpus-based extracted ngram-based mapping. In: Proceedings of TREC (2015)
20. Waltz, D.L.: An English language question answering system for a large relational database. Commun. ACM. **21**, 526–539 (1978)
21. Wang, D., Nyberg, E.: CMU OAQA at TREC 2015 LiveQA: discovering the right answer with clues. In: Proceedings of TREC (2015)
22. Wang, D., Nyberg, E.: CMU OAQA at TREC 2016 LiveQA: an attentional neural encoder-decoder approach for answer ranking. In: Proceedings of TREC (2016)

Linguistic Features as Evidence for Historical Context Interpretation

Jyi-Shane Liu[1(✉)], Ching-Ying Lee[2], and Hua-Yuan Hsueh[3]

[1] Department of Computer Science, Center for Creativity and Innovation Studies,
National Chengchi University, Taipei, Taiwan
liujs@nccu.edu.tw
[2] Department of Applied Foreign Languages, University of Kang Ning,
Taipei, Taiwan
[3] Department of History, National Chengchi University, Taipei, Taiwan

Abstract. Inspired by the great potential of linguistic features in preserving and revealing writers' state of mind and conception in certain space and time, we use linguistic features as a vehicle to extract pieces of significant information from a large set of text of known origin so as to construct a context for personal inspection on the writer(s). In this research, we choose a set of linguistic features, each of a grammatical function or a grammatical association pattern, and each represents a different perspective of contextual annotation. In particular, the selected grammatical items include personal pronoun, negation, noun chunk, and are used as text slicing tubes for extracting a certain aspect of information. The initial results show that some selected grammatical constructions are effective in extracting descriptive evidence for construing historical context. Our study has contributed to exploring an effective avenue for innovative history studies by means of examining linguistic evidence.

Keywords: Linguistic feature · Historical context

1 Introduction

Linguistic features are variant and invariant units of language use, and have been considered as a basis of distinction for a variety of purposes in language analysis. In particular, corpus linguistics exploits linguistic features as some of the identifiable patterns in natural texts for both quantitative and qualitative analytic frameworks. Simple occurrence count of a linguistic item or a grammatical construction presents an objective reference for some lexical and grammatical studies. More complex association patterns of linguistic and non-linguistic features provide additional in-depth information that is essentially indicative to the investigation of the use of a linguistic feature and the varieties of texts, e.g., registers, dialects, historical periods [1]. For example, some of the lexical and grammatical choices are identified as register features in the context of schooling that requires authoritativeness and explicit logical presentation [12].

© Springer International Publishing AG 2017
K. Ekštein and V. Matoušek (Eds.): TSD 2017, LNAI 10415, pp. 299–307, 2017.
DOI: 10.1007/978-3-319-64206-2_34

Authorship attribution is another area of study motivated by identifying unique linguistic features for the purpose of determining the authorship of an anonymous text [11]. An ideal context of authorship attribution is to have a limited closed set of candidates and a sufficiently large collection of training text for each known author. Recent research categorizes authorship attribution in real-life scenarios into three variants: (1) the profiling problem, where there is no candidate set available and the task is to gain the unknown author's demographic or psychological information, (2) the needle-in-a-haystack problem, where the size of the candidate set is large and the writing sample is very limited, and (3) the verification problem, where the question is whether the suspect is the author [10]. Authorship attribution problems have recently received much attention with the encouraging results in forensic linguistics [4], as well as in literary research [3,7]. An interesting application in Chinese literature is to help resolve the long-standing debate of authorship in *Dream of the Red Chamber*, one of *China's Four Great Classical Novels*, written sometime in the middle of the 18th century during the *Qing Dynasty*. Research has provided statistical evidence that reveals linguistic differences between the first 80 chapters, by the known author, *Cao Xueqin*, and the later 40 chapters, presumably by a second author [8,13].

Inspired by the great potential of linguistic features in preserving and revealing writers' state of mind and conception in certain space and time, we use linguistic features as a vehicle to extract pieces of significant information from a large set of text of known origin so as to construct a context for personal inspection on the writer(s). Our research interest is motivated by the need in historical study, with the opportunity of abundant text dataset, to gather as much descriptive evidence as possible to better understand an iconic figure, a social group, or an era that may either shed light on new conjecture or verify previous speculation. From the perspectives of systemic functional linguistics [5], there exists some obligatory linguistic features which appear regularly for underlying certain functions in text. In response to different contexts, writers consciously and subconsciously make lexical and grammatical choices to fulfill different functional purposes. Linguistic features in terms of lexical and grammatical association patterns, therefore, reflect the context of a text's production. By analyzing linguistic features in historical text, we attempt to establish a link between what were prominently expressed in a focused list of words and what were indicative in the historical context.

In this research, we choose a set of linguistic features, each of a grammatical function or a grammatical association pattern, and each represents a different perspective of contextual annotation. In particular, the selected grammatical items include personal pronoun, negation, noun chunk, and are used as text slicing tubes for extracting a certain aspect of information. Our research purpose is two folds. First, we examine whether certain types of linguistic-based contextual information from historical text can be useful in complementing subject understanding in historical study. Second, we verify the language behavior of certain types of linguistic features in different registers of actual historical text. In other words, we attempt to establish a two-way link connecting linguistic features and historical context and expect its future expansion. The initial results show that some selected grammatical constructions are effective in extracting descriptive evidence for construing historical context.

2   Data and Methodology

For the purpose of historical study, we focus on one particular individual, *Lei Chen* (*Lei* henceforth), based on his iconic status as a political leader and the availability of his writing in digital format. *Lei* was a political pioneer in advocating democracy and freedom in the era of Taiwan's authoritarian regime. He founded and edited the periodical *Free China* in 1949 and published articles that criticized government and demanded reform in political and juridical systems. He also attempted to form a political party in opposition to the authoritative ruling party, and in 1960 was charged with treason and jailed for 10 years. After his release from prison, *Lei* was still under government surveillance until his pass away in 1979.

The text data (in Mandarin Chinese) that involves *Lei's* writing include *Lei's* personal diary and the periodical *Free China Journal*. We differentiate *Lei's* article in *Free China* with articles by other authors and form three sets of text. The resulted background information are shown in Table 1. The organization of the text sets involves variables in genre, authorship, and register. First, the same author's writing in different genres, e.g., *Lei's* personal dairy vs. *Lei's* public journal articles, can be compared. Second, authorship in the same register, albeit one versus group, e.g., *Lei's* journal articles vs. Non-*Lei's* journal articles, can be inspected. Third, *Lei's* political life ended by imprisonment brings into three different writing registers in personal diary for further observation.

Table 1. Background information on text data

Text set	# of word tokens	# of entry/articles
Lei's diary	1,233,433	9,444
Part 1, politically active, 1948–60	627,464	3,649
Part 2, imprisonment, 1961–70	459,489	4,208
Part 3, politically withdrawn, 1971–79	146,480	1,587
Lei's articles in Free China, 1949–60	379,121	87
Non-Lei's articles in Free China, 1949–60	7,867,019	3,063

The set of linguistic features we selected are intended for providing a vantage point that allows strategical retrieval of some contextual interpretation from authorship production. In this regard, we consider a few representative grammatical construction based on their functional purposes in instantiating/revealing author's social context in content subject (field) and role relationships between writer/reader (tenor) [6].

Personal pronoun includes first person singular, first person plural, second person singular, second person plural, third person singular, and third person plural. Writers make pronominal choices to instantiate self-representation (personal self) and to interpret own position in social context (social self) [2]. Thus, personal pronoun us-age is proposed as an index of a writer's social functioning status.

Negation expression is a universal property of natural language. Many languages use negative markers to reverse the truth value of affirmative sentences and no language explicitly marks affirmative sentences [14]. In this regard, negative markers serve as a reference point to the subjects of negation. Thus, the compound of negative marker and the subject of negation may provide an index of a writer's judgmental and affective status.

Noun chunk is a sequential combination of words with part-of-speech of noun and other grammatical types to form a large semantic unit in a sentence. In this research, we consider two types of paired word chunk, noun-noun and noun-verb. Noun and verb have been central to the semantic content of expression. By retrieving noun-noun and noun-verb chunks, we hope to capture an essential part of a writer's content subject.

We apply the set of linguistic features to the text set and retrieve the matched lexical items. Each lexical item retrieved is accumulated with a frequency of occurrence. The absolute frequency in each text is then converted to the relative frequency per million words (tokens). In other words, occurrence frequency of lexical item is normalized so as to be comparable among text of different size. Lexical items in each featured type of grammatical construction are then ranked by the normalized frequency so that the most significant lexical items in each category are identified.

3   Analytical Results

The use of personal pronoun reflects writers' psychological proximity between personal self and others in an interactional context [9]. The relative frequency among the pronominal choice indicates the primary and secondary interactional roles in the social embedding of the writer's text production.

Figure 1 show the normalized frequency of the personal pronoun usage in the three text set. The list of pronominal choices are arranged from '*I*' at right to '*they*' at left in increasing psychological proximity distance to self (the first '*you*' is the singular form, the second '*you*' is the plural form, the second '*he*' is another less used third person singular form in Mandarin Chinese). All three texts show multi-polar interactional roles. The primary interactional roles in the diary involve self and the third personal singular (he), with a much higher frequency than in other two texts. This indicates that the diary has a high concentration on reflecting the writer's own status and the interactional process with a third person. In contrast, both *Lei's* and non-*Lei's* articles in *Free China* have three primary interactional roles, '*we*', '*he*', '*they*' that are not as polarized as the diary. This confirms that the articles are of public advocate nature and are intended to involve the addressee in a community sense.

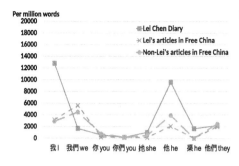

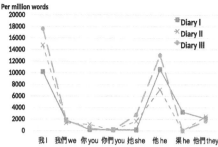

Fig. 1. Comparison of personal pronoun usage among texts

Fig. 2. Comparison of personal pronoun usage within diary

Figure 2 further examines the diary in three parts. The first part of the diary has the least frequency of '*I*' and higher aggregated frequency of third persons. At this stage in life, *Lei* was politically active and engaged in much social interaction. The second part of the diary shows the next highest frequency of '*I*' and least frequency of third person. At this stage, *Lei* was in prison with constrained and limited social interaction. The psychological attention was on person self. The third part of the diary has both the highest frequency of '*I*' and '*he*'. This also seems to concur with the more reflective nature in *Lei's* mostly family living life of political withdrawn.

Table 2. Top 3 indicative lexical items following the negation marker 不 bu4 (no, not) in normalized frequency (per million words)

Lei's diary		Lei's articles in Free China		Non-Lei's articles in Free China	
贊成(agree)	285	信任(trust)	148	合理(reasonable)	43
滿意(satisfy)	115	合理(reasonable)	78	容易(easy)	42
舒服(comfort)	83	公平(fair)	73	承認(admit)	41

Next, we examine the most frequent lexical items subject to the negation marker of '不' bu4 (no, not). The most negated lexical item in both texts of *Free China* is '是', which is a copula in grammatical sense as '*be*' with a five to seven times ratio to the second in rank. In contrast, the most negated lexical item in the diary is '好' (good), with a two times ratio to the next frequent item, '是' (copula/be). Other lexical items being frequently negated in all three texts are '*know*', '*wish*', '*should*', '*dare*'. Table 2 shows more indicative lexical items being negated with frequency somewhat lower than these modality items. This seems to support that diary is more judgmental in personal context and articles in *Free China* are more argumentative of certain affairs.

For comparison within the diary, as shown in Table 3, the subject of negation in part one is projected toward outside affairs. When *Lei* was in confinement,

Table 3. Top 3 indicative lexical items following the negation marker 不 bu4 (no, not) in normalized frequency (per million words) in *Lei's Dairy*

Part 1		Part 2		Part 3	
贊成(agree)	489	舒服(comfort)	174	願意(consent)	96
滿意(satisfy)	191	願意(consent)	111	贊成(agree)	89
參加(participate)	57	高興(happy)	72	舒服(comfort)	68

the subject of negation turns to his own physical and mental status. In the last stage of his life, *Lei's* frequency of negation reduced and involved both nature of opinion and physical condition.

Lastly, we observe the most frequent noun-noun chunks and noun-verb chunks in the texts. In Table 4, the top frequency noun subject in *Lei's* diary is *Mr. Hu*, which refers to *Hu Shih*, a pioneering intellect and a key contributor to Chinese liberalism and language reform. *Hu* was president of Beijing University and later president of the Academia Sinica in Taipei. *Hu* was an ally of *Lei's* political movement and the diary reveals *Hu's* significance in *Lei's* own account with the highest frequency. 3 pm (in the) afternoon reveals the peak hour of notable activities that worth recording. It may also indicate *Lei's* (and others') habit of substantial noon break at that time. Noun subjects in both texts of *Free China* are consistent with the content nature of advocating political reform. However, *Lei's* articles show a higher concentration in reviewing and disputing political issues than non-*Lei's* articles.

Table 4. Top 3 lexical items of noun-noun chunk in normalized frequency (per million words)

Lei's diary		Lei's articles in Free China		Non-Lei's articles in Free China	
(Mr. Hu)	923	(democracy politics)	702	(democracy politics)	186
(3 pm afternoon)	566	(public opinion institution)	424	(democracy country)	145
(family)	463	(democracy country)	413	(imperialism)	129

For comparison within the diary in Table 5, in part one, *Mr. Hu* was mentioned with an even higher frequency while *Lei* was politically active, followed by temporal annotation of activity. In part 2 and part 3, the central attention shifted toward family. A prison officer surnamed *Dong* appears to be influential in *Lei's* prison life. *Cheng Shewo* was a journalist and publisher, who owned a newspaper. *Lei's* frequent interaction with *Cheng* in the last stage of his life seems to indicate some kind of writing agenda. *Lei* also faced physical health issue in later years with frequent visit to a clinic. Another prominent noun-noun chunk item, *secret service agency*, also come up in the top ten list of *Lei's* third-part diary, indicating his awareness of being constant surveillance.

Table 5. Top 3 lexical items of noun-noun chunk in normalized frequency (per million words) in *Lei's Dairy*

Part 1		Part 2		Part 3	
(Mr. Hu)	1508	(family)	825	(family)	737
(3 pm afternoon)	631	(prison officer Dong)	655	(Cheng Shewo)	471
(today morning)	475	(3 pm afternoon)	594	(central clinic)	410

In terms of noun-verb chunk, Table 6 offers a few interesting observation. First, writing and visiting are *Lei's* primary activities. Second, the central themes in *Free China* are indeed political and judicial reform. Third, the higher frequency of primary noun-verb chunks in *Lei's* articles than non-*Lei's* articles supports that *Lei* did play the role of prime advocate.

Table 6. Top 3 lexical items of noun-verb chunk in normalized frequency (per million words)

Lei's diary		Lei's articles in Free China		Non-Lei's articles in Free China	
(today write)	275	(speech freedom)	181	(local autonomy)	75
(today sunny)	266	(local autonomy)	108	(democracy freedom)	69
(morning went)	122	(judicial independence)	92	(speech freedom)	68

In Table 7, noun-verb chunks are compared within the diary. Part one shows active social activity, while part 2 and part 3 are mostly individual activity, especially in writing. The relative frequency indicates that *Lei's* activities in part one were more diverse, in contrast to more concentration in part 2 and part 3. *Lei's* frequent hill-climbing in the morning also indicates that he had paid more attention to his physical condition in his later years.

Table 7. Top 3 lexical items of noun-verb chunk in normalized frequency (per million words) in *Lei's Dairy*

Part 1		Part 2		Part 3	
(morning went)	239	(today write)	690	(home write)	867
(airport send)	207	(today sunny)	520	(morning write)	382
(morning visit)	127	(last night sleep)	222	(morning hill-climbing)	375

4   Discussion and Conclusion

Text production involves essential linguistic focus on lexical choices and grammatical patterns, especially those that deliver viewpoints of a writer. Extralinguistic context entails the relation between the micro-structure of text and the macro-structure of society. Corpus linguistic and language use analysis pave the way for studying the empirical basis of society in these micro-macro relations. Integrating theoretical knowledge of systematic functional linguistics with computational text analysis, our research presents a case of applying the linguistic feature approach to extract significant information for construing historical contexts, and to further identify writing variations across genres and registers. Instead of assuming all words are equal and relying on statistical properties of common n-gram models, this approach exploits existed linguistic knowledge so as to effectively target meaningful components of expression. The language use behavior is also observed in the complete corpus of the subject of study. This allows us to use simple normalized frequency as adequate indicator of significance in presenting important contextual information.

The research implications comprise two folds. First, linguistic features attested in the historical text provide objective micro-evidences for better understanding historical issues. This study confirms that factual data produced by a comparative analysis on *Lei's* diary and journal articles are aligned with expert interpretations on prior subjective knowledge. Second, this study demonstrates the possibility for historians exploiting linguistic features to leverage large-sized historical text for new insights. This bilateral interaction may lead to fruitful endeavor in systematically digesting a sea of historical text for research advances. Our future research includes more thorough examination on sources of linguistic feature occurrences as well as replicating the inter-disciplinary collaboration in other historical texts and accumulating linguistic evidence for historical interpretation. Words and texts convey thoughts and express feelings. Our empirical study has contributed to exploring an effective avenue for innovative historical studies by means of examining linguistic evidence.

References

1. Biber, D., Reppen, R.: Corpus Linguistics. Sage, London (2012)
2. Brewer, M.B., Gardner, W.: Who is this "we"? Levels of collective identity and self representations. J. Pers. Soc. Psychol. **71**(1), 83 (1996)
3. Burrows, J.: 'Delta': a measure of stylistic difference and a guide to likely authorship. Literary Linguist. Comput. **17**(3), 267–287 (2002)
4. Chaski, C.E.: Who's at the keyboard? Authorship attribution in digital evidence investigations. Int. J. Digit. Evid. **4**(1), 1–13 (2005)
5. Eggins, S.: Introduction to Systemic Functional Linguistics. A&C Black, London (2004)
6. Halliday, M., Matthiessen, C.M., Matthiessen, C.: An Introduction to Functional Grammar. Routledge, London (2014)
7. Hoover, D.L.: Frequent word sequences and statistical stylistics. Literary Linguist. Comput. **17**(2), 157–180 (2002)

8. Hu, X., Wang, Y., Wu, Q.: Multiple authors detection: a quantitative analysis of dream of the red chamber. Adv. Adapt. Data Anal. **6**(04), 1450012 (2014)
9. Íñigo-Mora, I.: On the use of the personal pronoun we in communities. J. Lang. Polit. **3**(1), 27–52 (2004)
10. Koppel, M., Schler, J., Argamon, S.: Computational methods in authorship attribution. J. Am. Soc. Inf. Sci. Technol. **60**(1), 9–26 (2009)
11. Love, H.: Attributing Authorship: An Introduction. Cambridge University Press, Cambridge (2002)
12. Schleppegrell, M.J.: Linguistic features of the language of schooling. Linguist. Educ. **12**(4), 431–459 (2002)
13. Tu, H.C.: Using a text mining approach to study the authorship controversy on the last 40 chapters of the dream of the red chamber. In: Hsiang, J. (ed.) Digital Humanities and Craft: Technological Change. Center of Humanities Research, Taiwan (2014)
14. Zeijlstra, H.: Negation in natural language: on the form and meaning of negative elements. Lang. Linguist. Compass **1**(5), 498–518 (2007)

Morphosyntactic Annotation of Historical Texts.
The Making of the Baroque Corpus of Polish

Witold Kieraś[1][✉], Dorota Komosińska[1], Emanuel Modrzejewski[2],
and Marcin Woliński[1]

[1] Institute of Computer Science, Polish Academy of Sciences, Warszawa, Poland
wkieras@ipipan.waw.pl
[2] Institute of Polish Language, Polish Academy of Sciences, Kraków, Poland
https://www.ipipan.waw.pl
https://www.ijp.pan.pl

Abstract. In the paper, we present some technical issues concerning processing 17th & 18th century texts for the purpose of building a corpus of that period. We describe a chain of procedures leading from transliterated source texts to morphological annotation of text samples that was implemented for building the Baroque Corpus of Polish, a relatively large historical corpus of Polish texts from 17th & 18th c. The described procedure consists of: automatic transliteration from original spelling to modern one, morphological analysis (including the construction of an inflectional dataset for Baroque Polish) and a tool for manual morphosyntactic annotation. The toolchain is being used to create a small manually validated subcorpus, which will serve as training data for a stochastic tagger. Then a larger corpus will be annotated automatically and made available via the Poliqarp corpus search tool.

Keywords: Historical corpora · Morphosyntactic annotation · Baroque Polish

1 Introduction

The purpose of this paper is to present technical details of processing text samples for the Baroque Corpus of Polish (BCP).[1] The process involves manual typing of historical texts (transliteration), automatic transcription from original to contemporary spelling, automatic tokenization and morphological analysis, and adjustment to manual morphosyntactic annotation (disambiguation) in a multi-access web application.

Processing historical texts is a much more laborious and time-consuming process than in the case of contemporary linguistic data. Historical texts need

---

[1] The work being reported was co-financed by a National Science Centre, Poland grant DEC-2014/15/B/HS2/03119 and a Ministry of Science and Higher Education National Programme for the Development of Humanities grant 0036/NPRH2/H11/81/2012.

K. Ekštein and V. Matoušek (Eds.): TSD 2017, LNAI 10415, pp. 308–316, 2017.
DOI: 10.1007/978-3-319-64206-2_35

to be transliterated properly from original typesetting and transcribed[2] to modern spelling. Text processing tools, from sentencer to morphological analyzer, need to be adjusted to capture the diversity of historical inflectional forms, variability, punctuation rules and all kinds of conventions that do not appear in modern texts. While processing historical texts one needs to cope with much less standardized lexis, inflection, and spelling. In the article, we present our own attempt at dealing with 17th & 18th century Polish texts for the purpose of building a corpus of that period. We will focus on technical details of automated text processing, leaving apart more theoretical and philological problems such as acquiring and selecting text samples and creating a sound and adequate morphosyntactic tagset, although we keep them in mind and solve them in parallel.

The final Baroque corpus will be about 12 million tokens large, manually transliterated and automatically transcribed to modern spelling. A small subcorpus (ca. 500,000 tokens) of manually disambiguated samples (ca. 200 words each) will serve as a training data set for a stochastic tagger with which the rest of the corpus will be automatically annotated.

2   Automatic Transcription

The source texts of BCP are not homogeneous regarding the form in which they were acquired. They divide into three types: (1) original editions (in original spelling), (2) 19th century editions of Baroque texts edited in 19th century spelling, (3) contemporary editions in contemporary spelling. Texts representing the last type require no adjustment before further processing but they are a minority. Thus most of the samples require transcription which needs to be conducted automatically. For the purpose of transcription, we use the converter created by Janusz Bień and his team for the IMPACT project – a rule-based tool[3] [7] for substituting letters or sequences of letters based on the context in which they appear. The procedure itself is simple, but it requires building a relatively large set of rules which are created manually.

Different sets of rules were used for texts in categories (1) and (2), but the main body of rules is common for both types (the total amount of rules is between 3000 and 4000). The structure of a rule is based on regular expressions, cf. the following examples:

---

[2] Although it is possible to omit the transcription step (either manual or automatic) and operate on transliterated historical texts directly, we have decided to transcribe all texts to standard modern spelling. This allows us to avoid including all possible orthographic variations in the inflectional dictionary, which would be costly in terms of work required and dictionary size. Based on our experiences from other projects we claim that in the case of Polish omitting the transcription step is possible for texts published after 1830, when the spelling rules became more standardized and more similar to modern ones.

[3] https://bitbucket.org/jsbien/pol.

	left context	match	right context	replace with	exceptions
1.	.*	ô	.*	go	null
2.	^	naiw	.*	najw	naiwn.*
3.	T	j	T	y	mjr

1. Replace every ô (which is used in the source corpus to denote o in the superscript) with go independent on the context, for example: $ie^o \rightarrow iego$[4] ('his, him').
2. Replace initial $naiw$ with $najw$ if the following character is not an n: $naiwięk$-$szy \rightarrow największy$ ('the biggest one'), but $naiwny \rightarrow naiwny$ ('naïve').
3. Replace every j between consonants (capital T denotes any consonant) with a y: $zwjczaj \rightarrow zwyczaj$ ('habit'), but mjr (abbreviation for $major$ 'squadron leader') remains unchanged as the only exception from this rule.

Original editions (1) required a special set of rules which was prepared for the purpose of unifying differences in transliteration standards of two sources of the used texts (these include representation of special characters such as historical ligatures). Additional rules were provided to normalize the differences in manually introduced combining characters (for example, there is a difference between grave accent, modifier letter grave accent, combining grave accent, and modifier letter middle grave accent). Most rules operate on prefixes, but some modify whole words; sometimes it is more efficient to provide a rule that can be applied to the only one string rather than to specify the contexts or enumerate the exceptions. Special care was taken of the cases where a word was incorrectly split but newly formed tokens were recognized as existing forms, e.g. the string *poidzie* could be tokenized as *po* (ambiguous preposition) and *idzie* ('goes'), while it should be transcribed and analyzed as *pójdzie* ('will go').

The goal of transcription was to normalize the texts to make them better suited for morphological analysis with the tools described in the next section. This goal has been achieved as can be seen from Table 1. The process of transcription slightly changes the number of tokens in the corpus, but more importantly it greatly reduces the number of token types that have to be catered for by morphological analyzer. Due to normalization the percentage of unrecognized

Table 1. Number of tokens in the two representations of the corpus

	Transliteration	Transcription
Token occurrences	12,814,830	12,832,214
Token types	646,410	476,733
Unrecognized token types	75.55%	48.08%
Unrecognized token occurrences	24.04%	5.4%

---

[4] Subsequently this form is changed to *jego* by another rule.

tokens drops from hardly acceptable 24% to modest 5.4% (which is about twice the number for contemporary text).

3   Morphological Analysis

A morphological analyzer is an essential tool in building corpora for inflectional languages. Since compiling extensive morphological data sets is extremely time and cost consuming, it is much more reasonable to modify existing resources than to create new ones from scratch. The obvious choice for morphological analysis is Morfeusz 2 [12], most widely used in Polish NLP and highly configurable analyzer. It allows to customize all linguistically sensitive parts of the analysis: inflectional dictionary, tokenization and tagset.

For the purpose of the Baroque analyzer's dictionary we have used all existing inflectional data that was available, namely (1) inflectional information from the Electronic Dictionary of 17th & 18th century Polish (e-SXVII) [6], and (2) contemporary data of Grammatical Dictionary of Polish (SGJP) [10,13] modified ("aged") to fit into the tagset for Baroque morphosyntax, as well as (3) some automatically obtained extensions of inflectional paradigms from e-SXVII.

The data obtained from e-SXVII is relatively small, the dictionary is still under development, currently it consists of ca. 25,000 lexical entries but many of them are just stubs containing little inflectional data. The dictionary notes only those word forms that were attested in source texts. As a consequence it contains only ca. 70,000 word forms, which gives a ratio of less than three forms per paradigm, meaning that larger part of inflectional paradigms in the dictionary are incomplete. Over a half of the lexical entries contains only one form from the whole paradigm.

A much larger linguistic resource is the contemporary inflectional data of SGJP which serves as a basis for standard modern version of Morfeusz 2. Its extensive lexical basis goes back as far as last decades of 18th century, which makes it a perfect resource also in applications concerning historical corpora. SGJP consists of over 330,000 lexemes and nearly seven million word forms, which makes it the largest and most widely used inflectional data source for Polish. However, the data needs at least some adjustments to apply them in the Baroque. The adjustments, which we call "aging" of SGJP, are twofold: (a) fitting existing paradigms of modern inflectional system into Baroque tagset and (b) automatically generating forms absent in contemporary Polish but highly regular and of significant frequency in 17–18th c.

The fact that the grammatical gender system looked slightly different in 17–18th c. than now can serve as an example of the former. A so called masculine personal gender was only evolving during that period so contemporary gender system needed to be mapped into historical one to fit into Baroque tagset.

The later can be exemplified by historical feminine accusative adjectival forms of non-complex declension such as *piękną* 'beautiful', *radosnę* 'cheerful', which can be obtained by simple substitution of the last character from modern form *piękną, radosną*. Same applies to superlative forms of adjectives and adverbs such as *napiękniejszy* 'most beautiful', *napiękniej* 'most beautifully' which can be produced from modern forms *najpiękniejszy, najpiękniej*. Also some regular verbal imperative forms of dualis (nonexistent in modern Polish) were massively reconstructed for most verbs, e.g. *zamorduj* 'murder' (modern imp. 1st pers. sing.), *zamordujta* (historical imp. 2nd pers. dual.), *zamordujwa* (historical imp. 1st pers. dual.). Such regularities apply to extensive classes of lexemes with almost no exceptions, leading to obvious overproduction of historical forms for lexemes that didn't exist in 17–18th c. This is however a minor problem as long as those forms are not systematically homonymous with some other forms existing in the dictionary and do not cause a significant increase of morphological ambiguity. However, since the analysis at this stage serves mainly as a supporting tool for human annotators, even moderate increase of homonymous interpretations in the dictionary is less problematic than no interpretation at all, thus a reasonable overgeneration of forms is to some extent useful.

Some effort was put into automatic reconstruction of inflectional paradigms of lexemes noted in e-SXVII. The applied procedure goes as follows. For every existing paradigm in e-SXVII containing at least two different forms, a set of pairs consisting of ending and morphological tag is created. Endings are obtained simply by cutting off the longest common prefix of all the forms in the paradigm. With the set of pairs for all paradigms created, the procedure looks for supersets of every given set of pairs. If found, it applies the endings to the stem of the original lexeme together with a morphological tag, which creates a new form in the dictionary. For example, given two nouns CELNICTWO 'customs administration' and BOGACTWO 'richness' in the dictionary and their respective sets of forms {*celnictwo, celnictwem, celnictwie*}, {*bogactwo, bogactwa, bogactwem, bogactwie, bogactw*} together with their morphological tags, by cutting off the longest common prefixes for both sets we obtain two sets of pairs:

	(-o, subst:sg:nom:n),
(-o, subst:sg:nom:n),	(-a, subst:sg:gen:n)
(-em, subst:sg:inst:n),	(-em, subst:sg:inst:n),
(-ie, subst:sg:loc:n)	(-ie, subst:sg:loc:n)
	(∅, subst:pl:gen:n)

It is easy to notice that one is a subset of the other. The difference is a set of two pairs which can now be applied to the stem of lexeme CELNICTWO, producing two new forms in the dictionary: *celnictwa* and *celnictw* with their morphological interpretations. If a given paradigm has several matching supersets the one corresponding to the longest common suffix of forms is chosen.

The procedure of reconstructing paradigms is very simple, general and easy to apply. It is also relatively reliable. It gives favor to regular series of lexemes of the same inflectional characteristics. On the other hand it also propagates irregularities: both inflectional idiosyncrasies and dictionary errors. Many such errors were corrected thanks to this procedure as they became easier to spot. On the other hand the procedure is only heuristic and produces some incorrect forms. Large amount of those forms were eliminated as they contain bigrams and trigrams of characters not appearing in the source dictionary. The procedure enriched the data set by over 210,000 new forms.

The procedure of "aging" SGJP and reconstructing forms unnoted in e-SXVII allowed to reduce the number of tokens unrecognized by Baroque analyzer in the corpus from 11% to 5,4%. The ratio of unknown tokens is still higher than in the case of modern texts, but "aging" significantly decreases the number of morphological interpretations that need to be typed in manually by human annotator. The analyzer's dictionary is in constant development and is being enhanced based on annotators' feedback. Based on the manual annotation of the subcorpus the analyzer's dictionary will be extended before the automatic annotation of the full BCP.

Apart from analyzer's dictionary, tokenization rules also needed to be adjusted. Baroque Polish, when compared to modern language, is characterized by wider use of joint spelling that includes preposition spelled together with the succeeding noun, pronoun, adjective or numeral; NIE 'no' particle spelled together with verbal forms; as well as numerous postpositional clitics of various grammatical characteristics: conditional mood particle (*by*), emphatic particles (*-ż, -że, -ć*), personal auxiliaries (*-m, -ś, -śmy, -ście, -śwa, -śta*), and their variants, which in Baroque Polish are much more mobile and can be attached to a wider range of forms than nowadays. Analyzer's tokenization rules needed to be significantly modified to cover all these phenomena.

4   Manual Annotation

The process of manual annotation of the corpus is performed using a web-based application called Anotatornia 2. A similar tool [9] was used for manual annotation of National Corpus of Polish (NKJP) [8]. Unfortunately, due to fast aging of toolkits used to build web applications it turned out very difficult to deploy this tool in a new project. More importantly, the old tool did not meet a crucial requirement of the current project – parallel access to both transliterated and transcribed layers of the text sample. For that reason we have decided to build a new tool for morphosyntactic annotation of corpora.

Anotatornia 2 is a web-based application that allows for the work of a group of annotators. We assume that the text being annotated is divided into samples that are annotated separately. A sample should consist of a few paragraphs (or parts thereof), which consist of sentences, which in turn consist of tokens. Each token has its text provided in the transliterated and transcribed form. Interpretation of a token consists of a lemma and a morphosyntactic tag. We also have to track the number of page given token appears on in the printed original. This element of text structure is absent in contemporary NKJP but it is commonly requested by the users of historical corpora. Unfortunately, page divisions cross all other levels of text structure, which makes it difficult to represent in XML. The input and the output of Anotatornia is expressed in TEI XML.

The annotator is responsible for several tasks. The first is correcting automatically determined sentence boundaries. If a sample starts or ends in the middle

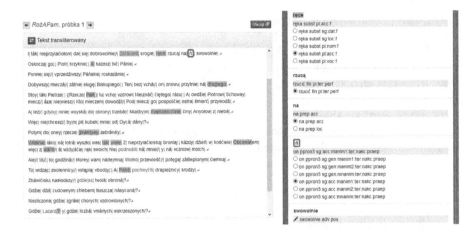

Fig. 1. A sample being annotated as seen by an annotator in Anotatornia 2

of an incorrectly determined sentence, such part of a sentence is excluded from annotation.

The next task is to check whether the tokens are determined correctly. As mentioned in the previous section, words are split into several tokens much more often in Baroque text than in a contemporary one. So the main point here is checking such split words.

The transcribed version of the text has to be checked for each token. Anotatornia by default displays the text sample in the transliterated form, that is the form closer to the printed original. However, the transcribed form is used on the list of morphosyntactic interpretations. This way both variants are shown simultaneously to the annotator and can be verified.

Finally, the annotator has to validate and complete inflectional interpretations. Anotatornia displays all interpretations generated by the morphological analyzer allowing the annotator to choose one of them with a single click. Moreover, the annotator can change all elements of a morphosyntactic interpretation or provide a new one. The system checks that modified morphosyntactic tags are consistent with the defined tagset. This is very important for the quality of annotation.

We adhere to the established best practice in manual corpus annotation: each text sample is annotated independently by two annotators, then the system compares their resulting annotation and, in case of conflicts, an adjudicator ("super-annotator") steps in to resolve them. In fact, there is an intermediate phase: when conflicts are detected the sample is shown once again to both annotators. The conflicting tokens are highlighted, but only the user's own annotation is shown. This way annotators are encouraged to check their work for simple errors but are not tempted to switch to other annotator's version. When allotting samples to annotators the system maximizes the number of different pairs of annotators working together. This is to minimize the biases introduced to annotation by particular annotators.

The adjudicator operates on a list of differences between annotations provided by the two annotators. He has to resolve all conflicts, which can be done by selecting one of the existing annotations or providing a new one. Moreover, the adjudicator has access to all elements of annotation of all tokens and can change any decision made by the annotators.

In the present project, a corpus of 500,000 tokens of BCP will be manually annotated. This work has only started, which means we have very limited data for evaluation (ca. 54,000 tokens). All tokens marked as a punctuation, foreign elements or structural markers were excluded for the purpose of the evaluation. The percentages of manual changes introduced by annotators are as follows: modification of transcription -2.43% of tokens, of tokenization -1.73%, morphological interpretation modified by the annotator -8.61%. If an annotator changes transcription or tokenization, the text is not passed again to the analyzer and an interpretation has to be constructed manually. This means that the number of modified interpretations includes two previous numbers. Currently annotators generate conflicts on 14.1% of tokens (this number is expected to drop in time since the annotators are still learning).

5   Poliqarp 2

Poliqarp is a corpus query engine developed and used for National Corpus of Polish. Its newer version, Poliqarp 2,[5] is much more flexible with respect to data and metadata that can be represented in the corpus. This version will be used to make BCP accessible for the users, which will allow to represent the text in both transliterated and transcribed forms and to address both forms in queries. Also metadata of BCP is more complicated than in the contemporary corpus since sometimes it is important to include information describing both the original publication and its later critical edition which was used as the base for including the text in the corpus.

6   Conclusions and Further Work

The BCP is a work in progress since the manual annotation of the subcorpus is ongoing. All components involved in processing text samples are being constantly enhanced according to the feedback from annotators and adjudicators, but it seems that the tools are already quite stable. Only minor changes are still introduced in transcription rules. Most frequent lexemes are already introduced in dictionary of the morphological analyzer.

When the manually annotated corpus is ready, a stochastic tagger [11] will be trained on the annotations and the rest of the BCP corpus will be automatically annotated. We also plan to check whether machine learning techniques can be

---

[5] https://sourceforge.net/projects/poliqarp2/.

used to train a better transcriber based on transcriptions manually corrected by the annotators.

The presented toolchain will also be used to prepare a similarly organized corpus of 19th century Polish texts.

References

1. Bronikowska, R., Gruszczyński, W., Ogrodniczuk, M., Woliński, M.: The use of electronic historical dictionary data in corpus design. Stud. Polish Linguist. **11**(2), 47–56 (2016)
2. Calzolari, N., Choukri, K., Declerck, T., Grobelnik, M., Maegaard, B., Mariani, J., Moreno, A., Odijk, J., Piperidis, S. (eds.): Proceedings of the Tenth International Conference on Language Resources and Evaluation, LREC 2016. ELRA, European Language Resources Association (ELRA), Portorož, Slovenia (2016)
3. Calzolari, N., Choukri, K., Declerck, T., Loftsson, H., Maegaard, B., Mariani, J., Moreno, A., Odijk, J., Piperidis, S. (eds.): Proceedings of the Ninth International Conference on Language Resources and Evaluation, LREC 2014. ELRA, Reykjavík, Iceland (2014)
4. Proceedings of the 24th International Conference on Computational Linguistics (COLING 2012) (2012)
5. Goźdź-Roszkowski, S. (ed.): Explorations across Languages and Corpora: PALC 2009. Peter Lang, Frankfurt am Main (2011)
6. Gruszczyński, W. (ed.): Elektroniczny słownik języka polskiego XVII i XVIII wieku. Kraków (2004). http://sxvii.pl/
7. Kresa, M., Szafran, K.: Przykład nowego zastosowania słownika polszczyzny historycznej. Prace Filologiczne LXIV, pp. 159–171 (2013)
8. Przepiókowski, A., Bańko, M., Górski, R., Lewandowska-Tomaszczyk, B. (eds.): Narodowy Korpus Języka Polskiego. Warszawa (2012)
9. Przepiórkowski, A., Murzynowski, G.: Manual annotation of the National Corpus of Polish with Anotatornia. In: Goźdź-Roszkowski [5], pp. 95–103
10. Saloni, Z., Gruszczyński, W., Woliński, M., Wołosz, R., Skowrońska, D.: Słownik gramatyczny języka polskiego, 3rd edn. (2015). http://sgjp.pl
11. Waszczuk, J.: Harnessing the CRF complexity with domain-specific constraints. The case of morphosyntactic tagging of a highly inflected language. In: COLING [4]
12. Woliński, M.: Morfeusz reloaded. In: Calzolari et al. [3], pp. 1106–1111
13. Woliński, M., Kieraś, W.: The on-line version of Grammatical Dictionary of Polish. In: Calzolari et al. [2], pp. 2589–2594

Last Syllable Unit Penalization in Unit Selection TTS

Markéta Jůzová[1], Daniel Tihelka[1(✉)], and Radek Skarnitzl[2]

[1] New Technologies for the Information Society (NTIS) and Department of Cybernetics, Faculty of Applied Sciences, University of West Bohemia, Pilsen, Czech Republic
{juzova,dtihelka}@ntis.zcu.cz
[2] Faculty of Arts, Institute of Phonetics, Charles University, Prague, Czech Republic
radek.skarnitzl@ff.cuni.cz

Abstract. While unit selection speech synthesis tries to avoid speech modifications, it strongly depends on the placement of units into the correct position. Usually, the position is tightly coupled with a distance from the beginning/end of some prosodic or rhythmic units like phrases or words. The present paper shows, however, that it is not necessary to follow position requirements, when the phonetic knowledge of the perception of prosodic patterns (mostly durational in our case) is considered. In particular, we focus on the effects of using word-final units in word-internal positions in synthesized speech, which are often perceived negatively by listeners, due to disruptions in local timing.

Keywords: Speech synthesis · Unit selection · Target cost · Word final lengthening

1 Introduction

Interactions between the segmental and prosodic levels of speech may be exemplified in several ways, but it is perhaps the temporal domain in which this interaction can most readily be observed. Temporal segmentation and the rhythm of speech have been shown to affect listeners to a great extent. Arhythmical or temporally unpredictable speech leads to longer reaction times in monitoring experiments [3,19] and, therefore, to increased cognitive processing of the incoming speech signal. In addition, speech with unnatural durational patterns has been shown to decrease intelligibility [18], or to induce negative perceptions regarding the speakers, making the speakers sound, for instance, more nervous and anxious [30] or less competent.

In the field of text-to-speech (TTS) synthesis, the durational patterns (as well as other prosody patterns) can either be modelled explicitly, i.e. by

This research was supported by the Czech Science Foundation (GA CR), project No. GA16-04420S, and by the grant of the University of West Bohemia, project No. SGS-2016-039.

© Springer International Publishing AG 2017
K. Ekštein and V. Matoušek (Eds.): TSD 2017, LNAI 10415, pp. 317–325, 2017.
DOI: 10.1007/978-3-319-64206-2_36

means of an estimator, assigning a particular duration value to each speech unit to be synthesized [8,12,20], or by a symbolic description, defining only deep-level description (discriminative features defining what the prosody should/should not be) and expecting that the appropriate surface-level duration patterns (the particular unit durations) will emerge as a result of this description. While the former approach is used mainly in generative approaches like HMM [11], DNN [33] or single instance speech synthesis [16,27], the latter is mostly employed in unit selection speech synthesis, where it provides more robust selection criteria when compared to the following of generated prosody contours [24]. The insidious part is, however, the definition of the discriminative features. The usual way is to describe the position of speech units within a prosodic pattern (whether a phrase, word or syllable) in the source speech recordings and to select each particular unit into the most similar position in the synthesized phrase. The side effect is the increased pressure on the selection criteria, trying to balance the trade-off for all the features used, which is in details described in Sect. 3. First, however, let us look at the specific cases where it is important to consider temporal/durational properties of speech units and why this is necessary.

2   The Special Status of Final Syllables

Concatenative speech synthesis may be regarded as one of the key stimulating factors in the research of the effects of the prosodic domain on the duration of speech segments. Klatt [13] showed how segmental duration interacts with linguistic characteristics at various levels and devised a series of rules which predicted the durations of speech sounds in American English through the multiplication of a base value by coefficients related to various segmental and prosodic attributes (see [5,22] for similar studies).

The above-mentioned and other studies document several ways in which segmental duration is affected by the prosodic structure of utterances. In many languages, accented syllables are realized as longer than unaccented ones, or syllables tend to shorten in longer words [9]. Probably the most salient reflection of prosodic structure in the segmental strand is called phrase-final lengthening, which appears to be related to the general declination observed on a number of levels: ends of prosodic phrases are thus marked by lower speech rate and a drop in fundamental frequency [10,14], as well as larger and longer articulatory gestures [4] and laxer phonation [17]. Phrase-final lengthening has been documented in many languages [9], including Czech [7,29], and is considered to be universal in speech. A number of studies (see the Refs. in [9]) have agreed that the rate lowering concerns the rhyme of the final syllable (i.e., the vowel nucleus plus any consonants in the coda following the vowel).

Lengthening is thus a well-documented phenomenon at the level of prosodic phrases. In this study, we are interested in temporal adjustments at the level of individual words, which have been researched considerably less (although mentioned in literature). When comparing the duration of the schwa vowel framed in sentences /Her pop[p][a] posed a problem/ and /Her pop [o]pposed the marriage/,

Beckman and Edwards [2] found the schwa in the first sentence significantly longer. Similar results were obtained by [6], who compared pairs like /*lettuce–let us*/ or /*inquires–in choirs*/, though the effect was relatively small. More recently, word-final lengthening is discussed in [31], who, however, did not reach a definitive conclusion, and in [32], who report significantly longer durations in word-final positions. All of these studies consider English, though.

Until recently, no data on word-final lengthening have been available for Czech. A new study on the acoustic characteristics of Czech lexical stress [23] showed, however, that duration is the only parameter which systematically varies with lexical stress – note that Czech is a language with stress fixed on the first syllable of the base rhythm unit, also called the *prosodic word* [21]. Figure 1 shows part of the duration data taken from multi-syllabic words in spontaneous speech; only words which did not appear as phrase-final were analysed in the study. The results differ from what is typical in most languages: the vowel in the stressed syllable tends to be shortest and that in the last syllable is always longest (though not always statistically significant).

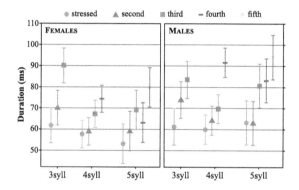

Fig. 1. Vowel duration in individual syllables of 3-, 4-, and 5-syllabic words; based on [23].

The account presented in this section indicates that phrase-final lengthening is essentially a ubiquitous and very salient phenomenon, but even word-final lengthening is non-negligible. The present study examines speech sounds in the last syllable of phonological words from the perspective of concatenative unit-selection synthesis of Czech. The results presented by [23] suggest that the word-final syllable may enjoy a special status in the prosodic hierarchy of Czech, and this may need to be reflected in the speech synthesis algorithm. Specifically, it is possible that units which appeared in the word-final position in the source recordings should not be selected for synthesis within words. This study therefore tests the hypothesis that the presence of word-final (source) units in word-internal positions in synthesized speech will be perceived negatively by listeners, since the lengthening will disrupt the local timing.

3   Handling of Unit Position in TTS Systems

The previous section indicates the great importance of the last syllable. We decided to check this in our TTS system ARTIC based on unit selection speech synthesis method. The highest influence of the unit position has the design of *target cost* (TC).

The target cost, as usually handled, is set to strive for putting the units into the same suprasegmental surroundings as they originally had in the source recordings, which is achieved when target features match. Thus, to make sure that the last-syllable position is handled properly, as described in Sect. 2, we could simply define an additional feature with *onset, nucleus, coda* symbolic values assigned to units recorded in the last syllable, and with *not-last* value for units from other syllables. Based on the match or mismatch of such feature values, an additional penalty would be added to the total TC, encouraging (in theory) the placement of units into the required position. However, in case of value mismatch, there is the same penalty no matter the value mismatch – i.e. the same penalty for e.g. *unknown↔nucleus* as for *coda↔onset* (unless a feature exchange matrix is defined somehow). An alternative way of defining the last-syllable position as a binary feature, i.e. a unit either *belongs* to the last syllable or *not*, is not going to improve the situation very much – there is better change to avoid mismatch of last/non-last syllable units, but units within the last syllable may still be interchanged between coda and onset parts (while both match the feature), which is not better either.

In our current TTS version, therefore, the computation of position within a prosodic word (hereafter referred to as *p-word*) is based on the assumption that there is no need to put a unit into the identical position in which the unit was originally recorded [26,28]. This is achieved through the definition of a *suitability* to the required position related to the beginning/middle/end of each p-word, represented by 3 values, each corresponding to one von Hann window spanned through the p-word, as illustrated in Fig. 2. Such a position feature can avoid a completely wrong placement (i.e. unit originating at the beginning of a p-word to be placed to its end), while it still allows some kind of flexibility in units interchanging, when the unit originates in a position which is "close" to the one in which it is to be placed. From the point of view of the selection

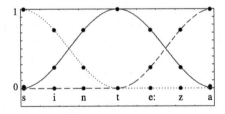

Fig. 2. The illustration of the correspondence of windowing functions to a prosodic word "synthesis". Individual windows are distinguished by line style, points correspond to the values describing the candidates.

algorithm, the set of candidates suitable for the given p-word position (as defined by the three values) is wider than if the position were defined as, for example, distance from the beginning and from the end of the p-word. And the wider set to select from gives the algorithm a better chance to choose a more appropriate unit, when the other target and concatenation features are taken into account as well.

On the other hand, such a position measure (and either of similar) is not able to handle a "last-syllable" occurrence feature reliably. We observed unnatural unit lengthenings close to the end of a p-word which occurred when a unit from the last syllable was placed to the penultimate syllable, since its position was suitable to the position required, as measured by the current approach.

4   New Unit Position Reflecting Final Syllable Status

Based on the findings from Sect. 2, we thus re-defined the unit position feature in a way that instead of measuring appropriateness for the given position, we strongly penalize placements into inappropriate positions, while expecting any other placements to be equally suitable, i.e. not penalizing them in any way. Let us emphasize that this completely replaces the original position function from Sect. 3, instead of just being added as an extra measure. The reason is that adding a feature would increase the pressure on the selection algorithm, lowering the set of candidates matching the feature and thus increasing the chance that a unit not matching the required position will be used after all. And moreover, ensuring a position placement is not that significant, as suggested in Sect. 2.

Let us also note that there was no exact syllabification used to identify the last p-word syllable (neither is it possible to detect syllables exactly [15]). Instead, we simply found the last vowel (or syllabic consonant) V in each p-word, and the diphones *[*-V]*, *[V-*]* as well as the remaining diphones until the end of p-word were considered to be the syllable constituents (cf. references in Sect. 2 which show that the rhyme of the final syllable is most affected by final lengthening).

Having the new positional feature, we synthesized more than a million phrases by the original (marked as TTS_{base} hereafter) and by the modified version of unit selection method (TTS_{syll}) embedded into our TTS system. The resulting sequences of units provided by TTS_{base} for each phrase were further analysed – both sequence of the phrase units and its corresponding selected units were passed through TTS_{syll} module which returned the high TC value for units with inappropriate syllable position. This number of position misplacements was used as the base score for the selection of phrases to be evaluated by listening tests.

4.1   Listening Tests Overview

To verify our presumption, we carried out a large 3-scale preference listening test, where two variants of the same phrase, synthesized by TTS_{base} and TTS_{syll}, were compared. To select the phrases for the evaluation, the set of all the synthesized phrases was first limited to those having 40 characters at most, since smaller differences in long phrases are rather hard for listeners to recall [1]. Then, only the phrases with two or more misses, as described in Sect. 4, were kept. The final set of 25 phrase pairs was selected randomly, while 4 different professional synthetic voices, two male and two female, were used – in total 100 phrases were thus compared.

22 listeners participated in the test, 7 of them being speech synthesis experts, 6 being phoneticians and 9 naive listeners. We intentionally did not inform any of them about the experiment's details, since the aim was to test the overall quality of the new TTS system. Furthermore, during the listening test, the order of the samples was randomized across the listening prompts, so the listeners did not know which one was synthesized by TTS_{base} and by TTS_{syll}. The listeners were able to listen to each stimulus repeatedly, they were instructed to use earphones, and had to choose one of the following choices: *sample A sounds better/samples are of the same quality/sample B sounds better*. The A/B assignments were then normalized to $A = 1$ where TTS_{syll} variant was preferred, $A = -1$ where TTS_{base} was preferred, and $A = 0$ otherwise. The final score s of the listening test T was then computed using the Eq. 1

$$s = \frac{\sum_{A \in T} A}{\sum_{A \in T} 1} \tag{1}$$

Thus, the positive value of s indicates the improvement of the overall quality when using the new last-syllable feature.

Let us note that for the purpose of this testing, we did not use the procedure designed in [25]. The main reason is that we do not have to rely on numbers of changed units when comparing the output sequences from the two system versions; instead we can simply detect wrong cases in TTS_{base}, as described in Sect. 4. Nevertheless, we plan to use the procedure before the final deployment of TTS_{syll} to the production-ready version of our TTS system.

5 Results

The results of the listening tests are shown in the Table 1. All the score values s are positive, indicating a considerable improvement of the quality of synthesized samples.

Since the score value for *male speaker 1* obtained from phonetics experts' answers seems to be low and not so conclusive, we decided to prove the statistical significance of this result. We have carried out the sign test with the null hypothesis H0:*the outputs of the both systems are of the same quality*, and alternative hypothesis H1:*the output of one system sounds better*. The computed p-value $= 0.0193$, so we can reject the null hypothesis H0 at $\alpha = 0.05$ significance level, concluding that the quality of TTS_{syll} system is really higher.

However, there is a considerable number of "*same quality*" evaluations. It suggests that not all candidates from the last syllable cause a speech artefact

Table 1. The results obtained from listening tests. The table contains the number of listener answers and the score values s computed by Eq. 1.

	Male spkr 1	Male spkr 2	Female spkr 1	Female spkr 2	All speakers
	Numbers of answers in percents				
TTS_{syl} better	58.7%	56.4%	65.1%	53.6%	58.5%
Same quality	20.0%	27.1%	21.8%	29.3%	24.5%
TTS_{base} better	21.3%	16.5%	13.1%	17.1%	17.0%
	Score value s				
All listeners	**0.375**	**0.398**	**0.520**	**0.365**	**0.415**
TTS experts	0.531	0.480	0.531	0.457	0.500
Phonetics experts	0.187	0.313	0.520	0.387	0.352
Naive listeners	0.378	0.391	0.511	0.280	0.390

when placed to non-last syllable position – note that, as described in Sect. 4, we *know* that there is a unit from the last syllable placed to non-last position in the TTS_{base}.

On the other hand, the listeners sometimes preferred the TTS_{base} variant. Therefore, we inspected the problematic prompts, trying to find the cause of the preference. The main problem was a disturbing artefact of another kind (not directly related to syllable position) in TTS_{syl} variant, while none of the syllable position misses in TTS_{base} was perceived negatively, nor was there another artefact. In a few cases, moreover, even TTS_{syl} still contained unnatural inter-phrase lengthenings. These originated from the failures of p-word tokenization in source recordings, and thus the system used the last-syllable unit into inappropriate position without being able to realize it.

6   Conclusion

The results of the listening tests clearly confirm the importance of the correct handling of the last syllable of a p-word. Let us also note that the change of paradigm, when instead of "forcing" units into the expected position we try to avoid their use in positions where they are known to cause audible artefacts, follows the principle of units synonymy/homonymy established in [24, 26]. We believe that this is the right direction towards the tuning of unit selection features, as it allows the use of units in a much wider range of placements (than the one in which the unit has been placed in the source recordings), and it also avoids the definition of a penalty function which would evaluate a distance of what the unit is (where it *is* placed) to what it is required to be (where we *try* to place it).

One of the possible further improvements will now be the focus of p-word tokenization which was found to be incorrect in some of the cases examined and proved to be the clear cause of TTS_{base} preference in these. Also, the findings in Sect. 5 suggest that there may be some durational (or prosodic, in general) patterns, allowing the exchange of the last syllable and non-last syllable units under some conditions. Answering this phenomena would even more relax the pressure on the selection algorithm, thus widening the set of candidates to be used.

References

1. Baddeley, A.: Human Memory: Theory and Practice. Psychology Press, East Sussex (1997). Revised edn
2. Beckman, M., Edwards, J.: Lengthenings and shortenings and the nature of prosodic constituency. In: Papers in Laboratory Phonology I: Between the Grammar and the Physics of Speech, pp. 152–178. Cambridge University Press, Cambridge (1990)
3. Buxton, H.: Temporal predictability in the perception of English speech. In: Cutler, A., Ladd, D.R. (eds.) Prosody: Models and Measurements, vol. 14, pp. 111–121. Springer, Heidelberg (1983)
4. Byrd, D., Saltzman, E.: The elastic phrase: modelling the dynamics of boundary-adjacent lengthening. J. Phonetics **31**, 149–180 (2003)
5. Crystal, T.H., House, A.S.: Segmental durations in connected-speech signals: current results. J. Acoust. Soc. Am. **83**, 1553–1573 (1988)
6. Cutler, A., Butterfield, S.: Syllabic lengthening as a word boundary cue. In: Proceedings of the 3rd Australian SST, pp. 324–328 (1990)
7. Dankovičová, J.: The domain of articulation rate variation in Czech. J. Phonetics **25**, 287–312 (1997)
8. Fernandez, R., Rendel, A., Ramabhadran, B., Hoory, R.: Prosody contour prediction with long short-term memory, bi-directional, deep recurrent neural networks. In: Proceedings of Interspeech, pp. 2268–2272. ISCA (2014)
9. Fletcher, J.: The prosody of speech: timing and rhythm. In: The Handbook of Phonetic Sciences, pp. 521–602. Blackwell Publishing Ltd. (2010)
10. Gussenhoven, C.: The Phonology of Tone and Intonation. Cambridge University Press, Cambridge (2004)
11. Hanzlíček, Z.: Czech HMM-based speech synthesis: experiments with model adaptation. In: Habernal, I., Matoušek, V. (eds.) TSD 2011. LNCS, vol. 6836, pp. 107–114. Springer, Heidelberg (2011). doi:10.1007/978-3-642-23538-2_14
12. Holm, B., Bailly, G.: Generating prosody by superposing multi-parametric overlapping contours. In: Proceedings of ICSLP, pp. 203–206 (2000)
13. Klatt, D.H.: Linguistic uses of segmental duration in English: acoustic and perceptual evidence. J. Acoust. Soc. Am. **59**, 1208–1221 (1976)
14. Ladd, D.R.: Intonational Phonology, 2nd edn. Cambridge University Press, Cambridge (2008)
15. Matoušek, J., Hanzlíček, Z., Tihelka, D.: Hybrid syllable/triphone speech synthesis. In: Proceedings of 9th Interspeech (Eurospeech), Lisbon, Portugal, pp. 2529–2532 (2005)
16. Matoušek, J., Romportl, J., Tihelka, D., Tychtl, Z.: Recent improvements on ARTIC: czech text-to-speech system. In: Proceedings of Interspeech, Jeju Island, Korea, pp. 1933–1936 (2004)
17. NíChasaide, A., Yanushevskaya, I., Gobl, C.: Prosody of voice: declination, sentence mode and interaction with prominence. In: Proceedings of 18th ICPhS (2015). Paper 476

18. Quené, H., van Delft, L.E.: Non-native durational patterns decrease speech intelligibility. Speech Commun. **52**(11–12), 911–918 (2010)
19. Quené, H., Port, R.: Effects of timing regularity and metrical expectancy on spoken-word perception. Phonetica **62**(1), 1–13 (2005)
20. Romportl, J., Kala, J.: Prosody modelling in Czech text-to-speech synthesis. In: Proceedings of the 6th ISCA SSW, Bonn, pp. 200–205 (2007)
21. Romportl, J., Matoušek, J., Tihelka, D.: Advanced prosody modelling. In: Sojka, P., Kopeček, I., Pala, K. (eds.) TSD 2004. LNCS, vol. 3206, pp. 441–447. Springer, Heidelberg (2004). doi:10.1007/978-3-540-30120-2_56
22. van Santen, J.P.H.: Assignment of segmental duration in text-to-speech synthesis. Comput. Speech Lang. **8**, 95–128 (1994)
23. Skarnitzl, R., Eriksson, A.: The acoustics of word stress in Czech as a function of speaking style. In: Proceedings of Interspeech (2017)
24. Tihelka, D.: Symbolic prosody driven unit selection for highly natural synthetic speech. In: Proceedings of 9th Interspeech (Eurospeech), pp. 2525–2528. ISCA, Bonn (2005)
25. Tihelka, D., Grůber, M., Hanzlíček, Z.: Robust methodology for TTS enhancement evaluation. In: Habernal, I., Matoušek, V. (eds.) TSD 2013. LNCS, vol. 8082, pp. 442–449. Springer, Heidelberg (2013). doi:10.1007/978-3-642-40585-3_56
26. Tihelka, D., Matoušek, J.: Unit selection and its relation to symbolic prosody: a new approach. In: Proceedings of 9th ICSLP, vol. 1, pp. 2042–2045. ISCA, Bonn (2006)
27. Tihelka, D., Méner, M.: Generalized non-uniform time scaling distribution method for natural-sounding speech rate change. In: Habernal, I., Matoušek, V. (eds.) TSD 2011. LNCS, vol. 6836, pp. 147–154. Springer, Heidelberg (2011). doi:10.1007/978-3-642-23538-2_19
28. Tihelka, D., Romportl, J.: Exploring automatic similarity measures for unit selection tuning. In: Proceedings of 10th Interspeech, pp. 736–739. ISCA, Brighton (2009)
29. Volín, J., Skarnitzl, R.: Temporal downtrends in Czech read speech. In: Proceedings of Interspeech, pp. 442–445 (2007)
30. Volín, J., Poesová, K., Skarnitzl, R.: The impact of rhythmic distortions in speech on personality assessment. Res. Lang. **12**, 209–216 (2014)
31. White, L., Turk, A.E.: English words on the procrustean bed: polysyllabic shortening reconsidered. J. Phonetics **38**(3), 459–471 (2010)
32. Windmann, A., Šimko, J., Wagner, P.: Polysyllabic shortening and word-final lengthening in English. In: Interspeech 2015, pp. 23–40 (2015)
33. Wu, Z., Watts, O., King, S.: Merlin: an open source neural network speech synthesis system. In: Proceedings of 9th ISCA SSW, pp. 218–223, September 2016

On Multilingual Training of Neural Dependency Parsers

Michał Zapotoczny, Paweł Rychlikowski, and Jan Chorowski[✉]

Institute of Computer Science, University of Wrocław, Wrocław, Poland
mzapotoczny@gmail.com, {pawel.rychlikowski,jan.chorowski}@cs.uni.wroc.pl

Abstract. We show that a recently proposed neural dependency parser
can be improved by joint training on multiple languages from the same
family. The parser is implemented as a deep neural network whose only
input is orthographic representations of words. In order to successfully
parse, the network has to discover how linguistically relevant concepts
can be inferred from word spellings. We analyze the representations of
characters and words that are learned by the network to establish which
properties of languages were accounted for. In particular we show that
the parser has approximately learned to associate Latin characters with
their Cyrillic counterparts and that it can group Polish and Russian
words that have a similar grammatical function. Finally, we evaluate the
parser on selected languages from the Universal Dependencies dataset
and show that it is competitive with other recently proposed state-of-the
art methods, while having a simple structure.

Keywords: Dependency parsing · Recurrent neural networks · Multitask training

1 Introduction

Parsing text is an important part of many natural language processing applications. Recent state-of-the-art results were obtained with parsers implemented
using deep neural networks [3]. Neural networks are flexible learners able to
express complicated input-output relationships. However, as more powerful
machine learning techniques are used, the quality of results will not be limited by the capacity of the model, but by the amount of the available training
data. In this contribution we examine the possibility of increasing the training
set by using treebanks from similar languages.

For example, in the upcoming Universal Dependencies (UD) 2.0 treebank
collection [27] there are 863 annotated Ukrainian sentences, 333 Belarusian, but
nearly 60k Russian ones (divided into two sets: a default one of 4.4k sentences and
SynTagRus with 55.4k sentences). Similarly, there are 7k Polish sentences and
a little over 100k Czech ones[1]. Since these languages belong to the same Slavic

---

[1] However, experiments use UD 1.3 dataset which does not include Belarusian and
Ukrainian.

© Springer International Publishing AG 2017
K. Ekštein and V. Matoušek (Eds.): TSD 2017, LNAI 10415, pp. 326–334, 2017.
DOI: 10.1007/978-3-319-64206-2_37

language family, performance on the low resource languages should improve by joint training the model also on a better annotated language [5]. In this paper, we demonstrate this improvement. Starting with a parser competitive with the current state-of-the-art, we are able to further improve the results for tested languages from the Slavic family. We train the model on pairs of languages through simple parameter sharing in an end-to-end fashion, retaining the structure and qualities of the base model.

2   Background and Related Work

Dependency parsers represent sentences as trees in which every word is connected to its head with a directed edge (called a dependency) labeled with the dependency's type. Parsers often contain parts that are learned on a corpus. In example, transition-based dependency parsers use the learned component to guide their actions, while graph-based dependency parser learn a scoring that measures the quality of inserting a *(head, dependency)* edge into the tree.

Historically, the learning algorithms were relatively simple ones, e.g. transition-based parsers used linear SVMs [25,26]. Recently, those simple learning models were successfully replaced by deep neural networks [3,8,14,31]. This trend coincides with successes of those models on other NLP tasks, such as language modeling [19,24] and translation [33].

Neural networks have enough capacity to directly solve the parsing task. For example a constituency parser can be implemented using a sequence-to-sequence network originally developed for translation [32]. Similarly, a graph-based dependency parser can be implemented by solving two supervised tasks: head selection and dependency labeling. Both are easily solved using neural networks [11,12,21,35]. Moreover, neural networks can extract meaningful features from the data, which may augment or replace manually designed ones, as it is the case with word embeddings [23] or features derived from the spelling of words [4,11,20].

Another particularly nice property of neural models is that all internal computations use distributed representations of input data that are embedded in highly dimensional vector spaces [18]. These internal representation can be easily shared between tasks [7]. Likewise, neural parsers can share some of their parameters to harness similarities between languages [2,5,13,17]. Creation of multilingual parsers is further facilitated by the introduction of standardized treebanks, such as the Universal Dependencies [27].

3   Model

Our multilingual parser can be seen as n identical neural dependency parsers for n languages, which share parameters. When all parameters are shared a single parser is obtained for all n languages. When only a subset of parameters is shared the model can be seen as a parser for a main language that is partially regularized using data for other languages.

Each of the n parsers is a single neural network that directly reads a sequence of characters and finds dependency edges along with their labels [11]. We can functionally describe four basic parts: *Reader, Tagger, Labeler/Scorer*, and an optional *POS Tag Predictor* (Fig. 1).

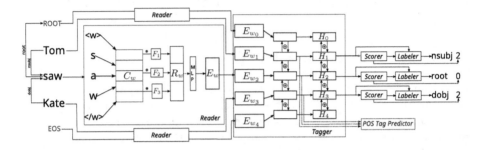

Fig. 1. The model architecture.

The **reader** is tasked with transforming the orthographic representation of a single word w into a vector $E_w \in \mathbb{R}^{\text{Edim}}$, also called the word w's embedding. First, we represent each word as a sequence of characters fenced with start-of-word and end-of-word tokens. We find low dimensional characters embeddings and concatenate them to form a matrix C_w. Next we convolve this matrix with a learned filterbank F

$$R_{w,i} = \max(C_w * F_i),\tag{1}$$

where F_i is the i-th filter and $*$ denotes convolution over the length of the word. Thanks to the start- and end-of-word tokens the filters can selectively target infixes, prefixes and suffixes of words. Finally, we max-pool the filter activations over the word length and apply a small feedforward network to obtain final word embedding $E_w = \text{MLP}(R_w)$.

The **tagger** processes complete sentences and puts individual word embeddings E_w into their contexts. We use a multi-layer bidirectional GRU Recurrent Neural Network (BiRNN) [9,28]. The output of the tagger is a sequence of the BiRNN's hidden states $H_0, H_1, \ldots, H_n$ with $H_i \in \mathbb{R}^{\text{Hdim}}$, where H_0 corresponds to a prepended ROOT word and n is the length of the sentence. Please observe that while the embedding E_i of the i-th word only depends on the word's spelling, the corresponding hidden state H_i depends on the whole sentence. We have also added an **auxiliary network to predict POS tags** based on hidden states H_i. It serves two purposes: first, it can provide extra supervision on POS tags known during training. Second, it helps to attribute errors to various parts of the network (c.f. Sect. 4.4). The POS tag predictor is optional: its output is not used during inference because the tagger communicates all information to the scorer and labeler through the hidden states H_i.

Finally, the network produces the dependency tree by solving two supervised learning tasks: using a **scorer** to find the head word, then using a **labeler** to find the edge label. The **scorer** determines whether each pair of hidden vectors (H_w, H_h) forms a dependency. We employ per-word normalization of scores: for a given word location $w \in 1, 2, \ldots, n$ scores are SoftMax-normalzied over all head locations $h \in 0, 1, 2, \ldots, n$. The **labeler** reads a pair of hidden vectors (H_w, H_h) and predicts the label of this dependency edge. During training we use the ground-truth head location, while during inference we use the location predicted using the *scorer*. We employ the following **training criterion**: $L = \alpha_h L_h + \alpha_l L_l + \alpha_t L_t$, where L_h, L_l, L_t are negative log-likelihood losses of the scorer, the labeler and POS tag predictor, respectively.

4   Experiment Details and Results

4.1   Model Hyperparameters

We have decided to use the same set of hyperparameters for all languages and multilingual parsers, which were a compromise in model capacity for languages that had small and large treebanks. The reported size of recurrent layers is slightly too big for low-resources single-language parser, but we have determined that it is optimal for languages with large treebanks and for multilingual training.

The *reader* embeds each character into vector of size 15, and contains 1050 filters (50·k filters of length k for k = 1, 2,..., 6) whose outputs are projected into 512-dimensional vector transformed by a 3 equally sized layers of feedforward neural network with ReLU activation. Unlike [11,20] we decided to remove Highway layers [30] from the *reader*. Their usage introduced a marginal accuracy gain, while nearly doubling the computational burden. The *tagger* contains 2 BiRNN layers of GRU units with 548 hidden states for both forward and backward passes which are later aggregated using addition [11]. Therefore the hidden states of the tagger are also 548-dimensional. The *POS tag predictor* consists of a single affine transformation followed by a SoftMax predictor for each POS category. The *scorer* uses a single layer of 384 tanh for head word scoring while the *labeller* uses 256 Maxout units (each using 2 pieces) to classify the relation label [16]. The training cost used the constants $\alpha_h = 0.6, \alpha_l = 0.4, \alpha_t = 1.0$.

We regularize the models using Dropout [29] applied to the *reader* output (20%), between the BiRNN layers of the *tagger* (70%) and to the *labeller* (50%). Moreover we apply mild weight decay of 0.95.

We have trained all models using the Adadelta [34] learning rule with epsilon annealed from 1e-8 to 1e-12 and adaptive gradient clipping [10]. Experiments are early-stopped on validation set Unlabeled Attachment Score (UAS) score. Unfortunately, due to limited computational resources we are only able to present the results for a subset of the UD treebanks that are shown in Table 1.

Multilingual models use the same architecture. We unify the inputs and outputs of all models by taking the union of all possible token categories (characters, POS categories, dependency labels). If some category does not exist within a particular language we use a special UNK token. All parsers are trained in parallel minimizing a sum of their individual training costs. We use early-stopping on the main (first) language UAS score. We equalize training mini-batches such that

Table 1. Baseline results of single language models from UD v1.3. Our models use only orthographic representations of tokenized words during inference and work without a separate POS tagger. Ammar et al. [2] uses version 1.2 of UD and uses gold language ids and predicted coarse tags. SyntaxNet [1,3] works on predicted POS tags, while ParseySaurus [1] uses word spellings.

Language	#sentences	Ours		SyntaxNet		Ammar et al.	ParseySaurus	
		UAS	LAS	UAS	LAS	LAS	UAS	LAS
Czech	87 913	**91.41**	**88.18**	89.47	85.93	-	89.09	84.99
Polish	8 227	90.26	85.32	88.30	82.71	-	**91.86**	**87.49**
Russian	5 030	83.29	79.22	81.75	77.71	-	**84.27**	**80.65**
German	15 892	82.67	76.51	79.73	74.07	71.2	**84.12**	**79.05**
English	16 622	87.44	83.94	84.79	80.38	79.9	**87.86**	**84.45**
French	16 448	**87.25**	**83.50**	84.68	81.05	78.5	86.61	83.1
Ancient Greek	25 251	**78.96**	**72.36**	68.98	62.07	-	73.85	68.1

each contains the same number of sentences from all languages. We determined the optimal amount of parameter sharing and show it in Table 2. Moreover, we never share the start-of-word and end-of-word tokens to indicate to the network which language is parsed.

4.2   Main Results

Our results on single language training are presented in Table 1. Our models reach better scores than the highly tuned SyntaxNet transition-based parser [3]

Table 2. Impact of parameter sharing strategies on main language parsing accuracy when multilingual training is used for additional supervision.

Shared parts	Main lang	Auxiliary lang	UAS	LAS
-	Polish	-	90.26	85.32
Parser	Polish	Czech	90.72	85.57
Tagger, Parser	Polish	Czech	91.19	86.37
Tagger, POS Predictor, Parser	Polish	Czech	91.65	86.88
Reader, Tagger, POS Predictor, Parser	Polish	Czech	**91.91**	**87.77**
Parser	Polish	Russian	90.31	85.07
Tagger, POS Predictor, Parser	Polish	Russian	**91.34**	**86.36**
Reader, Tagger, POS Predictor, Parser	Polish	Russian	89.16	82.94
-	Russian	-	83.29	79.22
Parser	Russian	Czech	83.15	78.69
Tagger, POS Predictor, Parser	Russian	Czech	83.91	79.79
Reader, Tagger, POS Predictor, Parser	Russian	Czech	**84.78**	**80.35**

and are competitive with the DRAGNN based ParseySaurus which also uses character-based input [1].

Multilingual training (Table 2) improves the performance on low-resource languages. We observe that the optimal amount of parameter sharing depends on the similarity between languages and corpus size – while it is beneficial to share all parameters of the PL-CZ and RU-CZ parser, the PL-RU parser works best if the reader subnetworks are separated. We attribute this to the quality of Czech treebank which has several times more examples than Polish and Russian datasets combined.

4.3 Analysis of Language Similarities Identified by the Network

We have first analyzed whether a PL-RU parser can learn the correspondence between Latin and Cyrillic scripts[2]. We have inspected the reader subnetworks of a PL-RU parser that shared all parameters. As described in Sect. 3, the model begins processing a word by finding the embedding of each character. For the analysis we have extracted the embeddings associated with all Polish and Russian characters. We have paired Polish and Russian letters which have similar pronunciations. We note that the pairing omits letters that have no clear counterparts (e.g. the Russian letter я correspond to the syllable "ja" in Polish).

a-а, b-б, c-ц, d-д, e-е, e-э, f-ф, g-г, h-х, i-и, j-й, k-к, l-л, m-м, n-н, o-о, p-п, r-р, s-с, t-т, u-у, w-в, y-ы, z-з, ł-л, ż-ж

Adapting the famous equation $king - man + woman \approx queen$ [23] we inspected to what extent our network was able to deduce Latin-Cyrillic correspondences. For all distinct pairs $(p_1 - r_1, p_2 - r_2)$ of letter correspondences we computed the vector $C(p_2) - C(p_1) + C(r_1)$, where C stands for char embedding, and found Russian letter which had the closest embedding vector. In 48.3% cases we choose the right vector. We found it quite striking given that the two languages have separated from their common root (Proto-Slavic) more than 1000 years ago. Moreover, relations between Polish and Russian letters are side effects, not the main objective of the neural network.

We have also examined word representations E_w computed for Polish and Russian by the shared reader subnetwork. As one could expect, the network was able to realize that in these languages morphology is suffix based. However, the network was also able to learn that words built from different letters can behave in similar way. We can observe it in both monolingual or multilingual context. Table 3 shows some Polish adjectives and the top-7 Russian words with the closest embedding. All Russian words which are not *italics* have the same morphological tags as the Polish word. In the first row we can observe 2 suffixes *-ской* (**skoy**) (skoy) and *-нной* (**nnoy**) (nnoy) quite distant from polish *-owej* (owej). In the second row we see that the model was able to correctly alias the Polish 3-letter suffix *-ych* with the Russian 2 letter suffix *-ых* which are pronounced the same way. The relation found by the network is purely syntactical – there is no easy-to-find connection between semantics of these words.

---

[2] Conveniently, the Unicode has separate codes for Latin and Cyrillic letters.

Table 3. The network learns to group Polish words with Russian words that have a similar grammatical function.

Polish word	Closest Russian embeddings
przedwrześniowej	адренергической тренерской таврической непосредственной археологической философской *верхнюю*
większych	автомобильных *трёхдневные* технических практических официальных оригинальных
policyjnym	главным историческим глазным непосредственным *косыми* летним двухсимвольным

4.4 Common Error Analysis

We have investigated two possible sources of errors produced by the parser. First, we verified if using a more advanced tree-building algorithm was better than using a greedy one. We have observed that the *scorer* produces very sharp probability distributions that can be transformed into trees using a greedy algorithm that simply selects for each word the highest scored head [11,12]. Counterintuitively, the Chu-Liu-Edmonds (CLE) maximum spanning tree algorithm [15] often makes the decoding results slightly worse. We have established that the network is so confident in its predictions that non-top scores do not reflect alternatives but are only noise. Therefore when the greedy decoding creates a cycle the CLE usually breaks it in a wrong place introducing another pointer error.

We have used the *POS predictor* to pinpoint which parts of the network (*reader/tagger* or *labeler/scorer*) were responsible for errors. Tests showed that if the predicted tag was wrong, the *scorer* and *labeler* will nearly always produce erroneous results too.

5 Conclusions and Future Works

We have demonstrated a graph-based dependency parser implemented as a single deep neural network that directly produces parse trees from characters and does not require other NLP tools such as a POS tagger. The proposed parser can be easily used in a multilingual setup, in which parsers for many languages that share parameters are jointly trained. We have established that the degree of sharing depends on language similarity and corpus size: the best PL-CZ parser and RU-CZ shared all parameters (essentially creating a single parser for both languages), while the best PL-RU parser had separate morphological feature detectors (i.e. *readers*). We have also determined that the network can extract meaningful relations between languages, such as approximately learning a mapping from Latin to Cyrillic characters or associate Polish and Russian words that have a similar grammatical function. While this contribution focused on improving the performance on a low-resource language using data from another languages, similar parameter sharing techniques could be used to create one universal parser [2].

We have performed qualitative error analysis and have determined to regions for possible future improvements. First, the network does not indicate alternatives to the produced parse tree. Second, errors in word interpretation are often impossible to correct by the upper layers of the network. In the future we plan to investigate training a better POS tagging subnetwork possibly using other sources of data.

Acknowledgments. The experiments used Theano [6], Blocks and Fuel [22] libraries. The authors would like to acknowledge the support of the following agencies for research funding and computing support: National Science Center (Poland) grant Sonata 8 2014/15/D/ST6/04402, National Center for Research and Development (Poland) grant Audioscope (Applied Research Program, 3rd contest, submission no. 245755).

References

1. Alberti, C., et al.: SyntaxNet models for the CoNLL 2017 shared task. arXiv:1703.04929, March 2017
2. Ammar, W., et al.: Many languages, one parser. Trans. Assoc. Comput. Linguist. **4**(0), 431–444 (2016)
3. Andor, D., Alberti, C., Weiss, D., Severyn, A., Presta, A., Ganchev, K., Petrov, S., Collins, M.: Globally normalized transition-based neural networks. arXiv:1603.06042 [cs], March 2016
4. Ballesteros, M., Dyer, C., Smith, N.A.: Improved transition-based parsing by modeling characters instead of words with LSTMs. arXiv preprint arXiv:1508.00657 (2015)
5. Bender, E.M.: On achieving and evaluating language-independence in NLP. Linguist. Issues Lang. Technol. **6**(3), 1–26 (2011)
6. Bergstra, J., et al.: Theano: a CPU and GPU math expression compiler. In: Proceedings of SciPy (2010)
7. Caruana, R.: Multitask learning. Mach. Learn. **28**(1), 41–75 (1997)
8. Chen, D., Manning, C.D.: A fast and accurate dependency parser using neural networks. In: EMNLP, pp. 740–750 (2014)
9. Cho, K., et al.: Learning phrase representations using RNN encoder-decoder for statistical machine translation. CoRR abs/1406.1078 (2014)
10. Chorowski, J., Bahdanau, D., Cho, K., Bengio, Y.: End-to-end continuous speech recognition using attention-based recurrent NN: first results. arXiv:1412.1602 [cs stat], December 2014
11. Chorowski, J., Zapotoczny, M., Rychlikowski, P.: Read, tag, and parse all at once, or fully-neural dependency parsing. CoRR abs/1609.03441 (2016)
12. Dozat, T., Manning, C.D.: Deep biaffine attention for neural dependency parsing. CoRR abs/1611.01734 (2016)
13. Duong, L., Cohn, T., Bird, S., Cook, P.: A neural network model for low-resource universal dependency parsing. In: EMNLP, pp. 339–348. Citeseer (2015)
14. Dyer, C., Ballesteros, M., Ling, W., Matthews, A., Smith, N.A.: Transition-based dependency parsing with stack long short-term memory. arXiv preprint arXiv:1505.08075 (2015)
15. Edmonds, J.: Optimim branchings. J. Res. Natl. Bur. Stand. B **71B**(4), 233–240 (1966)

16. Goodfellow, I., Warde-Farley, D., Mirza, M., Courville, A., Bengio, Y.: Maxout networks. In: ICML, pp. 1319–1327 (2013)
17. Guo, J., Che, W., Yarowsky, D., Wang, H., Liu, T.: Cross-lingual dependency parsing based on distributed representations. In: ACL, vol. 1, pp. 1234–1244 (2015)
18. Hinton, G.E., McClelland, J.L., Rumelhart, D.E.: Paralell Distributed Processing: Explorations in the Microstructure of Cognition: Foundations, vol. 1. MIT Press/Bradford Books, Cambridge (1986)
19. Jozefowicz, R., Vinyals, O., Schuster, M., Shazeer, N., Wu, Y.: Exploring the limits of language modeling. arXiv:1602.02410 [cs], February 2016
20. Kim, Y., Jernite, Y., Sontag, D., Rush, A.M.: Character-aware neural language models. arXiv preprint arXiv:1508.06615 (2015)
21. Kiperwasser, E., Goldberg, Y.: Simple and accurate dependency parsing using bidirectional LSTM feature representations. arXiv:1603.04351 [cs], March 2016
22. van Merriënboer, B., et al.: Blocks and fuel: frameworks for deep learning. arXiv:1506.00619 [cs stat], June 2015
23. Mikolov, T., Sutskever, I., Chen, K., Corrado, G.S., Dean, J.: Distributed representations of words and phrases and their compositionality. In: NIPS, pp. 3111–3119 (2013)
24. Mikolov, T., Karafiát, M., Burget, L., Cernocky, J., Khudanpur, S.: Recurrent neural network based language model, Makuhari, Chiba, Japan, September 2010
25. Nivre, J.: Algorithms for deterministic incremental dependency parsing. Comput. Linguist. **34**(4), 513–553 (2008)
26. Nivre, J., et al.: MaltParser: a language-independent system for data-driven dependency parsing. Nat. Lang. Eng., 1 (2005)
27. Nivre, J., et al.: Universal dependencies 1.2. http://universaldependencies.github.io/docs/
28. Schuster, M., Paliwal, K.K.: Bidirectional recurrent neural networks. IEEE Trans. Signal Process. **45**(11), 2673–2681 (1997)
29. Srivastava, N., Hinton, G., Krizhevsky, A., Sutskever, I., Salakhutdinov, R.: Dropout: a simple way to prevent neural networks from overfitting. JMLR **15**, 1929–1958 (2014)
30. Srivastava, R.K., Greff, K., Schmidhuber, J.: Highway networks. arXiv:1505.00387 [cs], May 2015
31. Titov, I., Henderson, J.: A latent variable model for generative dependency parsing. In: Proceedings of IWPT (2007)
32. Vinyals, O., Kaiser, L., Koo, T., Petrov, S., Sutskever, I., Hinton, G.: Grammar as a Foreign language. arXiv:1412.7449 [cs stat], December 2014
33. Wu, Y., et al.: Google's neural machine translation system: bridging the gap between human and machine translation. arXiv:1609.08144, September 2016
34. Zeiler, M.D.: Adadelta: an adaptive learning rate method. arXiv:1212.5701 (2012)
35. Zhang, X., Cheng, J., Lapata, M.: Dependency parsing as head selection. CoRR abs/1606.01280 (2016)

Meaning Extensions, Word Component Structures and Their Distribution: Linguistic Usages Containing Body-Part Terms Liǎn/Miàn, Yǎn/Mù and Zuǐ/Kǒu in Taiwan Mandarin

Hsiao-Ling Hsu[1(✉)], Huei-ling Lai[2], and Jyi-Shane Liu[3]

[1] Graduate Institute of Linguistics, National Chengchi University,
Taipei City, Taiwan
heidimavis@hotmail.com
[2] Department of English, National Chengchi University, Taipei City, Taiwan
hllai@nccu.edu.tw
[3] Department of Computer Science, National Chengchi University,
Taipei City, Taiwan
liujs@nccu.edu.tw

Abstract. This study analyzes and compares the linguistic expressions of three sets of body-part terms extracted from the largest, balanced and widely-used Mandarin Chinese corpus, and aims to find their actual usage patterns in the real world context of Mandarin Chinese. It is found that PERSON and EMOTION are the most prevalent metonymic meaning in the six body part terms. As for the metonymic and metaphorical meanings in the six body-part terms and their corresponding word component structures, it is found that when the body-part terms denote PERSON, the most dominant word component structure is $[NN]_N$; when they denote EMOTION, $[NN]_N$ and $[VN]_V$ are the most dominant structures. In addition, the $[NN]_N$ structure shows the highest frequency of occurrences in all the six body part terms when they are used metaphorically.

Keywords: Meaning extensions · Word component structures · Tagging · Metonymies · Metaphors

1   Introduction

Metaphor and metonymy are considered two important mechanisms that people apply to understand abstract concepts [4]. Human body parts are regarded as the most primary source domain which people map to understand abstract concepts in the target domain, for we are all born with our body parts and they are the foremost way we interact with the world. In English, people often use the linguistic expression – *lose face* to mean 'lose one's own dignity or self-respect', and a similar expression – *diūliǎn* 丢臉 'lose face' is also found in Mandarin Chinese [11].

© Springer International Publishing AG 2017
K. Ekštein and V. Matoušek (Eds.): TSD 2017, LNAI 10415, pp. 335–343, 2017.
DOI: 10.1007/978-3-319-64206-2_38

Prolific findings based on systematic cognitive networks have been proposed for the analysis of metonymic and metaphorical expressions with body-part terms. However, studies of body-part term metaphors are mostly based on linguistic data limited to reliable and stable dictionaries or underdeveloped corpora, leading to their findings more of a theoretically-based reasoning than an empirically-based treatment (e.g. [4, 8, 10–16]). Particularly, corpora with a multitude of linguistic expressions have shown that language use is not always stable and fixed but dynamic [1]. Novel usages often emerge from various contexts, and continue to change for different communicative purposes. For instance, *dǎliǎn* 打臉 'hit face' has recently been used metonymically and metaphorically among young generations and even among news reporters or anchors. Nowadays, the word is used not only to express their literal meaning 'to hit one's face' but also to be extended to express 'to cause someone to feel embarrassed or ashamed'.

In addition, viewing the components of words in terms of their form class identity (i.e. syntactic form class identity or 'part of speech') can tell us what systematic knowledge native speakers possess to understand and use the words [6]. From usage-based linguistic evidence, it is highly expected that generalized patterns can be found based on the interaction between the form class of the word constituents and their senses.

To better capture such dynamicity and the generalized patterns exhibited by language use, this study, incorporating computational linguistics approaches, corpus-based approaches, and cognitive semantic theories, aims to find actual usage patterns in the real world context of Mandarin Chinese. By combining these approaches, we analyze and compare the linguistic expressions of three sets of body-part terms in Mandarin Chinese – *face* (臉 *liǎn*/ 面 *miàn*), *eye* (眼 *yǎn*/目 *mù*), and *mouth* (嘴 *zuǐ*/口 *kǒu*). The three sets of body-part terms are selected because they are considered the most distinctive body-parts on which people depend to understand and interact with the outside world: people recognize others through their faces, perceive the outside world through their eyes and express themselves through their mouths. Therefore, these three prominent body parts are regarded as a good starting point to understand human cognition and the six body part terms are the source domain vocabularies this paper uses to identify metonymies and metaphors in the corpus.

2   Related Work

2.1 Metaphors and Metonymies with Body Part Terms in Mandarin Chinese: An Overview for 臉 *liǎn*, 面 *miàn*, 眼 *yǎn*, 目 *mù*, 嘴 *zuǐ*, and 口 *kǒu*

Body-parts are everything that human beings are made of and are keys to human existence in spatial and temporal domains. We depend on our body-parts to interact with the outside world. The important status of face in human body parts has attracted many scholars to investigate how *liǎn* and its counterpart *miàn* are used in Mandarin Chinese. It is suggested that four non-literal meanings can be found in *liăn* and *miàn*. They would be used to express "PER-SON, CHARACTER, EMOTION, or DIGNITY" [11,12,16]. Qin [7] studies the conceptual metaphors of *yăn* & *mù* and proposes that the metaphorical use of *yăn* & *mù* consists of four major experiential domains: knowledge/intellection; emotion/attitude; social relationship/relationship between entities; and shape/time. Zhao [14] in his dessertation re-examines the metaphoric and metonymic expressions containing *yăn* or *mù* in Chinese, and proposes that at least six metonymies: PERSON, EMOTION, EYESIGHT/ABILITY, INTELLECTION, FOCUS, and QUANTITY, and three metaphors: NEAR TIME/SPACE, TINY HOLE, and CENTER can be identified. Zhao [14] has found four metonymic usages of *zuĭ*: EMO-TION, FLAVOR, CHARACTER, and UTTERANCES, five in *kŏu*: PERSON/QUANTITY, EMOTION, FLAVOR, CHARACTER, and UTTERANCES, and two metaphoric usages of *kŏu*: MOUTH-SHAPED and GATEWAY. He has also found that *zuĭ* is harder to be extended into metaphoical meanings.

2.2 Describing Word Components: Form Class Description

Many ways have been proposed to characterize and describe the word components: relational description, modification structure description, semantic description, syntactic description and form class description [6:21–33]. Among these descriptions, Packard [6] favors the "form class" method of word component analysis, for this approach avoids problems that cannot be solved under other approaches. Form class description is the method used by Lu [5] in his work *Word Formation in Chinese*. Under this approach, a word's component morphemes are described in terms of their syntactic form class identity, or "part of speech", and they are often coded in this way: $[M1\ M2]_{w1}$, M1 M2 represent the form class identity of a word's component morphemes and w1 refers to the form class identity of the word [6]. Some advantages are proven by adopting this approach. One is that if word's components are viewed in terms of their form class identity, they can be easily and systematically categorized. Also, "it allows us to account for different types of systematic knowledge that native speakers possess regarding the composition of words [6:32]."

3 Method

3.1 The Data

The data are all extracted from Academic Sinica Balanced Corpus of Modern Chinese 4.0 (simplified as Sinica Corpus 4.0) Sinica Corpus is considered the first, the largest and the most representative word-based Chinese Corpus established by Institute of Information Science and CKIP group in Academic Sinica Taiwan [3]. The words are segmented in Sinica Corpus 4.0 and tagged with their grammatical functions. By incorporating computational linguistics approaches, instances containing either one of the six target body-part terms *liăn/miàn*, *yăn/mù* and *zuĭ/kŏu* in Sinica Corpus 4.0 are extracted based on the reliable word segmentation results. The target body-term *liăn/miàn*, *yăn/mù* and *zuǐ/kǒu* are first searched in the corpus. Then, once the target body-part terms are located in the texts, the words containing the target body-part terms are extracted according to the CKIP segmentation. Lastly, the raw data are exhibited in the excel templates with the following information: word, part of speech, and frequency.

(i) Word: It is the target instance containing the target body-part terms which is segmented according to the CKIP segmentation. In this study, only disyllabic words are analyzed.

(ii) Part of Speech: It shows the part of speech of the "word" (not the word components) containing the target body-part terms in the texts collected by the corpus according to CKIP parsing results. The information of the words' part of speech is important for our form class identity (i.e. word component structure) analyses. According to Packard [6], the internal structure of word components may influence the output of the word, namely, the part of speech of the words.

(iii) Frequency: It represents the number of the occurrence that the word containing the target body-part terms is used in the texts collected by the corpus. In this present study, it is important information to show the distribution of the meanings and word component structures.

3.2　Coding Schema

Based on the raw data, we develop several steps to analyze the forms and meanings of the words containing the target body-part terms. The form class identity of word components and the meanings of words are analyzed according to the following steps.

(i) Step One (class identification): The form class of the target body-part terms are first identified as they are listed in authoritative *Chinese Dictionary* established by *Ministry of Education, R.O.C* (http://dict.revised.moe.edu.tw/cbdic/index.html) and *Chinese Wordnet* (http://cwn.ling.sinica.edu.tw).

(ii) Step Two (structural analysis): Once the form class of the target body-part terms are identified, the form classes of the other components that co-occur with target body-part terms in disyllabic words are analyzed based on the components' role and position within that word and the form classes as they are listed in the *Chinese Dictionary* and *Chinese Wordnet* [7]. For example, a disyllabic word like *diūliǎn* 丟臉 'lose face' is analyzed as a verb $[]_V$ composed of a verb and a noun elements ($[VN]_V$); *yǎnjìng* 眼鏡 'glasses' is analyzed as a noun $[]_N$ constructed by two noun elements ($[NN]_N$); *zhèngmiàn* 正面 'front' is analyzed as a noun $[]_N$ composed of an adjective and a noun ($[AN]_N$).

(iii) Step Three (meaning identification): The various meanings of the six target body-part terms are analyzed and categorized based on their linguistic context and their usages in the real world context, with previous findings as references (e.g. [14]). For instance, the disyllabic word like *liǎnjiá* 臉頰 'cheek' is analyzed and categorized into "body-part terms that denote literal meanings", while word like *shǎngliǎn* 賞臉 'to honor someone' is analyzed and categorized into "body-part terms that denote non-literal meanings", and here, it denotes the non-literal meaning DIGNITY.

4   Results

The results are displayed in the following tables. In total, 1076 types of disyllabic words containing the target body part terms were analyzed: 66 for *liǎn*; 334 for *miàn*; 192 for *yǎn*; 122 for *mù*; 66 for *zuǐ*; and 296 for *kǒu* (Table 1). As for the distribution of meanings in *liǎn*, *miàn*, *yǎn*, *mù*, *zuǐ*, and *kǒu*, it is found that the counterparts of *liǎn*, *yǎn*, and *zuǐ* – *miàn*, *mù*, and *kǒu* are much more frequently used to denote non-literal (metonymic and metaphorical) meanings (Table 1).

Table 1. Number of types and the distribution of literal and non-literal meanings in *liǎn*, *miàn*, *yǎn*, *mù*, *zuǐ*, and *kǒu*

Body part terms	Number of types	%	Literal	Metonymic	Metaphorical	Total (%)
Liǎn	66	6.13%	48.63%	51.37%	0.00%	100.00%
Miàn	334	31.04%	3.37%	**24.97%**	**71.66%**	100.00%
Yǎn	192	17.84%	42.83%	37.27%	19.91%	100.00%
Mù	122	11.34%	0.49%	**6.27%**	**93.24%**	100.00%
Zuǐ	66	6.13%	66.28%	27.93%	5.79%	100.00%
Kǒu	296	27.51%	6.09%	**49.01%**	**44.90%**	100.00%
Total	1076	100%				

The distribution of metonymic meanings in *liǎn*, *miàn*, *yǎn*, *mù*, *zuǐ*, and *kǒu* is given in Table 2. Based on the data extracted from Sinica Corpus 4.0, the six body part terms all denote the metonymic meanings PERSON and EMO-TION. Compare *liǎn*, *yǎn*, and *zuǐ* with their counterparts – *miàn*, *mù*, and *kǒu*, and it is found that they exhibit a very similar pattern: the metonymic mean-ings that *liǎn*, *yǎn*, and *zuǐ* denote are nearly the same as their correspond-ing counterparts. For instance, *liǎn* and *miàn* both denote PERSON (e.g. 頭臉 *tóuliǎn* 'people'; 人面 *rénmiàn* 'people'), EMOTION (e.g. 翻臉 *fānliǎn* 'suddenly get mad'; 面容 *miànróng* 'countenance'), CHARACTER (e.g. 黑臉 *hēiliǎn* 'appear as the hatchet man'; 假面 *jiǎmiàn* 'masked'), DIGNITY (e.g. 丟臉 *diūliǎn* 'lose face'; 情面 *qíngmiàn* 'face-saving'), and APPEARANCE (e.g. 露臉 *lóuliǎn* 'show up'; 露面 *lóumiàn* 'show up'). *Yǎn* and *mù* also denote the same metonymic meanings: they are both used to denote PERSON (e.g. 眼眉 *yǎnméi* 'spy'; 頭目 *tóumù* 'leader of a tribe'), EMOTION (e.g. 眼神 *yǎnshén* 'expression in one's eyes'; 瞪目 *dàngmù* 'stare in anger'), ABILITY (e.g. 鷹眼 *yīgyǎn* 'sharp eyes'; 鷹目

Table 2. The distribution of metonymic meanings in *liǎn*, *miàn*, *yǎn*, *mù*, *zuǐ*, and *kǒu*

Terms/Freq	Metonymies							Total
Liǎn (%)	PERSON	EMOTION	CHARACTER	DIGNITY	APPEARANCE			
	0.51	**71.18**	9.26	13.38	5.66			100.00
Miàn (%)	PERSON	EMOTION	CHARACTER	DIGNITY	APPEARANCE	CONFRONT	QUANTITY	
	11.56	4.95	0.21	3.97	17.12	**62.07**	0.13	100.00
Yǎn (%)	PERSON	EMOTION	ABILITY	INTELLECTION	FOCUS			
	0.23	**38.37**	13.17	30.61	17.62			100.00
Mù (%)	PERSON	EMOTION	ABILITY	INTELLECTION	FOCUS			
	5.52	13.25	1.36	34.92	**44.94**			100.00
Zuǐ (%)	PERSON	EMOTION	FLAVOR	UTTERANCES	CHARACTER			
	3.55	14.20	15.38	**49.70**	17.16			100.00
Kǒu (%)	PERSON	EMOTION	FLAVOR	UTTERANCES	QUANTITY			
	33.48	0.50	16.36	**47.82**	1.84			100.00

yīngmù 'sharp eyes'), INTELLECTION (e.g. 明眼 *míngyǎn* 'discerning eye'; 目光 *mùguāng* 'vision'), and FOCUS (e.g. 耀眼 *yàoyǎn* 'dazzling'; 奪目 *duómù* 'to catch one's eyes'). *Zuǐ* and *kǒu* both denote PERSON (e.g. 名嘴 *míngzuǐ* 'pundit'; 戶口 *hùkǒu* 'number of households'), EMOTION (e.g. 嘟嘴 *dūzuǐ* 'pout'; 劈口 *pīkǒu* 'yell at someone suddenly'), FLAVOR (e.g. 挑嘴 *tiāozuǐ* 'be fussy about food'; 口感 *kǒugǎn* 'texture') and UTTERANCES (e.g. 頂嘴 *dǐngzuǐ* 'talk back'; 口才 *kǒucái* 'eloquence'). In terms of the percentage of frequency, *liǎn* (71.18%) and *yǎn* (38.37%) are used to denote EMOTION the most frequently; *miàn* (62.07%) denotes CONFRONT (e.g. 面對 *miànduì* 'face (v.)') the most frequently; *mù* (44.94%) denotes FOCUS the most frequently; and both *zuǐ* (49.70%) and *kǒu* (47.82%) denote UTTERANCES the most frequently.

The distribution of metaphorical meanings in *miàn*, *yǎn*, *mù*, *zuǐ*, and *kǒu* is displayed in Table 3. *Liǎn* denotes no metaphorical meanings. Three metaphorical meanings are found in *miàn*: SURFACE (e.g. 地面 *dìmiàn* 'ground'), CONDITION (e.g. 市面 *shìmiàn* 'market conditions') and SIDE (e.g. 北面 *běimiàn* 'the north'), with SIDE showing the highest frequency. *Yǎn* and *mù* both denote NEAR TIME/SPACE (e.g. 眼前 *yǎnqiàn* 'at present'; 目前 *mùqiàn* 'at present'), TINY HOLE (e.g. 網眼 *wǎngyǎn* 'tiny hole of a net'; 網目 *wǎngmù* 'tiny hole of a net'), and CENTER (e.g. 詞眼 *cíyǎn* 'motif of lyrics'; 綱目 *gāngmù* 'outline'), and they are used to denote NEAR TIME/SPACE the most frequently. *Mù* also denote ENTRY (e.g. 書目 *shūmù* 'bibliography'), which is extended from the metaphorical meaning TINY HOLE. Both *zuǐ* and *kǒu* are used to denote MOUTH-SHAPED (e.g. 壺嘴 *hùzuǐ* 'spout'; 口袋 *kǒudài* 'pocket'), while *kǒu* has two more metaphorical meanings GATEWAY (e.g. 進口 *jìnkǒu* 'import') and NEAR TIME/SPACE (e.g. 當口 *dāngkǒu* 'at present'). *Zuǐ* is only used to denote the metaphorical meaning MOUTH-SHAPED (100.00%), and *kǒu* shows the highest percentage of frequency when it denotes the metaphorical meaning GATEWAY (87.39%).

Table 3. The distribution of metaphorical meanings in *miàn, yǎn, mù, zuǐ,* and *kǒu*

Terms/Freq	Metaphors				Total
Miàn %	SURFACE	CONDITION	SIDE		
	22.55%	18.04%	**59.41%**		100.00%
Yǎn %	NEAR TIME/SPACE	TINY HOLE	CENTER		
	84.19%	1.03%	14.78%		100.00%
Mù %	NEAR TIME/SPACE	TINY HOLE	CENTER	ENTRY	
	52.56%	0.13%	22.95%	24.36%	100.00%
Zuǐ %	MOUTH-SHAPED				
	100.00%				100.00%
Kǒu %	MOUTH-SHAPED	GATEWAY	NEAR TIME/SPACE		
	11.90%	**87.39%**	0.71%		100.00%

In general, the six body-part terms are all used to denote the metonymic meanings PERSON and EMOTION. Because only these two metonymic meanings are shared by all the six body-part terms, here, only their corresponding word component structures are compared. It is found that when the six target body part terms denote the metonymic meaning PERSON, their word component structures show the highest frequency of occurrences in $[NN]_N$, such as 頭臉 *tóuliǎn* 'people'; 人面 *rénmiàn* 'people'; 眼眉 *yǎnméi* 'spy'; 頭目 *tóumù* 'leader of a tribe'; 名嘴 *míngzuǐ* 'pundit'; and 戶口 *hùkǒu* 'number of households'. When *liǎn, miàn* and *yǎn* denote the metonymic meaning EMOTION, the $[NN]_N$ structure also shows the highest frequency of occurrences, such as 嘴臉 *zuǐliǎn* 'facial expressions'; 面容 *miànróng* 'countenance'; and 淚眼 *lèiyǎn* 'tearful eyes', while $[VN]_V$ shows the highest frequency of occurrences when *mù, zuǐ* and *kǒu* denote EMOTION, such as 瞪目 *dàngmù* 'stare in anger'; 嘟嘴 *dūzuǐ* 'pout'; and 劈口 *pīkǒu* 'yell at someone suddenly'.

As for the metaphorical meanings in the six body part terms and their corresponding word component structures, the $[NN]_N$ structure shows the highest frequency of occurrences in all the six body part terms when they are used metaphorically, such as 眼前 *yǎnqiàn* 'at present' and 目前 *mùqiàn* 'at present'. The only difference is that when *kǒu* is used to denote the metaphorical meaning GATEWAY, the $[VN]_V$ structure is preferred rather than the $[NN]_N$ structure, such as 進口 *jìnlǒu* 'import'.

5 Discussions

This corpus-based study attempts to investigate the exact usages of body-part terms in Mandarin Chinese and the syntagmatics in their meaning extensions.

Recently, how to identify and quantify metonymies and metaphors in corpora has been a trend [9]. Several studies have been done to identify the grammar in metaphors in English, and it is found that when source domain nouns are used to denote metaphorical meanings, they bear different part-of-speech: they are used as verbs, e.g. *dog* [2]. However, still little research has been done to quantify metonymies and metaphors in corpora, and in particular, little research has been done to investigate the correlation between word component structures and meanings in Mandarin Chinese. This paper provides a systematic way to identify the word component structures and metonymic and metaphorical expressions with body-part terms in disyllabic words found in Sinica Corpus 4.0 with quantification and statistical evaluation.

Although previous studies have identified some metonymic and metaphorical meanings that *liǎn*, *miàn*, *yǎn*, *mù*, *kǒu*, and *zuǐ* may denote, what exact metonymic and metaphorical meanings the six body part terms denote in people's everyday usages still cannot be discovered from the traditional and introspective approach. By analyzing words that contain body part terms extracted from Sinica Corpus 4.0, a more comprehensive and complete picture is provided to exhibit how people used the six body part terms. It is found that *miàn*, *mù*, and *kǒu* are more frequently used to denote non-literal meanings than their counterparts – *liǎn*, *yǎn*, and *zuǐ* (Table 1). This may be the reason why there are two candidates which refer to the same body part in Mandarin Chinese: both of the two candidates refer to the same body part, but one is "literal-meaning-oriented" and the other is "non-literal-meaning-oriented". In addition, it is found that PERSON and EMOTION are the most prevalent metonymic meanings in the six body-part terms. This shows that when body parts are extended to denote non-literal meanings, denoting PERSON and EMOTION is the initial stage in their non-literal meaning extensions, that is, other metonymic meanings are extended on the basis of either PERSON and EMOTION. As for their corresponding word component structures, it is found that when the body-part terms denote PERSON, the most dominant word component structure is $[NN]_N$; when they denote EMOTION, $[NN]_N$ and $[VN]_V$ are the most dominant structures. And, the $[NN]_N$ structure shows the highest frequency of occurrences in all the six body part terms when they are used metaphorically.

It is highly expected that the results may provide some insights for future study to refine procedures to help identify metaphors and metonymies in corpora and then accelarate quantification and statistical evaluation.

References

1. Cameron, L., Deignan, A.: The emergence of metaphor in discourse. Appl. Linguist. **27**(4), 671–690 (2006)
2. Deignan, A.: The grammar of linguistic metaphors. Trends Linguist. Stud. Monogr. **171**, 106 (2006)
3. Huang, C., Chen, K., Chang, L., Hsu, H.: An introduction to academia Sinica balanced Corpus. In: Proceedings of ROCLING VIII, pp. 81–99 (1995). (in Chinese)
4. Hung, Y.Y., Gong, S.P.: A cross-linguistic approach to body part metonymy in English and Chinese. In: Effects of an On-line Syntactic Analysis Strategy Instruction on University Students' Reading Comprehension of English Science Texts, vol. 16, p. 146 (2011)
5. Lakoff, G., Johnson, M.: Metaphors We Live By. University of Chicago Press, Chicago (1980)

6. Lu, Z.: Word Formation in Chinese. Scientific, Beijing (1964)
7. Packard, J.L.: The Morphology of Chinese: A Linguistic and Cognitive Approach. Cambridge University Press, Cambridge (2000)
8. Qin, X.: The conceptual metaphors of "eye" - a comparative study based on the corpus between English and Chinese. J. Foreign Lang. **31**(5), 37–43 (2008)
9. Stefanowitsch, A., Gries, S.T. (eds.): Corpus-Based Approaches to Metaphor and Metonymy, vol. 171. Walter de Gruyter, Berlin (2007)
10. Tsai, L.C.: The Metaphor of Body Parts in Chinese. Unpublished Master of Arts Thesis, National Tsing-Hua University, Taiwan (1994)
11. Wen, X., Wu, S.Q.: A study on the cognitive features of the metaphoric expressions involving the English and Chinese FACE. J. Southwest Univ. **6**, 28 (2007). (Social Sciences Edition)
12. Xie, H.Z.: Comparative study on metaphorical meaning of vocabulary with the denotation of "face" between Chinese and English. Unpublished Master of Arts Thesis, Central China University, China (2011)
13. Yu, N.: Metaphor from body and culture. In: The Cambridge Handbook of Metaphor and Thought, pp. 247–261 (2008)
14. Yu, N.: Speech organs and linguistic activity/function in Chinese (a revised and expanded version of a 2009 book chapter). In: Maalej, Z., Ning, Y. (eds.) Embodiment via Body Parts: Studies from Various Languages (Human Cognitive Processing series, 31), pp. 117–148. John Benjamins, Amsterdam (2011)
15. Zhai, L.F.: The interaction of metaphor and metonymy in the Chinese expressions of body-part terms Yan and Mu. Canadian Soc. Sci. **4**(1), 57–62 (2009)
16. Zhao, X.D.: A cognitive study on meaning transference of body-part terms. Unpublished Doctoral Dissertation, Fudan University, China (2010)

An Unification-Based Model for Attitude Prediction

Manfred Klenner[(✉)]

Computational Linguistics, University of Zurich, Zürich, Switzerland
klenner@cl.uzh.ch
https://www.cl.uzh.ch

Abstract. Attitude prediction strives to determine whether an opinion holder is positive or negative towards a given target. We cast this problem as a lexicon engineering task in the context of deep linguistic grammar formalisms such as LFG or HPSG. Moreover, we demonstrate that attitude prediction can be accomplished solely through unification of lexical feature structures. It is thus possible to use our model without altering existing grammars, only the lexicon needs to be adapted. In this paper, we also show how our model can be combined with dependency parsers. This makes our model independent of the availability of deep grammars, only unification as a processing mean is needed.

Keywords: Sentiment · Opinion inference · Lexical functional grammar

1 Introduction

Attitude prediction comprises the identification of an opinion holder, an opinion target and the positive or negative attitude of the holder towards the target. It is a variant of stance detection where the targets are not known in advance and the writer is not necessarily the only opinion holder. Take *Peres accused Syria to support Hezbollah.* Here, the writer claims that Peres (an opinion holder) has a negative attitude towards Syria (the target). However, the sentence also implies that Peres is against the Hezbollah (another target). Moreover, Hezbollah is the (potential) beneficiary of a support event which - as the reason of the accusation - is (contextually) perceived as being negative. Being against a event means that one is also against any beneficiaries of that event. Correspondingly, one has a positive attitude towards victims of any disapproved event (e.g. in *A0 complained that A1 was hurt*, A0 is a proponent of the victim A1). As discussed in [7], the truth commitment that comes with particular verbs, its interaction with negation in the course of the determination of event factuality [9] are crucial components of any attitude prediction. So far, attitude prediction has been cast as logic-based inference, either with machine learning (Probabilistic Soft Logic, see [2]) or Description Logics (cf. [7]). In this paper, we introduce a lean model for attitude prediction. Instead of deduction, unification is used and instead of rules, verb classes are specified on the basis of attribute value pairs (features).

© Springer International Publishing AG 2017
K. Ekštein and V. Matoušek (Eds.): TSD 2017, LNAI 10415, pp. 344–352, 2017.
DOI: 10.1007/978-3-319-64206-2_39

Our model can easily be combined with existing deep grammars such as the Pargram LFG grammars [1]. Only the lexicon needs to be adapted leaving the grammar rules as they are. However, our model is independent of the existence of such grammars. We only need unification, not unification grammars. Actually, we show how to combine our model with dependency parsers. This makes our approach widely applicable. Our current model is for German.

2   Unification-Based Grammars

The core of unification or constraint-based grammars are feature structures (f-structure, henceforth) and unification. A feature structure is a list of attribute value pairs, where the attributes are, among others, names of grammatical functions or morphological categories. Values are atomic or complex, i.e. feature structures again. Unification of two feature structures is successful if the unification of corresponding attributes is successful. Attribute values unify if both are atomic and identical or, in the case of complex values, both recursively unify. We work with LFG [3] and the Xerox Linguistic Environment (XLE) grammar engineering tool [4]. In addition to the standard unification principle, LFG requires feature structures to be coherent (a governable function must have a governor) and complete (any governed function must be realized). In XLE so-called templates are available. A template is a means to build classes and to save specification effort (which is interesting from an engineering perspective).

3   Attitudes and Polar Effects

If we assume that the writer is committed to the truth of his text, we can infer the attitudes among the entities referred to, but we also are able to find out whether the scenario is good or bad *of* or *for* a referent. If *A0 injures A1* this is bad *of* A0 and bad *for* A1. We call these resulting states a predicate produces the *polar effects* and use *polar roles* like *PFOR* (positive for) to denote that a referent occupies such a role. Polar effects are crucial, they allow to understand the role a referent plays in a text. They allow to understand how an entity is perceived (or cast) by the text author. For instance, we found that the German Chancellor, Angela Merkel, takes quite often the role of a negative actor (a *NOF* (negative of) role) given the texts of the AfD, a German right-wing party.

Obviously verbs are most crucial for these tasks. For instance, we know that *cheat* means that A0, the actor, acts in a negative way and thus should be regarded as a villain, while A1, the patient, is a victim. If we know the polar effect of a verb (on the target), we know the attitude (of the opinion holder towards the target). The (situation-specific) attitude of A0 expressed by *cheat* towards A1 is negative. The point is that these polar effects and - to a certain degree also the attitudes - depend on the factuality status of the sentence.

We refer to the concept of event factuality as discussed in [9]. If an event is factual, polar effects take place, if an event is counterfactual, the inverted effects might (depending on the verb) take place. If the event is nonfactual, no effect is cast.

The truth of *Italy has helped the migrants to survive* makes *Italy* a benefactor and the *migrants* a beneficiary. The negated form, *Italy not helped the migrants to survive*, turns *Italy* into a villain and the *migrants* into a victim. Finally the modal, nonfactuality indicating version *Italy might help the migrants to survive* blocks any inferences.

Such verbs do have a truth commitment wrt. their complement clause. An affirmative and truth committed subclause gives rise to event factuality. Thus *cheat* denotes a true event given that *A0 regrets that he cheated A1* is true. If the subclause is negated, the truth commitment makes the event counterfactual: *A0 regrets that he not has told the truth*. In this case, the inverse event is true (*he lied*). We also have to take the affirmative status of the matrix verb into account, which might alter the truth commitment.

According to [7], we need to distinguish for each verb a signature for the affirmative and for the negated usage (see also [5]). We propose to have three categories: 'T' (truth committing), 'F' (falsehood committing) and 'N' (no commitment). For instance, *regret* as a factive verb has 'T'-'T' (affirmative-negated), while *force* has 'T'-'N' and *refuse* has 'F'-'N'. *A0 refuses to cheat* means that *cheat* is counterfactual, while in *A0 not refuses to cheat* we cannot tell whether there will be a cheating event. Finally, non-factive verb like *hope* have 'N'-'N'. It crucial not to confuse truth commitment and factuality. The first one holds for the whole (even negated) subclause, while the second relates to the event expressed by the verb.

4 Unification-Based Attitude Prediction and Role Assignment

We have specified a reasoning scheme on the basis of unification. Constraining equality and existential constraints are building blocks for this. Constraining equality (notation = c) is a global constraint mechanism that demands that a particular value is introduced by a defining equation (notation =) somewhere in the lexicon or grammar. Existential constraints pose the restriction that a particular attribute must or must not be present.

Figure 1 shows the f-structures of two sentences: *Peres accuses Syria to support Hezbollah* (left-hand side) and *Peres regrets that Syria supports Hezbollah* (right-hand side). The predicted attitudes are captured under the attribute REL (either *pro* or *con*), the polar roles here are *PFOR* and *POF* (in the *regret* sentence). For instance, the *accuse* version gives rise to the negative attitude of Peres towards Syria, represented by (dropping the grammatical functions) *con<Peres,Syria>* in Fig. 1. Only in the *regret* version, we find a pro relation, *pro<Syria,Hezbollah>* (since *support* then is factual) and a *POF* filled by *Syria* and a *PFOR* role filled by *Hezbollah*.

A verb subcategorizing for a complement clause gets a verb signature ('T', 'F' or 'N') depending on the affirmative status (affirmative or negated) of the verb. We introduce an attribute *TCOM* and embed it under the subcategorized clausal complement (either COMP or XCOMP). In Fig. 1 *support* has *TCOM N* in the

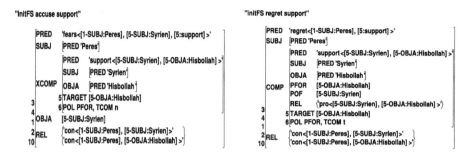

Fig. 1. F-structures for Peres accuse/regret Syria support Hezbollah

case of *accuse* while *TCOM T* with *regret*. Figure 2 shows the (partial) entry for *beschuldigen (accuse)*, a verb that has signature 'T-N'. We use XLE notation: verb form followed by part of speech, the morphology indicator * and the feature equations[1]. Line 1 gives the semantic form of the verb (with the subcategorized grammatical functions). Line 2 is a control equation (*beschuldigen* has object control). Line 3 is the disjunctive condition for truth commitment (realization of 'T-N'). The attribute *NEG* exists if the verb is negated, otherwise the attribute is not present at all. The existential constraint (↑NEG) thus checks whether the clause is negated. It is a kind of precondition for the assignment of the subclause's truth commitment): 'T' if affirmative (~(↑NEG)), 'N' if negated (↑NEG).

beschuldigen (accuse) V *
1   (↑PRED)='accuse<(↑SUBJ)(↑OBJA)(↑XCOMP)>'
2   (↑XCOMP SUBJ)=(↑OBJA)
3   {~(↑NEG) (↑XCOMP TCOM) = T | (↑NEG) (↑XCOMP TCOM) = N}
4   @(nforverb OBJA)
5   @(disapprove XCOMP)

Fig. 2. Lexical entry for the verb *beschuldigen (accuse)*

Lines 4 and 5 are template invocations that relate to the two main tasks, the prediction of polar effects and attitudes. Most of the time, the subject is the opinion source. The Target role is more flexible, either *OBJA, OBJD or OBJP* might occupy it (depending on the verb). We use the polar roles *PFOR* (positive for), *NFOR* (negative for), *POF* (positive of) and *NOF* (negative of) to model the verb-specific polar effects. For instance, *criticize* has a *PFOR* role (if factual) and the direct object *OBJA* as the target. Given the verb *enjoy*, it is the subject that takes the *PFOR* role. These polar roles are, thus, abstract semantic roles with a polarity load. The template *@nforverb* (line 4 from Fig. 5) is invoked with the target role, *ROLE*, which for *accuse* is *OBJA*.

---

[1] An up arrow inserts a feature into the feature structure defined by the equation.

Before we introduce *@nforverb* we briefly discuss (the template for) factuality:

factual =
{˜(↑NEG) (↑TCOM)=c T| (↑NEG) (↑TCOM)=c F|˜(COMP↑) ˜(XCOMP↑) ˜(↑MOD) ˜(↑NEG)}.

Factuality holds, if the verb is affirmative and has T as signature, or if negated and the signature is F, or if it is not embedded (inside-out determination) and there is neither modality nor negation present. ˜(COMP↑) prohibits (inside-out) embedding under COMP. This realises factuality determination of the outmost matrix verb.

Figure 3 shows the definition of (the verb class) *@nforverb*. *@nforverb* and *@pforverb* carry out the assignment of polar roles under factuality and counterfactuality. They also establish the *inner* attitude prediction, i.e. in cases where source and target have the same verbal head (the templates *@direct_con* and *@direct_pro* from Fig. 3). Line 1 from Fig. 3 sets the target role. The rest of the definition depends on the factuality status of the verb. If the verb is factual (line 2) then the target role is set as *NFOR*. If the verb is counterfactual (line 3) the polar role is inverted, e.g. set to *PFOR*. The verb *criticize*, e.g., is a *@nforverb* verb. If A0 criticizes A1, this is negative for A1 (*NFOR*) and the attitude of A0 towards A1 is negative (*@direct_con*). IF *criticize* ist nonfactual (line 4 and 5), e.g. embedded into *hope*, then no polar role is set. However, in order to determine the *outer* attitude, namely the one of the opinion source of the matrix clause towards the referents of the embedded verb in nonfactual cases, we need to know the polar role profile of the verb. Is the target someone who would benefit (*PFOR*) or suffer (*NFOR*) from a situation where the event denoted by the verb was true? To provide this information is the function of the attributes *NFORView* and *PFORView*: they define the role profile without instantiating roles.

nforverb(ROLE) =
1   (↑TARGET)=(↑ROLE)
2   {@(factual)   (↑NFOR)=(↑ROLE) @direct_con (↑NFORView)=+
3 |@(counterfactual) (↑PFOR)=(↑ROLE) @direct_pro (↑PFORView)=+
4 |@nonfactual
5   {(↑NEG)(↑PFORView)=+ | (↑NEG) (↑NFORView)=+} }.

Fig. 3. Template definition

Figure 4 shows the definition of the *@disapprove* template responsible for attitudes between the matrix clause and the subclause. Again, the factuality status is crucial, but this time it is the one of the matrix clause.

First of all, given a verb like *accuse* of type *@disapprove*, the opinion holder of the matrix clause is against the subject of the subclause if the matrix clause is factual: this is the function of the template call @(subj_con C), line 1.

disapprove(C) =
1   {@factual @(subj_con C)
2   {(↑C PFORView) 'con<(↑SUBJ)(↑C TARGET)>' ∈ (↑REL)
3 |(↑C NFORView) 'pro<(↑SUBJ)(↑C TARGET)>' ∈ (↑REL)|@(NOView C)}
4 |@counterfactual @(subj_pro C)
5   {(↑C PFORView) 'pro<(↑SUBJ)(↑C TARGET)>' ∈ (↑REL)
6 |(↑C NFORView) 'con<(↑SUBJ)(↑C TARGET)>' ∈ (↑REL)|@(NOView C)}
7 |@nonfactual}.

Fig. 4. Template definition

The variable C indicates the subclause type (XCOMP, COMP). If the subclause obeys to a *PFORView* (*accuse that A0 has helped A1*) (see line 2) then a *con* relation is set. *REL* is an attribute that takes a set as its value (since the opinion source might have more than one attitude): ∈ is used, thus, instead of (↑=↓). If the matrix is factual but the subclause turns out to have a *NFORView* (*accuse that A0 not has helped A1*) then the relation is *pro* (line 3). The corresponding definition holds for counterfactuality (see line 4–6). In case that the sublclause is a verb without a polar viewpoint, *@NOView* applies (it just verifies that the verb has no polar view).

In our model, direct inferences are restricted to a single level of embedding. We claim that this is sufficient. Inferences for deeper nested structures like in the sentence *X criticizes that A0 has not helped to free A1* where X has a positive attitude towards A1 can be drawn with transitive rules like: con<X,A0> ∧ con<A0,A1> → pro<X,A1>.

5   Model Initialisation from a Dependency Parse

We used the verb resource[2] of [6] in our implementation. It was automatically mapped to XLE specifications like the one shown in Fig. 2.

Our model for attitude prediction is purely lexicalistic. A lexical entry of a verb fully specifies its behavior either as an embedding or embedded verb in a simple or complex sentence. It thus can be combined with any existing deep linguistic grammar. But we also can combine it with dependency parsers. We only need to determine the grammatical functions of the involved verbs at the right embedding level. This information is available from the dependency tree. Figure 5 shows an example. The right-hand side is the result of the mapping: a feature structure of type DepFS with one embedding level. All information stems from the dependency tree. The label *obji* is mapped to *XCOMP*. We need a single XLE grammar rule (slightly simplified) in order to parse this: G → DepFS V V: {(↑COMP)=↓ |(↑XCOMP)=↓}.

The expression to be parsed is: *parse {initFS accuse support}* (in general: initFS matrix_verb subclause_verb1 subclause_verb2 ...). First *initFS* is identified as *DepFS*, its f-structure is unified with those of the matrix verb (*accuse*).

---

[2] Available from https://pub.cl.uzh.ch/projects/opinion/lrec_data.txt.

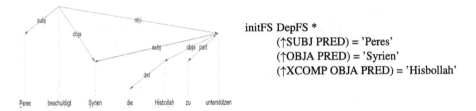

initFS DepFS *
($\uparrow$SUBJ PRED) = 'Peres'
($\uparrow$OBJA PRED) = 'Syrien'
($\uparrow$XCOMP OBJA PRED) = 'Hisbollah'

Fig. 5. Peres beschuldigt (accuses) Syrien die Hisbollah zu unterstützen (to support)

The subclause verb is unified with the result, its f-structure gets embedded under *XCOMP*. See Fig. 1 for the resulting f-structure.

6   Empirical Evaluation

It comes as a surprise that the interannotator agreement for the task at hand is low. In the empirical evaluation reported in [7] comprising 160 sentences it is 43% (Cohens $\kappa = 0.19$). One of the reasons is that humans seem to be "selective annotators and focus on the most striking attitudes more than on the more hidden ones" (see [7, p. 83]). They also report that the two annotators often produced different and - as it turned out - complementary annotations for a given sentence. Thus, reconciliation was unproblematic, i.e. both accepted their respective additional attitude annotations. This indicates that attentiveness might be a problem, but also that factors might play a role that are beyond the verb semantics, e.g. stemming from world knowledge.

We used the data from [7]. Note that our two models are similar[3], but our model is leaner and realized with a totally different framework. Also, our notion of factuality differs (see Related Work). However, our attitude labels are easily mapped onto theirs.

Their data consists of (a) 80 complex made-up sentence (their precision was 83.89%, recall is 93.72%) and (b) 80 newspaper sentences (59.04% precision and 71.15% recall). Our system achieved 85.12%, recall 91.52%) for (a) and 65.24% precision and 75.13% recall for (b). It is obvious that this is but a first evaluation. A larger data set is needed - which is not available yet. We also need to clarify how humans actually perform attitude prediction. This is future work.

7   Related Work

The goal of the rule-based approach of [2] is to detect entities that are in a positive (PosPair) or negative (NegPair) relation to each other. Rules are realized in the framework of Probabilistic Soft Logic, where the rule weights depend on the output of the preprocessing pipeline made out of two SVM classifiers and

---

[3] We also use the parser in [10].

three existing sentiment analysis systems. The model of [2] also copes with event-level sentiment inference, however factuality is not taken into account. Also, polar roles do not play any role in their framework.

[7] stress the point that factuality determination is a crucial part of sentiment inferences. They introduce a rule-based system for German realized with Description Logic and SWRL. The rules also are taking the affirmative and factuality status of the sentence into account. The goal is to instantiate relations (con and pro) expressing the attitudes of entities towards each other. We agree that factuality is a crucial part of such a model. However, we use a tripartite distinction while their factuality labels are binary.

Recently, [8] have presented an elaborate model that is meant to explicate the relations between all involved entities: the reader, the writer, and the entities referred to by a sentence. Also, the internal states of the referents and their values are part of the model. The underlying resource, called connotation frames, was created in a crowd sourcing experiment, the model parameters (e.g. values for positive and negative scores) are average values. Our verb resource is, on the contrary, specified by experts. Again, factuality is not taken into account in their model.

8 Conclusion

In this paper, a purely unification-based approach for sentiment reasoning is introduced. The approach is independent of any existing deep linguistic grammar, but can be coupled easily with it. Only the verb lexicon needs to be augmented with additional verb-specific features. This would result in a system that carries out attitude prediction etc. while parsing instead of afterwards (like current systems do). In the current paper, we pursued another possibility. Namely, to couple the model with a dependency parser. Only the embedding skeleton needs to be derived from the dependency tree. Then unification with verb entries carries out the whole inference process. To the best of our knowledge, our work is the first that exploits the idea of feature unification on top of a dependency parse tree in order to solve a sophisticated problem. Feature structures and unification are an elegant representational scheme and provide powerful processing means. We have shown how to reap the benefits of this. This might stimulate other researcher to also pose their problems in terms of such a framework.

Our system realizes a linguistically informed approach to solve the problem of attitude prediction and the assignment of polar roles. Future work will focuses on a broader evaluation in the context of stance detection.

References

1. Butt, M., King, T.H., Masuichi, H., Rohrer, C.: The parallel grammar project. In: Carroll, N.J., Sutcliffe, R. (eds.) Proceedings of the Workshop on Grammar Engineering and Evaluation, COLING 2002, pp. 1–7 (2002)

2. Deng, L., Wiebe, J.: Joint prediction for entity/event-level sentiment analysis using probabilistic soft logic models. In: Proceedings of the 2015 Conference on Empirical Methods in Natural Language Processing (EMNLP), Lisbon, Portugal, pp. 179–189 (2015)
3. Kaplan, R.M., Bresnan, J.: Lexical-functional grammar: a formal system for grammatical representation. In: The Mental Representation of Grammatical Relations. The MIT Press, Cambridge (1982)
4. Kaplan, R.M., King, T.H., Maxwell III, J.T.: Adapting existing grammars: the XLE approach. In: Carroll, N.J., Sutcliffe, R. (eds.) Proceedings of the Workshop on Grammar Engineering and Evaluation, COLING 2002, pp. 29–35 (2002)
5. Karttunen, L.: Simple and phrasal implicatives. In: Proceedings of the First Joint Conference on Lexical and Computational Semantics, pp. 124–131. Association for Computational Linguistics, Stroudsburg (2012)
6. Klenner, M., Amsler, M.: Sentiframes: a resource for verb-centered German sentiment inference. In: Calzolari, N., Choukri, K., Declerck, T., Goggi, S., Grobelnik, M., Maegaard, B., Mariani, J., Mazo, H., Moreno, A., Odijk, J., Piperidis, S. (eds.) Proceedings of the Tenth International Conference on Language Resources and Evaluation (LREC), Portoro, Slovenia, pp. 2888–2891 (2016)
7. Klenner, M., Clematide, S.: How factuality determines sentiment inferences. In: Gardent, C., Raffaella Bernardi, I.T. (eds.) Proceedings of *SEM 2016: The Fifth Joint Conference on Lexical and Computational Semantics, Berlin, Germany, pp. 75–84, August 2016
8. Rashkin, H., Singh, S., Choi, Y.: Connotation frames: a data-driven investigation. In: Proceedings of the 54th Annual Meeting of the Association for Computational Linguistics (ACL), Berlin, Germany, pp. 311–32, August 2016
9. Saurí, R., Pustejovsky, J.: FactBank: a corpus annotated with event factuality. Lang. Resour. Eval. **43**(3), 227–268 (2009)
10. Sennrich, R., Schneider, G., Volk, M., Warin, M.: A new hybrid dependency parser for German. In: Proceedings of the German Society for Computational Linguistics and Language Technology (GSCL), Potsdam, Germany, pp. 115–124 (2009)

Optimal Number of States in HMM-Based Speech Synthesis

Zdeněk Hanzlíček[✉]

Faculty of Applied Sciences, NTIS - New Technology for the Information Society,
University of West Bohemia, Univerzitní 22, 306 14 Plzeň, Czech Republic
zhanzlic@ntis.zcu.cz
http://www.ntis.zcu.cz/en

Abstract. This paper deals with using models with a variable number of states in the HMM-based speech synthesis system. The paper also includes some implementation details on how to use these models in systems based on the HTS toolkit, which cannot handle the models with an unequal number of states directly. A bypass to enable this functionality is proposed here. A data-based method for the determination of the optimal number of states for particular models is proposed here and experimentally tested on 4 large speech corpora. The preference listening test, focused on local differences, proved the preference of the proposed system to the traditional system with 5-state models, while the size of the proposed system (the total number of states) is lower.

Keyword: HMM-based speech synthesis

1 Introduction

In the statistical parametric approaches to speech synthesis, hidden Markov models (HMMs) are traditionally used for the acoustic modeling of speech [7]. Recently, deep neural networks (DNNs) have been successfully applied in many areas of speech processing, including the speech synthesis [2]. Today, HMMs seem to be outperformed, especially their potential for the further quality increasing is infinitesimal.

However, from a practical perspective, tunning of the system parameters to achieve the possibly best results is always very important. This paper deals with a fundamental feature of HMM-based TTS systems – the number of states. In traditional systems, this number is equal for all models; it is usually set to 5. The main reason is probably the limitation of the HTS toolkit that is commonly used to train the models; this toolkit cannot handle models with a variable number of states directly.

This research was supported by the Czech Science Foundation (GA CR), project No. GA16-04420S. Access to computing and storage facilities owned by parties and projects contributing to the National Grid Infrastructure MetaCentrum, provided under the programme CESNET LM2015042, is greatly appreciated.

© Springer International Publishing AG 2017
K. Ekštein and V. Matoušek (Eds.): TSD 2017, LNAI 10415, pp. 353–361, 2017.
DOI: 10.1007/978-3-319-64206-2_40

Setting the optimal number of states for particular models could have the positive effect on the speech quality – system size ratio. 5-state models seem to represent the optimal choice in traditional systems; the benefit of more states for all models is dubious. However, maybe some models don't need 5 states, whereas some other models could take advantage of more than 5 states. The fundamental issue of this work is: Are there any benefits of using a variable number of states in a HMM-based TTS system?

This paper is organized as follows, Sect. 2 describes briefly the baseline HMM-based TTS system with one equal number of states for all models and the modification to enable the use of the variable number of states. The procedure for determination of the optimal number of states for particular phones is proposed in Sect. 3. The performed experiments and the evaluation are described here as well. Finally, Sect. 4 concludes this paper and outlines the future work.

2   System Description

2.1   Baseline System

This section gives only a brief description of the baseline HMM-based TTS system. We focus primarily on the differences from the standard HMM-based TTS system [7] and on features important for the subsequent description of our experimental system.

In the baseline system, speech is described by a sequence of parameter vectors containing 40 mel-generalized cepstral coefficients (MGC), pitch ($\log F_0$) and 21 band a periodicity coefficients (BAP). MGCs and BAPs were obtained by the STRAIGHT analysis method [1]; the pitch was extracted by using the PRAAT software[1]. The speech parameter vectors are modeled by a set of multi-stream context dependent HMMs by using the HTS toolkit[2].

In the HMM-based speech synthesis framework, the prosody is modeled implicitly by using models with large context description, i.e. individual models are defined for various phonetic, prosodic and linguistic context. Then, a speech unit is given as a phone with its phonetic, prosodic and linguistic context information. In our baseline system, a context-depended unit is represented by the following string template

$$p_1 - p_2 + p_3 \, @P: \, p_{w1} - p_{w2} \, @S: \, s_{w1} \mid s_{h1} - s_{w2} \mid s_{h2} \, @W: \, w_{h1} - w_{h2} \, / \, P_x$$

where all subscripted bold letters are contextual factors defined in Table 1 and the remaining characters in the template help to refer to particular factors, which is important during the model clustering stage.

[1] Praat: doing phonetics by computer, www.praat.org.
[2] HMM-based Speech Synthesis System (HTS), http://hts.sp.nitech.ac.jp.

Table 1. Definition of contextual factors. Note: All the position-related factors are forward and backward (fw and bw). Their values are limited to 5, since we assume that the marginal positions are most prominent and the following positions are less relevant.

Factors		Values
p_1, p_2, p_3	Previous, current and next phoneme	Czech phoneme set (see Table 2)
p_{w1}, p_{w2}	Position of phone in prosodic word (fw, bw)	1–5
s_{w1}, s_{w2}	Position of syllable in prosodic word (fw, bw)	
s_{h1}, s_{h2}	Position of syllable in phrase (fw, bw)	
w_{h1}, w_{h2}	Position of prosodic word in phrase (fw, bw)	
P_x	Prosodeme (type of phrase)	P0, P1.1, P2.2, P3.1 (see [4] for more details)

The standard training procedure involves 3 main stages[3]:

1. initialization and training of models for single phones (i.e. disregarding the context)
2. training of full-context models (initialized from the corresponding phone models)
3. model clustering

The clustering of full-context models substantially improves their robustness, reduces the total number of models/states/parameters and the clustering trees built during this process enable to derive models for contexts unseen in the training data.

2.2 Experimental System

The main issue is that the HTS toolkit does not support models with a variable number of states. This is probably the principal reason why the equal number of states is so typical in HMM-based speech synthesis. A possible solution, so-called hard state skipping, was proposed by X. Shao [5]: all models have the same number of states, but some states are marked as "unproductive" and are skipped. However, a number of algorithmic changes are required in both training and synthesis stages.

Since we did not want to perform any complex HTS modification, we proposed a simple bypass that allows to use a variable number of states in the current system: We split all models into particular states and obtained a sequence of 1-state models this way[4]. The indices of default states are turned into additional

---

[3] The detailed scheme of the training procedure is more complex, e.g. the reestimation and clustering of models are usually repeated twice.

[4] A bug had to be fixed in HTS toolkit ver.2.2 (file `HFB.c`) to allow using the 1-state models or else it did not work properly.

contextual factors and the derived 1-state models are represented by the following string template

$$p_1 - p_2 \ ^\wedge i_1 \ ^\wedge i_2 + p_3 \ @P: p_{w1-} p_{w2} \ @S: s_{w1} \mid s_{h1} {}_- s_{w2} \mid s_{h2} \ @W: w_{h1} {}_- w_{h2} \ / \ P_x$$

where i_1 and i_2 are the forward and backward state indices, respectively. For example, the following 3-state model

$$\text{\#-a+b@P:1\_5@S:1|1\_3|5@W:1\_5/P0}$$

is turned into the following sequence of 1-state models

$$\text{\#-a\^1\^3+b@P:1\_5@S:1|1\_3|5@W:1\_5/P0}$$
$$\text{\#-a\^2\^2+b@P:1\_5@S:1|1\_3|5@W:1\_5/P0}$$
$$\text{\#-a\^3\^1+b@P:1\_5@S:1|1\_3|5@W:1\_5/P0}$$

Using the state indices as the contextual factors enables to define the related clustering questions. Then, states with different indices can be clustered together, which was not possible in the baseline system. To increase the flexibility of clustering process, forward and backward state indices were introduced similarly as for other positional factors.

The modified training procedure involves the following stages[5]:

1. individual initialization of phone models (only HInit and HRest)
2. splitting of phone models into 1-state phone models and their reestimation (first use of HERest)
3. training of 1-state full-context models (initialized by 1-state phone models)
4. clustering of 1-state full-context models

It is obvious that the numbers of states have to be specified before the transition to 1-state models, i.e. we can define different number of states for particular phones. It is also possible to start directly with the 1-state models; however, the proposed procedure allows simply to use a phone-level segmentation for the model initialization.

3   Experiments and Results

3.1   Experimental Data

For our experiments, 4 large Czech[6] speech corpora recorded for the purposes of speech synthesis [3] were used: 2 male voices (denoted as M_1 and M_2) and 2 female voices (denoted as F_1 and F_2). From each corpus 10,000 declarative sentences (about 14 h of speech) were selected.

---

[5] Names of HTS tools are stated here to specify the point of transition to 1-state models as precisely as possible.

[6] However, proposed methods are certainly not language-dependent.

3.2   Optimal Number of States

The system proposed in the Sect. 2.2 is capable to use models with a variable number of states. This section deals with the issue how to determine the optimal number of states for particular models. We proposed a simple procedure based on the Viterbi algorithm.

1. The available data are divided into 2 parts: training and alignment data.
2. Sets of full-context models with one general number of states are successively estimated from training data. The number of states ranged between 1 and 7 in our experiments.
3. The alignment data is successively time-aligned with all model sets (by using the Viterbi algorithm).
4. The average likelihood per frame is calculated for each phone and number of states; an example of the dependence between the likelihood and the number of states is presented in Fig. 1.
5. The optimal number of states is determined for each phone.

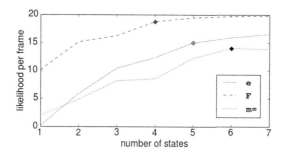

Fig. 1. Dependence of the likelihood-per-frame on the number of states: examples for speaker M_1. Selected numbers are marked with "diamonds".

The optimal number of states is not necessarily related to the maximum likelihood since the difference around the maximum seems to be insignificant in some cases and it is probably better to use the lower value. Moreover, for some longer and frequent phones, e.g. long vowels and diphthongs, the likelihood tends to continue rising even for a higher number of states than 7. To obtain reasonable values, we decided to select the number of states from which the further likelihood increase is lower than 1.0.

Results for particular speakers are summarized in Table 2. The selected speech data was divided in half, i.e. 5,000 utterances were using for training and remaining 5,000 for alignment. Results seem to be partly speaker-dependent; therefore, an individual setting was used for each speaker in the following experiments.

Two systems were created for each speaker: the baseline system with 5-state models and the system with the optimal number of states, referred to as **5s** and **Vs**,

Table 2. The optimal number of states for particular phones and speakers. SAMPA alphabet [6] is used for the transcription of particular phones. Examples in brackets are not phonetically transcribed, but the related characters are highlighted.

SAMPA (example)	M_1	M_2	F_1	F_2	SAMPA (example)	M_1	M_2	F_1	F_2
a (máma)	5	6	5	5	a: (táta)	4	5	4	5
a_u (auto)	5	5	5	5	b (bod)	4	5	4	5
c (kutil)	4	5	4	4	d (den)	5	5	6	6
d_z (leckdo)	4	4	5	5	d_Z (léčba)	4	5	5	5
e (pes)	5	5	5	4	e: (lépe)	5	5	5	4
e_u (euro)	5	6	5	7	f (facka)	4	3	4	3
F (nymfa)	4	3	5	3	g (guma)	4	6	3	5
G (bych ho)	3	4	5	3	h\ (hák)	4	4	3	3
i (pivo)	5	5	5	5	i: (víno)	4	4	4	5
j (voják)	4	4	3	3	J (laň)	5	4	5	4
J\ (děti)	5	6	6	5	k (oko)	6	5	4	5
l (los)	3	5	3	4	l= (vlk)	4	4	4	4
m (mír)	5	5	5	4	m= (osm)	6	4	3	4
n (nos)	4	4	6	5	N (banka)	4	4	3	4
o (bok)	5	4	5	6	o: (jód)	4	3	3	4
o_u (pouto)	5	6	5	4	p (pak)	5	4	6	4
P\ (moře)	6	4	4	4	Q\ (tři)	5	4	4	4
r (rak)	4	5	6	5	r= (bratr)	6	6	5	7
s (osel)	4	4	3	3	S (pošta)	4	4	3	3
t (otec)	5	6	5	6	t_s (ocel)	4	5	4	5
t_S (oči)	5	5	5	4	u (rum)	5	5	5	5
u: (růže)	4	5	4	4	v (vlak)	4	5	4	4
x (chyba)	4	3	3	3	z (koza)	4	3	4	3
Z (žena)	4	4	4	4					

respectively. Table 3 presents the number of clustered states in particular systems; it is directly proportional to the final number of model parameters in each system. For all speakers, **Vs** contains about 10% less clustered states than **5s**. It is important for the further comparison since **Vs** is not advantaged by using more states than **5s**.

3.3 Listening Test

To ascertain the benefit of using models with a variable number of states, a preference listening test was carried out.

Table 3. Number of clustered states for particular speakers. Note that particular parameter streams are clustered independently; therefore the number of clustered states is different for each stream.

Param. stream	M_1		M_2		F_1		F_2	
	#	%	#	%	#	%	#	%
5s unclust.	721,065	100.00	692,025	100.00	666,960	100.00	754,580	100.00
mgc	4,094	0.57	4,481	0.65	3,555	0.53	3,528	0.46
lf0	13,175	1.83	9,731	1.41	11,398	1.71	14,633	1.94
bap	2,954	0.41	3,336	0.48	2,974	0.45	2,832	0.38
Vs unclust.	635,438	100.00	650,256	100.00	598,097	100.00	684,907	100.00
mgc	3,641	0.57	3,666	0.56	2,933	0.49	3,138	0.46
lf0	12,473	1.96	8,341	1.28	10,084	1.69	12,840	1.87
bap	2,607	0.41	2,606	0.40	2,398	0.40	2,498	0.36

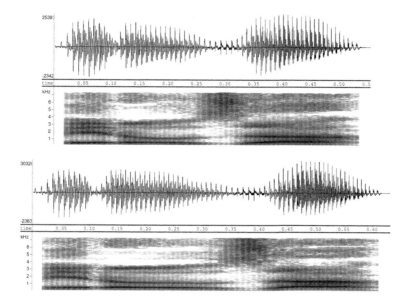

Fig. 2. The word "viróza" [viro:za] synthesized with M_1 voice by 5s and Vs system, respectively. Differences in waveform (both shape and duration) and spectrum of vowels o: and a are evident.

However, we noticed that the baseline and experimental systems produce often audibly almost identical utterances. The assumed reason is that the 5-state models in the baseline system are rich enough to capture all relevant speech properties in most situations. The quality seems to culminate around this number, only several phones have higher optimal number of states. This is why the 5-state models are employed in standard HMM-based speech synthesis systems.

However, all utterances were not equal, some pairs were definitely different. They did not vary globally, the differences were usually limited to one phone and its vicinity – see example in Fig. 2. To find the suitable sentences for the listening comparison, 500 utterances were synthesized by both **5s** and **Vs** systems. Then, the DTW alignment was applied on parameter trajectories. Sentences with the highest value of maximum local difference were selected for the test.

Test contained 70 pairs of utterances – only for speakers M_1 and F_1; speech corpora for M_2 and F_2 were processed additionally and therefore they could not be included in the test. 10 listeners participated in this test; they were instructed to focus on local differences and used a 4-point evaluation scale:

- the 1st utterance is preferred over the 2nd
- no preference – utterances are identical
- no preference – utterances are different
- the 2nd utterance is preferred over the 1st

Detailed results of the test are presented in Table 4 and Fig. 3. The default system with 5-state models was preferred in about 11% cases, proposed system in almost 45% cases and remaining 44% cases were evaluated as of equal quality. Results for particular speakers were consistent.

Table 4. Results of the preference listening test.

Speaker	5s preferred	w/o preference		Vs preferred
		Equal	Different	
M_1	8.94%	26.02%	21.95%	43.09%
F_1	13.22%	14.94%	25.29%	46.55%
Avg.	11.08%	20.48%	23.62%	44.82%

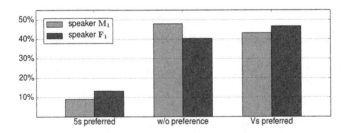

Fig. 3. Results of the preference listening test.

4   Conclusion

This paper presented experiments on using models with a variable number of states in HMM-based speech synthesis. We described a simple bypass to enable the use of such models in systems based on the HTS toolkit. We also proposed and compared the optimal number of states for 4 different speakers. A preference listening test, focused on local differences, proved the preference of the proposed system, while the size of the system (the number of clustered states) is slightly lower.

The proposed approach could be also used to minimize the system, i.e. for a consistent quality-balanced reduction of the number of states, for example, when less training data is available. Results seem to be partly speaker-dependent, however, a general speaker-independent setting would be useful for a quick creation of new voices.

References

1. Kawahara, H., Masuda-Katsuse, I., de Cheveigne, A.: Restructuring speech representations using a pitch-adaptive time-frequency smoothing and an instantaneous-frequency-based F0 extraction: possible role of a repetitive structure in sounds. Speech Commun. **27**, 187–207 (1999)
2. Ling, Z.H., Kang, S.Y., Zen, H., Senior, A., Schuster, M., Qian, X.J., Meng, H.M., Deng, L.: Deep learning for acoustic modeling in parametric speech generation: a systematic review of existing techniques and future trends. IEEE Signal Process. Mag. **32**(3), 35–52 (2015)
3. Matoušek, J., Tihelka, D., Romportl, J.: Building of a speech corpus optimised for unit selection TTS synthesis. In: Proceedings of LREC (2008)
4. Romportl, J., Matoušek, J., Tihelka, D.: Advanced prosody modelling. In: Sojka, P., Kopeček, I., Pala, K. (eds.) TSD 2004. LNCS (LNAI), vol. 3206, pp. 441–447. Springer, Heidelberg (2004). doi:10.1007/978-3-540-30120-2_56
5. Shao, X., Pollet, V., Breen, A.: Refined statistical model tuning for speech synthesis. In: Proceedings of the 7th ISCA Workshop on Speech Synthesis, pp. 284–287 (2010)
6. Wells, J.: SAMPA computer readable phonetic alphabet. In: Handbook of Standards and Resources for Spoken Language Systems, pp. 684–732. Mouton de Gruyter, Berlin (1997)
7. Zen, H., Tokuda, K., Black, A.W.: Statistical parametric speech synthesis. Speech Commun. **51**(11), 1039–1064 (2009)

Temporal Feature Space for Text Classification

Stefano Giovanni Rizzo[(⊠)] and Danilo Montesi

Department of Computer Science and Engineering, University of Bologna,
Mura Anteo Zamboni 7, Bologna, Italy
{stefano.rizzo8,danilo.montesi}@unibo.it
http://www.smartdata.cs.unibo.it/

Abstract. In supervised learning algorithms for text classification the
text content is usually represented using the frequencies of the words it
contains, ignoring their semantic and their relationships. Words within
temporal expressions such as *"today"* or *"last February"* are particularly affected by this simplification: the same expression can have a
different semantic in documents with different timestamps, while different expressions could refer to the same time. After extracting temporal
expressions in documents, we model a set of temporal features derived
from the time mentioned in the document, showing the relation between
these features and the belonging category. We test our temporal approach on a subset of the New York Times corpus showing a significant
improvement over the text-only baseline.

Keywords: Temporal features · Automatic text classification · Semantic annotation

1 Introduction

The goal of text classification tasks is to assign the category of a text document
(such as an email or a web page) given the features that represent its content.
Usually, a document is represented using a bag-of-word: a boolean vector with
one element for each word in the document collection. In the bag-of-word representation, the feature (i.e. an element of the boolean vector) denotes the presence
or the absence of each word. A text classifier, trained using these features, will
estimate the category of a document given the presence or absence of the more
representative words.

Intuitively, the better the features can describe documents with respect to
their categories, the higher will be the accuracy of a model trained with such features. Processing the text features using NLP techniques with the goal of enhancing accuracy has a long history in Information Retrieval and Text Classification
tasks. In practice, most of the attempts made using a richer, linguistic representation of text, instead of traditional word tokens, were ineffective, bringing no evidence to justify a text processing much more complex than the simple tokenization
[10]. More recently however, the use of semantic annotations in addition to tokens
has improved the accuracy of retrieval tasks significantly [6,9,14]. Moreover, a new

© Springer International Publishing AG 2017
K. Ekštein and V. Matoušek (Eds.): TSD 2017, LNAI 10415, pp. 362–370, 2017.
DOI: 10.1007/978-3-319-64206-2_41

trend is emerging which gives particular attention on the temporal dimension of text [4], and for which temporal semantic annotation and extraction is a crucial step [1,3].

The particular interest towards content-level temporal information resides in its great variance in expressing the same object, because synonymy and polysemy relations in temporal expressions (timexes) are more complex than in other named entities:

- Synonyms of absolute timexes: any absolute mention of a time interval such as "*4 October 2016*", can have a number of variations for each mentionable time interval (e.g. "*4/10/2016*", "*4 October 2016*", "*the fourth of October 2016*" etc.).
- Synonyms of relative timexes: the same moment in time can be mentioned using a relative time expression such as "*tomorrow*" or "*one year ago*", depending on the time of writing. What was "*today*" for a philosopher of the ancient Greece becomes "*two thousand years ago*" in the present time.
- Hypernyms of relative timexes: the same time expression can refer to any interval, depending on the time of writing: "*next year*" could refer to 1950 if written in 1949, or could refer to 2017 if written in 2016.

Improving text categorization by means of semantic annotations of named entities has been considered in the past [2], however to the best of our knowledge no work has been done to exploit content-level time for text categorization.

In this paper we propose a temporal features space, in addition to the traditional text features space, to improve text classification tasks such as topic categorization or new event detection. The set of novel temporal features for documents, derived from time mentioned in text, captures the temporal peculiarities of the documents related to their category in a low-dimensional representation. These features take in consideration how much time is mentioned in a document, the central time of the narrated events, and the span of the intervals cited.

After formally defining how these features are built, we show the results of ANOVA (analysis of variance) to assure correlation between temporal features and categories on the New York Times dataset, a well-known corpus of manually categorized documents. Finally, we evaluate how much accuracy improvement is obtained using both temporal and text features sets.

2 Temporal Feature Space

Each document, in its textual content, cites a number of absolute and relative dates (*content-level time*). For instance, a certain document can contain a temporal expression such as "On *2016 Christmas eve*" referring to the absolute date 2016-12-24. The same document could also contain a relative temporal expression such as "the match we watched *yesterday*", referring to a relative date, which depends on the creation time of the document (a timestamp known as document creation time or DCT). All the temporal expressions, both absolute and relative, can be annotated and normalized into exact timestamp intervals using a temporal annotator, such as Heideltime [13]. We define this set of all the mentioned intervals as the *temporal scope* of the document.

The temporal features we define are all derived from the temporal scope of documents. From this set of intervals, we extract different time features that are able to finely describe the document characteristic in the temporal space, without using a plain *bag-of-chronons* that would require a very high-dimensionality representation:

1. **Temporality feature:** some texts are more time-related than others (e.g. news article vs philosophy argument). In the same way, documents belonging to different categories can have a significantly different *temporality*. Temporality is an indicator of how much time there is in a document.
2. **Focus and mean time features:** these two features denote the central scope of the intervals mentioned in the text, with two different notions on what is the central time window of the narrated events.
3. **Interval size feature:** depending on the topic, the mentioned intervals can be short, such as one day, or longer such as years and millenniums. The interval size considers the span's length of the mentioned intervals.

2.1   Temporality

We define the temporality of a document as the number of timexes in a document. Despite its simplicity, this feature captures a property that can strongly discern some topics and categories. This is due to the fact that the subjects of some categories rely on many time mentions, while others hardly make use of time in their narrative.

Definition 1 (Temporality). *Given the temporal scope T_D of a document as the set of all the mentioned intervals in its content, the* **temporality** *is the cardinality of T_D.*

$$time_{temporality}(T_D) = |T_D| \tag{1}$$

2.2   Mean Time and Focus Time

The set of time expressions in a document often revolves around a central time window, such as the time of the main event described. We provide two different central time definitions: mean time and focus time.

Definition 2 (Mean time window). *Given the temporal scope of a document as the set of all the mentioned intervals in its content, the* **mean time window** *is an interval $[t_s, t_e]$ where t_s is the mean of all the start times in the temporal scope and t_e is the mean of all the end times in the temporal scope.*

$$time_{window}(T_D) = [\frac{1}{|T_D|} \sum_{t \in T_D} t_s, \frac{1}{|T_D|} \sum_{t \in T_D} t_e] \tag{2}$$

The mean time window, aggregating all the mentioned intervals, gives a rich information on the time extent of the document. However, averaging the intervals can lose a very crucial information, which is what the "focus" of the document is.

Different works in literature have different conceptions on what the focus time of a document is. For Strotgen et al. [12] the focus time is the most frequent time in a document, while in more complex approaches [8] the focus time is the one with which the document's terms are mostly associated in the corpus. Following the former notion of focus time [12], we define our focus time as the mode of the frequency distribution, that is, the interval which is most frequently mentioned in the document.

Definition 3 (Focus time). *Given the temporal scope of a document as the set of all the mentioned intervals in its content, the **focus time** is an interval* $[m_s, m_e]$ *where m_s is the mode of all the start times in the temporal scope and m_e is the mode of all the end times in the temporal scope.*

$$time_{focus}(T_D) = [mode(t_s), mode(t_e)] \tag{3}$$

In order to illustrate how well the focus time can approximate the time of a document, and the difference between the focus time and the mean time window, we picked two very different documents.

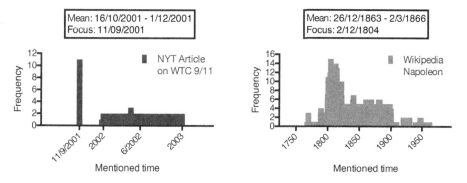

Fig. 1. Temporal scope for two documents: a New York Times article on the WTC terrorist attack and the Wikipedia main article on Napoleon. Top frame shows Mean time and Focus time.

In Fig. 1 we represent the content time of the two different documents as a frequency distribution of each interval. On the left, a New York Times article on the WTC 2001 attack, written in 2002, shows a peak on the day of the attack, but it also shows other mentioned intervals about events happened after the attack, along the year 2002. For this reason, the mean time window for this article is the period from 16/10/2001 to 1/12/2001, while the focus time identifies only the main event time as the 11th of September 2001. The same results are obtained on a totally different kind of text document: the Wikipedia article on Napoleon

Bonaparte. The focus time is 2 December 1804, the date of his incoronation, while the mean takes into account all the related events.

2.3 Interval Size

The temporal expressions found in a document, once normalized to time intervals, can have a different size depending on the span of the cited time. For instance, in the gregorian calendar, this time span can be of 7 days if a week is mentioned, 28 to 31 days if a month is mentioned, 365 days for the year and so on. Moreover, there can be found a smaller percentage of irregular intervals, in temporal expressions such as *"I will train for 10 days"* or *"The Great War lasted from 28 July 1914 to 11 November 1918"*. Put together, all these intervals compose a set that can be very diversified, but can also follow some patterns depending on the topic of a document and, therefore, on its category.

Definition 4 (Intervals size). *The intervals size of a document is the mean size of all the intervals of its temporal scope.*

$$time_{size} = \frac{\sum_{x \in T_D} x_e - x_s}{|T_D|} \tag{4}$$

This feature, although simple in its definition, is valuable in discriminating documents from different categories, as we show in the next section.

3 Experimental Results

We evaluate the proposed features on a random subset[1] of 75 thousand documents from the The New York Times Annotated Corpus[2], which is the most used corpus for temporal related tasks [1,11] because of its temporal richness both in content (we extracted 15 million temporal expressions over 1.8 million documents) and in production time variance (it spans over 20 years of articles). Timexes have been identified and normalized using Heideltime [13], considered the state of the art tagger able to recognize even BC period dates. The category annotation of New York Times articles is provided by the New York Times Newsroom, the New York Times Indexing Service and the online production staff at nytimes.com.

After annotating all the temporal expressions in the corpus and extracting the normalized time intervals, we randomly sampled 5,000 articles for each one of the 15 most occurrent online section categories: *Arts, Business, Magazine, New York and Region, Obituaries, Opinion, Paid Death Notices, Real Estate, Sports, Style, Technology, Travel, U.S., Week in Review, World.*

---

[1] Dataset with precomputed features available at https://smartdata.cs.unibo.it/data/ TFTC/.

[2] NYT Corpus available at https://catalog.ldc.upenn.edu/LDC2008T19.

3.1 Significance Study

The ANOVA test (analysis of variance) is a statistical model to analyze the difference for a specific variable (feature) over a set of groups (categories). The ANOVA test has been widely used for feature selection because it gives a measure of the *reliability* of a feature [7]. All the proposed features have significantly different averages across different categories (p-value ≪ 0.01), meaning that the difference between the mean values of the categories, for each defined feature, is not due to chance.

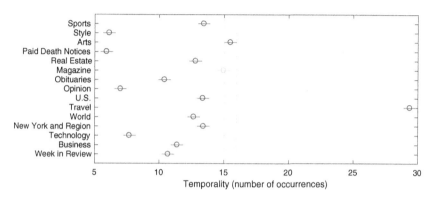

Fig. 2. Mean (circle) and variance (line) of **temporality** for each category in NYT Section dataset. (Color figure online)

In Fig. 2 we show the mean temporality for each category and the variance within the same category. Circle points denote the mean temporality for the category, while horizontal lines over each point denote the extension of the variance.

By looking at the variance lines overlaps, it is possible to visually identify categories with a significantly different temporality: if the variance line for two or more categories overlaps, these categories do not significantly differ, thus a classifier trained only on this feature will not be able to discriminate between them. As an example, in Fig. 2 we highlight in blue the *Arts* category, while in red are shown 13 categories that significantly differs from *Arts*. In gray, the *Magazine* category is the only category which cannot be distinguished from *Arts* with significant confidence.

Among the fifteen categories, the mean time is less scattered than the other features, as shown in Fig. 3. Despite this most categories are significantly different from each other. The highlighted examples in blue is the category *Real Estate*: the average of its mean time is totally indistinguishable from the category *Travel*, however it is well distinguishable from the other 13 categories.

The means of interval sizes in Fig. 4 are also quite diverse among categories. The category *U.S.*, highlighted in blue, is an unlucky example: the sizes of its mentioned intervals are similar to *World, New York and Region* and *Opinion*, while significantly different from the other 11 categories.

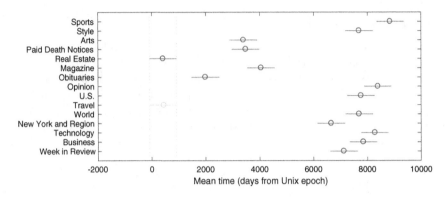

Fig. 3. Mean (circle) and variance (line) of **mean time** for each category in NYT Section dataset. (Color figure online)

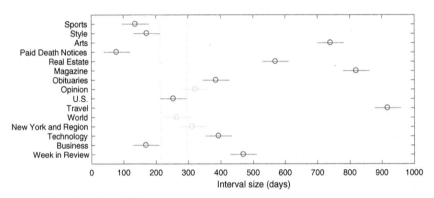

Fig. 4. Mean (circle) and variance (line) of **interval size** for each category in NYT Section dataset. (Color figure online)

It is noteworthy that, for each pair of categories in this corpus there exists at least one of the defined features for which they significantly differ.

3.2   Classification Accuracy

To test the improvement in accuracy we combine two simple kNN (k Nearest Neighbor) classifiers: one trained on the text features (*tf.idf* vector), the other trained using the novel temporal features. We set k as 35 for both classifiers since this is a known good choice to yield stable effectiveness [16]. Once we evaluate the accuracy for the single classifier, we use this accuracy as the weight in the linear combination of the two classifier [15]. We obtain an F1 macro-averaged accuracy of 0.247 using the temporal features only and 0.681 using the traditional *tf.idf* features, while combining both temporal and text features we obtain 0.694.

We further test the improvement using a state-of-the-art classifier, the tree boosting model XGBoost [5], on the union set of both feature spaces, this time using a single classifier. In this setting we obtain 0.417 using only time features, 0.852 using text features and 0.864 using both.

Applying the t-test on the predicted categories of all tests it results that the improvement obtained adding the temporal features is not due to chance (p-value ≪ 0.01).

4   Conclusion

In this paper we have shown how the time mentioned in documents is category-dependent, thus can be exploited to recognize the category of documents with higher accuracy. We defined a set of time features to synthesize the temporal scopes of documents by their centrality, extent and size. Analyzing the variance of these features among different categories, it results that most categories are far apart from a temporal point of view while smaller groups of categories share similar temporal aspects. The experimental evaluation confirms that while using solely the proposed temporal features does not lead to adequate accuracy, their combination with traditional terms-based features yields a significant improvement over the text-only baseline.

References

1. Berberich, K., Bedathur, S., Alonso, O., Weikum, G.: A language modeling approach for temporal information needs. In: Gurrin, C., He, Y., Kazai, G., Kruschwitz, U., Little, S., Roelleke, T., Rüger, S., Rijsbergen, K. (eds.) ECIR 2010. LNCS, vol. 5993, pp. 13–25. Springer, Heidelberg (2010). doi:10.1007/978-3-642-12275-0_5
2. Bloehdorn, S., Hotho, A.: Boosting for text classification with semantic features. In: Mobasher, B., Nasraoui, O., Liu, B., Masand, B. (eds.) WebKDD 2004. LNCS, vol. 3932, pp. 149–166. Springer, Heidelberg (2006). doi:10.1007/11899402_10
3. Brucato, M., Montesi, D.: Metric spaces for temporal information retrieval. In: Rijke, M., Kenter, T., Vries, A.P., Zhai, C.X., Jong, F., Radinsky, K., Hofmann, K. (eds.) ECIR 2014. LNCS, vol. 8416, pp. 385–397. Springer, Cham (2014). doi:10. 1007/978-3-319-06028-6_32
4. Campos, R., Dias, G., Jorge, A.M., Jatowt, A.: Survey of temporal information retrieval and related applications. ACM Comput. Surv. (CSUR) 47(2), 15 (2015)
5. Chen, T., Guestrin, C.: XGBoost: a scalable tree boosting system. In: Proceedings of the 22nd ACM SIGKDD International Conference on Knowledge Discovery and Data Mining, pp. 785–794. ACM (2016)
6. Fernández, M., Cantador, I., López, V., Vallet, D., Castells, P., Motta, E.: Semantically enhanced information retrieval: an ontology-based approach. Web Semant. Sci. Serv. Agents World Wide Web 9(4), 434–452 (2011). JWS special issue on Semantic Search
7. Guyon, I., Elisseeff, A.: An introduction to variable and feature selection. J. Mach. Learn. Res. 3, 1157–1182 (2003)
8. Jatowt, A., Au Yeung, C.M., Tanaka, K.: Estimating document focus time. In: Proceedings of the 22nd ACM International Conference on Conference on Information and Knowledge Management, pp. 2273–2278. ACM (2013)

9. Kara, S., Alan, Ö., Sabuncu, O., Akpinar, S., Cicekli, N.K., Alpaslan, F.N.: An ontology-based retrieval system using semantic indexing. Inf. Syst. **37**(4), 294–305 (2012). Semantic Web Data Management

10. Moschitti, A., Basili, R.: Complex linguistic features for text classification: a comprehensive study. In: McDonald, S., Tait, J. (eds.) ECIR 2004. LNCS, vol. 2997, pp. 181–196. Springer, Heidelberg (2004). doi:10.1007/978-3-540-24752-4_14

11. Radinsky, K., Agichtein, E., Gabrilovich, E., Markovitch, S.: A word at a time: computing word relatedness using temporal semantic analysis. In: Proceedings of the 20th International Conference on World Wide Web, WWW 2011, pp. 337–346. ACM, New York (2011)

12. Strötgen, J., Alonso, O., Gertz, M.: Identification of top relevant temporal expressions in documents. In: Proceedings of the 2nd Temporal Web Analytics Workshop, pp. 33–40. ACM (2012)

13. Strötgen, J., Gertz, M.: Heideltime: High quality rule-based extraction and normalization of temporal expressions. In: Proceedings of the 5th International Workshop on Semantic Evaluation, pp. 321–324. Association for Computational Linguistics (2010)

14. Vallet, D., Fernández, M., Castells, P.: An ontology-based information retrieval model. In: Gómez-Pérez, A., Euzenat, J. (eds.) ESWC 2005. LNCS, vol. 3532, pp. 455–470. Springer, Heidelberg (2005). doi:10.1007/11431053_31

15. Wu, S., Crestani, F.: Data fusion with estimated weights. In: Proceedings of the Eleventh International Conference on Information and Knowledge Management, pp. 648–651. ACM (2002)

16. Yang, Y., Liu, X.: A re-examination of text categorization methods. In: Proceedings of the 22nd Annual International ACM SIGIR Conference on Research and Development in Information Retrieval, pp. 42–49. ACM (1999)

Parkinson's Disease Progression Assessment from Speech Using a Mobile Device-Based Application

T. Arias-Vergara[1(✉)], P. Klumpp[2], J.C. Vásquez-Correa[1,2],
J.R. Orozco-Arroyave[1,2], and E. Nöth[2]

[1] Faculty of Engineering, Universidad de Antioquia UdeA,
Calle 70 No. 52-21, Medellín, Colombia
tomas.arias@udea.edu.co
[2] Pattern Recognition Lab, Friedrich-Alexander-Universität Erlangen-Nürnberg,
Erlangen, Germany

Abstract. This paper presents preliminary results of individual speaker models for monitoring Parkinson's disease from speech using a smartphone. The aim of this study is to evaluate the suitability of mobile devices to perform robust speech analysis. Speech recordings from 68 PD patients were captured from 2012 to 2016 in four recording sessions. The performance of the speaker models is evaluated according to two clinical rating scales: the Unified Parkinson's Diseae Rating Scale (UPDRS) and a modified version of the Frenchay Dysarthria Assessment (m-FDA) scale. According to the results, it is possible to assess the disease progression from speech with Pearson's correlations of up to $r = 0.51$. This study suggests that it is worth to continue working on the development of mobile-based tools for the continuous and unobtrusive monitoring of Parkinson's patients.

Keywords: Parkinson's disease · Monitoring · Mobile device · Speaker model

1 Introduction

Parkinson's disease (PD) is a neurological disorder caused by the progressive loss of dopamine in the substantia nigra of the midbrain [1]. PD produces motor and non-motor impairments. The primary motor symptoms include resting tremor, slowness of movement, postural instability, rigidity and several dimensions of speech are affected including phonation, articulation, prosody, and intelligibility [2,3]. Currently the neurologist relies on medical history, physical and neurological examinations to assess the patients. However, the motor skills of patients with PD are impaired, thus to visit a hospital to perform medical screenings and/or assessments is not a straightforward task [4]. There have been proposed many approaches focused on the telemonitoring of PD from speech using portable devices [5–8]. Most of these devices consider sustained phonations to evaluate disorders in the vocal folds vibration. The set of features used include the

© Springer International Publishing AG 2017
K. Ekštein and V. Matoušek (Eds.): TSD 2017, LNAI 10415, pp. 371–379, 2017.
DOI: 10.1007/978-3-319-64206-2_42

fundamental frequency and its variability, the sound pressure level, and the vocal formants, however, more speech tasks need to be included to perform more complete screenings. Furthermore, these portable devices are original designs, thus they are not open source systems. Other studies consider mobile device applications for the at-home monitoring of PD patients. In [9] it is presented a smartwatch-based system to monitor speech and voice impairments of PD patients. The system consists of the combination of a tablet and smartwatch to perform the data collection. Speech problems are analyzed considering the sustained phonation of vowel /a/. The speech recordings are sent to a cloud-based server to store and perform the speech analyses; however, those analyses are limited to only phonation measures. In [10] the authors present a portable system for the automatic recognition of the syllables /pa-ta-ka/. The proposed approach consists of a tablet and a headset to capture the speech signals. The system was trained using speech recordings from two group of speakers: patients with traumatic brain injuries and PD patients. The automatic recognition of /pa-ta-ka/ is performed in the mobile device using an Automatic Speech Recognition (ASR)-based system. Speech impairments are assessed using a single metric, which consists of the syllable error rate. Another platform to monitor PD using a smartphone is presented in [11]. The application includes several tests to evaluate different PD symptoms related to dysphonia, postural instability, bradykinesia, and tremor. The monitoring and assessment of PD symptoms is performed with a defined protocol used to measure different motor impairments in voice, gait, dexterity, and balance. Although several motor impairments are considered, it is not clear whether the proposed system is suitable to assess each patient individually. There is an increasing interest in the research community to develop methodologies for the at-home monitoring of PD using mobile devices-based systems. This paper presents a methodology for the automatic monitoring of the progression of PD from speech using a smartphone-based application. The method is based on the Gaussian Mixture Model–Universal Background Model (GMM–UBM) approach, which was already used in [12] to assess the PD progression using individual patient/speaker models. This methodology is implemented in the smartphone application *Apkinson* which was introduced during the Third Frederick Jelinek Memorial Summer Workshop at Johns Hopkins University[1]. This initial version was upgraded later on in [13]. The system provides a mobile solution for the continuous and unobtrusive monitoring of PD from speech. The main objective of this paper is to evaluate the suitability of *Apkinson* for the assessment of PD using a robust-in-phone approach.

2   Methods

Articulatory model
Speech problems are modeled using the articulation approach proposed in [14]. Voiced and unvoiced segments are extracted and grouped separately. Hamming windowing with 30 ms length and a time-shift of 10 ms is applied.

---

[1] https://www.clsp.jhu.edu/workshops/16-workshop/remote-monitoring-of-neurode generation-through-speech/.

Voiced/unvoiced frames are modeled with 12 MFCCs and the log-energy of the signal distributed in 17 Bark bands, forming a 29-dimensional feature vector.

Gaussian Mixture Model-Universal Background Model

The Gaussian Mixture Models (GMM)-based systems are capable to represent arbitrary probabilistic densities. In speech processing GMMs are used to represent the distribution of feature vectors extracted from a single speaker or a group of speakers. If the GMM is trained using features extracted from a large sample of speakers, the resulting model is called Universal Background Model (UBM). The training process of the UBM consists of calculating the weights ω, the mean vectors μ, and the covariance matrices Σ that represent the population of speakers. The speaker model is found by adapting the parameters of the UBM using the training data from the speaker to be modeled [15].

Distance computation

The speech of PD patients can be assessed using the individual speaker models obtained from the GMM-UBM approach. The resulting models are based on probabilistic representations of the articulatory model, i.e., voiced/unvoiced segmentation. Hence, the Bhattacharyya distance d_{Bha} is used to detect changes in speech of PD patients. This methodology was first presented in [12] and showed promising results for tracking PD progression from speech. Thus, we implemented this approach to evaluate the progression of PD patients using a mobile application.

3   Apkinson

The mobile application used in this paper was originally introduced during the Third Frederick Jelinek Memorial Summer Workshop at Johns Hopkins University [16] and further upgraded in [13]. *Apkinson* is designed as a portable solution for the continuous and unobtrusive monitoring of PD from speech. The key feature of the application is that allows to capture speech from phone calls to perform the analyses, therefore the speech of the patients can be analyzed without interrupting their daily routine, i.e., the patients do not notice that they are been monitored. *Apkinson* comprises four main processes on its background: (1) phone call detection, (2) recording, (3) collection of meta data, and (4) signal analysis. The phone call detection is constantly running in the background. In order to preserve the privacy of the people, *Apkinson* only records the speech of the patient under monitoring, who has to sign an informed consent before start using the App. The mobile use-patterns and meta data provide useful information about the patient, for instance date, time, contact who called or was called, and the duration of the conversations can show important information about the mood state of the patient. These data can help to understand non-motor symptoms also developed by PD patients like depression [17]. The speech signal analysis is meant to be implemented in a separate plug-in in order to reduce the overall size of the application. This is a key feature in *Apkinson* because it allows to other developers to create additional plug-ins for speech analysis without applying major changes in the main application. In this study the speaker model-based approach presented in [12] is implemented in the speech analysis stage of *Apkinson* in order to assess the suitability of the mobile devices to perform robust processing.

4   Experimental Setup

Clinical scales

All of the patients considered in this study were evaluated by clinical experts according to two scales: the Movement Disorders Society – Unified Parkinson's Disease Rating Scale (MDS-UPDRS) [18] and a modified version of the FDA scale [19]. The MDS-UPDRS is a perceptual scale used to assess motor and non-motor abilities of PD patients. In this study only the third section (MDS-UPDRS-III) is considered because it evaluates the motor capability of the patients. The modified version of the FDA scale, i.e., m–FDA was introduced during the Third Frederick Jelinek Memorial Summer Workshop at Johns Hopkins University [16]. The original version of FDA requires the patient to be with the examiner, which is not possible for most of the PD patients due to their reduced mobility. The m–FDA can be applied considering speech recordings of the patient, thus the patient is not required to visit the clinician to be administered.

Data description

Recordings of 68 PD patients (35 male and 33 female) collected in five recording sessions from 2012 to 2016 are considered. The original sampling frequency was 44.1 kHz, thus the signals were down-sampled to 16 kHz in order to meet with *Apkinson's* requirements. A subset of seven patients participated in four sessions, thus their speech samples are considered here as the test set. For the experiments presented here, only the repetition of /pa-ta-ka/ was considered. The MDS-UPDRS-III and m–FDA scores of the seven patients used to test the speaker models are provided in Table 1. The MDS-UPDRS-III labels of the third recording session were not available.

Speaker model on a mobile device

The training process is performed off-line, i.e., the parameters $\lambda = \{\omega, \mu, \Sigma\}$ are pre-calculated off-line. The set of features used to train the UBM are extracted from the voiced/unvoiced segments as in [20]. The trained parameters are stored in the smart-phone as a plain text file in the *Apkinson's* folder. The speech recordings of the test patients are locally stored in the smart-phone prior to the adaptation process. Voiced/unvoiced segmentation and feature extraction are performed in the smart-phone. The pre-computed parameters $\{\omega, \mu, \Sigma\}$ are considered to obtain the individual speaker models by using the Maximum A Posterior (MAP) adaptation approach. Finally, the Bhattacharyya distance (d_{Bha}) between the adapted model and the UBM is calculated. Figure 1 summarizes the methodology.

Table 1. Demographic information of the patients considered in the test set (those who participated in five recording sessions). S-i, $i \in \{1, \ldots, 5\}$, and indicates the recording session.

Patient	Sex	Age	MDS-UPDRS-III					m–FDA				
			S-1	S-2	S-3	S-4	S-5	S-1	S-2	S-3	S-4	S-5
P1	M	64	28	19	-	13	-	31	15	17	16	-
P2	F	57	41	35	-	33	33	29	39	24	21	40
P3	F	51	38	49	-	44	45	14	20	1	12	15
P4	F	55	43	10	-	19	-	13	13	13	21	15
P5	M	59	6	8	-	24	21	21	36	12	13	17
P6	M	68	14	25	-	7	-	37	22	18	23	31
P7	F	55	29	26	-	26	30	23	26	16	16	14

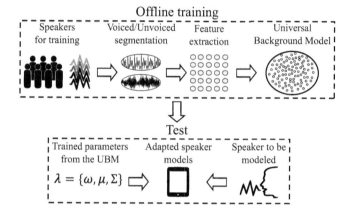

Fig. 1. General methodology.

Validation of models

The performance of the speaker models is evaluated estimating the Pearson's correlation coefficient r between the estimated distances and (1) the MDS-UPDRS-III, and (2) the m–FDA scores. It is important to note that in the beginning, the Spearman's correlation was also considered; however, since this measure works with ranked variables, the obtained results were biased due to the low amount of data.

5 Experiments and Results

Three different versions of the UBM were considered for training: using only features from the voiced segments, only with features from unvoiced segments, and merging the set of features from both segments. The number of Gaussians used to train the UBM ranges from 2 to 16 in steps of 1. Table 2 shows the Pearson's correlation between the Bhattacharyya distances (d_{Bha}) and the MDS-UPDRS-III scores. Average values among the seven patients are also included. It can be observed that the highest correlations ($r = 0.51$) are obtained when only features from the voiced segments are considered for training. When only features from unvoiced segments are considered, the correlation is $r = 0.33$. When both sets of features are merged for training, the correlation does not improve.

Table 2. Pearson's correlation between d_{Bha} and the MDS-UPDRS-III. **M:** Number of Gaussians used to train the model. **AVG:** Average of the correlations per model.

Segment	M	Patient 1	Patient 2	Patient 3	Patient 4	Patient 5	Patient 6	Patient 7	AVG
Voiced	11	0.94	0.83	0.28	0.39	−0.54	0.99	0.67	0.51
Unvoiced	4	0.26	−0.18	0.71	0.55	0.61	−0.30	0.66	0.33
Fusion	6	0.96	0.94	−0.48	0.44	0.41	0.17	0.37	0.40

Table 3 shows the results obtained when the performance of the models is evaluated considering the m–FDA scores. Similarly to the MDS-UPDRS-III, the best results were obtained with voiced segments with correlations of up to $r = 0.40$. It can be observed that these results are lower than those obtained with MDS-UPDRS-III. This result was not expected because the m–FDA is designed to assess specific speech problems developed by Parkinson's patients; however, it may be considered that this experiment only included the repetition of /pa-ta-ka/. A fair evaluation of the m–FDA scale needs to include other speech tasks like sustained phonation of vowel /a/, read text, and monologue. Previous studies have shown that it is possible to model the speech problems of the patients according to the m–FDA scale with correlations of up to 0.72 [21].

Table 3. Pearson's correlation between d_{Bha} and m–FDA. **M:** Number of Gaussians used to train the model. **AVG:** Average value of the correlations per trained model.

Segment	M	Patient 1	Patient 2	Patient 3	Patient 4	Patient 5	Patient 6	Patient 7	AVG
Voiced	16	0.78	0.55	0.01	−0.58	0.98	0.12	0.93	0.40
Unvoiced	6	−0.69	0.11	0.43	−0.20	0.41	0.97	0.09	0.16
Fusion	14	0.92	0.57	0.32	−0.12	0.65	−0.79	−0.10	0.21

Figures 2A and B show the best results obtained considering features extracted upon voiced segments. The x-axis represents the recording sessions and the y-axis indicates the normalized values of the d_{Bha} and the clinical scores. As the ranges of the two scales are different, the normalization (using the z-score approach) is necessary for displaying purposes, i.e., to make the curves comparable in the same picture. The distances computed from each speaker model represent the progression of the disease. Note that in Fig. 2A and B the trends of the computed distances follow the trend of the neurological and dysarthria scales. This behavior can be observed clearly for patients P1, P2, and P6. For patients P5 and P7 this behavior is only observed in Fig. 2B. These results indicate that, to some extent, the MDS-UPDRS-III scale is reflecting speech impairments in PD patients; however, as it was shown in [21] when more speech tasks are considered, the predictions of the m–FDA scores exceed those obtained with the neurological scale. Figure 2C compares the real labels of the MDS-UPDRS-III and m–FDA scales in the same recording sessions for patient. It can be observed that both scales do not follow the same trend, indicating that although the speech production is a motor process, a general neurological scale could not be the most suitable to assess speech impairments.

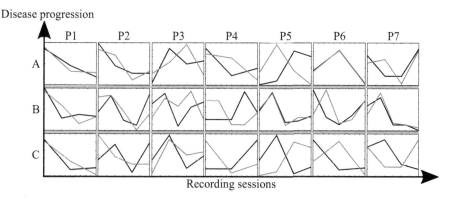

Fig. 2. Parkinson's disease progression curves per patient. **A.** d_{Bha} (Gray line) vs. MDS-UPDRS-III labels (Black line). **B.** d_{Bha} (Gray line) vs. m–FDA (Black line). **C.** m–FDA (Black line) vs. MDS-UPDRS-III labels (Gray line).

6   Conclusions

In this study a methodology to assess the PD progression from speech using a mobile device application is presented. The method is based on the GMM-UBM approach to obtain individual speaker models. The training of the universal background model was performed off-line. The obtained parameters are stored in the smart-phone and used afterwards to obtain the speaker models. The speech problems of the patients are modeled using an articulation approach based on the voiced/unvoiced segmentation. The performance of the models was evaluated according to the MDS-UPDRS and the m–FDA scales. According to the results it is possible to model the variation in the speech of the patients with correlations of up to $r = 0.51$ with the neurological scale. The results reported here are promising but still not satisfactory. In order to improve them, especially those related with the correlations with the dysarthria scale, future versions of the mobile application need to include more speech tasks. This paper is a step forward in the development of mobile-based tools for the continuous and unobtrusive monitoring of people with Parkinson's disease. Currently, data collection using *Apkinson* is ongoing, thus in the near future we expect to develop speaker models using recordings of real phone calls and/or dedicated tests captured with smart-phones.

Acknowledgments. This work was financed by COLCIENCIAS through the project Nº 111556933858. The work reported here was started at JSALT 2016, and was supported by JHU via grants from DARPA (LORELEI), Microsoft, Amazon, Google and Facebook. Thanks also to CODI from University of Antioquia by the grant Numbers 2015-7683 and PRV16-2-01.

References

1. Hornykiewicz, O.: Biochemical aspects of Parkinson's disease. Neurology **51**(2), S2–S9 (1998)
2. Ho, A.K., Iansek, R., Marigliani, C., Bradshaw, J.L., Gates, S.: Speech impairment in a large sample of patients with Parkinsons disease. Behav. Neurol. **11**(3), 131–137 (1999)
3. Darley, F.L., Aronson, A.E., Brown, J.R.: Differential diagnostic patterns of dysarthria. J. Speech Lang. Hear. Res. **12**(2), 246–269 (1969)
4. Theodoros, D.G., Constantinescu, G., Russell, T.G., Ward, E.C., Wilson, S.J., Wootton, R.: Treating the speech disorder in Parkinson's disease online. J. Telemedicine Telecare **12**(Suppl. 3), 88–91 (2006)
5. Rubow, R., Swift, E.: A microcomputer-based wearable biofeedback device to improve transfer of treatment in Parkinsonian dysarthria. J. Speech Hear. Disord. **50**(2), 178–185 (1985)
6. Wirebrand, M.: Real-time monitoring of voice characteristics using accelerometer and microphone measurements. Master's thesis, Linkping University, Linkping, Sweden (2011)
7. Vásquez-Correa, J.C., et al.: New computer aided device for real time analysis of speech of people with Parkinson's disease. Revista Facultad de Ingeniería Universidad de Antioquia **1**(72), 87–103 (2014)
8. Carullo, A., Vallan, A., Astolfi, A.: Design issues for a portable vocal analyzer. IEEE Trans. Instrum. Meas. **62**(5), 1084–1093 (2013)
9. Dubey, H., et al.: EchoWear: smartwatch technology for voice and speech treatments of Patients with Parkinson's disease. In: Proceedings of the Conference on Wireless Health, pp. 15:1–15:8. ACM (2015)
10. Tao, F., Daudet, L., Poellabauer, C., Schneider, S., Busso, C.: A portable automatic PA-TA-KA syllable detection system to derive biomarkers for neurological disorders. In: Proceedings of the Seventeenth Annual Conference of the International Speech Communication Association, pp. 362–366 (2016)
11. Zhan, A., et al.: High frequency remote monitoring of Parkinson's disease via dmartphone: platform overview and medication response detection. arXiv preprint arXiv:1601.00960 (2016)

12. Arias-Vergara, T., et al.: Parkinson's disease progression assessment from speech using GMM-UBM. In: Proceedings of the Seventeenth Annual Conference of the International Speech Communication Association, pp. 1933–1937 (2016)
13. Klumpp, P.: Implementation of a mobile monitoring application for patients with Parkinson's disease. Master's thesis, Friedrich-Alexander-Universität Erlangen-Nürnberg (2017). https://sourceforge.net/projects/apkinson/files/Documentation/Master
14. Orozco-Arroyave, J., et al.: Voiced/unvoiced transitions in speech as a potential bio-marker to detect Parkinson's disease. In: Sixteenth Annual Conference of the International Speech Communication Association, pp. 95–99 (2015)
15. Reynolds, D.A., Quatieri, T.F., Dunn, R.B.: Speaker verification using adapted Gaussian mixture models. Digit. Signal Proc. **10**(1), 19–41 (2000)
16. Nöth, E., et al.: Remote monitoring of Neurodegeneration through Speech. In: Final Presentation of the Third Frederick Jelinek Memorial Summer Workshop (JSALT), August 2016
17. Tan, L.C.: Mood disorders in Parkinson's disease. Parkinsonism Relat. Disord. **18**(Suppl. 1), S74–S76 (2012)
18. Goetz, C.G., et al.: Movement disorder society-sponsored revision of the Unified Parkinson's disease rating scale (MDS-UPDRS): scale presentation and clinimetric testing results. Mov. Disord. **23**(15), 2129–2170 (2008)
19. Enderby, P.M., Palmer, R.: FDA-2: Frenchay Dysarthria Assessment: Examiner's Manual. Pro-ed, Austin (2008)
20. Orozco-Arroyave, J., et al.: Automatic detection of Parkinson's disease in running speech spoken in three different languages. J. Acoust. Soc. Am. **139**(1), 481–500 (2016)
21. Vasquez-Correa, J.C., et al.: Multi-view representation learning via GCCA for multimodal analysis of Parkinson's disease. In: Proceedings of 2017 IEEE International Conference on Acoustics, Speech, and Signal Processing, ICASSP 2017 (2017)

A New Corpus of Collaborative Dialogue Produced Under Cognitive Load Using a Driving Simulator

George Christodoulides[(⊠)]

Centre Valibel, IL&C, Université catholique de Louvain,
Louvain-la-Neuve, Belgium
george@mycontent.gr

Abstract. We present an experiment designed to collect both monologue and dialogue speech, produced under conditions that will tax the
attentional resources of the speaker. Pairs of participants (native speakers of French) were asked to perform tasks involving use of the auditory
memory, complex memorisation and recall, and collaborative exchange
of information. Following the dual-task paradigm, we induced continuous
attentional load to one of participants, using the Continuous Tracking
and Reaction (ConTRe) task, in a driving simulator. In this article, we
present the corpus and an initial analysis of the prosodic characteristics (silent pauses, filled pauses, disfluencies) in speech produced under
cognitive load.

Keywords: Dialogue · Cognitive load · French · Attention · Prosody ·
Disfluencies · Hesitation · Dual-task paradigm

1 Introduction

Cognitive load reflects the mental demand placed by a task on the person performing it, and is derived from the limited capacity of cognitive systems, such
as working memory and attention [4,12,13]. Speech production (conceptuali-
sation of a message, formulation and articulation) and speech perception and
comprehension are processes that engage cognitive resources, including working
memory, to a different degree. It is expected that situations of high cognitive
load will cause detectable effects, among others, in the prosodic structuring of
speech.

The objective of the present study was to collect both monologue and dialogue speech produced under conditions that will tax the attentional resources
of the speaker. Furthermore, we sought to create a realistic communicative situation that would encourage participants to produce long stretches of speech.
We elicited both monologue and collaborative dialogue speech between pairs
of participants, while one of them was using a driving simulator. We used the
dual task paradigm to induce cognitive load: one of the two participants ("the
driver") is constantly performing a secondary task, which demands attention

© Springer International Publishing AG 2017
K. Ekštein and V. Matoušek (Eds.): TSD 2017, LNAI 10415, pp. 380–392, 2017.
DOI: 10.1007/978-3-319-64206-2_43

and co-ordination, in the driving simulator. The other participant ("the passenger") is not engaging in any secondary task. During the same experimental session, participants switched roles: a first recording was performed with the first participant in the role of the driver, and a second recording was performed with the second participant in the role of the driver (a design that facilitates within-subject comparisons). In this article, we present the experimental design, the structure and annotations of the corpus collected, and an initial study on prosodic features of speech produced under varying levels of cognitive load.

2 Experimental Design

The experimental sequence for a given pair of participants ran in three phases:

1. Syntactically Unpredictable Sentences (SUS) speech perception test,
2. Radio News collaborative dialogue task, and
3. Taboo Game task.

Each task was repeated in two conditions: a "slow" and a "fast" driving condition, by changing the configuration of the task performed on the driving simulator (see below). The objective of this manipulation was to induce higher attentional load on the participant by increasing the difficulty of the secondary task (driving).

2.1 Perception of Syntactically Unpredictable Sentences

In the SUS speech perception test participants listened to short sentences presented over their headphones, and were asked to repeat them as faithfully as possible. The sentences were selected from the French Syntactically Unpredictable Sentences corpus [1,11]. The SUS corpus contains sentences following one of the following syntactic forms:

Adverb det. Noun1 Verb-t-pron. det. Noun2 Adjective?
Determiner Noun1 Adjective Verb determiner Noun2
Determiner Noun1 Verb preposition determiner Noun2

The content words are singular, monosyllabic (unless a final schwa was uttered) and have a high frequency of use according to the BRULEX lexicon; prepositions and determiners are also monosyllabic; adjectives normally placed before a noun in French were avoided. The choice of words is such that the sentences are definitely meaningless: it is thus not possible to use the context or logical induction to infer a word that was not perceived correctly. The SUS list was read by a professional male speaker in a soundproof booth, and the recordings were sampled at 16 kHz (16 bits, mono) in the Wave format. The SUS Phrase Audio corpus was further refined by optimising "for homogeneity in terms of phoneme-distribution as compared to average French, and for word occurrence frequency of the employed monosyllabic keywords as derived from French language databases" [11, 2028]. Twenty lists of 12 sentences each are publicly available by the Groupe Audio-Acoustique of the LIMSI research centre, as part of the SUS calibrated audio corpus. The sentence lists used in the experiment can be found in the Annex. These sentences were mixed with multi-talker babble noise, to create the stimuli used in the speech perception experiment. The signal-to-noise ratio ranged from 0 dB (no noise) to −20 dB, in −2 dB intervals. A small computer programme was written to control the presentation of stimuli. The experimenter would listen to the sentence as repeated by the participant; if more than half of the content words were correctly repeated, the experimenter decreased the SNR (i.e. the next stimulus would be presented in louder babble noise), otherwise the experimented increased the SNR.

2.2   Recitation and Collaborative Dialogue in the Radio News Task

The second, and main, phase of the experiment consisted of the Radio News collaborative task. In this task, the driver listened to four consecutive radio news items, extracted from the corpus described in [7]. The news items were recorded during 2006–2008 from the Belgian radio stations "La Première" and "Bel-RTL". Only the driver listened to the radio news items, through their headphones. A list of four questions related to the content of the news items was simultaneously presented to the passenger's computer screen. After the playback of the radio news items was finished, the driver was asked to give a summary of the news, containing as much information as possible (in order to help the passenger to answer the questions, but without knowledge of the questions at this point). After the driver completed the summary, the passenger would attempt to answer the comprehension questions in sequence, and giving their answers aloud. If there was missing information (as was frequently the case) the driver and the passenger engaged in dialogue in order to find the best answer. After each set of four comprehension questions, a series of four open-ended questions was presented to the passenger, one at a time, as a means to stimulate discussion on a subject related to the news items. The passenger posed the question to the driver, and the two participants exchanged views on the subject. The questions were selected to stimulate debate about topical, societal issues, and to encourage participants to express their personal opinion. Two sets of four items were presented per driving simulator condition to each pair of participants; therefore, for each pair of participants we recorded 4 driver summaries (2 per condition), 4 driver-passenger exchanges on the comprehension questions (2 per condition) and 8 short dialogues (4 per condition).

2.3   The Taboo Task

In the third and final phase of the experiment, the participants engaged in fast-paced collaborative dialogue in a game of Taboo. A target word was presented on the computer screen of the passenger (invisible to the driver) along with a list of "forbidden" words. The task of the passenger was to help the driver guess the target word, giving clues but without using the forbidden words.

The objective was to have the driver guess as many words as possible in one minute. The passenger had the possibility to skip a word. The presentation of the next word (when the driver had correctly guessed or when a forbidden word was used) was controlled by the experimenter. We recorded one game per driver simulator condition.

After completing the three phases, driver and passenger exchanged roles, and the experiment was repeated. Different SUS sentences, radio news items and related questions, as well as the Taboo words were used (i.e. participants were confronted with new material, regardless of whether they were taking the role of the driver in the first run or in the second run). At the end of the experiment, participants were asked to fill in a questionnaire with basic demographic information and a subjective rating of the perceived difficulty of each task.

2.4 Driving Simulator and the Continuous Tracking and Reaction Task

The driving simulator used was the OpenDS Pro, version 3.5, a Java-based open source system (the Pro version used is commercially licensed). The participant was controlling the simulator by means of the Thrustmaster T500 RS steering wheel and foot pedals. The driving task used was the ConTRe (Continuous Tracking and Reaction) task (see Fig. 1).

(a) Cylinders   (b) Traffic lights

Fig. 1. ConTRe task screenshots

The driving task is described as follows in [9]: "The driver's primary task in the simulator is comprised of actions required for normal driving: operating the brake and acceleration pedals, as well as turning the steering wheel. System feedback, however, differs from normal driving. In the ConTRe task, the car moves autonomously with a constant speed through a predefined route on a unidirectional straight road consisting of two lanes. Neither operating the acceleration or brake pedal, nor changing the direction of the steering wheel does have an effect on speed or direction of the vehicle. Accordingly, motion rather feels like a video clip. Steering, braking and using the gas pedal do not actually control the car, but instead manipulate a moving cylinder which is rendered in front of the car. On the road ahead, the driver perceives two such cylinders, which continuously move at a constant longitudinal distance (20 m) in front of the car. The two cylinders differ only in colour: one is blue and the other one is yellow. The latter is called the reference cylinder, as it moves autonomously cording to an algorithm. The movement direction and the movement speed of the reference cylinder are neither controlled nor predictable by the user, except that the cylinder never exceeds the roadsides. In contrast, the driver controls the lateral position of the blue cylinder with the help of the steering wheel, trying to keep it overlapping with the reference cylinder as well as possible. As the user turns the steering wheel, the controllable cylinder moves to the left or to the right, depending on the direction of the steering wheel and its angular velocity (i.e. the steering wheel controls the cylinder's lateral acceleration). Effectively, this corresponds to a task where the user has to follow a curvy road or the exact lateral position of a lead vehicle, although it is more strictly controlled and thus with less user-dependent variability. Furthermore, there is a traffic light placed on top of the reference cylinder containing two lights: The lower one can be lighted green, whereas the top light shines red when it is switched on. Either none or only one of these lights appears at a time. The red light requires an immediate brake reaction with the brake pedal, whereas green indicates that an immediate acceleration with the gas pedal is expected. As soon as the driver reacts correctly, the light is turned off". (pp. 2–3).

The task's parameters were configured to provide two levels of task difficulty. In the EASY condition, the speed of the lateral movement of the reference cylinder was 0.4 m/s, while in the DIFFICULT condition, the speed was 1.0 m/s; the maximum speed of the controllable cylinder was 4 m/s. These values were selected after a pilot study involving 10 participants (cf. [5] for a study using three levels of difficulty) (Fig. 2).

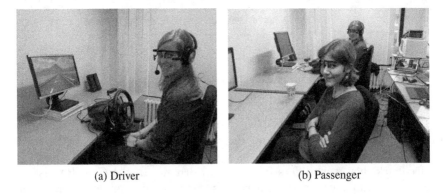

(a) Driver   (b) Passenger

Fig. 2. Participants in the driving simulation collaborative dialogue study

3   Participants and Method

3.1   Participants

A pilot study was run with 10 participants (all female, all university students). The pilot study subjects participated in the experiment for course credit and were not otherwise remunerated. The main study, reported here, was conducted with 28 participants, in 14 pairs (6 male participants and 22 female). The average age of the main study participants was 22.3 years (standard deviation: 3.2; min: 19; max: 36). Most participants were university students, and were recruited through the participant pool of the Faculty of Psychology at Université catholique de Louvain. All participants reported French as their mother tongue; 23 participants were born and raised in Belgium and 5 participants in Metropolitan France. 26 participants reported knowledge of one or more second languages. None of the participants reported any auditory problems and 9 participants had musical training. Regarding their driving experience, 20 participants reported having a driver's licence (obtained on average 2.93 years before the experiment) and 16 participants reported driving regularly. With regards to the relationship between the participants, 16 did not know each other before the experiment; 10 participants were friends; and 2 were in an intimate relationship.

The demographic data was collected by means of a questionnaire, given to both participants at the end of the experiment (both recording sessions). The subjective task difficulty ratings were also collected using the same questionnaire. One participant did not complete the questionnaire.

3.2   Data Acquisition

Data from multiple sources was collected throughout the experiment. Both participants were wearing a Sennheiser ME3 head-worn microphone and Philips FX4M headphones. They could listen to each other, and they were relatively isolated from outside noise. Their speech was recorded throughout the experiment. The microphones were connected to a Focusrite 18i8 audio interface (through two microphone preamplifiers), which was used to control the interconnection between the driver's microphone and the passenger's headphones and vice versa; the audio signal was then routed to a Zoom Z24 multi-track recorder. The stimuli were presented using a dedicated laptop PC running Windows; the stimuli were presented through the special interface developed for the SUS phase of the experiment, and using Winamp for the Radio News phase of the experiment. The audio signal of the stimuli presented to the participants was also routed to the Zoom multi-track recorder. Two audio synchronisation signals, from the eye-tracker and the driving simulator, were also connected to the Zoom recorder; finally the internal microphone of the recorder was used to capture ambient noise in the room and to serve as a backup recording. Using this configuration, we collected an 8-track time-synchronised recording of the driver's and passenger's speech, the stimuli presented to the driver and passenger, the events of the eye tracker (start and stop) and of the driving simulator (start, stop, red light, green light, coded as short pure tones with different frequencies) and of any interaction with the experimenter. The audio recordings were performed using a sampling rate of 44.1 kHz and 16-bit resolution.

In addition to the audio recordings, eye tracker data was collected for the driver only, using the portable head-mounted Pupil eye-tracker [8]. In this study, we used the second-generation model, recording both eyes at a sampling rate of 60–80 Hz. The OpenDS driving simulator system, which was running on a dedicated laptop Windows PC, also recorded driving behaviour data: the precise time at which the driver pressed the acceleration or brake pedal in response to the traffic lights (and a calculated response time), as well as steering wheel position and the deviation of the controllable cylinder from the reference cylinder. The driving simulator data were stored in a MySQL database.

3.3   Speech Corpus Annotation and Analysis

The speech recordings of the driver and passenger were split in sections depending on the task performed and the driving simulator condition, using the Ardour audio editor. The following sections were separated: SUS/Driver and SUS/Passenger, EASY and DIFF conditions; Radio News part 1 and part 2 (8 + 8 questions) Driver and Passenger, EASY and DIFF conditions; Taboo game, EASY and DIFF conditions. The total duration of speech recorded (i.e. including both recording channels, the Driver and the Passenger) was 47.6 h. The SUS task sub-corpus contains 555.6 min (9.2 h) of recordings; the Radio News task sub-corpus contains 1915.9 min (31.9 h) of recordings; and the Taboo Game sub-corpus contains 385 min (6.4 h) of recordings. The total number of section recordings is 442. Due to the large amount of data collected, we focus our analysis only on the Radio News task sub-corpus.

The Radio News task sub-corpus was further processed and its recordings split into corpus samples, using the following breakdown: Summary by the driver; Exchange between the driver and the passenger on the basis of the comprehension questions; Dialogue between the driver and the passenger (each question/topic is a different corpus sample). The sub-corpus contains 657 sections (in 1314 audio files since the driver and passenger speech were recorded in different channels). The total duration of the sub-corpus, after removing the regions of recordings where the driver was listening to the radio news stimuli (and therefore there is no speech to analyse) is 1424.5 min (23.7 h).

Samples were selected for manual transcription. The selection was representative of all tasks performed under both conditions (e.g. we selected 8 question-based dialogues for each pair of participants, 4 in the EASY and 4 in the DIFF driving conditions, out of the original 16 dialogues), giving priority to dialogues longer than one minute. This selection of samples was then transcribed under Praat, and the transcriptions imported in Praaline. The final corpus analysed is 8.7 h long and its contents are presented in Table 1.

A cascade of automatic analysis tools was applied: the orthographic transcriptions were converted into phonetic transcriptions and aligned with the speech signal at the phone and syllable level using EasyAlign [6]. A morphosyntactic

Table 1. Radio news sub-corpus (manually transcribed and analysed) contents

Condition/Task	Driver or Passenger		Total (both)	
	Count	Duration (min)	Count	Duration (min)
DIFF	116	150,46	176	265,79
Summary	28	35,14	28	35,14
Exchange	27	31,80	54	63,61
Dialogue	61	83,52	122	167,04
EASY	114	144,34	172	256,22
Summary	28	32,46	28	32,46
Exchange	30	35,88	60	71,76
Dialogue	56	76,00	112	152,00
Grand total	230	294,80 (4,9 h)	348	522,01 (8,7 h)

analysis of the corpus was performed using the DisMo annotator [3]. The Prosogram script [10] was applied to the entire corpus (using the automatic annotation functionality in Praaline [2]), to obtain detailed prosodic information on each syllable, based on pitch stylisation. The corpus contains 54.002 tokens and 73.381 syllables.

4   Results

4.1   Subjective Ratings of Task Difficulty

This section presents the analysis of the subjective ratings of task difficulty, as reported by the participants. The questionnaire used a 7-point Likert scale for task difficulty rating. The results of the subjective ratings are shown in Table 2 and Fig. 3. The most difficult tasks were, as expected, listening to the radio news (comprehension) and summarising them (production). Tasks performed under the fast driving condition were systematically rated more difficult than task performed under the slow driving condition. The passenger's participation in the Radio News and Taboo tasks was rated as the easiest tasks; however, repeating the SUS sentences as a passenger was rated as more difficult than participating in the free exchange dialogue or playing the Taboo game while driving in the slow condition.

4.2   Prosodic Analysis of the Corpus

We focus on the temporal characteristics of the speech of the participants, while performing different tasks and under different conditions (driver in the EASY or DIFF driving simulator, passenger). We first turn our attention to articulation ratio and silent pause ratio: as can be seen on Fig. 4, the task associated with the highest cognitive load (S: summary) is associated with the highest silent pause

Table 2. Summary of subjective ratings of task difficulty

Task/Condition		Mean	Std Dev
RListen_DrvFast	Listening to news, fast driving	5,11	1,52
RSummary_Drv	Summarising (driver)	5,11	1,42
RListen_DrvSlow	Listening to news, slow driving	4,30	1,30
SUS_DrvFast	Repeating SUS, fast driving	4,15	1,10
Taboo_DrvFast	Taboo game, guessing, fast driving	3,37	1,22
RDialogue_DrvFast	Open-ended dialogue, fast driving	3,26	1,35
SUS_DrvSlow	Repeating SUS, slow driving	3,19	1,16
SUS_Pass	Repeating SUS, passenger (no dual task)	3,11	1,45
Taboo_DrvSlow	Taboo game, guessing, slow driving	2,81	1,06
RDialogue_DrvSlow	Open-ended dialogue, slow driving	2,74	1,07
Taboo_Pass	Taboo game, describing (no dual task)	2,38	1,18
RAnswers_Pass	Answering questions (no dual task)	1,74	0,97
RQuestions_Pass	Asking questions (no dual task)	1,26	0,52
RDialogue_Pass	Open-ended dialogue, passenger (no dual task)	1,22	0,42

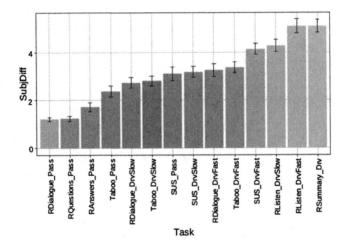

Fig. 3. Subjective difficulty ratings of tasks in this study (mean and standard error)

ratio, and the lowest articulation ratio. Among the dialogue tasks, exchanging information on the comprehension questions (E: exchange) has a higher articulation ratio and lower silent pause ratio compared to the free dialogue (Q: questions).

A more detailed analysis of the duration of silent pauses, as a mixture of 2 log-normal distributions is presented in Fig. 5 and Table 3. It can be observed that the long silent pause component distribution has a higher weight factor

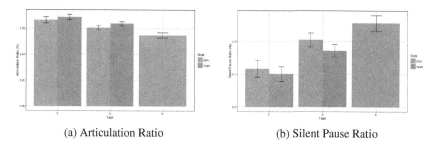

(a) Articulation Ratio   (b) Silent Pause Ratio

Fig. 4. Articulation Ratio (left) and Silent Pause Ratio (right) per Task and Speaker Role (means and 95% CI)

across tasks (the relevant figures are in boldface); whereas speech produced by the passenger (under no cognitive load) has a more balanced distribution between short and long silent pauses. However, there is no clear correlation between the short/long pause balance and the task that the driver is executing. The task of driving is altering the silent pause distribution coarsely, with higher cognitive load associated with a higher proportion of long pauses in speech. Furthermore, a Wilcoxon rank sum test indicates a significant difference in mean silent pause duration produced by the Driver and the Passenger ($p < 0.001$).

Table 3. Analysis of silent pause length as a mixture of 2 component log-normal distributions, per task, driving condition and speaker role

Task/Condition/Speaker role	Component weight λ		Mean value μ		Std deviation σ	
	Comp1	Comp2	Comp1	Comp2	Comp1	Comp2
S EASY	25,6%	74,4%	1,824	2,828	0,349	0,288
S DIFF	29,7%	70,3%	1,844	2,838	0,352	0,314
E EASY driver	20,6%	79,4%	1,743	2,767	0,262	0,325
E DIFF driver	27,8%	72,2%	1,666	2,748	0,282	0,309
E EASY passenger	57,3%	42,7%	2,285	2,781	0,620	0,229
E DIFF passenger	44,5%	55,5%	2,241	2,698	0,557	0,275
Q EASY driver	17,5%	82,5%	1,610	2,714	0,255	0,340
Q DIFF driver	20,8%	79,2%	1,723	2,726	0,355	0,341
Q EASY passenger	60,7%	39,3%	2,362	2,730	0,587	0,269
Q DIFF passenger	43,2%	56,8%	2,265	2,768	0,583z	0,250

Regarding filled pauses, Fig. 6 presents two measures calculated per task type (S: summary, E: exchange and Q: dialogue on questions) and speaker role (Driver vs. Passenger): the filled pause ratio, i.e. the percentage of speech time covered by filled pauses, and the filled pause rate, i.e. the number of filled pauses per second. A Wilcoxon rank sum test indicates that filled pause ratio is significantly higher for the driver compared to the passenger ($p < 0.0001$), while filled pause

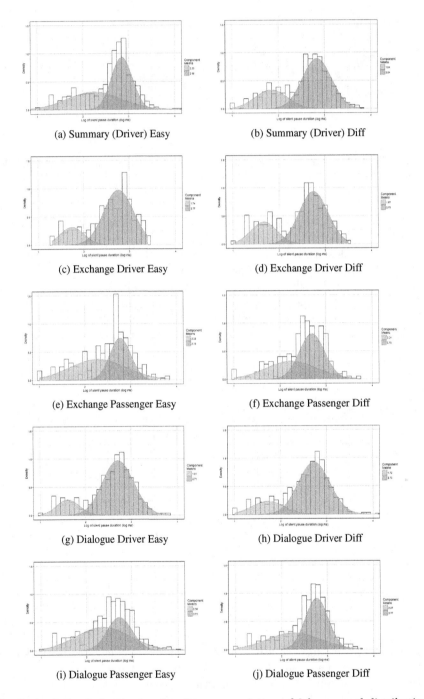

Fig. 5. Analysis of silent pause durations as a mixture of 2 log-normal distributions, per task, driving condition and speaker role

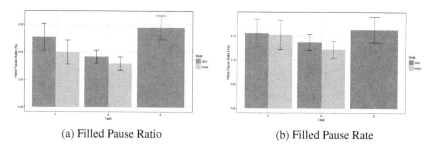

(a) Filled Pause Ratio

(b) Filled Pause Rate

Fig. 6. Filled pause ratio (left) and filled pause rate (right) per task and speaker role

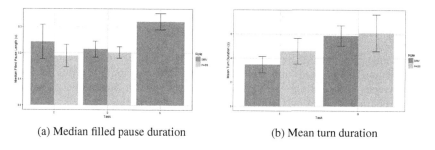

(a) Median filled pause duration

(b) Mean turn duration

Fig. 7. Median filled pause duration per task and speaker role (left) and Mean turn duration for the two dialogue tasks, per speaker role (right)

rate is not significantly different for the two speaker roles. In other words, the additional attentional requirements of the secondary task (driving) do not lead to more filled pauses, but to globally longer filled pauses (cf. Fig. 7a: median filled pause length per task and speaker role). This finding is consistent with the previously mentioned observations on silent pauses and articulation ratio.

Finally, with respect to dialogue dynamics, Fig. 7b presents the mean turn duration (in seconds) per task type and speaker role for the two tasks involving dialogue (E: exchange of factual information on the radio news and Q: open-ended dialogue based on questions). The difference in means is statistically significant between the two task types (Wilcoxon rank sum test, $p < 0.0001$) but not between driver and passenger in each of the two tasks: this measure is affected by the communicative context with the E task being more dynamic as expected.

5   Conclusion

We have presented an experiment to collect monologue and dialogue speech, produced under varying levels of cognitive load. An initial study on the prosodic characteristics of the corpus indicated that tasks associated with a higher cognitive load had a higher silent pause ratio and a lower articulation ratio; that the additional attentional requirements of the secondary task (driving) did not lead to more filled pauses, but to globally longer filled pauses; and that with respect to dialogue dynamics, the communicative context had a stronger effect than the induced cognitive load. This study has produced a large new corpus of French dialogue, which is still being annotated and analysed, with a view to making it available to the community.

References

1. Boula de Mareüil, P., d'Alessandro, C., Raake, A., Bailly, G., Garcia, M.N., Morel, M.: A joint intelligibility evaluation of French text-to-speech synthesis systems: the EvaSy SUS/ACR campaign. In: Proceedings of the 5th International Conference on Language Resources and Evaluation, pp. 2034–2037 (2006)
2. Christodoulides, G.: Praaline: integrating tools for speech corpus research. In: Proceedings of the 9th International Language Resources and Evaluation Conference, pp. 31–34 (2014)
3. Christodoulides, G., Avanzi, M., Goldman, J.P.: DisMo: a morphosyntactic, disfluency and multi-word unit annotator: an evaluation on a corpus of French spontaneous and read speech. In: Proceedings of the 9th International Language Resources and Evaluation Conference, pp. 3902–3907 (2014)
4. Cowan, N.: Working memory and attention in language use. In: Guendouzi, J., Loncke, F., Williams, M.J. (eds.) The Handbook of Psycholinguistic and Cognitive Processes, pp. 75–98. Psychology Press, New York (2011)
5. Engonopoulos, N., Sayeed, A., Demberg, V.: Language and cognitive load in a dual task environment. In: Proceedings of the 35th Annual Meeting of the Cognitive Science Society, pp. 2249–2254 (2013)
6. Goldman, J.P.: Easyalign: an automatic phonetic alignment tool under Praat. In: Proceedings of Interspeech 2011, vol. 3233–3236 (2011)
7. Hupin, B., Simon, A.C.: Analyse phonostylistique du discours radiophonique. Recherches en communication **28** (2007)
8. Kassner, M., Patera, W., Bulling, A.: Pupil: an open source platform for pervasive eye tracking and mobile gaze-based interaction: Technical report. http://arxiv.org/abs/1405.0006
9. Mahr, A., Feld, M., Mehdi Moniri, M., Math, R.: The ConTRe (Continuous Tracking and Reaction) task: a flexible approach for assessing driver cognitive workload with high sensitivity. In: Kun, A.L., Boyle, L.N., Reimer, B., Riener, A. (eds.) Adjunct Proceedings of the 4th International Conference on Automotive User Interfaces and Interactive Vehicular Applications, pp. 88–91 (2012)
10. Mertens, P.: The Prosogram: semi-automatic transcription of prosody based on a tonal perception model. In: Bel, B., Marlien, I. (eds.) Proceedings of Speech Prosody 2004, pp. 549–552 (2004)
11. Raake, A., Katz, B.F.: SUS-based method for speech reception threshold measurement in French. In: Proceedings of the 5th International Conference on Language Resources and Evaluation, pp. 2028–2033 (2006)
12. Shriffin, R., Schneider, W.: Controlled and automatic human information processing II: perceptual learning, automatic attending and a general theory. Psychol. Rev. **84**(2), 127–190 (1977)
13. Wickens, C.D., Hollands, J.G., Banbury, S., Parasuraman, R.: Engineering Psychology and Human Performance, 4th edn. Psychology Press, Hove (2012). International edition

Phonetic Segmentation Using Knowledge from Visual and Perceptual Domain

Bhavik Vachhani, Chitralekha Bhat$^{(\boxtimes)}$, and Sunil Kopparapu

TCS Innovation Labs, Mumbai, India
{bhavik.vachhani,bhat.chitralekha,sunilkumar.kopparapu}@tcs.com
http://www.tcs.com

Abstract. Accurate and automatic phonetic segmentation is crucial for several speech based applications such as phone level articulation analysis and error detection, speech synthesis, annotation, speech recognition and emotion recognition. In this paper we examine the effectiveness of using visual features obtained by processing the image spectrogram of a speech utterance, as applied to phonetic segmentation. Further, we propose a mechanism to combine the knowledge from visual and perceptual domains for automatic phonetic segmentation. This process can be considered analogous to manual phonetic segmentation. The technique was evaluated on TIMIT American English Corpus. Experimental results show significant improvements in phonetic segmentation, especially for lower tolerances of 5, 10 and 15 ms, with an absolute improvement of 8.29% for TIMIT database for a 10 ms tolerance is observed.

Keywords: Unsupervised phonetic segmentation · Edge detection · Multi-taper · Visual phonetic segmentation

1 Introduction

In the terms of speech signal processing, a phone is defined as a distinct speech sound independent of the spoken language. Phonetic segmentation is the process of identifying the boundaries between phones within a given speech utterance. Accurate phonetic segmentation forms the basis for speech tasks such as phone level articulation analysis, speech synthesis, transcription, annotation, speech recognition, emotion recognition etc.

Manual segmentation and labeling is carried out by an expert who marks the phonetic boundaries in the speech utterance by referring to speech spectrograms, energy, duration of various speech sound units and pitch. In practice, accurate phone segmentation is obtained manually by using both perceptual and visual cues for the task. This process is described as human spectrogram reading and ensures perceptual validity. Studies have examined inter- and intra-transcriber consistencies in manual segmentation [11,14,19], where it has been observed that no two human annotators are likely to produce the same phonetic boundary for a given speech utterance. Owing to large sizes of speech corpora and transcriber

© Springer International Publishing AG 2017
K. Ekštein and V. Matoušek (Eds.): TSD 2017, LNAI 10415, pp. 393–401, 2017.
DOI: 10.1007/978-3-319-64206-2_44

fatigue, manual segmentation turns out be tedious, time-consuming and less efficient. Hence, research into automatic segmentation methods becomes inevitable. Techniques for phonetic segmentation are categorized into (1) Supervised and (2) Unsupervised methods. Supervised methods are model based and require extensive training on speech corpora, that have been transcribed accurately. Generally, these methods employ a two-stage process for automatic phonetic segmentation, with the automatic speech recognizer (ASR) providing an initial estimate of phone boundaries, followed by a boundary refinement process as described in [1,9,10,16,20]. A variety of boundary refinement techniques have been used in literature such as specific boundary level acoustic models [20], regression tree [1], acoustic-phonetic knowledge [10], phone-transition-dependent SVM classifiers [9]. Yet another supervised technique uses a discriminative learning [6]. In [5], author describes a phonetic segmentation method using a combination of auditory attention and phone posterior features. Unsupervised methods on the other hand are heuristic based and make use of speech features such as spectral measures, energy, duration and pitch and peak-picking algorithms for boundary detection. In [4] authors use energy changes and sub-band analysis in the short-term Fourier transform (STFT) of a speech utterance. Support Vector Machine (SVM)-based technique wherein a biomimetic model of the human auditory processing is used for automatic phonetic segmentation [7]. In [8], authors present an image processing approach to speech phoneme segmentation. Despite excellent results, manually annotated data is considered more reliable for speech processing tasks.

In this paper we propose a method for automatic phonetic segmentation (APS) that is very similar to the manual segmentation done by experts who use both visual and perceptual cues to segment the phone boundaries. The main contribution of this paper is a method that makes use of the visual cues obtained by treating the spectrogram of a speech utterance as an image in combination with perceptual cues obtained from Perceptual Linear Prediction Cepstral Coefficients (PLPCC) along with Spectral Transition Measure (STM) [18]. Both the visual and perceptual cues make use of a multi-taper based spectral estimation [17].

The rest of the paper is organized as follows. In Sect. 2, we describe the methodology used to obtain phonetic segmentation, Sect. 3 describes the experimental set up and the data used, In Sect. 4 we describe the results for various tolerance levels and we conclude in Sect. 5.

2   Proposed Technique

In this paper, we propose a technique for automatic phonetic segmentation to be analogous to manual segmentation process wherein both visual and perceptual cues are used for segmentation process. Figure 1 shows an overview of the proposed technique.

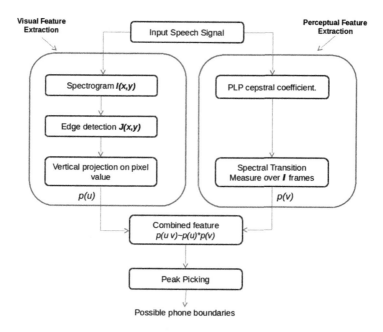

Fig. 1. Proposed technique for phonetic segmentation

2.1   Visual Domain Phonetic Segmentation

Conventionally spectral estimation of speech is done by applying a Hamming-window or a single taper for speech signal processing. A major limitation of the single taper method is that, by using one taper a significant portion of the signal is discarded and the data points at the extremes are down-weighted, giving a high variance for the direct spectral estimate [12]. Hence, a multi-taper method is used so that the statistical information lost by using just one taper is partially recovered by using multiple windows for the same duration. The multi-taper spectrum is a weighted sum of the several tapered periodograms. Spectral estimation of a signal s using multi-taper method is as follows,

$$S_{(m,k)} = \frac{1}{M} \sum_{p=0}^{M-1} \lambda(p) \sum_{j=0}^{N-1} w_p(j)s(m,j)e^{-i2\pi\frac{k}{N}j} \tag{1}$$

where $w_p(j)$ is the pth data taper function, M is the number of tapers and $\lambda(p)$ is the weight corresponding to the pth taper, N is the speech frame length and k is the FFT points. In practice, weights are designed so as to compensate for increased energy loss at higher order tapers. The spectrograms used for both visual cues as well as perceptual were derived from multi-taper spectral estimation described above.

Visual cues are obtained by treating the spectrogram of a speech utterance as a $2D$ image, where frequency bins are plotted along the y-axis and time frames along the x-axis with the intensity of the image indicating the energy at a particular frequency. A visual inspection of a spectrogram clearly displays the transition pattern between phones as shown in Fig. 2. This visual information is used by annotators in addition to the Perceptual cues for manual phonetic segmentation.

For a given utterance $s(t)$ Multi-taper based spectral estimation is used to compute $2D$ spectrogram image $I(x,y)$ using Eq. 1. Edge detection algorithm is then applied on the spectrogram image $I(x,y)$ of the speech utterance to obtain an initial estimate of the phonetic boundaries. In this paper, we use Canny edge detection algorithm from MATLAB toolbox.

Figure 2 shows the correspondence between phonetic segmentation and the edges detected in the spectrogram of the speech utterance $s(t)$.

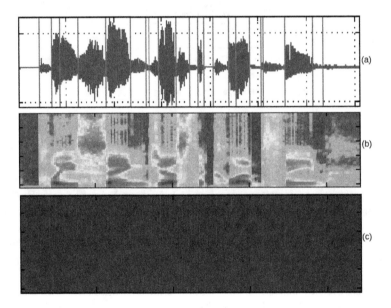

Fig. 2. (a) Speech signal $s(t)$ with manual segment boundaries, (b) Multi-tapered spectrogram $I(x,y)$, (c) Edge detection output $J(x,y)$

$$STM_v(x) = \sum_{i=1}^{n} J(x, y_i) \qquad (2)$$

where, $x = 1, 2, ...m$ (total number of frames) and $y = 1, 2, ...n$ (number of FFT bins).

A vertical projection based on pixel value of image $J(x,y)$ is computed using Eq. 2, wherein the pixel values are summed over a column of the edge detected image $J(x,y)$. This projection profile can be considered as the contour (STM_v) whose peaks indicate the location of edges along the time axis. A threshold determined dynamically from the projection profile contour STM_v, i.e. a threshold specific to a given speech utterance is used to eliminate over-segmentation. Of the statistical measures such as mean and median, the mean of the projection profile was found to be the most suitable as threshold τ_M using Eq. 4.

$$STM_{vt}(x) = \begin{cases} STM_v(x), & \text{if } STM_v(x) > \tau_M \\ \tau_M, & \text{otherwise} \end{cases} \tag{3}$$

$$\tau_M = \frac{1}{m} \sum_{x=1}^{m} STM_v(x) \tag{4}$$

It was observed that this technique achieved high recall of approximately 87% (refer Table 2) indicating a high percentage of correct boundaries within a given tolerance level. However, the process was also pervaded with low precision, indicating high over-segmentation. The phonetic boundaries thus obtained using visual cues were further refined using perceptual cues as described in Sect. 2.2.

2.2 Combining Knowledge from Perceptual and Visual Domains

The process for phonetic segmentation using multi-taper based PLPCC-STM has been described in [18]. The STM (Spectral Transition Measure) contour obtained from this process is used to refine the phonetic boundaries from the visual domain. Given that both the processes provide a contour whose peaks indicate the presence of a phonetic boundary, we treat the contour value as the probability of occurrence of a peak. Let $p(u)$ be the probability of occurrence of a boundary in the visual contour (STM_{vt}) and $p(v)$ be the probability of occurrence of a boundary in the perceptual contour (PLPC-STM). We compute the mixed probability $p(uv) \approx p(u) \times p(v)$ as a transition measure/contour. This is done in order to maximize the probability of a phone boundary when using both visual and perceptual cues. A peak picking algorithm is used on the contour provided by $p(uv)$ to obtain the final phonetic boundaries. Figure 3 shows the speech utterance with boundaries marked manually along with automatically obtained phonetic segmentation using Visual (STM_{vt}), Perceptual (PLPCC-STM) and the proposed technique.

3   Experimental Setup

All experiments were carried out on TIMIT American English corpus [3]. TIMIT contains 2,34,925 between-phone boundaries manually determined by experts.

3.1   Multi-taper Spectral Estimation

Multi-taper spectral estimation was done using Discrete Prolate Spheroidal sequences (DPSS) also known as Thomson or Slepian tapers [17] with 6 orthonormal tapers.

$$w_p(j) = \frac{\sin[\omega_c T(p-j)]}{(p-j)}, \qquad j = 0, 1, \ldots, N-1 \tag{5}$$

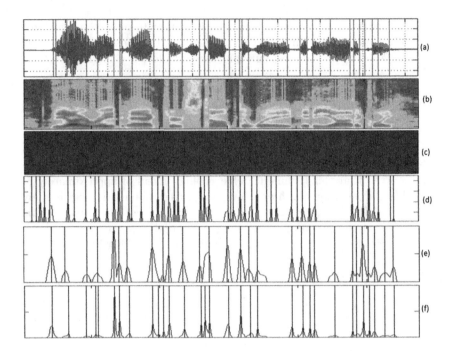

Fig. 3. (a) Speech signal with manually segmented boundaries, (b) Multi-taper spectrogram $I(x,y)$, (c) Edge detection output $J(x,y)$, (d) $p(u)$ computed using visual features STM_{vt}, (e) $p(v)$ computed using perceptual features PLPLCC-STM, and (f) $p(uv)$ computed using combining perceptual+visual evidences

where N denotes the desired window length in samples, ω_c is the desired mainlobe cut-off frequency in radians per second, and T is the sampling period in seconds. Twelve dimensional PLPCC features were computed using Thomson multi-taper spectral estimation with a 30 ms window and a 10 ms shift rate. For spectrogram computation we have used 256 FFT bins with a 30 ms window and a 10 ms shift rate.

4 Results and Analysis

In Table 1, we compare the performance of our proposed technique with known unsupervised techniques in the literature, for TIMIT clean speech for a tolerance level of 20 ms. Automatic phonetic segmentation was carried out as described in Sect. 2. In this paper we have used Precision, Recall and F-score as a performance measures to evaluate the phonetic segmentation results. For evaluation of automatic phonetic segmentation performance we also use various tolerance levels such as 20 ms, 15 ms, 10 ms and 5 ms. Table 2 gives an indication of the performance of phonetic segmentation from visual domain only. It can be seen that F-score is impacted by a low precision, caused by over-segmentation. The perceptual cue-based phonetic segmentation serves in reducing the over-segmentation

Table 1. Unsupervised APS methods comparison for TIMIT corpus

Method	Precision (Pr)	Recall (Re)	F-score (Fs)
Dusan et al. [2] [2006]	72.73	75.2	73.94
Qiao et al. [13] [2008]	78.76	77.5	78.13
Shah et al. [15] [2014]	60.69	80.2	69.1
Baseline [18][2016]	84.6	75.8	80
Visual domain	67.40	86.64	75.82
Proposed (visual+perceptual)	82.52	81.11	**81.81**

Table 2. APS using visual features only for TIMIT corpus

Database	TIMIT					
Tolerance (ms)	Visual-Spectrogram			Visual-MT Spectrogarm		
	Pr	Re	Fs	Pr	Re	Fs
5	29.00	40.14	33.67	30.68	39.44	34.51
10	49.18	68.05	57.10	51.53	66.23	57.96
15	59.46	82.29	69.04	61.92	79.60	69.66
20	64.33	89.02	74.69	**67.41**	**86.64**	75.82

to a great extent as can be seen in Table 3. The overall effect of the proposed technique is to improve the F-score over the baseline for all tolerance levels of 20 ms, 15 ms, 10 ms and 5 ms. However, significant improvements were seen at lower tolerance levels. The process of combining the two probability contours as discussed in Sect. 2.2 serves to relocate phonetic boundaries so as to be closer to the ground truth as compared to the boundaries obtained by the visual and perceptual domains alone.

Table 3. APS using baseline [18] and proposed technique for TIMIT corpus

Database	TIMIT					
Tolerance (ms)	Baseline - Perceptual			Proposed - Visual+Perceptual		
	Pr	Re	Fs	Pr	Re	Fs
5	31.75	27.60	29.53	37.74	37.09	37.41 (+7.88)
10	58.32	50.70	54.24	63.02	61.94	**62.47 (+8.29)**
15	76.67	66.65	71.31	75.74	74.44	75.08 (+3.77)
20	84.64	75.82	79.99	82.52	81.11	**81.81 (+1.82)**

5 Conclusion

Accurate and automatic phonetic segmentation is very helpful in several speech based applications like phone level articulation analysis, speech synthesis, phoneme level speech corpus annotation, automatic speech recognition and even speech emotion recognition. The best accuracies are achieved by expert human annotators who look at the spectrogram of the speech utterance and listen to the utterance for perceptual cues to manually mark phone boundaries. In this paper, motivated by this observation, we have proposed a novel and robust technique that makes use of the visual and perceptual cues to identify phone boundaries. The initial estimate of the phone boundary segments are obtained by performing simple image processing technique of edge detection followed by contour building on the 2D image spectrogram of the utterance. The obtained phone boundaries are refined using perceptual cues obtained from Perceptual Linear Prediction Cepstral Coefficients (PLPCC) along with Spectral Transition Measure (STM). This approach of mimicking the manual phone segmentation process adopted by an expert human is the main contribution of this paper. The use of multi-taper based spectral estimation to construct the spectrogram of an utterance instead of a single taper results in better estimate of phone boundaries, this is yet another contribution of this work. The usefulness and robustness of proposed technique was evaluated on the standard TIMIT American English Corpus. Experimental results show an absolute improvement of 8.29% for TIMIT database for a 10 ms tolerance. The improvement is visible especially for lower tolerances of 5, 10 and 15 ms and for tolerance of 20 ms our approach performs as well as the state of the art phone segmentation. As a way forward, an exploration into identifying a suitable pre-processing technique to enhance the edges of the spectrogram image or a post-processing technique to reduce over-segmentation will render the proposed technique well suited for unsupervised phonetic segmentation.

References

1. Adell, J., Bonafonte, A.: Towards phone segmentation for concatenative speech synthesis. In: Proceedings of the 5th ISCA Speech Snthesis Workshop, pp. 139–144 (2004)
2. Dusan, S., Rabiner, L.R.: On the relation between maximum spectral transition positions and phone boundaries. In: INTERSPEECH- ICSLP, Ninth International Conference on Spoken Language Processing, 17–21 September 2006, Pittsburgh, PA, USA (2006)
3. Garofolo, J.S.: Getting started with the darpa timit cd-rom: an acoustic phonetic continuous speech database. In: National Institute of Standards and Technology (NIST) (1988)
4. Golipour, L., O'Shaughnessy, D.D.: A new approach for phoneme segmentation of speech signals. In: INTERSPEECH, pp. 1933–1936. ISCA (2007)
5. Kalinli, O.: Automatic phoneme segmentation using auditory attention features. In: Proceedings of the INTERSPEECH, pp. 2270–2273 (2012)
6. Keshet, J., Shalev-Shwartz, S., Singer, Y., Chazan, D.: Phoneme alignment based on discriminative learning. In: INTERSPEECH 2005, pp. 2961–2964 (2005)
7. King, S., Hasegawa-Johnson, M.: Accurate speech segmentation by mimicking human auditory processing. In: IEEE International Conference on Acoustics, Speech and Signal Processing, ICASSP 2013, 26–31 May, Vancouver, BC, Canada, pp. 8096–8100 (2013)
8. Leow, S.J., Chng, E.S., Lee, C.H.: Language-resource independent speech segmentation using cues from a spectrogram image. In: 2015 IEEE International Conference on Acoustics, Speech and Signal Processing (ICASSP), pp. 5813–5817, April 2015
9. Lo, H.Y., Wang, H.M.: Phonetic boundary refinement using support vector machine. In: IEEE International Conference on Acoustics, Speech and Signal Processing (ICASSP), vol. 4, pp. 933–936, April 2007
10. Patil, V., Joshi, S., Rao, P.: Improving the robustness of phonetic segmentation to accent and style variation with a two-staged approach. In: INTERSPEECH, pp. 2543–2546 (2009)

11. Pitt, M.A., Johnson, K., Hume, E., Kiesling, S., Raymond, W.: The buckeye corpus of conversational speech: labeling conventions and a test of transcriber reliability. Speech Commun. **45**, 89–95 (2005)
12. Prieto, G.A., Parker, R.L., Thomson, D.J., Vernon, F.L., Graham, R.L.: Reducing the Bias of Multitaper Spectrum Estimates, vol. 171, pp. 1269–1281. Oxford University Press, Oxford (2007)
13. Qiao, Y., Shimomura, N., Minematsu, N.: Unsupervised optimal phoneme segmentation: objectives, algorithm and comparisons. In: ICASSP, pp. 3989–3992 (2008)
14. Raymond, W.D., Pitt, M.A., Johnson, K., Hume, E., Makashay, M.J., Dautricourt, R., Hilts, C.: An analysis of transcription consistency in spontaneous speech from the buckeye corpus. In: INTERSPEECH (2002)
15. Shah, N.J., Vachhani, B.B., Sailor, H.B., Patil, H.A.: Effectiveness of PLP-based phonetic segmentation for speech synthesis. In: Proceedings of the ICASSP, Florence, Italy, pp. 270–274 (2014)
16. Stolcke, A., Ryant, N., Mitra, V., Yuan, J., Wang, W., Liberman, M.: Highly accurate phonetic segmentation using boundary correction models and system fusion. In: Proceedings of the ICASSP, Florence, Italy, pp. 5552–5556 (2014)
17. Thomson, D.: Spectrum estimation and harmonic analysis. Proc. IEEE **70**, 1055–1096 (1982)
18. Vachhani, B., Bhat, C., Kopparapu, S.: Robust phonetic segmentation using multitaper spectral estimation for noisy and clipped speech. In: 2016 24th European Signal Processing Conference (EUSIPCO), pp. 1343–1347, August 2016
19. Wesenick, M.B., Kipp, A.: Estimating the quality of phonetic transcriptions and segmentations of speech signals. In: Proceedings of the Fourth International Conference on Spoken Language 1996, ICSLP 1996, vol. 1, pp. 129–132. IEEE (1996)
20. Yuan, J., Ryant, N., Liberman, M., Stolcke, A., Mitra, V., Wang, W.: Automatic phonetic segmentation using boundary models. In: Proceedings of the INTERSPEECH, pp. 2306–2310 (2013)

The Impact of Inaccurate Phonetic Annotations on Speech Recognition Performance

Radek Safarik[✉] and Lukas Mateju

Institute of Information Technology and Electronics,
Technical University of Liberec, Studentska 2, 461 17 Liberec, Czech Republic
{radek.safarik,lukas.mateju}@tul.cz
https://www.ite.tul.cz/speechlab/

Abstract. This paper focuses on impact of phonetic inaccuracies of acoustic training data on performance of automatic speech recognition system. This is especially important if the training data is created in automated way. In this case, the data often contains errors in a form of wrong phonetic transcriptions. A series of experiments simulating various common errors in phonetic transcriptions based on parts of GlobalPhone data set (for Croatian, Czech and Russian) is conducted. These experiments show the influence of various errors on different languages and acoustic models (Gaussian mixture models, deep neural networks). The impact of errors is also shown for real data obtained by our automated ASR creation process for Belarusian. The results show that the best performance is achieved by using the most accurate data; however, certain amount of errors (up to 5%) does have relatively small impact on speech recognition accuracy.

Keywords: Speech recognition · Gaussian mixture models · Deep neural networks · Phonetic annotations · Phoneme corruption

1 Introduction

Modern Automatic Speech Recognition (ASR) systems are composed of two different parts, language dependent and language independent. To adapt the ASR system to a new language, the language dependent components have to be altered while the language independent components remain unchanged. The language dependent components are pronunciation lexicon, Language Model (LM) and Acoustic Model (AM). The orthographic components (lexicon and LM) are built from large corpus of target language. The corpus can be easily gathered by downloading texts from various Internet sources. While the creation of LM model is fairly straightforward the building of AM is much more complicated. It needs tens of hours of recorded speech from many speakers as well as phonetic annotations. It is necessary to spend a lot of time and human effort to create such a speech database. It is also possible to buy already created databases (e.g. GlobalPhone [12]), but these tend to be quite expensive.

© Springer International Publishing AG 2017
K. Ekštein and V. Matoušek (Eds.): TSD 2017, LNAI 10415, pp. 402–410, 2017.
DOI: 10.1007/978-3-319-64206-2_45

Therefore, we proposed an approach for automatic creation of all necessary parts with only minimal human intervention based on free Internet resources. In [9], we showed how this approach was used to create ASR system for all South Slavic languages in almost automated way. The approach for building AM is based on scanning web pages and downloading large amount of audio/video data with any related text. The downloaded data is transcribed by an existing ASR system (of target or closely related language). The output is then compared with the related text to find matching segments. Finally, these segments are cut out and used for further processing. If the match between the recognized text and the related text is 100%, the segments are automatically used for training of new AM. If the match is higher than 70% but not perfect, the segments are transcribed with the new improved AM and so on. The transcription can be also manually tuned by using our effective tool. It can show the recognized and reference text, play the audio as well as highlight mismatched words.

This approach guarantees very accurate phonetic transcriptions suitable for AM training. However, the amount of obtained training data is significantly lower than the amount of downloaded data. For this very reason, an extensive amount of data needs to be downloaded and processed to obtain enough training data. Therefore, it is vital to know the influence of various amount of errors in training data on speech recognition performance. Another important thing is if the influence is similar on different languages or not. For this purpose, we evaluate all experiments on three different Slavic languages (Croatian, Czech, Russian). Note that we focused on Polish language in [11].

The amount of related works focusing on influence of inaccurate training data on speech recognition performance is scarce. However, in [13], the authors showed these effects using Gaussian Mixture Models (GMM). The errors were simulated by substituting phonetically similar and dissimilar phonemes. The result showed that even 20% word error rate did not significantly influence the performance. Similar works but from different fields focused on dealing with errors in training data in image processing [4], class imbalances in training data [1] and identifying and filtering mislabeled training data [2].

2   ASR System

Our ASR system has been in development for more than 15 years and at this point it is capable of real-time transcription using a lexicon of size over a half million words. The original application was aimed for Czech language. However, now we a have working solution for almost all Slavic languages [9]. The ASR system has been used in various application over the years: voice dictation, transcription of historical audio archive [10] or broadcast monitoring [8]. The ASR system is modular and separates the language independent parts (signal processing, decoder) from the language dependent ones (acoustic model, lexicon and language model). This allows us to easily apply it for different languages or modify it for different application.

Our system supports two types of acoustic models: GMM-HMM (Gaussian Mixture Models - Hidden Markov Models) and DNN-HMM (Deep Neural Networks - Hidden Markov Models). The former ones are triphone multi-gaussian models [5] with 32 mixtures to represent the phonemes. The models are speaker-independent and context-dependent. 39-dimensional Mel-Frequency Cepstral Coefficients (MFCC) are extracted from the input signal. Additionally, Cepstral Mean Subtraction (CMS) and HLDA transformation are applied. Note that the frame length is 25 ms and the time shift is 10 ms. The training is done using HTK speech recognition toolkit[1].

The latter acoustic model follows the DNN-HMM architecture presented in [3] with the exception of no pre-training. To represent the input signal, 39-dimensional log filter banks are computed. These features are also locally normalized within 1 s long window. To provide context, the final input feature vector is a concatenation of 5 previous frames, current frame and 5 following frames. To speed-up the training process, the hyper-parameters of DNN are slightly toned down from the ones we normally use [7]. The hyper-parameters used in this work are as follows. The networks have 5 hidden layers each consisting of 768 neurons. The employed activation functions are ReLU and softmax for hidden layers and output layer, respectively. The networks are trained within 15 epochs using the mini-batches of size 1024 with the learning rate set to 0.08. Note that torch library[2] is used to train the networks.

The orthographic part of the ASR system is composed of lexicon and language model. These parts are gathered from a large amount of text data (corpus) of given language. The lexicon is created by selecting the most frequent words in the corpus. For these words, a phonetic transcription is generated by G2P system or in special cases by hand. Each of the words can also have multiple pronunciation variants. The employed language models are based on N-grams. Due to the large size of the lexicon, the recognizer uses only bigrams. The unseen bigrams are backed-off by the Knesser-Ney smoothing algorithm [6].

3 Experimental Work

The main focus of the experiments was to evaluate the influence of inaccurate training data on speech recognition performance. We simulated the inaccuracies by random substituting phonemes and by random deleting or repeating words. The experiments were conducted using 3 Slavic languages to compare the influence of errors on different phonetics (and different sized phoneme sets). The Sect. 4 describes another experiment on Belarusian data with real inaccuracies. It studied the relation between the amount of training data and the quality of phonetic annotations. The evaluation was done using GMM-HMM and DNN-HMM acoustic models.

---

[1] http://htk.eng.cam.ac.uk/.

[2] http://torch.ch.

3.1   Train and Test Data

The experiments were done for 3 Slavic languages (Croatian, Czech, Russian) on data from GlobalPhone database. For each language, approximately one hundred utterances from one hundred speakers were available. Hence, the results are comparable among the evaluated languages. Each language data was divided into train and test data set. The first 10 speakers were used for evaluation and the remaining 90 speakers were utilized for training of acoustic models. The distribution of data is shown in Table 1. Note that the experiments could be easily replicated as the GlobalPhone database is accessible.

Table 1. The distribution of train and test sets of GlobalPhone for different languages in hours.

Language	Train	Test	Phonemes
Croatian	14.0	2.0	35
Czech	27.0	3.8	40
Russian	23.9	2.5	52

3.2   Experimental Setup

The conducted experiments can be divided into two main groups. The first one focused on the effect of phoneme substitutions while the latter one studied the effect of inserting repeating or removing words from training data.

To study the effect of phoneme substitution on speech recognition performance, we had to alter the phoneme transcriptions of GlobalPhone data. For each phoneme a probability was rolled. If this number was lower than a certain threshold, the phoneme was substituted. The percentage of substituted phonemes ranged from 1% to 50% (half of the phonemes corrupted) and differed for each experiment. The replacement phonemes were randomly selected from the phoneme set of given language. The same procedure was then repeated for clusters of phonemes of size 2 and 3.

Quite often the speakers accidentally repeat or omit words from their utterances. To simulate this situation, we randomly (with probability ranging from 1% to 40%) repeated (insertions) or removed (deletions) words from data.

3.3   Results

The experiments are evaluated using standard Word Error Rate (WER) metric:

$$WER[\%] = \frac{I + S + D}{N} * 100, \tag{1}$$

where I, D and S are the number of insertions, deletions and substitutions, respectively. N is the total number of words in the reference text.

Figure 1 shows the comparison of different languages within the scope of experiment with a random substitution of phonemes using DNN models. It is evident that WER grew with the amount of substituted phonemes. However, this grow was not so significant, e.g. WER only got worse by 5% for 20% of substituted phonemes. The increase in WER was more noticeable with additional substitutions. Nevertheless the performance was still decent even with a half of the data corrupted.

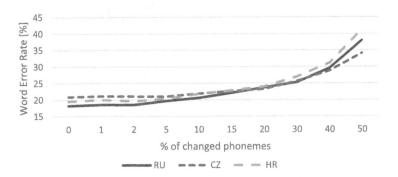

Fig. 1. Performance comparison on various languages based on amount of phoneme substitutions.

As the Fig. 1 shows, the behaviour of WER on all languages was fairly comparable. The different steepness of increase in WER was due to different amount of training data for each language (larger amount of data (Czech) had enough correct data to train better model than languages with less resources). Due to the high similarity between the results of different languages and to increase the readability of figures, we further evaluate experiments only on Russian data.

In the second experiment, we focused on the influence of substituted clusters. Such clusters can be created e.g. when voice assimilation wrongly affects the whole cluster of consonants during automatic creation of pronunciation in lexicon. The Fig. 2 depicts the results of experiments based on substitution of one phoneme, two consecutive phonemes and three following phonemes in training data. The amount of substituted phonemes is comparable in all cases.

The models based on clustered substitutions of bigger size outperformed the single phoneme substitutions. This was caused by employment of triphone model. When errors were randomly distributed in the train set, each substituted phoneme affected three triphones (the current one and the two surrounding ones) meaning that three random substituted phonemes altered nine triphones. In case of altering clusters of size 3, only five triphones were affected (the three substituted ones and the two surrounding ones).

The following experiment was about the influence of deleting or inserting repeating words. It simulated situations when the speaker repeats some words or completely omits them during recording. It may also happen in our automatic mining process if the match between the recognized text and related text

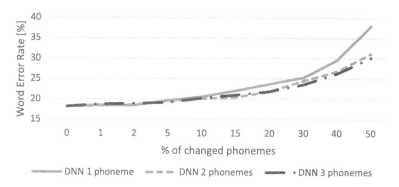

Fig. 2. The influence of substituted cluster sizes on speech recognition performance.

is not 100%. That means that there may be some wrong, missing or added words. The results in Fig. 3 show the influence of repeating and deleting words from training data on performance of GMM-HMM and DNN-HMM models. The results showed that repetition of words is not a significant problem and that the performance was only slightly worsened. On the other hand, the word deletions significantly impacted the performance, especially for DNN. It was caused by assignment of phonemes to frames during training. The repeated words got assigned to the lowest count of frames possible while the remaining frames were assigned mostly correctly. When a word was deleted from a transcription, all of its corresponding frames had to be reassigned to surrounding phonemes. This resulted in larger amount of influenced frames.

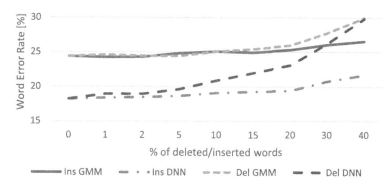

Fig. 3. Influence of insertions and deletions on performance of GMM-HMM and DNN models.

Finally, the last experiment compared the behaviour of GMM-HMM and DNN-HMM trained on corrupted training data. The results are shown in Fig. 4. Note that all GMM-HMM and DNN-HMM pairs were always trained on the same corrupted data. As expected, DNN-HMM model outperformed GMM-HMM on clear data. However, it was significantly more vulnerable to errors in training data. It was due to different training process and assignment of phonemes to frames of audio signal. GMM-HMM reassigned phonemes in each iteration of training by using new model each time, while DNN used the same assignment all along the training.

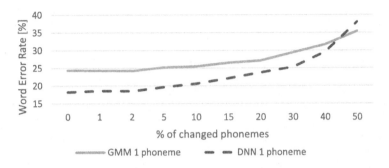

Fig. 4. Performance comparison between GMM-HMM and DNN-HMM models.

4   Practical Experiment with Automatically Created Training Data

In this experiment, we focused on finding a relation between amount of training data and its accuracy. As we are currently working on Belarusian language for our ASR system, we chose it for this experiment as well. All acoustic data downloaded from public TV and radio stations' web pages were processed automatically and only segments with 100% match between recognized text and related text were used for training. We studied the influence of adding segments with more than 95%, 90% and 80% match to training set. This means that the most refined model had the least amount of data and vice versa. Note that the errors were mostly caused by missing words, inserted words or wrong phoneme transcriptions of words.

Table 2. Results on Belarusian language with various amount of errors in training data.

Data precision	Hours	WER
100%	16.6	35.95
95%+	19.2	38.24
90%+	20.7	39.32
80%+	27.2	41.80

To test the models, our TV/radio data set for Belarusian[3] was employed. The data set annotated by native speakers consisted of TV and radio shows in total length of hour and half. The results, as summarized in Table 2, show the importance of clean annotations over the amount of training data.

5   Conclusions

In this paper, we focused on importance of precise phonetic transcriptions for speech recognition task. We presented several experiments focused on different phonetic corruption of annotations. The results (performance wise) can be summarized as follows:

- up to 5% of substitutions, the reduction in performance was fairly low;
- the differences between languages were negligible;
- the bigger clusters of substituted phonemes outperformed single substitutions;
- repetitions of words did not worsen the performance significantly;
- deletions on the other hand dropped in performance noticeably;
- the performance of DNN-HMM models dropped significantly more than GMM-HMM models (especially with more errors).

We can say that if maximum precision is not the main goal, certain amount of phonetic inaccuracies (up to 5%) may still result in high-performing speech recognition system. On the other hand, by using the most precise annotations, slight reduction in WER can be achieved. However, obtaining such clean data can be time consuming as well as costly. We confirmed our hypothesis by the experiments conducted on Belarusian data set where the performance dropped by 2% by adding data with up to 5% of various errors.

Acknowledgements. This work was supported by the Technology Agency of the Czech Republic (Project No. TA04010199) and by the Student Grant Scheme 2017 of the Technical University in Liberec.

References

1. Batista, G.E., Prati, R.C., Monard, M.C.: A study of the behavior of several methods for balancing machine learning training data. ACM SIGKDD Explor. Newsl. **6** (2004)
2. Brodley, C.E., Friedl, M.A.: Identifying mislabeled training data. J. Artif. Intell. Res. **11**, 131–167 (1999)
3. Dahl, G.E., Yu, D., Deng, L., Acero, A.: Context-dependent pre-trained deep neural networks for large-vocabulary speech recognition. Trans. Audio, Speech Lang. Proc. (2012)
4. Hansen, M.S., Kozerke, S., Pruessmann, K.P., Boesiger, P., Pedersen, E.M., Tsao, J.: On the influence of training data quality in k-t BLAST reconstruction. Magn. Reson. Med. **52**(5), 1175–1183 (2004)

---

[3] https://gitlab.ite.tul.cz/SpeechLab/EastSlavicTestData.

5. Huang, X., Acero, A., Hon, H.W.: Spoken Language Processing A Guide to Theory, Algorithm, and System Development, 1st edn. Prentice Hall, Upper Saddle River (2001)
6. Kneser, R., Ney, H.: Improved backing-off for m-gram language modeling. In: Proceedings of the IEEE International Conference on Acoustics. Speech and Signal Processing, Detroit, Michigan, vol. I, pp. 181–184, May 1995
7. Mateju, L., Cerva, P., Zdansky, J.: Investigation into the use of deep neural networks for LVCSR of Czech. In: 2015 IEEE International Workshop of Electronics, Control, Measurement, Signals and their application to Mechatronics (ECMSM), pp. 1–4 (2015)
8. Nouza, J., Zdansky, J., Cerva, P.: System for automatic collection, annotation and indexing of Czech broadcast speech with full-text search. In: 2010 15th IEEE Mediterranean Electrotechnical Conference, Melecon 2010, pp. 202–205, April 2010
9. Nouza, J., Safarik, R., Cerva, P.: Asr for South Slavic languages developed in almost automated way. In: INTERSPEECH, pp. 3868–3872 (2016)
10. Nouza, J.e.a.: Speech-to-text technology to transcribe and disclose 100,000+ hours of bilingual documents from historical Czech and Czechoslovak radio archive. In: INTERSPEECH, pp. 964–968. ISCA (2014)
11. Safarik, R., Mateju, L.: Impact of phonetic annotation precision on automatic speech recognition systems. In: 2016 39th International Conference on Telecommunications and Signal Processing (TSP), pp. 311–314, June 2016
12. Schultz, T.: Globalphone: A multilingual speech and text database developed at Karlsruhe university. In: Proceedings of the ICSLP, pp. 345–348 (2002)
13. Sundaram, R., Picone, J.: Effects on transcription errors on supervised learning in speech recognition. In: IEEE International Conference on Acoustics, Speech, and Signal Processing, 2004 Proceedings, ICASSP 2004, vol. 1, p. I-169. IEEE (2004)

Automatic Detection of Parkinson's Disease: An Experimental Analysis of Common Speech Production Tasks Used for Diagnosis

Anna Pompili[1]([✉]) [ID], Alberto Abad[1] [ID], Paolo Romano[1] [ID],
Isabel P. Martins[2,4] [ID], Rita Cardoso[4,5,6] [ID], Helena Santos[4,5,6],
Joana Carvalho[4,5,6], Isabel Guimarães[3,4,5] [ID], and Joaquim J. Ferreira[4,5,6] [ID]

[1] INESC-ID/IST, Lisbon, Portugal
{anna,alberto.abad}@l2f.inesc-id.pt
[2] Laboratório de Estudos de Linguagem, Faculty of Medicine,
University of Lisbon, Lisbon, Portugal
[3] Department of Speech Therapy, Escola Superior de Saúde do Alcoitão,
SCML, Estoril, Portugal
[4] Instituto de Medicina Molecular, Lisbon, Portugal
[5] Laboratory of Clinical Pharmacology and Therapeutics,
Faculty of Medicine, University of Lisbon, Lisbon, Portugal
[6] CNS - Campus Neurológico Sénior, Torres Vedras, Portugal

Abstract. Parkinson's disease (PD) is the second most common neu-rodegenerative disorder of mid-to-late life after Alzheimer's disease. During the progression of the disease, most individuals with PD report impairments in speech due to deficits in phonation, articulation, prosody, and fluency. In the literature, several studies perform the automatic classification of speech of people with PD considering various types of acoustic information extracted from different speech tasks. Nevertheless, it is unclear which tasks are more important for an automatic classification of the disease. In this work, we compare the discriminant capabilities of eight verbal tasks designed to capture the major symptoms affecting speech. To this end, we introduce a new database of Portuguese speakers consisting of 65 healthy control and 75 PD subjects. For each task, an automatic classifier is built using feature sets and modeling approaches in compliance with the current state of the art. Experimental results permit to identify reading aloud prosodic sentences and story-telling tasks as the most useful for the automatic detection of PD.

Keywords: Parkinson's disease · Phonation · Articulation · Prosody

1 Introduction

Parkinson's disease (PD) is a progressive degenerative disorder of the central nervous system characterized by motor and non-motor symptoms. The cardinal motor signs of PD include the characteristic clinical picture of resting tremor, rigidity, bradykinesia, and impairment of postural reflexes, while non-motor

K. Ekštein and V. Matoušek (Eds.): TSD 2017, LNAI 10415, pp. 411–419, 2017.
DOI: 10.1007/978-3-319-64206-2_46

symptoms include cognitive disorders, sleep and sensory abnormalities. These symptoms slowly worsen during the disease with a nonlinear progression. Motor symptoms of PD influence also the speech production of language. Dysarthria is typically observed in PD patients, it is characterized by weakness, paralysis, or lack of coordination in the motor-speech system, affecting respiration, phonation, articulation and prosody. The main deficits of PD speech are: loss of intensity, monotony of pitch and loudness, reduced stress, inappropriate silences, short rushes of speech, variable rate, imprecise consonant articulation and harsh and breathy voice. The standard method to evaluate and rate the neurological state of Parkinson's patients is based on the revised version, provided by the Movement Disorders Society, of the unified Parkinson's disease rating scale (MDS-UPDRS) [1]. The motor part of the MDS-UPDRS (Sect. 3) addresses speech evaluating volume, prosody, clarity and repetition of syllables. There are several speaking tasks that could be used to evaluate the extent of speech and voice disorders in PD. The most traditional of them are the sustained vowel phonation, rapid syllable repetition (diadochokinesis), and variable reading of short sentences, longer passages or freely spoken spontaneous speech [2].

Recently, the automatic detection of Parkinson's disease through speech has gained the interest of the scientific community. Current state of the art includes plenty of works targeting either the discrimination of PD patients from healthy controls (HC) [3–5] or the evaluation of the correlation among different speech characteristics and the severity of the disease [6–8]. These works differ on many aspects: on the set of features considered, on the speech tasks used for the analysis, and on the statistical approach used in the characterization of the problem. Nevertheless, there are few studies that specifically focus on comparing common speech production tasks typically used for diagnosis in terms of their utility for automatic PD discrimination [3].

The main goal of this work is to investigate the role of common speech production tasks used for the automatic detection of PD. To this end, we consider a new database of Portuguese speakers consisting of 65 healthy control (HC) and 75 PD subjects, each one of them performing 8 different speech tasks that are typically used in Speech and Language Therapy clinical evaluations. For each one of these tasks, we report individual automatic PD detection experiments based on a conventional machine learning approach. Feature extraction is based on a selection of the most representative measures typically considered in studies assessing how the symptoms of this disease affect the speech production. Thus, we are not interested in comparing the large amount of different acoustic measures and learning approaches that have emerged along the years, but rather in defining a feature set and classification strategy, based on the literature review, that can be suitable for assessing the different speech tasks. In the following, Sect. 2 reports on some recent works in the area of automatic detection of the disease. Section 3 describes the data and the speech tasks, while Sect. 4 explains the approach followed in this study. Results are presented and discussed in Sect. 5. Finally, the conclusions are provided in Sect. 6.

2   Related Work

In the last years there has been an increasing number of research works aiming at the automatic characterization and assessment of dysarthria in PD using speech. The primary focus of these studies is either discerning PD patients from HC or attempting to monitor the disease progression through the estimation of the UPDRS scale.

For instance, in Bocklet et al. [3], the authors investigate four different systems in order to assess acoustic, prosodic and voice-related features. The corpus used is composed of 88 German speaking subjects affected by PD and 88 HC subjects performing eight tasks. The proposed systems deal with different acoustic and prosodic features (MFCCs, F_0, energy, duration, pauses, jitter, and shimmer), and with the estimation of the parameters of the physical glottis. Another system performs feature extraction with openSMILE [9] considering the 1582 acoustic features of the INTERSPEECH 2010 Computational Paralinguistic Challenge (ComParE2010) baseline [10]. With the combination of all the speech tasks and the fusion of the four systems, the authors report a recognition result of 79% in discriminating HC subjects from PD patients.

In Bayestehtashk et al. [6], the authors investigated the automatic evaluation of the severity of PD from speech. The corpus used is composed of 168 English speaking patients at different stages of the disease. The recordings include three different tasks: sustained phonation of the English vowel /a/, diadochokinesis (DDK) evaluation and reading text. The classifiers are based on the ComParE2010 feature set. According to the results, the reading text and DDK tasks are the most effective to perform the evaluation of the extent of the disease. The authors reported a mean absolute error of 5.5 using the motor sub-scale of UPDRS that takes values in the [0, 108] range.

In Orozco-Arroyave et al. [5], the authors explore different acoustic measures on a set of recordings composed of the five vowels existing in the Spanish language. The corpus is composed of 50 subjects affected by PD and 50 healthy subjects, both groups are balanced by gender and age. The analysis includes several acoustic measures, among these the first two formants (F1 and F2), the pitch, the jitter, the shimmer, the vowel articulation index (VAI), the triangular vowel space area (tVSA), and three new measures based on the tVSA. Results, in agreement with previous studies, have shown that measures of the variability of the pitch are among the most important features. Also, combining articulation and phonation features led to an improvement in the results, achieving 81.3% of classification accuracy.

Finally, regarding PD for European Portuguese, Proença et al. [11] investigated acoustic and phonetic-prosodic characteristics of speech produced by PD patients while reading phonetically rich sentences and isolated words. The corpus is composed of 22 patients (12 females, 10 males) with different degrees of PD severity. Only vowels in continuous speech context were analyzed. First and second formant frequencies, vowel space area (VSA), VAI, MFCC, spectral and prosodic parameters were calculated for each speaker. Results have shown a centralization of vowel formant frequencies for PD speech, besides exploiting acoustic, spectral and prosodic features for classifying PD speech have shown that dynamic features are of highest importance in this task.

3   Corpus Description

The FraLusoPark database [12] contains 140 European Portuguese speakers. The control group, composed of 65 healthy volunteers, is age-matched and sex-matched with the PD group, composed of 75 subjects. Patients were recorded twice, OFF medication (i.e.: at least 12 h after withdrawal of all anti-Parkinsonian drugs), and ON medication (i.e.: following at least 1 h after the administration of the usual medication).

Subjects were recorded in a quiet room, with a specialized speech recording equipment (Marantz PMD661 MKII recorder), using a unidirectional headset microphone sampled at 48 kHz with 16-bit resolution.

Participants were required to perform several speech production tasks with an increasing complexity in a fixed order: (1) three repetitions of the sustained phonation of the vowel /a/, (2) two repetitions of the maximum phonation time (vowel /a/ sustained as long as possible), (3) oral diadochokinesia (repetition of the pseudo-word *pataka* at a fast rate for 30 s.), (4–5) reading aloud of 10 words and 10 sentences, (6) reading aloud of a short text ("The North Wind and the Sun"), (7) storytelling speech guided by visual stimuli, and (8) reading aloud of a set of sentences with specific prosodic properties.

The total duration of the recordings is approximately 6 h and 31 min for the control group, and 7 h and 30 min for the PD group. Demographics data of the corpus are presented in Table 1.

Table 1. Demographic and clinical data for patient and control groups.

	PD patients		Controls	
	M	F	M	F
Gender	38	37	34	31
Age	64.6 ± 11.9	66.9 ± 8.5	62.4 ± 12.4	66.6 ± 14.4
Years diagnosed	6.7 ± 4.5	10.8 ± 5.6	–	–
MDS-UPDRS-III	32.1 ± 12.9	38.3 ± 14.5	–	–

4   Methodology

In this work, we use the database described in the previous section to conduct an analysis of the performance of automatic PD classification for each one of the 8 speech production tasks. For the purpose of this study, only recordings ON medication were considered. Data was manually preprocessed in order to remove the therapist's speech. Additionally, each spontaneous intervention introduced by the subject, that was not directly related with the task, was removed as well.

After that, recordings were down-sampled to 16 kHz. The selected feature set, described in detail in the next Sect. 4.1, has been extracted with the openSMILE toolkit [9]. The selected model is a Random Forest classifier as implemented in the WEKA toolkit [13]. This implementation relies on bootstrap aggregating, also known as bagging, a machine learning ensemble meta-algorithm designed to improve the stability and accuracy of machine learning algorithms used in statistical classification and regression. Bagging reduces variance and helps to avoid over-fitting. A stratified k-fold cross validation per speaker strategy is used for training and evaluation of each speech task separately, with k being equal to 5. Thus, we ensure that the train and the test sets at each iteration do not contain the same speakers. Also, the percentage of speakers of each class is balanced in the two data sets at each iteration.

4.1   Features Selected for PD Detection

Motor symptoms of PD affect also the motor-speech system, influencing the production of language at various dimensions: phonation, articulation, and prosody. Phonation problems are related with vocal fold bowing and incomplete closing of vocal folds [14,15]; articulation deficits are manifested as reduced amplitude and velocity of the articulatory movements in the lips, tongue and jaw [16]; prosody impairments comprise changes in loudness, pitch, and timing, which overall contribute to the resulting intelligibility of speech [17].

In the literature the most traditional measures used in examining phonation include measurement of F_0, jitter, shimmer, and Harmonics to Noise Ratio (HNR) [5,17]. Articulation is typically assessed considering differences in vocal tract resonances. The first and second formant frequencies and the vowel space area are frequently studied [11,18]. Prosodic analysis includes measurements of F_0, intensity, articulation rate, pause, and rhythm [3,17,19].

In this study we consider some of the measures that are repeatedly referred in the majority of the works examined. In particular, our custom set of features is reported in Table 2. First, these features are initially computed at the frame level, the so-called low-level descriptors, which are obtained based on a sub-set of the Geneva Minimalistic Acoustic Parameter Set (GeMAPS) [20] and the MFCC pre-built configuration files. Then, in a second step, two functionals (mean and standard deviation) are applied in order to obtain a feature vector of constant length for the whole utterance. For some features, (F_0 and loudness), mean and standard deviation of the slope of rising/falling signal parts were also computed. Finally, we obtain a 114-dimensional feature vector composed of 78 MFCC based features and 36 GeMAPS based features. Notice that some other features also frequently mentioned in the literature (i.e.: the articulation rate, pause analysis, or VSA) were not considered in order to build a general purpose feature set, which could be suitable for each task under assessment.

Table 2. Description of the acoustic features based on 53 low-level descriptors plus 6 functionals.

Descriptors	Functionals
Logarithmic F_0 (1), Loudness (1)	mean and stdev, mean and stdev of the slope of rising/falling signal parts (x6)
Jitter (1), Shimmer (1), Formant 1 bandwidth (1), Formant 1, 2, 3 frequency (3), amplitude (3), Harmonic to Noise Ratio (1), Harmonic difference: H1-H2 (1), H1-A3 (1), MFCC [1–12] (12), LOGenergy (1), First and second derivative of MFCC and Log-energy (26)	mean and stdev (x2)

4.2 Sentence-Level vs. Segmental Feature Extraction

On a first attempt, the recordings of every speech production task for each speaker have been processed as described previously to obtain a feature vector of 114 elements. We refer to this approach as sentence-level feature extraction. This strategy results in a single feature vector per speaker and task. In other words, cross-validation experiments for each task are limited to only 140 sample vectors, which will probably result in poorly trained models and less reliable results. Alternatively, in order to increase the number of samples, we have also performed a segmental feature extraction strategy. In this case, we obtain a feature vector as previously described for each audio sub-segment of fixed length equal to 4 s with a time shift of 2 s. This approach permits increasing the amount of training samples for the cross-validation experiments, besides extracting more detailed information of the speech productions.

5 Experimental Results and Analysis

Table 3 shows classification accuracy (%) results for each speech production task following the two feature extraction strategies described previously: sentence-level and segmental. As expected, the former approach led to poorer results, mostly motivated by the reduced number of training samples (only 112 in each task at each cross-validation iteration). However, we also believe that we are losing valuable information when applying the functionals to long speech segments as the ones corresponding to each speech production. On the other hand, the segmental feature extraction strategy leads to very remarkable improvements in terms of classification accuracy. In particular, the reading words task achieves a maximum of 40.6% relative improvement, followed by the reading sentences task with 31.5% relative improvement.

Overall, from these results we can observe that the reading prosodic sentences task achieved the best recognition accuracy (85.10%). In fact, this is the best

Table 3. Task-dependent recognition results on the 2-class detection task (PD vs. HC).

Feature extraction results - accuracy (%)		
Task	Sentence level	Segmental
Sustained vowel phonation (/a/)	55.00	58.14
Maximum phonation time (/a/)	60.00	75.65
Rapid syllables repetitions	60.71	73.28
Reading of word	54.29	76.35
Reading of sentences	62.14	81.74
Reading of text	65.00	79.86
Storytelling guided by visual stimuli	66.43	82.32
Reading of prosodic sentences	70.71	85.10

performing task also in the case of sentence-level feature extraction. This observation confirms the relevance of this task, which was carefully designed in order to explore language-general and language-specific details of PD dysprosody. The second most discriminant task in terms of automatic PD classification is the storytelling one (82.32%). As a matter of fact this task corresponds to the production of spontaneous speech, since the subject has to create a story based on temporal events represented in a picture. Although its overall duration is extremely variable and dependent on the speaker, this task definitely contains many important acoustic and prosodic information. This result is very encouraging for the development of tele-monitoring applications that may use spontaneous speech recorded over the telephone. The next most discriminant tasks are those consisting of reading short passages of text and sentences. Again, we believe that these productions are richer in terms of acoustic and prosodic information, which makes them more convenient for automatic PD detection in contrast to less informative rapid syllables repetitions or maximum phonation time of vowel /a/. In general, it is likely that more complex tasks will contain more linguistics phenomena, like for instance co-articulations, that may provide important cues for discrimination. Moreover, these more complex tasks consist generally of longer speech productions, which is expected to be beneficial for the segmental feature extraction approach. Nevertheless, we also note that both feature extraction strategies provide coherent results in terms of identifying the top-4 most significant speech production tasks. Finally, we observe that the sustained phonation of vowels /a/ is the task that achieved the worst results with the segmental approach by a large margin (58.14%). From a quick analysis we notice that this task is the one with shortest speech productions, resulting in less speech segments. Anyway, this result deserves a deeper analysis, since it is in contradiction with some previous observations for other languages.

6   Conclusions

In this work, we have analyzed the potential discriminative ability of a large set of common speech production tasks for the automatic detection of PD. For this purpose, we considered a database containing European Portuguese PD and HC speakers performing 8 tasks designed to assess speech disorders at various dimensions. For each task, automatic classification experiments have been conducted using a Random Forest classifier and a custom set of acoustic features carefully selected based on the study of the state of the art. The experimental results have shown that the most important production tasks are reading of prosodic sentences and storytelling, achieving a PD classification accuracy of 85.10% and 82.32%, respectively. Future work includes the analysis of conversational speech production tasks, besides the use of i-vector based classifiers, which is the current state of the art in speech recognition tasks such as speaker or language recognition.

Acknowledgments. This work was supported by Portuguese national funds through – Fundação para a Ciência e a Tecnologia (FCT), under Grants SFRH/BD/97187/2013 and Projects with reference UID/CEC/50021/2013 and CMUP-ERI/TIC/0033/2014.

References

1. Movement disorder society task force on rating scales for Parkinson's disease. The Unified Parkinson's Disease Rating Scale (UPDRS): Status and recommendations (2003)
2. Goberman, A.M., Coelho, C.: Acoustic analysis of Parkinsonian speech I: speech characteristics and L-Dopa therapy. NeuroRehabilitation **17**(3), 237–246 (2002)
3. Bocklet, T., Steidl, S., Nöth, E., Skodda, S.: Automatic evaluation of Parkinson's speech-acoustic, prosodic and voice related cues. In: Interspeech, pp. 1149–1153 (2013)
4. Orozco-Arroyave, J.R., Hönig, F., Arias-Londoño, J.D., Vargas-Bonilla, J.F., Skodda, S., Rusz, J., Nöth, E.: Voiced/unvoiced transitions in speech as a potential bio-marker to detect Parkinson's disease. In: Interspeech, pp. 95–99 (2015)
5. Orozco-Arroyave, J.R., Belalcázar-Bolaños, E.A., Arias-Londoño, J.D., Vargas-Bonilla, J.F., Haderlein, T., Nöth, E.: Phonation and articulation analysis of Spanish vowels for automatic detection of Parkinson's disease. In: Sojka, P., Horák, A., Kopeček, I., Pala, K. (eds.) TSD 2014. LNCS, vol. 8655, pp. 374–381. Springer, Cham (2014). doi:10.1007/978-3-319-10816-2_45
6. Bayestehtashk, A., Asgari, M., Shafran, I., McNames, J.: Fully automated assessment of the severity of Parkinson's disease from speech. Comput. Speech Lang. **29**(1), 172–185 (2015)
7. Arias-Vergara, T., Vasquez-Correa, J., Orozco-Arroyave, J.R., Vargas-Bonilla, J.F., Nöth, E.: Parkinson's disease progression assessment from speech using GMM-UBM. In: Interspeech, pp. 1933–1937 (2016)
8. Orozco-Arroyave, J.R., Vasquez-Correa, J., Hönig, F., Arias-Londoño, J.D., Vargas-Bonilla, J.F., Skodda, S., Rusz, J., Noth, E.: Towards an automatic monitoring of the neurological state of Parkinson's patients from speech. In: 2016 IEEE International Conference on Acoustics, Speech and Signal Processing (ICASSP), pp. 6490–6494. IEEE (2016)

9. Eyben, F., Wöllmer, M., Schuller, B.: Opensmile: the Munich versatile and fast open-source audio feature extractor. In: Proceedings of the 18th ACM International Conference on Multimedia, MM 2010, pp. 1459–1462. ACM, New York (2010)
10. Schuller, B., Steidl, S., Batliner, A., Burkhardt, F., Devillers, L., Müller, C., Narayanan, S.: The INTERSPEECH 2010 paralinguistic challenge. In: Interspeech (2010)
11. Proença, J., Veiga, A., Candeias, S., Perdigão, F.: Acoustic, phonetic and prosodic features of Parkinson's disease speech. In: STIL-IX Brazilian Symposium in Information and Human Language Technology, 2nd Brazilian Conference on Intelligent Systems, Brazil (2013)
12. Pinto, S., Cardoso, R., Sadat, J., Guimarães, I., Mercier, C., Santos, H., Atkinson-Clement, C., Carvalho, J., Welby, P., Oliveira, P., D'Imperio, M., Frota, S., Letanneux, A., Vigario, M., Cruz, M., Martins, I.P., Viallet, F., Ferreira, J.J.: Dysarthria in individuals with Parkinson's disease: a protocol for a binational, cross-sectional, case-controlled study in French and European Portuguese (FraLusoPark). BMJ Open 6(11), e12885 (2016)
13. Hall, M., Frank, E., Holmes, G., Pfahringer, B., Reutemann, P., Witten, I.H.: The WEKA data mining software: an update. SIGKDD Explor. Newsl. 11(1), 10–18 (2009)
14. Hanson, D.G., Gerratt, B.R., Ward, P.H.: Cinegraphic observations of laryngeal function in Parkinson's disease. Laryngoscope 94(3), 348–353 (1984)
15. Perez, K.S., Ramig, L.O., Smith, M.E., Dromey, C.: The Parkinson larynx: tremor and videostroboscopic findings. J. Voice 10(4), 354–361 (1996)
16. Skodda, S., Visser, W., Schlegel, U.: Vowel articulation in Parkinson's disease. J. Voice 25(4), 467–472 (2011)
17. Rusz, J., Cmejla, R., Ruzickova, H., Ruzicka, E.: Quantitative acoustic measurements for characterization of speech and voice disorders in early untreated Parkinson's disease. J. Acoust. Soc. Am. 129(1), 350–367 (2011)
18. Vásquez-Correa, J., Orozco-Arroyave, J.R., Arias-Londoño, J.D., Vargas-Bonilla, J.F., Nöth, E.: Design and implementation of an embedded system for real time analysis of speech from people with Parkinson's disease. In: Symposium of Signals, Images and Artificial Vision - 2013, STSIVA - 2013, pp. 1–5, September 2013
19. Skodda, S., Schlegel, U.: Speech rate and rhythm in Parkinson's disease. Mov. Disord. 23(7), 985–992 (2008)
20. Eyben, F., Scherer, K.R., Schuller, B.W., Sundberg, J., Andr, E., Busso, C., Devillers, L.Y., Epps, J., Laukka, P., Narayanan, S.S., Truong, K.P.: The Geneva Minimalistic Acoustic Parameter Set (GeMAPS) for voice research and affective computing. IEEE Trans. Affect. Comput. 7(2), 190–202 (2016)

Unified Simplified Grapheme Acoustic Modeling for Medieval Latin LVCSR

Lili Szabó[1]([✉]), Péter Mihajlik[2,3], András Balog[2], and Tibor Fegyó[1,3]

[1] SpeechTex Ltd., Budapest, Hungary
{lili,tfegyo}@speechtex.com
[2] THINKTech Research Center, Budapest, Hungary
[3] Budapest University of Technology and Economics, Budapest, Hungary
{mihajlik,abalog}@thinktech.hu
http://www.speechtex.com

Abstract. A large vocabulary continuous speech recognition (LVCSR) system designed for dictation of medieval Latin language documents is introduced. Such language technology tool can be of great help for preserving Latin language charters from this era, as optical character recognition systems are often challenged by these historical materials. As corresponding historical research focuses on the Visegrad region, our primary aim is to make medieval Latin dictation available for texts and speakers of this region, concentrating on Czech, Hungarian and Polish. The baseline acoustic models we start with are monolingual grapheme-based ones. On one hand, the application of medieval Latin knowledge-based grapheme-to-phoneme (G2P) mapping from the source language to the target language resulted in significant improvement, reducing the Word Error Rate (WER) by 13.3%. On the other hand, applying a Unified Simplified Grapheme (USG) inventory set for the three-language acoustic data set complemented with Romanian speech data, resulted in a further 0.7% WER reduction - without using any target or source language G2P rules.

Keywords: G2P · Medieval Latin · Under-resourced speech recognition · Unified simplified grapheme modeling

1 Introduction

The pronunciation of Latin texts mainly depends on the era and region of their origin [3]. Apart from the two widely studied classical and ecclesiastical pronunciation styles [1], many other regional pronunciations exist that emerged after the classical era. One of these pronunciation groups is the East-Central European [3] one, described in detail in Sect. 3.2. Although the target pronunciation is considered to be uniform for this group, it also has to be taken into account, that the acoustic base of the different source languages varies, which can lead to various accents. It also has to be noted, that apart from the variations in the pronunciations, orthographic and grammatical variations of Latin are also exhibited through regions.

© Springer International Publishing AG 2017
K. Ekštein and V. Matoušek (Eds.): TSD 2017, LNAI 10415, pp. 420–428, 2017.
DOI: 10.1007/978-3-319-64206-2_47

This raises the question of how to create a speech recognition system which has to deal with pronunciation variations of native speakers of different languages reading linguistically different texts. We propose a system built for the recognition of medieval Latin speech spoken by speakers from the Visegrad region. It is also important to collect in-domain textual/language data for the language model from the relevant geographican regions and time. We describe the data acquisition process in Sect. 2.1.

Our baseline system consists of separately trained grapheme-based acoustic models for three of the Visegrad languages (Czech, Hungarian, Polish) complemented with the Romance language Romanian. We apply two different acoustic/pronunciation modeling techniques to develop models that are superior to the baseline. The first one, discussed in detail in Sect. 3.2, is a knowledge-based pronunciation modeling technique, where the source language phonemes are mapped to the target language phonemes. The second method applied is a Unified Simplified Grapheme (USG) acoustic modeling approach, where a joint grapheme inventory is established for all the languages participating in the joint acoustic model training, described in Sect. 3.3. The evaluation of all systems is presented in Sect. 4.

Related Work. Different adaptation techniques have been proposed in [5] and [2] to train acoustic models from multiple source languages for a single target language where training data was limited. Similar work has been done for multi-dialectal languages such as Arabic in [12] where jointly trained acoustic models were outperformed by methods that unify dialect specific-acoustic models using knowledge distillation and multitask learning. However, no approach is known for the authors where the graphemes of multiple languages are merged successfully and applied for acoustic modeling of a different language. To our knowledge, no previous work has been done on medieval Latin speech recognition, nor on classical Latin for that matter.

2 Data

2.1 Textual Data

As part of our inquiry was to cover linguistic variability across the Visegrad region, aquiring textual data posed a few challenges. First of all, textual data are scarce for medieval Latin, and texts originating from this geographical region are even more difficult to obtain in electronic format. Additionally, most of the available sources mix local languages and Latin, with no metadata to separate them. For the scope of this paper, we collected monolingual (Latin) texts only.

Training Data. A smaller amount of in-domain data (medieval charters) were collected from [10] (Monasterium), with an overall of 480k tokens. These documents are originating from the Hungarian Kingdom, from 1000 to 1524 AD. To increase the vocabulary size of the language model, we collected a relatively larger (but still small, compared to state-of-the-art language models used in speech recognition) 1.3M-token corpus from [11] (LatinLibrary). This corpus consists of literary and historical texts from the post-classical era. In spite of our efforts, at the time of writing this paper, we could not gather a measurable amount of textual data from the age and area of the Kingdoms of Bohemia and Poland.

Test Data. Using independent sources, three charters were selected from the Kingdoms of Bohemia (CZ), Hungary (HU) and Poland (PL), from around 1200–1300 AD, as test data for evaluating the language model, and to test the performance of the LVCSR approaches. The test sets were read out loud by historians fluent in medieval Latin.

Alternate Spellings. One interesting feature of the acquired corpora was that they contained a significant number of spelling variants. Having spelling variants in the corpus with identical pronunciation introduces noise, and thus has a negative effect on recognition results. We merged the spelling for the variants by favouring the more frequent variant in the corpus (e.g. *maiestati* to *majestati*). Resolving spelling variants resulted in a more consistent corpus in terms of perplexity (reducing it from 775 to 672), and reduced the OOV rate by 0.8%.

Language Model. The word trigram language models we built from the two corpora were estimated with the SRI Language Modeling toolkit (SRILM) [6] using modified Kneser-Ney smoothing method. After estimating the mixture parameter, linear interpolation was used to merge the two language models.

The perplexity measures on the test data showed that the Monasterium corpus originating from the time and era of the Hungarian Kingdom was indeed best fitting with the Hungarian subset of the test data with a perplexity of 82, and an OOV rate of 0.9%. The perplexities measured on the Czech and Polish origin text sets were ranging from 500 to 3200. Adding the LatinLibrary corpus increased the perplexity significantly (up to 672), but reduced the OOV rate by 7% on the overall test data, as well as the WER, so we decided to use the interpolated language model.

2.2   Speech Data

Training Data. For Czech, the read part of Speecon database [9] was used, recorded with medium distance microphone, 76 h in sum. The 567-hour database for Hungarian consisted of the Speecon [8] database, manually transcribed broadcast news (112 h) and conversational speech data. With the exception of the Hungarian knowledge-based model (described in Sect. 3.2), the 112-hour broadcast news set was used for training. For Polish, only broadcast news data [7] was available, comprising 31 h of manually transcribed speech. The Romanian speech database used for the experiments was originally collected for [7] consisting of 35 h of broadcast news.

Test Data. Native speakers of Czech, Hungarian, Polish and Slovak - all of whom have experince with medieval Latin - were asked to record the three test sets described in Sect. 2.1. The recording conditions were accurately controlled: close-talking microphones, quiet, non reverberant acoustic environment, fluent, flawless speech, and at least 16 kHz, 16 bit (linear PCM) encoding. No instructions were given regarding the pronunciation, the speakers were using their expertise on medieval Latin pronunciation - affected certainly by their native language. The overall length of the recorded test speech was around 30 min. The recordings can be found at the project webpage.[1]

3   Acoustic Modeling

Building an acoustic model for speech recognition requires long hours of transcribed speech. As of today (medieval) Latin is not spoken natively, and as to our knowledge, there is no recorded speech database. One obvious way to handle this problem is by creating a medieval Latin database; a proposition that requires lot of time, resources and trained speakers of medieval Latin. Another way of circumventing the lack of available speech data is to use speech data of spoken languages, preferably those ones whose native speakers are going to use the system.

For all the different pronunciation modeling methods, the acoustic models were trained as follows. Mel-Frequency Cepstrum+Energy features were used with Linear Discriminant Analysis (LDA)+Maximum Likelihood Linear Transformation (MLLT), with a splice context of ± 4 frames, 10 ms of frame shift. 9×40 dimensional spliced up feature vectors served as input to the feed-forward, 6 hidden-layer neural network with p-norm [4] activation function. Prior to DNN training, a Gaussian Mixture Model (GMM) pre-training was performed. Clustering and Regression Tree (CART) [4] was applied to obtain acrossword context dependent shared state phone (or graph) models and their time alignment. The number of senones (and so the size of the DNN softmax output layer) was between 7.000 and 11.000 depending on the nature of the training data. The size of the hidden layers was kept constantly on 2.000. A minibatch size of 512, an initial learning rate of 0.1, and final learning rate of 0.01 was applied in 20 epochs using the Kaldi toolkit [4].

3.1   Grapheme-Based Pronunciation Modeling

For our three separately trained baseline systems, grapheme-based acoustic models were used where pronunciation is modeled using graphemes as subword units. The language-specific graphemes (e.g. ö, ń) that are not part of the Latin alphabet were trained, but not used in the recognition phase.

---

[1] http://medilatin.speechtex.com.

3.2   Source-Target Grapheme to Phoneme Mapping (G2P)

This method utilizes already trained acoustic models where the source language phonemes are mapped to the target language phonemes using expert knowledge. The source language acoustic models are trained with G2P mapping from orthographic transcriptions to native phonemes. In our experiments we used Czech and Hungarian as (separately trained) source languages for the target language Latin. After mapping source language phonemes to Latin phonemes, Latin-specific pronunciation rules were implemented. These include a set of context independent digraph mappings and context dependent rewrite rules, summarized in Tables 1 and 2 respectively, for both Czech and Hungarian. Both languages fully cover the phoneme inventory of medieval Latin which is of size 24.

Table 1. Latin digraph context-insensitive rewrite rules.

	Digraph			
	ae	oe	ph	qu
CZ	e	oe	f	kv
HU	e	ø	f	kv

Table 2. Latin context-sensitive rewrite rules. V: vowel, VP: palatal vowel, ˆVP: everything but a palatal vowel, C: consonant, ∗: zero or any, ˆ: beginning of word, [ˆstx]: not s, t or x.

GR	c	c	ch	ch	gu	gu	ti	ti
PH	ts	k	h	k	gv	gu	tsi	ti
Rule	cVP	cˆVP	VC*ch	ˆC*ch	guV	guC	[ˆstx]tiV	tiC

3.3   Unified Simplified Grapheme Acoustic Modeling

The second method used for improving speech recognition of medieval Latin - this time in a fully data driven way - was the Unified Simplified Grapheme (USG) acoustic modeling technique. Our motivation with using this technique was three-fold:

1. Develop a target language acoustic model using available language resources.
2. Support recognition of medieval Latin spoken by speakers of diverse native language background.
3. As the writing systems in the Visegrad region are originating from medieval Latin, we were aiming to validate the intuition that by unifying and simplifying the native graphemes, the deviations from the common ancestor may cancel out.

We experimented with joint three- and four-language USG acoustic models of any combination of the four languages (Czech, Hungarian, Polish and Romanian). The joint acoustic model requires a unified grapheme inventory for the training. Our proposal was to simplify all special characters, i.e. those graphemes that had a diacritic mark (acute, caron, etc.) on them, were mapped back to their normalized form. Table 3 contains examples of the unification/simplification process for all four languages. For the four languages an overall of 32 of such unifications/simplifications were made, reducing the unified grapheme inventory set from 58 to 26. Further than that, those graphemes that are non-native to Latin, and can straightforwardly mapped to a native Latin grapheme(s), were also replaced. These included mappings from x to ks, y to i and w to v. As a result, a unified and simplified grapheme inventory set was produced, formally compatible with medieval Latin. The USG units were then used as acoustic model units in the multiple language training.

Table 3. Simplification examples for the unified model.

Language	CZ	HU	PL	RO
Orthographic form	řekl	őz	miś	apă
USG transcription	rekl	oz	mis	apa

4   Experimental Results

We conducted experiments on medieval Latin, spoken by native speakers of four languages (Czech, Hungarian, Polish and Slovak), where the test texts were originating from different regions, as described in Sects. 2.1 and 3. The best performing monolingual grapheme-based model results were that of Hungarian, with 34.6% overall WER (see in Table 4), possibly because of the larger training data - this was the reference value when comparing the results. On a related note, we also found that only the Hungarian monolingual grapheme-based acoustic model had the best performance over its own test sets.

Table 4. Word Error Rate (WER[%]) results for monolingual grapheme-based acoustic models of Czech, Hungarian, Polish and Romanian (CZ, HU, PL, RO).

AM language	Speaker				
	CZ	HU	PL	SK	$\sum$
CZ	53.6	73.8	62.9	45.7	59.0
HU	33.7	28.6	47.1	29.1	**34.6**
PL	65.0	67.6	46.4	51.1	57.5
RO	53.6	69.1	44.7	43.8	52.8

Table 5. WER[%] for Czech-Latin source-target G2P model. Acoustic model training set: 76 h.

	Latin test text			
Speaker	CZ	HU	PL	$\sum$
CZ	43.8	28.2	49.1	40.4
HU	48.7	40.0	58.7	49.1
PL	53.3	18.2	53.2	41.6
SK	30.3	30.0	44.0	34.8
$\sum$	43.9	28.9	50.8	41.2

Table 6. WER[%] for Hungarian-Latin source-target G2P model. Acoustic model training set: 567 h.

	Latin test text			
Speaker	CZ	HU	PL	$\sum$
CZ	19.4	**6.4**	28.0	17.9
HU	25.0	25.4	20.2	23.5
PL	28.9	15.4	41.3	28.5
SK	20.4	**9.1**	22.9	17.5
$\sum$	22.6	12.5	28.1	**21.1**

4.1 Source-Target G2P Mapping Results

The results on the experiments with the knowledge-based pronunciation modeling technique, where the native phonemes of the source phoneme-based acoustic models were mapped to the target phonemes in the pronunciation dictionary, are in Table 5 for the source language Czech, and in Table 6 for the source language Hungarian. The Hungarian knowledge-based acoustic model - possibly due to the larger data set - outperforms the (Hungarian grapheme-based) baseline significantly, with an 21.1% overall WER. It is worth mentioning that the Czech and Slovak speaker test sets achieve a surprisingly low 6.4% and 9.1% WER respectively on the Hungarian text test set.

4.2 USG Results

The results for the three-language joint acoustic models are in Table 7. Among the three-language USG models, the Czech-Hungarian-Romanian model had the best performance with a competitive overall 21.9% WER. Complemented with Polish, we got the best experimental results of 20.4% with the four-language USG model (see in Table 8). We also measured the WER on any combination of three of the four languages, and found that each language contributed to the four-language model.

It is worth mentioning, that compared to the knowledge-based Hungarian model (Table 6), the results on the Polish speaker test set improved by an absolute 6.5%. This could be due to the ability of the four-language model to generalize better over different speaker test tests. This generalizing ability intensifies when adding training data of a new language, as the models of similar graphemes are merged, and work better on different native language speaker test sets.

When comparing Tables 6, 7 and 8, it is somewhat surprising that the Hungarian speaker test set had an absolute 10.8% less WER on the Hungarian text test set using the knowledge-based G2P model. We had expected the Hungarian speakers to perform better with the Hungarian knowledge-based model and

Table 7. WER[%] for all the three-language USG models.

	Speaker				
AM language	CZ	HU	PL	SK	$\sum$
CZ+HU+PL	28.2	28.2	27.7	22.4	26.6
CZ+HU+RO	23.3	21.4	23.9	19.2	**21.9**
CZ+PL+RO	24.6	33.1	25.6	19.8	25.8
HU+PL+RO	24.8	21.5	25.7	20.7	23.2

Table 8. WER[%] for USG model of Czech, Hungarian, Polish and Romanian (CZ+HU+PL+RO).

	Latin Test Text			
Speaker	CZ	HU	PL	$\sum$
CZ	20.4	11.8	30.7	21.0
HU	21.1	14.6	25.7	20.5
PL	23.0	**10.0**	33.0	22.0
SK	14.5	12.7	24.8	17.3
$\sum$	19.9	12.2	29.0	**20.4**

Hungarian text test set setting, but in fact the phoneme mapping masked the difference between mid-front /e:/ and open-front /ɛ/ in the pronunciation of the Hungarian speakers. In addition to that, they were pronouncing the named entities using their native pronunciation, which also increased the WER in case of the knowledge-based approach. Similarly, we see an absolute 3% WER improvement on the whole Hungarian speaker test set whith the four-language USG model when comparing it to the Hungarian G2P results.

Finally, the results show that the experiments conducted on the Hungarian origin text test set yielded to the best results with all models. This is due to the fact that the in-domain part of the language model training data was originating from the Hungarian language region, see Sect. 2.1.

5 Conclusions

In this paper, we introduced two acoustic modeling techniques for a target language independent medieval Latin speech recognizer to elevate the efforts of digitizing medieval Latin charter data. Our goal was to build an acoustic model for medieval Latin, borrowing speech data from different source languages (Czech, Hungarian, Polish and Romanian). Our test set consisted of medieval Latin charters originating from different regions read by native speakers of the above languages. With the objective of building an acoustic model without source language speech data, we presented two approaches: knowledge-based G2P modeling, and USG modeling.

The results showed that both methods outperform by far the best baseline system. We found that the best model was the four-language USG model. When comparing it to the knowledge-based Hungarian phoneme-based model, which was using expert knowledge to map words to phoneme sequences, and trained on larger amount of data, it seemed that the four-language USG model was better in evening out the inconsistencies of the pronunciations in different speaker test sets.

Future research directions include acquiring a considerable amount of medieval speech and textual data, as well as implementing a more refined G2P modeling using a unified phoneme inventory set. Furthermore, adding more data when using the USG approach may result in even higher recognition accuracy, allowing dictational applications.

References

1. Allen, W.S.: Vox Latina: A Guide to the Pronunciation of Classical Latin. Cambridge University Press, Cambridge (1978). [Eng.], 2nd edn., New York
2. Besacier, L., Barnard, E., Karpov, A., Schultz, T.: Automatic speech recognition for under-resourced languages: a survey. Speech Commun. **56**, 85–100 (2014)
3. Encyclopedia of Caribbean Literature, Latin Regional Pronunciation (2007)
4. Povey, D., Ghoshal, A., Boulianne, G., Burget, L., Glembek, O., Goel, N., Hannemann, M., Motlicek, P., Qian, Y., Schwarz, P., Silovsky, J., Stemmer, G., Vesely, K.: The Kaldi speech recognition toolkit. In: IEEE 2011 Workshop on Automatic Speech Recognition and Understanding. IEEE Signal Processing Society (2011)
5. Schultz, T., Waibel, A.: Language-independent and language-adaptive acoustic modeling for speech recognition. Speech Commun. **31**, 31–51 (2001)
6. Stolcke, A.: SRILM - an extensible language modeling toolkit. In: Proceedings of the 7th International Conference on Spoken Language Processing (ICSLP), pp. 901–904 (2002)
7. Tarjan, B., Mozsolics, T., Balog, A., Halmos, D., Fegyo, T., Mihajlik, P.: Broadcast news transcription in Central-East European languages. In: 3rd IEEE International Conference on Cognitive Infocommunications, pp. 59–64 (2012)
8. Hungarian speecon database (2003). http://catalog.elra.info/product_info.php?products_id=1093
9. Czech speecon database (2004). http://catalog.elra.info/product_info.php?products_id=1095
10. Monasterium.net archive. http://monasterium.net/mom/HU-PBFL/archive
11. Latin library archive. http://www.thelatinlibrary.com/medieval.html
12. Waters, A., Bastani, M., Elfeky, M.G., Moreno, P., Velez, X.: Towards acoustic model unification across dialects. In: 2016 IEEE Workshop on Spoken Language Technology (2016)

Experiments with Segmentation in an Online Speaker Diarization System

Marie Kunešová[1,2]([✉]), Zbyněk Zajíc[1], and Vlasta Radová[1,2]

[1] NTIS - New Technologies for the Information Society, Faculty of Applied Sciences, University of West Bohemia, Univerzitní 8, 306 14 Plzeň, Czech Republic
{mkunes,radova}@kky.zcu.cz, zzajic@ntis.zcu.cz
[2] Department of Cybernetics, Faculty of Applied Sciences, University of West Bohemia, Univerzitní 8, 306 14 Plzeň, Czech Republic

Abstract. In offline speaker diarization systems, particularly those aimed at telephone speech, the accuracy of the initial segmentation of a conversation is often a secondary concern. Imprecise segment boundaries are typically corrected during resegmentation, which is performed as the final step of the diarization process. However, such resegmentation is generally not possible in online systems, where past decisions are usually unchangeable. In such situations, correct segmentation becomes critical. In this paper, we evaluate several different segmentation approaches in the context of online diarization by comparing the overall performance of an i-vector-based diarization system set to operate in a sequential manner.

Keywords: Speaker diarization · Speaker change detection · i-vectors · Convolutional neural network

1 Introduction

Speaker diarization is a speech processing task which aims at categorizing different speech sources in a conversation of two or more speakers, such that utterances produced by the same speaker are assigned the same label. In other words, we are trying to determine "Who speaks when?", typically without any prior knowledge about the number and identities of the speakers.

Speaker diarization systems can be divided into two main categories: *offline* and *online*. Offline systems process a given audio recording as a whole, requiring that the entirety of the data is available at the beginning of the process. This allows these systems to use all available information for their decisions. *Online* systems, by contrast, operate in a strict left-to-right manner and can process an incoming audio stream in real-time. The decisions made by these systems can be based only on previously seen data, independent of future information, and once made, cannot be changed.

The most common diarization approach, used by both offline and online systems, consists of two main steps: segmentation and clustering. The input signal

© Springer International Publishing AG 2017
K. Ekštein and V. Matoušek (Eds.): TSD 2017, LNAI 10415, pp. 429–437, 2017.
DOI: 10.1007/978-3-319-64206-2_48

is split into short intervals and these are then merged into clusters corresponding to the individual speakers. Common algorithms include clustering based on the Bayesian information criterion (BIC) [11] or on distances between i-vectors [12]. In the case of offline systems, there is often an additional resegmentation step, which refines the original segment boundaries.

An alternative diarization approach combines segmentation and clustering into a single iterative process, often with the use of Hidden Markov Models (HMMs) or related concepts [1,9]. However, this approach is not typically used in online diarization.

There are many possible ways to perform segmentation. Ideally, we want to have segments which only contain a single speaker. This is best achieved with speaker change detection (SCD) - identifying possible speaker boundaries and then splitting the conversation there. Common approaches include the Bayesian Information Criterion (BIC), Generalized Likelihood Ratio (GLR) [11], Support Vector Machines (SVM) [5] or Deep Neural Networks (DNNs) [7,15].

However, the SCD approach is problematic in spontaneous telephone conversations. These typically contain very short speaker turns and frequent overlapping speech, which makes it difficult to correctly detect speaker turns. For this reason, most authors in the telephone domain (e.g. [6,12,13]) choose to simply cut the conversation into very short intervals of fixed length. It is assumed that any inaccuracies can be resolved during the later stages of the diarization process, typically by performing a final resegmentation step. This was also confirmed in our recent paper [16], which compared the fixed length approach with SCD-based segmentation using GLR distance. There, we showed that even though the SCD approach led to better initial clusters, the final results of both options after resegmentation were comparable.

Unfortunately, no such resegmentation is possible in online systems. This means that a proper initial segmentation again becomes important.

This paper is in part inspired by the recent work of Zhu and Pelecanos [19], who have proposed an incremental adaptation process for online i-vector based speaker diarization. Their original paper focuses mainly on the clustering step and sidesteps the question of segmentation by utilizing *oracle* segmentation based on reference transcripts (although this has very recently been extended with ASR-based segmentation [3]). In our work, we follow up on their results by implementing the suggested online approach in our own diarization system, while using a different segmentation.

The main goal of this paper is then to compare multiple different segmentation options in the context of online speaker diarization. For this purpose, we use the aforementioned i-vector-based system and evaluate its performance on telephone data from the CALLHOME American English corpus [2]. As most telephone conversations involve only two individuals, our system explicitly assumes the presence of only two speakers and we limit our experiments to the two-speaker subset of the corpus.

2 Offline Diarization System

For our online diarization experiments, we have re-purposed an originally *offline* state-of-the-art diarization system which is based on i-vectors. In this section, we describe this base offline system, while the subsequent adjustments to a more sequential approach will be presented in Sect. 3.

The basic structure of the system is based on the i-vector approach which has recently become standard in speaker diarization [12,19]. The specific implementation largely follows the descriptions presented in our previous papers [8,16] and a diagram of the main steps can be seen in Fig. 1. The diarization process starts with the extraction of acoustic features from the conversation, followed by its splitting into short segments. This segmentation step can use one of multiple possible approaches, some of which will be explored in Sect. 4.

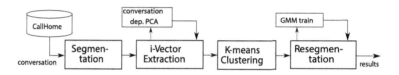

Fig. 1. Diagram of the offline diarization system.

As the next step of the process, we obtain a simplified representation of the individual segments. For each segment of the conversation, we first accumulate a supervector of statistics [17], from which we subsequently extract an i-vector via Factor Analysis [4]. The size of the i-vectors is further reduced with the aid of a conversation-dependent Principal Component Analysis (PCA) transformation [14].

Following this, the i-vectors are clustered in order to determine which parts of the conversation were produced by the same speaker. As we limit our data to conversations between only two speakers, we can use a simple k-means algorithm based on cosine distance between i-vectors [14].

Finally, we perform a frame-wise iterative resegmentation based on Gaussian Mixture Models (GMMs) trained on the data from each cluster. This serves to refine the speaker boundaries and correct mistakes caused by imprecise segmentation.

3 Online System

As our main goal was simply to investigate the sequential segmentation and clustering process, without the need for actual real-time output, we decided against implementing a complete, fully online diarization system. Rather, we have simply adjusted the original offline process which was described in Sect. 2 so that each of the steps separately operates in a left-to-right manner. As such, the initial steps of both systems are identical. However, the original k-means clustering is replaced by a sequential algorithm, while both the conversation-dependent PCA reduction of i-vectors and the final resegmentation step, which are not possible to perform online, are removed entirely.

As the clustering step, we employ the i-vector adaptation process proposed by Zhu and Pelecanos [19], which is given by

$$T_n = \alpha V_n V_n^T + (1 - \alpha_n)I \,, \qquad \alpha_n = \frac{n}{n + R} \,, \tag{1}$$

where n is the number of i-vectors which have been processed so far, V_n is the first principal component of the i-vectors, T_n is an i-vector transformation matrix and R is the relevance factor which controls the rate of the adaptation.

The resulting sequential clustering then works as follows: For each new i-vector (which corresponds to a new segment), we first update the transformation matrix T_n using the formula in (1) and use it to transform all i-vectors seen up to this point. Then we calculate the cosine distance between the new transformed i-vector and all existing clusters, where the distance to a cluster is calculated as the average of the distances to all of its i-vectors. If the distance to the closest cluster is lower than a threshold (we will designate this threshold as θ) or the maximum number of clusters is reached (in our case, this number is 2), the new i-vector is assigned to this cluster. Otherwise, a new cluster is created.

Because all decisions made by the system are final and unchangeable, an incorrect decision at an early point in a recording can significantly impact the rest of the clustering process. In this regard, extremely short segments, particularly those under 0.5 s are the most problematic, as they typically do not contain sufficient information about the speaker in order to be correctly clustered.

As some of the segmentation approaches which we compared may produce such short segments, it was necessary to slightly adjust the clustering algorithm in order to avoid this issue. We achieve this by excluding any segments under 1 s in length from the regular clustering process. Instead, the corresponding i-vectors (which we do not consider to be representative of any speakers) are simply labeled as the nearest existing cluster (they are never used to create a new one), but they are not included in the calculation of T_n in (1) and we also do not consider them in later distance calculations.

4   Segmentation

In this paper, we compare several different segmentation approaches. For these experiments, we chose to use the segmentation algorithms which were previously described in our two recent papers [8,16] in the context of *offline* speaker diarization. All of these segmentation approaches assume the possibility of their use in online diarization, i.e. they operate sequentially or can be relatively easily adjusted in such manner.

Some of the described approaches rely on information about the presence of silence and speech which would under real conditions be provided by a voice activity detector (VAD). However, in order to avoid any specific VAD method from influencing the results of the segmentation, we chose to use *oracle* VAD obtained from the reference transcripts and we discuss the possible dependence on VAD in the description of each segmentation method.

When performing segmentation, it is also important to consider the length of the resulting segments. In particular, we need to have a sufficient amount of information in order to be able to extract an i-vector from each segment. Typically, a segment length of at least 1–2 s is considered to be the minimum in order to obtain i-vectors which are representative of the speakers.

4.1   Fixed Length Segments

The simplest segmentation option is to split the input stream into short intervals of equal length, without considering any potential speaker boundaries. In our system, we follow the example of [12] by using overlapping segments. This allows us to increase the amount of information contained in a single i-vector while retaining a higher precision of the segmentation. Specifically, we chose to use segment length of 2 s with a 1 s overlap between neighboring segments.

4.2   GLR-Based Speaker Change Detection

As the first speaker change detection approach, we used the Generalized Likelihood Ratio (GLR)-based algorithm described in our previous paper [16]. This is a two-pass algorithm, which means that it is not suitable for true online diarization in its current form. However, we believe that it should be possible to implement a relatively similar algorithm in a strictly left-to-right form.

In the original two pass approach, which we used here, the algorithm first calculates the GLR distance between two sliding windows over the entirety of the conversation. A smaller number of the most likely speaker change points are found as the local maxima whose topographical prominence exceeds a set threshold. During the second pass, segments above a specific length are split according to the algorithm suggested in [16].

Finally, any segments which contain only a small percentage of speech frames (as determined by VAD), are labeled as silence and subsequently discarded. This means that the performance of this approach depends on VAD implementation and also gives it an advantage over the other approaches when reference VAD is used (as was done in our experiments).

4.3   CNN-Based Speaker Change Detection

The second SCD-based segmentation method which we considered uses a Convolutional Neural Network (CNN) as a regressor. We employ a CNN which was trained on spectrograms of acoustic signal using the method described in [8]. The reference training labels were in the form of a fuzzy labeling function L. Figure 2 depicts an example of a spectrogram, the values of the labeling L and the CNN output as a probability of speaker change P. Speaker changes are then identified

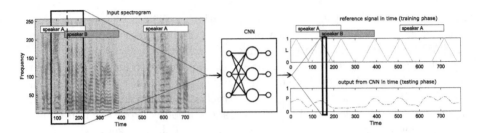

Fig. 2. The input speech as spectrogram is processed by the CNN into the output function P (a probability of change in time). The L-function (the reference speaker change) for the CNN training is depicted on top.

as peaks in the signal P using non-maximum suppression with a window size of 5 samples (0.5 s). We also apply a threshold of 0.5 on the detected peaks in order to remove insignificant local maxima. The signal between two detected speaker changes is considered as one segment.

Additionally, we also utilize the information about the change P from the CNN for weighting of the acoustic data in a segment, in order to refine the statistics accumulation process used for the subsequent i-vector generation [18].

In the offline version of this segmentation approach, we discard any segments under 1 s in length, as they are considered unreliable. They are only processed later during resegmentation. However, in the online variant, which does not have resegmentation, we keep all segments, regardless of length.

Processing the spectrogram window using a CNN takes only a very short time, which makes this approach suitable for online diarization. It should also be noted that the network is trained to detect all types of speaker boundaries and as such, does not need any information from a voice activity detector.

4.4   Oracle Segmentation

For comparison purposes, we also implemented oracle segmentation. In this approach, the conversations are split according to the reference transcripts: each individual record from the transcript becomes a single segment. As many of these segments are very short (often under 1 s), we adjust them slightly by joining any two segments from the same speaker which are separated by a silence of less than 0.5 s (this does not, however, eliminate all short segments). Otherwise, the segments are kept exactly as recorded in the transcripts, including any partial overlaps.

5   Results

For the evaluation of our system, we used the CALLHOME American English corpus of telephone speech [2], with both channels mixed into a single one. As 35 of the recorded conversations had been used for training the CNN which we use for one of the segmentation approaches, we limited our experiments to the remaining 77 conversations with only two participants.

The results are evaluated using the Diarization Error Rate (DER), as defined by NIST [10], with the customary tolerance collar of 0.25 s around speaker boundaries. Contrary to a common practice in telephone speech diarization, we do not ignore overlapping segments during the evaluation. However, our listed error rates only include two of the three components of DER: missed speech (speech incorrectly labeled as silence) and speaker error (speech labeled as the wrong speaker). False alarm (silence incorrectly labeled as speech) is removed before evaluation with the help of the reference transcripts.

In Table 1, we present the results achieved with the four segmentation methods for a fixed decision threshold $\theta = 0.6$ and different values of the relevance factor R, which controls the rate of the adaptation (see Sect. 3). This includes $R = \infty$, which is equal to not using adaptation. We may notice that the adaptation process proposed in [19] can improve the final DER in all four cases, but the individual segmentation approaches have different optimal values of R.

For comparison, we also show the results of the offline system (adapted from our previous works [8, 18]).

Table 1. Offline and online diarization results for different segmentation approaches, measured in terms of DER [%]. R is the relevance factor of the i-vector adaptation, with the value of ∞ being equal to no adaptation. Decision threshold for the online approach was $\theta = 0.6$. Offline results (except oracle) were adapted from [8, 18].

	Offline	Online								
R	–	∞	8192	...	1024	512	256	128	64	32
Fixed length	9.23	18.62	**18.47**	...	18.88	19.34	19.80	20.43	–	–
GLR	11.98	15.04	–		14.29	14.15	14.12	**13.74**	14.23	14.29
CNN	7.84	15.16	–		**14.77**	14.95	14.91	15.93	–	–
oracle	6.80	10.98	–		–	9.60	9.58	**9.30**	9.78	10.71

The table shows that in the offline scenario, the naïve fixed length segmentation produces reasonable results (likely due to resegmentation [16]), although it is surpassed by the CNN-based approach. However, in our online system, this simple option is no longer sufficient, achieving nearly double the error of the oracle option. This suggests that correct segmentation is much more important in online systems.

Of the two SCD-based approaches, the GLR-based method scored better. However, this may be influenced by its reliance on VAD (as discussed in Sect. 4.2). As our experiments used oracle VAD from reference transcripts, this gives the approach an advantage compared to the CNN-based option, which did not use any information from VAD.

6 Conclusion

In this paper, we compared several different segmentation approaches in an i-vector-based speaker diarization system operating in a left-to-right manner. We have found that the final system performance highly depends on the quality of the segmentation step. In particular, the simple naïve splitting by fixed length, which is commonly used in offline systems, does not appear to be sufficient for an online approach. Instead, more sophisticated methods are required, such as one of the other approaches which we explored here.

Acknowledgments. This research was supported by the Ministry of Culture of the Czech Republic, project No. DG16P02B009.

References

1. Bozonnet, S., Evans, N.W., Fredouille, C.: The LIA-EURECOM RT 2009 speaker diarization system: enhancements in speaker modelling and cluster purification. In: Proceedings ICASSP, pp. 4958–4961. IEEE (2010)
2. Canavan, A., Graff, D., Zipperlen, G.: CALLHOME American English speech, LDC97S42. In: LDC Catalog, Linguistic Data Consortium, Philadelphia (1997)
3. Church, K., Zhu, W., Vopicka, J., Pelecanos, J., Dimitriadis, D., Fousek, P.: Speaker diarization: a perspective on challenges and opportunities from theory to practice. In: Proceedings ICASSP, pp. 4950–4954 (2017)
4. Dehak, N., Kenny, P., Dehak, R., Dumouchel, P., Ouellet, P.: Front-end factor analysis for speaker verification. IEEE Trans. Audio Speech Lang. Process. **19**(4), 788–798 (2011)
5. Fergani, B., Davy, M., Houacine, A.: Speaker diarization using one-class support vector machines. Speech Commun. **50**(5), 355–365 (2008)
6. Garcia-Romero, D., Snyder, D., Sell, G., Povey, D., McCree, A.: Speaker diarization using deep neural network embedings. In: Proceedings ICASSP, pp. 4930–4934 (2017)
7. Gupta, V.: Speaker change point detection using deep neural nets. In: Proceedings ICASSP, pp. 4420–4424 (2015)
8. Hrúz, M., Zajíc, Z.: Convolutional neural network for speaker change detection in telephone speaker diarization system. In: Proceedings ICASSP, pp. 4945–4949 (2017)
9. Lapidot, I., Bonastre, J.F.: On the importance of efficient transition modeling for speaker diarization. In: Proceedings Interspeech, 08–12 September 2016, pp. 2190–2193 (2016)
10. NIST: The 2009 (RT-09) rich transcription meeting recognition evaluation plan (2009). http://www.itl.nist.gov/iad/mig/tests/rt/2009/docs/rt09-meeting-eval-plan-v2.pdf
11. Rouvier, M., Dupuy, G., Gay, P., Khoury, E., Merlin, T., Meignier, S.: An open-source state-of-the-art toolbox for broadcast news diarization. In: Proceedings Interspeech, pp. 1477–1481 (2013)
12. Sell, G., Garcia-Romero, D.: Speaker diarization with PLDA i-vector scoring and unsupervised calibration. In: IEEE Spoken Language Technology Workshop, pp. 413–417 (2014)

13. Senoussaoui, M., Kenny, P., Stafylakis, T., Dumouchel, P.: A study of the cosine distance-based mean shift for telephone speech diarization. IEEE/ACM Trans. Audio Speech Lang. Process. **22**(1), 217–227 (2014)
14. Shum, S., Dehak, N., Chuangsuwanich, E., Reynolds, D., Glass, J.: Exploiting intra-conversation variability for speaker diarization. In: Proceedings Interspeech, pp. 945–948 (2011)
15. Wang, R., Gu, M., Li, L., Xu, M., Zheng, T.F.: Speaker segmentation using deep speaker vectors for fast speaker change scenarios. In: Proceedings ICASSP, pp. 5420–5424 (2017)
16. Zajíc, Z., Kunešová, M., Radová, V.: Investigation of segmentation in i-vector based speaker diarization of telephone speech. In: Ronzhin, A., Potapova, R., Németh, G. (eds.) SPECOM 2016. LNCS (LNAI), vol. 9811, pp. 411–418. Springer, Cham (2016). doi:10.1007/978-3-319-43958-7_49
17. Zajíc, Z., Machlica, L., Müller, L.: Initialization of fMLLR with sufficient statistics from similar speakers. In: Habernal, I., Matoušek, V. (eds.) TSD 2011. LNCS (LNAI), vol. 6836, pp. 187–194. Springer, Heidelberg (2011). doi:10.1007/978-3-642-23538-2_24
18. Zajíc, Z., Ilrúz, M., Müller, L.: Speaker diarization using convolutional neural network for statistics accumulation refinement. In: Proceedings Interspeech (2017, in press)
19. Zhu, W., Pelecanos, J.: Online speaker diarization using adapted i-vector transforms. In: Proceedings ICASSP, pp. 5045–5049. IEEE (2016)

Spatiotemporal Convolutional Features for Lipreading

Karel Paleček[✉]

Institute of Information Technology and Electronics,
Technical University of Liberec, 461 17 Liberec, Czech Republic
karel.palecek@tul.cz

Abstract. We propose a visual parametrization method for the task of lipreading and audiovisual speech recognition from frontal face videos. The presented features utilize learned spatiotemporal convolutions in a deep neural network that is trained to predict phonemes on a frame level. The network is trained on a manually transcribed moderate size dataset of Czech television broadcast, but we show that the resulting features generalize well to other languages as well. On a publicly available OuluVS dataset, a result of 91% word accuracy was achieved using vanilla convolutional features, and 97.2% after fine tuning – substantial state of the art improvements in this popular benchmark. Contrary to most of the work on lipreading, we also demonstrate usefulness of the proposed parametrization in the task of continuous audiovisual speech recognition.

Keywords: Audiovisual speech recognition · Deep learning · Spatiotemporal convolutional network · Lipreading

1 Introduction

Automatic lipreading is a task of recognizing speech purely from a video of a talking face. It is an inherently difficult task due to limited information content, speech ambiguities and high speaker-dependence. Majority of the work in this area has therefore traditionally concentrated on simplified applications such as phrase or continuous digit recognition. An area closely related to lipreading is audiovisual speech recognition (AVSR), where the main role of lipreading lies in enhancing performance of acoustic decoders, typically under noisy conditions. However, mainly due to the lack of publicly available data, large vocabulary lipreading is scarcely investigated even when coupled with acoustic decoders. For an overview of pre-deep learning advances in lipreading and AVSR see [14].

With its rapid advancement over the past decade, deep learning has gradually found its way into visual speech recognition as well as other applied research areas. One of its first applications in lipreading was in [6], where Ngiam et al. employed deep autoencoder trained via layer-by-layer stacked Restricted Boltzmann Machines. Purpose of the resulting deep belief network was to create joint audio-video bottleneck parametrization, which Ngiam et al. subsequently

© Springer International Publishing AG 2017
K. Ekštein and V. Matoušek (Eds.): TSD 2017, LNAI 10415, pp. 438–446, 2017.
DOI: 10.1007/978-3-319-64206-2_49

used in a support vector machine (SVM) to classify videos of isolated letters and digits on two popular datasets AVletters and CUAVE. Another example is [7], where Noda et al. trained a convolutional neural network (CNN) on still images of mouths to predict phonemes and classified 300 Japanese isolated words using a hidden Markov model (HMM) trained on deep bottleneck features. A fully end-to-end approach was taken by Wand et al. [12], where 51 isolated words were classified using long short term memory (LSTM) recurrent network. Most recently, two advanced end-to-end deep learning systems for lipreading sentences were presented in [1,2]. Assael et al. [1] trained the system to recognize structured sentences of the GRID corpus by optimizing connectionist temporal classification (CTC) criterion and significantly improved state of the art word error rate (WER) from 13.6% to 4.8% in a multi-speaker split, albeit with still only 51 word vocabulary. Chung et al. [2] designed a first end-to-end trained truly large vocabulary deep learning system for lipreading sentences in the wild. To this end, they utilized watch, listen, attend, and spell framework instead of CTC, and were able to push the results on GRID even further down to 3.3%. Their system was, however, pre-trained on a large proprietary dataset of BBC television broadcast with over 100 thousands audiovisual utterances, not available to other researchers.

In this work, we exploit the predictive power of deep learning methods from a perspective of feature extraction for lipreading, similarly to e.g. [7]. We employ spatiotemporal convolutional network in order to predict frame-level labels, therefore also utilizing speech dynamics. As a basic visual speech unit, we choose phonemes, as it has repeatedly been shown, see e.g. [11], that even for purely visual tasks, visemes suffer from several key problems such as low ratio of between to inner class variance and strong context dependence. We then utilize the unnormalized log-probability, the network last layer's output, as a visual parametrization for a traditional HMM-based decoding system. In the experiments section we show that despite the network being trained to predict Czech phoneme set (PACZ) [8], the features achieve state of the art results on English data too, demonstrating their generalization ability and robustness. Advantages of this approach mainly include easy integration into existing frameworks, where the visual information can be plugged into as an additional modality. The proposed visual parametrization is evaluated in both lipreading and continuous speech AVSR.

2   Convolutional Features

Spatiotemporal convolution has been proven successful in video classification tasks, see e.g. [3]. It generalizes classic 2D convolution by also considering the time dimension, i.e.

$$(\boldsymbol{x} \circledast \boldsymbol{w})_{t_0,u_0,v_0,c_0} = \sum_{t=0}^{l}\sum_{u=0}^{m}\sum_{v=0}^{n}\sum_{c=1}^{C} w_{c_0,t,u,v,c} x_{t+t_0,u+u_0,v+v_0,c} + b_{c_0} \qquad (1)$$

where x represents a 4D input video sequence of l frames with $m \times n$ pixels and C channels (e.g. RGB images), w is a bank of d 4D convolution kernels, and b is an added bias.

Here, we exploit its representational power to classify short video chunks x into one of 48 Czech phoneme classes including silences. The chunks consist of 7 or 11 frames of 64×64 RGB region of interest (ROI) that cover the speaker's mouth and its closest surroundings. We stack four blocks of spatiotemporal convolutions (1), batch normalization, spatiotemporal max-pooling and rectified linear unit, with each new layer having twice more convolution kernels than the previous one. In order to produce probability for each class, a linear layer with output dimension equal to the number of phonemes is added after the last convolution. See Fig. 1 for the details.

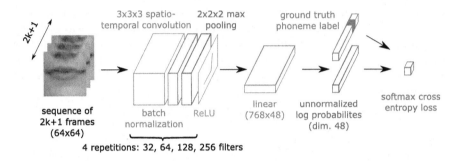

Fig. 1. Spatiotemporal convolutional network architecture

The optimal parameters θ^* (i.e. weights and biases) of the network are estimated by minimizing a softmax cross entropy loss of the network output vector $f(x; \theta)$ against ground truth labels r, i.e.

$$\underset{\theta}{\text{minimize}} \sum_{x,r \in \mathcal{X}} -f(x; \theta)_r + \log \sum_j e^{f(x;\theta)_j} \qquad (2)$$

where $\mathcal{X}$ is a training set consisting of labeled pairs (x, r). For the optimization, we employ the mini-batch stochastic gradient descent algorithm with a momentum of 0.9.

After the network is trained, we use its output vector $f(x_i; \theta^*)$, whose j-th element represents an unnormalized logarithmic probability of the j-th phoneme, as a robust visual parametrization for the i-th frame. In order to deal with borderline cases, the input video is padded with the first and last frames on its respective ends.

The features are post-processed by concatenating parametrization vectors for several consecutive frames and reducing their dimensionality via linear discriminant analysis (LDA). Delta features computed as a difference between two frame feature vectors may be appended to the resulting parametrization. The length of the neighborhood for concatenation or inclusion of delta features are treated as hyperparameters and are cross validated as per splits in Table 1.

Table 1. Overview of datasets used in this work

Dataset	# speakers	Voc. size	Training set	Validation set
TV-data	~50	26197	6.47 h	2.2 h
OuluVS	20	10	19 speakers (~29 min)	1 speaker (~1.5 min)
TULAVD-isol	54	366	45 speakers (~1 h)	9 speakers (~12 min)
TULAVD-cont	54	366	45 speakers (~3.1 h)	9 speakers (~40 min)

3 Data

Deep convolutional networks usually require large amount of data to train from scratch. In areas such as image classification, the preferred way often is to use a pre-trained model and then fine-tune it for specific purposes. In our case, this is not an option, since there are no publicly available spatiotemporal models that would be close to our application.

3.1 TV-data

In order to train the network, we utilize a manually transcribed dataset of Czech television broadcast. The complete dataset contains over 800 h of video content, most of which, however, is not suitable for training of our lipreading system. We process it in following way. Faces in each frame of every video are detected using histogram of oriented gradients (HOG) based deformable part model as implemented in the dlib library [5]. Also, for every frame precise locations of 68 landmarks describing facial shape of each face are predicted using the Ensemble of Regression Trees (ERT) algorithm [4]. We then pick a small subset of the complete data such that the resulting sequences are at least 3 s long and have only a single frontal facing talking speaker for the whole length. In order to achieve scale invariance we define the size of the region of interest (ROI) relative to the normalized mean facial shape. The coordinates of the ROI in the input image are then found by computing Euclidean transformation between the normalized shape and the detected one via least squares minimization.

All videos have been manually checked whether the visible face really is the speaker, or, for example, only serves for illustration purposes, such as when the studio host actually talks over the phone. The complete clean subset then contains approximately 11 h of audiovisual data, which have been split into training, validation, and test set in 0.6 : 0.2 : 0.2 respective ratio. See Table 1 for details. The phoneme class labels for each frame are produced by force-aligning a pre-trained acoustic model on mel frequency cepstral (MFCC) parametrization of the synchronized audio stream.

3.2   OuluVS

We demonstrate the performance of our proposed visual features on more restricted datasets that are better suited to video-only lipreading. OuluVS [13] is a popular and widely used publicly available dataset containing 20 speakers (17 male, 3 female), each of which utters 10 different short phrases five times. Examples of such phrases are for instance "Hello!" or "How are you?". The videos were recorded at 25 fps with resolution of 720×576 pixels in an interlaced mode. Even though OuluVS ships with four different kinds of pre-extracted ROIs, we use our own extraction procedure similar of Sect. 3.1.

3.3   TULAVD

TULAVD [9] contains data from 54 speakers, of which 23 are female and 31 male with age ranging from 20 to 70 years. Each speaker uttered 50 isolated words and 100 sentences in Czech language, which were automatically selected according to phonetic balance. First 50 sentences are common for all speakers, whereas the second 50 differ. Audiovisual utterances were captured by two Logitech C920 FullHD webcams and Microsoft Kinect, which also offers depth stream that is fully synchronized with the video. In this work, we only consider 640×480 pixels RGB data from the Kinect. The visual pre-processing pipeline is similar to the other two considered datasets.

4   Experiments

First we evaluate our proposed spatiotemporal convolutional features for lipreading in simpler and more restricted task of purely visual isolated unit recognition. To this end, videos are parametrized using the 48-dimensional output of our spatiotemporal convolution network, as trained on the TV-data dataset. On top of these features, a hidden Markov model with Gaussian mixture emissions (HMM/GMM) is constructed for each possible unit, i.e. a word on the TULAVD dataset, or a phrase in case of OuluVS. In case of TULAVD, we perform a 6-fold cross validation and report the average score. With OuluVS, we follow a speaker-independent leave-one-out cross validation scheme that is compatible with most existing research on this dataset.

Table 2 then presents the achieved word accuracy and compares the proposed parametrization to other popular methods, namely discrete cosine transformation (DCT) of the ROI, principal component analysis (PCA), active appearance model (AAM), spatiotemporal local binary pattern descriptor (LBPTOP) [13], dynamic histogram of oriented gradients [9], and random forest manifold alignment [10]. Input chunks of length 7 ($k = 3$) and 11 ($k = 5$) were evaluated in the experiments. As we can see, the former performs much better than the latter. Although longer sequences offer more context and discriminative information, they also represent more specific cases, and training models on longer inputs therefore is more prone to overfitting and sensitive to mismatch between training and testing sets.

Table 2. Recognition of isolated words and phrases

Parametrization	TULAVD [%]	OuluVS [%]
	50 words	10 phrases
DCT	72.5	79.2
PCA	73.9	77.9
AAM	74.1	82.1
LBPTOP [13]	74.2	82.5
HOGTOP [9]	86.4	85.7
Random forest manifold alignment [10]	-	89.7
phoneme-3D-CNN-k3	90.6	91.0
phoneme-3D-CNN-k5	74.4	-
phoneme-3D-CNN-k3 (fine-tuned)	**91.0**	**97.2**

Interestingly, parametrization trained on Czech data also performs very well in recognition of English phrases. We achieved state of the art result on the OuluVS of 91% correctly recognized phrases in speaker-independent lipreading. Moreover, the model could be further fine-tuned to different phoneme set by appending additional linear layer on top of the original output to produce unnormalized log-probabilities of each of the 33 English phonemes that are uttered within this dataset. Note that 20 different models had to be fine-tuned for every respective training set of the leave-one-out protocol in order to preserve the speaker independence. By applying the fine-tuned model, the best score reached as high as 97.1%. It is fair to note however, that the result mainly is due to low language variability caused by a small vocabulary, i.e. phonemes only have limited context options, making the chunks easier to classify. Improvement of fine-tuning on TULAVD is marginal, as the score increased by a mere 0.4%.

We also performed experiments continuous speech recognition on the TULAVD dataset. To this end, spatiotemporal features were linearly up-sampled to 100 Hz and concatenated with 39 mel-frequency cepstral coefficients (MFCC) that had been extracted from the audio stream. An HMM model was trained for each phoneme of the PACZ set. Simple bi-gram language model (LM) pre-trained on an external Czech newspaper corpus was applied and filtered for the 366 words in the test utterances. Again the proposed convolutional features perform the best, albeit with much smaller margin than in previous experiments. This is to be expected, as generally visual signal contains much less information than the acoustic one, which will be exploited by the decoder.

Table 3 presents the results of recognition of 50 test sentences of the TULAVD dataset with vocabulary size of 366 words for selected visual features. Results are represented by word accuracy (WAcc) and word correctness (WCor), which differ in that WCor ignores insertions errors. We can observe that under good acoustic conditions, visual parametrization adds relatively little improvement, which is of no surprise. The only two parametrizations increasing the resulting score are LBPTOP and our proposed spatiotemporal convolutional features.

Table 3. Recognition of continuous speech on TULAVD dataset

Parametrization	WAcc [%]	WCor [%]
MFCC	76.7	84.0
MFCC+DCT	42.7	67.6
MFCC+AAM	74.4	83.8
MFCC+LBPTOP [13]	77.7	86.5
MFCC+HOGTOP [9]	75.4	86.2
MFCC+phoneme-3D-CNN-k3	**79.2**	**86.9**

4.1 Error Analysis

It is interesting to look at the errors the network makes on the character level. When taking the maximum of the log-probabilities, the average phoneme accuracy on the validation set of the TV-data reached 25.2%. But the correct label is among the top 5 scoring classes more than 60% of the time, and among the top 10 almost 80% of the time. This suggests that even though the correct class on average still achieves high log-probability, which is important for purposes of parametrization, there are lots mismatches between phonemes of similar corresponding visemic classes. Indeed, by examining the confusion matrix we observed that the most common errors occur e.g. between long and short 'o' (0.65), 'ts' (pronunciation of 'c' in Czech) and 's' (0.49), or 'f' and 'v' (0.35). Note that, however, switching to prediction of visemes does not help due to problems mentioned in Sect. 1.

Figure 2 shows per-phoneme accuracy for 28 phonemes (excluding noise) that attained at least 1% frame-wise relative frequency in the validation set of the TV-data. It can be seen that clearly visible and on average longer vowels such as 'aa', 'o', or 'u', or visually distinctive consonants such as 'm', 'p', 's', or 'v' are

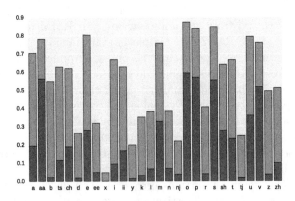

Fig. 2. Per-phoneme accuracy of the spatiotemporal network. The phoneme label convention follows PACZ [8]. Blue: top-1, red: top-5. (Color figure online)

predicted by the network with higher accuracy than e.g. 'x' (PACZ [8] label for Czech digraph 'ch', pronounced as voiceless velar fricative), 'nj', or 'tj'. As many of the easily distinguishable phonemes are shared ·between Czech and English, by capturing characteristic lip motion in its output the network generalizes well to the latter language as well.

5   Conclusion

We have presented a new method of feature extraction for lipreading suitable for both recognition of isolated unit as well as continuous speech. The features are based on learned spatiotemporal convolutions in a deep network trained to predict phonemes on a frame level. The proposed parametrization method can be utilized in traditional or hybrid decoders based on hidden Markov model as well as end-to-end trained deep learning systems. Compared to traditional parametrization methods, we have achieved superior performance on two datasets with different languages, demonstrating feature representativeness and robustness against input variability. On the popular OuluVS speaker-independent benchmark, state of the art word error rate was substantially reduced from 10.3% to 2.8%. Also, unlike many other popular parametrization methods, our features also generalize well to recognizing continuous speech jointly with audio, even in clean environment without acoustic noise. Our efforts in the nearest future will focus on developing end-to-end learned lipreading system for spontaneous speech using spatiotemporal features and attention mechanism.

References

1. Assael, Y.M., Shillingford, B., Whiteson, S., de Freitas, N.: LipNet: sentence-level lipreading. CoRR abs/1611.01599 (2016). http://arxiv.org/abs/1611.01599
2. Chung, J.S., Senior, A.W., Vinyals, O., Zisserman, A.: Lip reading sentences in the wild. CoRR abs/1611.05358 (2016). http://arxiv.org/abs/1611.05358
3. Karpathy, A., Toderici, G., Shetty, S., Leung, T., Sukthankar, R., Fei-Fei, L.: Large-scale video classification with convolutional neural networks. In: Proceedings of the 2014 IEEE Conference on Computer Vision and Pattern Recognition, CVPR 2014, pp. 1725–1732. IEEE Computer Society, Washington, DC (2014)
4. Kazemi, V., Sullivan, J.: One millisecond face alignment with an ensemble of regression trees. In: 2014 IEEE Conference on Computer Vision and Pattern Recognition, CVPR 2014, Columbus, OH, USA, 23–28 June 2014, pp. 1867–1874 (2014)
5. King, D.E.: Dlib-ml: a machine learning toolkit. J. Mach. Learn. Res. **10**, 1755–1758 (2009)
6. Ngiam, J., Khosla, A., Kim, M., Nam, J., Lee, H., Ng, A.Y.: Multimodal deep learning. In: Proceedings of the 28th International Conference on Machine Learning, ICML 2011, Bellevue, Washington, USA, 28 June–2 July 2011, pp. 689–696 (2011)
7. Noda, K., Yamaguchi, Y., Nakadai, K., Okuno, H., Ogata, T.: Lipreading using convolutional neural network. In: International Speech and Communication Association, pp. 1149–1153 (2014)

8. Nouza, J., Psutka, J., Uhlíř, J.: Phonetic alphabet for speech recognition of Czech (1997)
9. Palecek, K.: Lipreading using spatiotemporal histogram of oriented gradients. In: EUSIPCO 2016, Budapest, Hungary, pp. 1882–1885 (2016)
10. Pei, Y., Kim, T., Zha, H.: Unsupervised random forest manifold alignment for lipreading. In: IEEE International Conference on Computer Vision, Sydney, Australia, pp. 129–136 (2013)
11. Ramage, M.D.: Disproving visemes as the basic visual unit of speech (2013). http://www.mramage.id.au/phd
12. Wand, M., Koutník, J., Schmidhuber, J.: Lipreading with long short-term memory. CoRR abs/1601.08188 (2016). http://arxiv.org/abs/1601.08188
13. Zhao, G., Barnard, M., Pietikäinen, M.: Lipreading with local spatiotemporal descriptors. IEEE Trans. Multimedia **11**(7), 1254–1265 (2009)
14. Zhou, Z., Zhao, G., Hong, X., Pietikinen, M.: A review of recent advances in visual speech decoding. Image Vision Comput. **32**(9), 590–605 (2014)

Could Emotions Be Beneficial for Interaction Quality Modelling in Human-Human Conversations?

Anastasiia Spirina[✉], Wolfgang Minker, and Maxim Sidorov

Ulm University, Ulm, Germany
{anastasiia.spirina,wolfgang.minker,maxim.sidorov}@uni-ulm.de

Abstract. There are different metrics which are used in call centres or Spoken Dialogue Systems (SDSs) as an indicator for problem detection during the dialogue. One of such metrics is emotional state. The measurements of emotions can be a powerful indicator in different task-oriented services. Besides emotional state, there is another widely used metric: customer satisfaction (CS), which has a modification called Interaction Quality (IQ). The both models of CS and IQ may include emotional state as a feature. However, is it an actually necessary feature? Some users/customers can be very emotional, while other can be insufficiently emotional in different satisfaction categories. That is why emotional state may be not an informative feature for IQ/CS modelling. Our research is dedicated to the definition of the emotions measurements role in IQ modelling task.

Keywords: Human-human interaction · Task-oriented dialogues · Performances

1 Introduction

In modelling SDSs or in the field of call centre service improvements, the user/customer is a main indicator for further development direction. There are different metrics, which allow to evaluate the quality of interaction/service.

One of them is emotional state. The negative emotions may be the result of a mismatch of user's (customer's) expectations and the reality. As a result of the further analysis, the specific reasons of such problem can be determined. But sometimes the negative emotions can confuse a research, due to such emotions could be caused by external factors. Moreover, some people can be overly emotional or not enough emotional. Thus, all these reasons can lead to the false view. Therefore, the use of only this metric is not enough. But nonetheless, according to [27] emotions may provide rich information for better prediction of further consumer's consumption behaviour. In our point of view, the use of other metrics in combination with emotions is important. In this paper we described our study, dedicated to emotions' role determination in IQ modelling for human-human conversations (HHC).

© Springer International Publishing AG 2017
K. Ekštein and V. Matoušek (Eds.): TSD 2017, LNAI 10415, pp. 447–455, 2017.
DOI: 10.1007/978-3-319-64206-2_50

The rest of the paper is organised as follows: A concise observation of related work is given in Sect. 2. The following Sect. 3 provides a brief description of corpus, which was used for our study. Meanwhile, information about the experimental setup can be found in Sect. 4. Afterwards, the obtained results are presented in Sect. 5 and discussed then in Sect. 6. Finally, conclusions and future work are performed in Sect. 7.

2   Related Work

One of the widely used significant metrics, which help to evaluate the service quality, is CS [11]. Mostly, it is measured at the end of the calls handling the customer feedbacks, which were obtained through the various surveys. In [12] the authors present an attempt at automatically CS prediction using different call's features. The CS model is based on structured, prosodic, lexical and contextual features. It should be pointed out, that lexical features include sentiment words, which may reflect the speaker's emotion or affect. According to the experimental results, described in [12], exclusion of this feature category leads to accuracy reduction in comparison with the results, which were obtained using all feature categories. Nevertheless, we can not make a conclusion of the emotions importance, because the lexical feature category includes more than only sentiment words. It should be mentioned, that [12] provides the results of experiments of CS prediction at the end and in the middle of the calls.

In contrast to the previous metric, Interaction Quality [17,18] allows to assess performance at any point during the interaction. It is important to note, that it was originally designed to control an SDS performance during ongoing spoken interaction as an analogue of CS. Later, it was adapted to HHC [22]. The IQ model for human-computer spoken interaction (HCSI) is based on the various features, including emotions. IQ modelling for HCSI was conducted on the LEGO corpus [19,25], which contains the following emotion set: angry, slightly angry, very angry, neutral, and friendly. All features, which are used for IQ modelling in [18], are subdivided into the following groups: the automatic speech recognition parameters (ASR), features from the language understanding module (SLU), the dialogue manager-related parameters (DM), information about dialogue acts (DACT), emotional state, and user-specific information. The results, describing in [18], shows, that emotional state contributes only 0.009 in comparison with the results on the combination of feature sets: ASR, SLU, and DM. This contribution is not statistically significant according to the Wilcoxon Signed-Rank Test [28].

3   Corpus Description

The speech data for IQ modelling experiments includes 53 dialogues between employees and customers, which were split into 1,165 exchanges [24], which consist of agent's and customer's turn and possible overlaps. In turns, each turn/overlap is described by approx. 400 features, including acoustic features

(384-dimensional feature vector, extracted by *OpenSMILE* [5])configuration for Interspeech 2009 Emotion Challenge [20]), speech duration, emotions, gender and other. Moreover, there are features, which describe exchange itself: duration, number of overlaps, "who starts the exchange" and other. Besides, the feature set comprises the window and dialogue parameter levels, which describe the last n exchanges and the complete dialogue up to the current exchange correspondingly. These levels contain some statistical information. In our study the window level covers the three last exchanges with respect to the current exchange.

3.1   Interaction Quality

All exchanges were annotated with two IQ score labels, which are based on the different IQ-labeling guidelines. The rules for both annotation approaches can be found in [22].

The first approach is based on an absolute scale and is similar to the IQ score annotation guideline for HCSI [18]. Thereby this scale consists of five classes (1-bad, 2-poor, 3-fair, 4-good, 5-excellent), but in the corpus only three classes are presented (with the IQ scores "3", "4", "5"), where the biggest class (the IQ score "5") covers 96.39% of all exchanges, the smallest class (the IQ score "3") includes only four observations.

In contrast to the first approach with the absolute scale, the second approach is based on a scale of changes, which then is transformed into an absolute scale. The scale of changes consists of the following scores: "−2","−1","0","1","2","1_abs" (the last score is in the absolute scale). Afterwards with the assumption that all dialogues start with the IQ score "5" in absolute scale (from the first approach), the obtained labels were transformed into an absolute scale. In our case there are four scores: "6", "5", "4", "3". Similar to the first approach the classes are also unbalanced with 88.24% of exchanges from the majority class "5". While the second biggest class "6" consists of 8.24% of all data. Concerning the minority class "3", it also contains four exchanges, as for the first approach.

We will refer to the first approach as *IQ1* and the second approach as *IQ2*.

3.2   Emotions

For each agent/customer speech fragment three different emotion sets were used. These sets were chosen from [21]. The first set (denote it *em1*) contains the following emotion categories: angry, sad, neutral, and happy. The next set *em2* consists of such categories as: anxiety, anger, sadness, disgust, boredom, neutral, and happiness. The third emotion set *em3* includes fear, anger, sadness, disgust, neutral, surprise, and happiness. It should be mentioned, that not all categories are presented in the corpus.

Afterwards, each original emotion set was subdivided into neutral and other emotions (denote them as *em{1, 2, 3}2*) and into negative, neutral, and positive emotions (denote them as *em{1, 2, 3}3*). It should be mentioned, that the sets *em1* and *em13* are equals. That is why we have excluded the emotion set *em13* from our experiments.

4 Experimental Setup

Due to the fact, that there are two approaches for IQ labelling and eight different emotion sets, we have generated sixteen datasets. Moreover, we have formed datasets without information about emotions for both IQ annotation approaches. Thus, the total number of different sets is eighteen.

The exchange's IQ score prediction task can be presented as a classification problem. For our experiments we have chosen the following classification algorithms: Kernel Naive Bayes classifier (NBK) [9], *k*-Nearest Neighbours algorithm [29] with the dimensionality reduction technique Principal Component Analysis [1] (*k*NN_PCA), L2 Regularised Logistic Regression (LR) [3], Support Vector Machines [4,26] trained by Sequential Minimal Optimization (SVM) [13], Multilayer Perceptron [16] (MLP), J48 algorithm (an open source Java implementation of the C4.5 algorithm) (J48) [6,14].

Partially, the choice of such algorithms as J48, NBK, LR, SVM can be justified by [12], where these algorithms were used for measuring CS. The following algorithm, namely (*k*NN_PCA), has shown encouraging results in [23]. For some algorithm we have utilized the default settings, which are implemented in *Rapidminer*[1] and *WEKA* [8]. For other we have optimised some parameters by the grid optimisation with F_1-score [7] maximization.

For the first four algorithms, which are presented in the list above, we have used the same parameter settings, as in [23]. Whereas, for MLP and J48 the settings can be found in Table 1.

To get a statistically reliable results we have accomplished 10-fold cross-validation. After splitting on training and testing sets we have applied one more

Table 1. The settings for the classification algorithms and parametric optimization.

Parameter	Parameter's value
MLP	
Hidden layers	2
Training cycles	500
Learning rate	0.3
Momentum	0.2
J48	
Confidence threshold for pruning	[0.1, 0.5], step 0.1
Minimum number of instances per leaf	[1, 2], step 1
Laplace smoothing for predicted probabilities	True/False

---

[1] http://rapidminer.com/.

10-fold cross-validation on training set for the grid parameter optimization with F_1-score [7] maximization.

5 Results

To assess classification performance of the employed methods we have relied on macro-average metrics: F_1-score, unweighted average recall (UAR) [15], and accuracy. It means, that the values of these metrics were averaged over ten computations on different train-test splits.

The obtained results in terms of selected classification performance metrics for both *IQ1* and *IQ2* can be found in Tables 2 and 3. The first mentioned table, namely Table 2, presents the box plots, where the central fifty percentages of the achieved results are in the boxes. These box plots are based on the best achieved results for each dataset. So, the best results in terms of F_1-score for *IQ1* (0.581) and *IQ2* (0.623) have been achieved with *k*NN_PCA on *em2* and *em33* respectively. Moreover, *k*NN_PCA has shown the best results for *IQ1* (0.606) on *em2* and *IQ2* (0.589) on *em3* in terms of UAR. Whereas, in terms of accuracy the best results for *IQ1* (0.976) on *em3* and *IQ2* (0.941) on *em2* have been achieved with SVM. In turn, the result for *IQ2* for all chosen performance metrics have been obtained with SVM. Unlike *IQ2*, the results for *IQ1* in terms of F_1-score and UAR have been reached with *k*NN_PCA. The exception was only accuracy, where the best result has been obtained using NBK.

Table 2. The best achieved results for each dataset with and without emotions in terms of F_1-score, UAR, and accuracy.

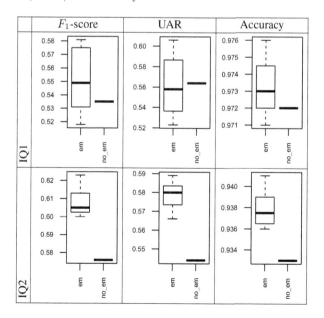

Table 3. The graphics of the obtained results for *IQ1* and *IQ2* in terms of F_1-score, UAR, and accuracy. The big black dots perform the results, which have been achieved on the dataset without emotion labels. Whereas the small grey dots mark the results, obtained with using information about emotions (eight emotion sets).

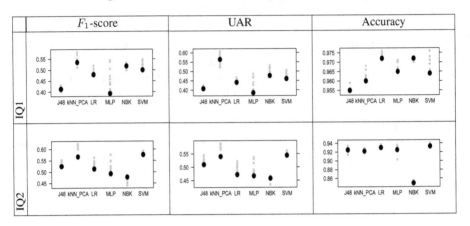

The second aforementioned table contains the graphics, which reflect the achieved results on different datasets (eight sets with emotions and one set without emotions) for each classification method. The big black dots highlight the results, which were achieved on the dataset without emotion labels. It was done to understand whether emotion sets are beneficial or not depending on classification methods.

6 Discussion

To determine the statistically significant differences between the achieved results we have applied the one-way analysis of variance (one-way ANOVA) [2] and the Tukey's honest significant difference (HSD) test [10] with the default settings, implemented in *R* programming language [31]. The one-way ANOVA determined, that there are no any statistically significant differences between means for *IQ1* and *IQ2* almost through all classification performance measures and all classification problems.

From the graphics from the Table 3 we can conclude, that almost for all utilized classification algorithms and for all chosen performance metrics the exclusion of emotional state has led to the performance reduction, but, as we found out from the statistical tests, these changes did not decrease the performance values dramatically. Moreover, from the Table 2 we can see, that in some cases the results, which were obtained on the dataset without emotions, have shown the worst result among the all datasets. But, nevertheless, for *IQ1* the dataset without emotions has shown the middle results in terms of different classification performance metrics.

Although, HHC is more complicated than HCSI and, thereby, they have differences, the obtained results have shown, that emotional state is not very important for modelling IQ for both HHC and HCSI.

But nonetheless, the obtained result is not enough to exclude emotional state from the further research in this field, because it can be only coincidence depending on corpora. To prove the state, that emotions may be excluded from the further IQ modelling process, an additional research is needed.

7   Conclusions and Future Work

In this paper we analysed the role of emotions in IQ modelling for HHC. The impact of emotions is not significant, similar to the results described in [18]. We could not reveal any statistically significant difference between the results, which were obtained on the datasets with and without emotions. Partially it might be explained by the fact, that the used corpus is highly unbalanced and all labels were annotated only by one expert rater. That is why, further research of this question is necessary. Moreover, we could not find any tendency in results with different emotion sets. It means, that there is no a definite emotions set, on which all the best results have been achieved.

As a future direction we plan to extend the list of classification algorithms for predicting an IQ score. Also, the techniques for unbalanced data should be performed. To obtain more objective corpus the number of expert rater and the number of dialogues in the corpus should be increased.

Acknowledgments. The work presented in this paper was partially supported by the DAAD (German Academic Exchange Service), the Ministry of Education and Science of Russian Federation within project 28.697.2016/2.2, and the Transregional Collaborative Research Centre SFB/TRR 62 "Companion-Technology for Cognitive Technical Systems" which is funded by the German Research Foundation (DFG).

References

1. Abdi, H., Williams, L.J.: Principal component analysis. WIREs Comput. Stat. **2**, 433–459 (2010)
2. Bailey, R.A.: Design of Comparative Experiments. Cambridge University Press, Cambridge (2008)
3. le Cessie, S., Houwelingen, J.C.: Ridge estimators in logistic regression. Appl. Stat. **41**(1), 191–201 (1992)
4. Cristianini, N., Shawe-Taylor, J.: An Introduction to Support Vector Machines and Other Kernel-based Learning Methods. Cambridge University Press, New York (2000)
5. Eyben, F., Weninger, F., Gross, F., Schuller, B.: Recent developments in opensmile, the Munich open-source multimedia feature extractor. In: Proceedings of ACM Multimedia (MM), pp. 835–838 (2013)
6. Gholap, J.: Performance tuning of J48 algorithm for prediction of soil fertility. Asian J. Comput. Sci. Inf. Technol. **2**(8), 251–252 (2012)

7. Goutte, C., Gaussier, E.: A probabilistic interpretation of precision, recall and F-Score, with implication for evaluation. In: Losada, D.E., Fernández-Luna, J.M. (eds.) ECIR 2005. LNCS, vol. 3408, pp. 345–359. Springer, Heidelberg (2005). doi:10.1007/978-3-540-31865-1_25

8. Hall, M., Frank, E., Holmes, G., Pfahringer, B., Reutmann, P., Witten, I.H.: The WEKA data mining software: an update. SIGKDD Explor. **11**(1), 10–18 (2009)

9. John, G.H., Langley, P.: Estimating continuous distribution in Bayesian classifiers. In: Eleventh Conference on Uncertainty in Artificial Intelligence, pp. 338–345 (1995)

10. Kennedy, J.J., Bush, A.J.: An Introduction to the Design and Analysis of Experiments in Behavioural Research. University Press of America, Lanham (1985)

11. Maar, B., Neely, A.: Managing and Measuring for Value: the Case of Call Centre Performance. Cranfield School of Management, cranfield (2004)

12. Park, Y., Gates, S.C.: Towards real-time measurement of customer satisfaction using automatically generated call transcripts. In: Proceedings of the 18th ACM conference on Information and knowledge management, pp. 1387–1396 (2009)

13. Platt, J.: Fast training of support vector machines using sequential minimal optimization. In: Schoelkopf, B., Burges, C., Smola, A. (eds.) Advances in Kernel Methods: Support Vector Learning. MIT Press, Cambridge (1999)

14. Quinkan, J.R.: C4.5: Programs for Machime Learning. Morgan Kaufmann Publishers, San Francisco (1993)

15. Rosenberg, A.: Classifying skewed data: importance to optimize average recall. In: Proceedings of INTERSPEECH 2012, pp. 2242–2245 (2012)

16. Rosenblatt, F.: Principles of Neurodynamics Perceptrons and the Theory of Brain Mechanisms. Spartan Books, Washingtion DC (1961)

17. Schmitt, A., Schatz, B., Minker, W.: Modeling and predicting quality in spoken human-computer interaction. In: Proceedings of the SIGDIAL 2011 Conference, pp. 173–184. Association for Computational Linguistics (2011)

18. Schmitt, A., Ultes, S.: Interaction quality: assessing the quality of ongoing spoken dialog interaction by experts - and how it relates to user satisfaction. Speech Commun. **74**, 12–36 (2015)

19. Schmitt, A., Ultes, S., Minker, W.: A parameterized and annotated corpus of the CMU let's go bus information system. In: International Conference on Language Resources and Evaluation (LREC), pp. 3369–3373 (2012)

20. Schuller, B., Steidl, S., Batliner, A.: The interspeech 2009 emotion challenge. In: Proceedings of INTERSPEECH 2009, pp. 312–315 (2009)

21. Sidorov, M., Brester, C., Schmitt, A.: Contemporary stochastic feature selection algorithms for speech-based emotion recognition. In: Proceedings of INTERSPEECH 2015, pp. 2699–2703 (2015)

22. Spirina, A., Sidorov, M., Sergienko, R., Schmitt, A.: First experiments on interaction quality modelling for human-human conversation. In: Proceedings of the 13th International Conference on Informatics in Control, Automation and Robotics (ICINCO), vol. 2, pp. 374–380 (2016)

23. Spirina, A., Vaskovskaia, O., Sidorov, M., Schmitt, A.: Interaction quality as a human-human task-oriented conversation performance. In: Ronzhin, A., Potapova, R., Németh, G. (eds.) SPECOM 2016. LNCS (LNAI), vol. 9811, pp. 403–410. Springer, Cham (2016). doi:10.1007/978-3-319-43958-7_48

24. Spirina, A.V., Sidorov, M.Y., Sergienko, R.B., Semenkin, E.S., Minker, W.: Human-human task-oriented conversations corpus for interaction quality modelling. Vestnik SibSAU **17**(1), 84–90 (2016)

25. Ultes, S., Platero Sánchez, M.J., Schmitt, A., Minker, W.: Analysis of an extended interaction quality corpus. In: Lee, G.G., Kim, H.K., Jeong, M., Kim, J.-H. (eds.) Natural Language Dialog Systems and Intelligent Assistants, pp. 41–52. Springer, Cham (2015)

26. Vapnik, V.N.: The Nature of Statistical Learning Theory. Springer, New York (1995)

27. Wang, J.: From customer satisfaction to emotions: alternative framework to understand customer's post-consumption behaviour. In: Proceedings of the 2012 International Joint Conference on Service Sciences, pp. 120–124 (2012)

28. Wilcoxon, F.: Individual comparisons by ranking methods. Biom. Bull. **1**, 80–83 (1945)

29. Witten, I.H., Frank, E., Hall, M.A.: Data Mining: Practical Machine Learning Tools and Techniques. Morgan Kaufmann, San Francisco (2011)

Multipoint Neighbor Embedding

Adrian Lancucki[(✉)] and Jan Chorowski

Institute of Computer Science, University of Wrocław, Wrocław, Poland
{adrian.lancucki,jan.chorowski}@cs.uni.wroc.pl

Abstract. Dimensionality reduction methods for visualization attempt to preserve in the embedding as much of the original information as possible. However, projection to 2-D or 3-D heavily distorts the data. Instead, we propose a multipoint extension to neighbor embedding methods, which allows to express datapoints from a high-dimensional space as sets of datapoints in a low-dimensional space. Cardinality of those sets is not assumed a priori. Using gradient of the cost function, we derive an expression, which for every datapoint indicates its remote area of attraction. We use it as a heuristic that guides selection and placement of additional datapoints. We demonstrate the approach with multipoint t-SNE, and adapt the $\mathcal{O}(N \log N)$ approximation for computing the gradient of t-SNE to our setting. Experiments show that the approach brings qualitative and quantitative gains, i.e., it expresses more pairwise similarities and multi-group memberships of individual datapoints, better preserving the local structure of the data.

Keywords: Manifold learning · Data visualization · t-SNE · Barnes-Hut algorithm

1 Introduction

The objective of dimensionality reduction is to construct a mapping from a high-dimensional dataset $\mathbf{X} = \{\mathbf{x}_1, \ldots, \mathbf{x}_N\}$ to a low-dimensional dataset $\mathbf{Y} = \{\mathbf{y}_1, \ldots, \mathbf{y}_N\}$ where typically $\mathbf{x}_i \in \mathbb{R}^S$, $\mathbf{y}_i \in \mathbb{R}^s$, and $S \gg s$. A special case of dimensionality reduction is visualization where $s = 2$ or $s = 3$.

Non-linear methods perform better than linear at visualization, because it is unlikely for high-dimensional data to lay in a linear subspace of such small dimensionality. The class of Neighbor Embedding (NE) algorithms [15] is especially useful in this task. We propose a multipoint extension to NE in which every datapoint $\mathbf{x}_i$ is embedded as a set of low-dimensional datapoints $Y_i = \{\mathbf{y}_i, \mathbf{y}'_i, \mathbf{y}''_i, \ldots\}$ of equal importance. Cardinality of Y_is is not assumed a priori, as the copies are added as necessary. Replication of each $\mathbf{y} \in Y_i$ is decided heuristically with a scoring function. We demonstrate the extension on the example of multipoint t-SNE, which is used during validation.

© Springer International Publishing AG 2017
K. Ekštein and V. Matoušek (Eds.): TSD 2017, LNAI 10415, pp. 456–464, 2017.
DOI: 10.1007/978-3-319-64206-2_51

1.1 Stochastic Neighbor Embedding

In Stochastic Neighbor Embedding (SNE) [7], the data is modeled with a distribution P, where $p_{i|j}$ is a conditional probability of $\mathbf{x}_i$ picking $\mathbf{x}_j$ as its neighbor, calculated by centering Gaussians at each $\mathbf{x}_i$. The embedding $\mathbf{Y}$ is modeled with a probability distribution Q. Similarly, it is constructed by centering Gaussians at each $\mathbf{y}_i$ [7]

$$p_{i|j} = \frac{\exp\left(-\|\mathbf{x}_i - \mathbf{x}_j\|^2/2\sigma_i^2\right)}{\sum_{k\neq i}\exp\left(-\|\mathbf{x}_i - \mathbf{x}_k\|^2/2\sigma_i^2\right)}, \quad q_{i|j} = \frac{\exp\left(-\|\mathbf{y}_i - \mathbf{y}_j\|^2\right)}{\sum_{k\neq i}\exp\left(-\|\mathbf{y}_i - \mathbf{y}_k\|^2\right)}. \quad (1)$$

The embedding is constructed by minimizing the cost defined as the sum of Kullback-Leibler divergences

$$C(Y) = KL(P\|Q) = \sum_i \sum_j p_{i|j} \log \frac{p_{i|j}}{q_{i|j}}. \quad (2)$$

In t-SNE [10] Gaussian distribution in Q is replaced with Student's t distribution.

2 Visualization with Multipoint Embeddings

We demonstrate and analyze our approach with multipoint t-SNE. Because we embed individual datapoints as sets of datapoints, we need to modify Q. We replace the Euclidean metric in Q with a distance function, which relates sets Y_i with Y_j through minimum distance between their elements $d_{ij} = \min\{\|\mathbf{y}_k - \mathbf{y}_l\| : \mathbf{y}_k \in Y_i \wedge \mathbf{y}_l \in Y_j\}$. Please note that it breaks the triangle inequality and is no longer a metric. Elements of Y_i are not weighted, and therefore are of equal importance to the embedding. The gradient can be derived with the chain rule [10], differing only in $\partial d_{ij}^2/\partial \mathbf{y}_i$ calculated using a subderivative of min.

NE methods have plausible interpretation as physical systems of springs [6]. The gradient of t-SNE can be decomposed into a difference of terms interpreted as attractive and repulsive forces [9,15]

$$\frac{\partial C}{\partial \mathbf{y}_i} = 4 \sum_j \frac{\partial d_{ij}^2}{\partial \mathbf{y}_i} \frac{p_{ij}}{(1 + d_{ij}^2)} - 4 \sum_j \frac{\partial d_{ij}^2}{\partial \mathbf{y}_i} \frac{q_{ij}}{(1 + d_{ij}^2)} = F_{attr} - F_{rep}. \quad (3)$$

Because of the heavy tail of Student's t distribution used to construct Q, the repulsive force F_{rep} acting on $\mathbf{y}_i$ comes from all datapoints. Conversely, because of the light tail of the Gaussian used to construct P, the attractive force F_{attr} comes only from those datapoints, which are close neighbors of $\mathbf{x}_i$ in $\mathbf{X}$.

Optimization of t-SNE cost function in Eq. (2) is hard, as the function is highly non-convex [7,10]. Minimization of the cost with gradient methods moves the datapoints slightly at every gradient update. In practice, a misplaced datapoint $\mathbf{y}_i$ might be repulsed by dissimilar datapoints with a force which balances

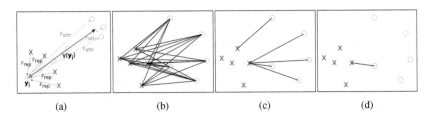

Fig. 1. Relationships between two sets Y_i **(blue** Xs**) and** Y_j **(green** Os**) which embed** $\mathbf{x}_i, \mathbf{x}_j \in \mathbf{X}$. Similar datapoints are in the same color. (a) A datapoint prevented from reaching the area of attraction (pointed by $\gamma(\mathbf{y}_i)$) by dissimilar datapoints, (b) Redundant pairwise relations, (c) Relations limited to, for every $y_i \in Y_i$, only the closest $y_j \in Y_j$, (d) The correct relation as the shortest distance between Y_i and Y_j. Best viewed in color.

the attractive force from a distant, similar datapoint (Fig. 1a). Thus the optimization might get stuck in a local optimum, where forces reach a spurious equilibrium.

Several strategies were proposed to remedy this problem, helping a datapoint to take a leap over dissimilar datapoints with a small increase in the cost in Eq. (2). For instance, a temporary dimension might be added to the embedding, weight of the datapoint might be lowered, or the whole embedding might be kept close early in the optimization [7,10].

2.1   Replication of Datapoints

Instead of encouraging the datapoints to move to remote areas, we propose a two-stage heuristic process, which happens during optimization. Firstly, we heuristically recognize obstructed datapoints, and place their copies in their remote areas of attraction. Secondly, we recognize and discard unused copies. This way a datapoint might be copied, but also moved by first being copied and then discarded.

Formally, for each $\mathbf{y}_i$, we compute its **potential** $\gamma(\mathbf{y}_i)$

$$\gamma(\mathbf{y}_i) = 4 \sum_{p_{ij} > q_{ij}} \frac{\partial d_{ij}^2}{\partial \mathbf{y}_i} (p_{ij} - q_{ij}). \tag{4}$$

We derive γ by multiplying the cost function of t-SNE in Eq. (3) by the $(1 + d_{ij}^2)$ term to cancel it out, because it promotes closer pairs of datapoints. Interestingly, γ becomes the gradient of the cost function of symmetric SNE (excluding the $p_{ij} > q_{ij}$ restriction). It also has a physical interpretation, under which $(p_{ij} - q_{ij})$ may be treated as a spring constant, and $\partial d_{ij}^2 / \partial \mathbf{y}_i$ as spring length [5]. We interpret vector $\gamma(\mathbf{y}_i)$ as the direction of a promising region, and $\|\gamma(\mathbf{y}_i)\|$ as the magnitude of its attraction.

We score each datapoint $\mathbf{y}$ with $\|\hat{\gamma}(\mathbf{y})\|$, and replicate a fraction of top scoring ones. The copies are initialized one by one, through line search along the direction of $\gamma(\mathbf{y}_i)$, so that they would minimize the cost in Eq. (2) (Fig. 1a). Placement may be approximate, as the optimization will be continued. For instance, 10% of datapoints might be replicated every 100 iterations up to total of 1000 iterations.

During **cleanup**, we discard misplaced datapoints, which experience little attractive force. Probability mass $\sum_j p_{ij}$ induces attractive force F_{attr} which acts on Y_i in Eq. (3). Because each p_{ij} is modeled by exactly one $\mathbf{y}_k \in Y_i$, we can distribute it among all elements in Y_i and normalize to a probability distribution r_{ki}. It can be interpreted as probability of $\mathbf{y}_k \in Y_i$ being the closest datapoint from Y_i to a random neighbor of Y_i. All datapoints with low r_{ki} should be discarded, as they experience small attractive force, and thus will likely be repelled from the embedding into infinity. Distribution r may also be interpreted as a weighting function for datapoints in Y_i.

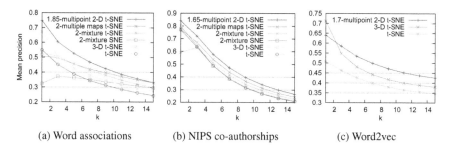

(a) Word associations　　(b) NIPS co-authorships　　(c) Word2vec

Fig. 2. Precision (in information retrieval sense) of reconstruction of k-NNs (higher is better). Decimal numbers denote how many maps/mixture components remained at the end (1.85 denotes ending up with 185% of initial datapoints etc.). For fairness of comparison, we select indexes of k highest entries in Q for a given $\mathbf{y}_i$, and k highest in P for a $\mathbf{x}_i$. The method is equivalent to selecting k closest datapoints w.r.t. the Euclidean distance for SNE and t-SNE, as their P and Q neighborhood functions monotonically decrease with the distance. Best viewed in color.

2.2   Fast Gradient Approximation

Barnes-Hut (B-H) algorithm for t-SNE [9, 15] approximates a low-dimensional n-body simulation [1], allowing to compute the gradient of the cost in $\mathcal{O}(N \log N)$ steps instead of $\mathcal{O}(N^2)$. The space is partitioned with a quad- or oct-tree, traversed in a depth-first manner. Nodes (cells) which meet the B-H condition [1] are treated as approximate summaries of its contents. Direct application of the B-H algorithm to multipoint t-SNE would overestimate F_{rep} by counting all pairwise relations (Fig. 1b) instead of only those between closest of datapoint replicas (Fig. 1d). We prevent it in two ways.

Firstly, we compute $\partial C / \partial \mathbf{y}_i$ for all $\mathbf{y}_i \in Y_i$ in parallel. For each pair (Y_i, Y_j) only the closest $\mathbf{y}_k \in Y_j$ should interact with Y_i. We traverse the quadtree in a depth-first manner and assign each cell to the closest $\mathbf{y}_i \in Y_i$ if it meets the B-H condition. Instead of immediately updating the gradient with those cells, we store them in lists, keeping one list for each $\mathbf{y}_i \in Y_i$. When traversal is over, we sort the list for each $\mathbf{y}_i$ based on the distance of those cells from $\mathbf{y}_i$. However, F_{rep} is still overestimated (Fig. 1c).

Secondly, we correct the number k_i of datapoints in the ith traversed cell. For each $\mathbf{y}_j$ in a cell, we would like to check if any of its replicas already interacts with Y_i, and if so, subtract 1 from k_i. We call such situation a collision. Let α_i denote the estimated number of collisions within a cell and $k_c = \sum_{j<i} k_j$ total number of datapoints in previously processed cells. Then[1] $\alpha_i = k_i + N\left(N - \frac{1}{N}\right)^{k_c}\left[\left(1 - \frac{1}{N}\right)^{k_i} - 1\right]$. Knowing the exact number of collisions $\sum_j k_j - N$, we normalize α_i as as $\hat{\alpha}_i = \frac{\alpha_i}{\sum_j \alpha_j}\left(\sum_j k_j - N\right)$. The proposed approximation is applicable when repulsive weights are equal, i.e., every datapoint repels the others with the same force.

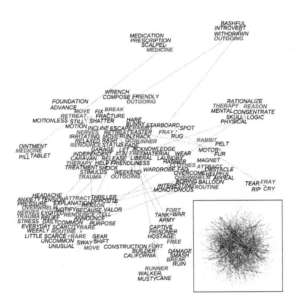

Fig. 3. Partial multipoint t-SNE embedding of word associations. Lines connect copies of the same datapoint (in red). 3-NNs were plotted for each copy. Overview of the complete embedding is shown in the lower right corner. Best viewed in color.

3 Related Work

To promote sparseness of the embedding, NE methods repulse all pairs of close datapoints. Weighted symmetric SNE (ws-SNE) [16] weights points in the repulsive term. Elastic Embedding (EE) [4] expresses the penalty with a simplified cost function.

---

[1] α_i amounts to expected number of collisions in a hash table of N locations when inserting k_i elements, after having inserted k_c elements.

Mixture SNE [7] embeds a single datapoint in multiple locations. The initial probability mass of a datapoint is distributed among its copies through weights. From the beginning of optimization, each datapoint has a fixed number of copies. Contrary, in our approach the copies are equally weighted, and their number varies.

Cook et al. [5] proposed visualization of pairwise similarities with mixtures of m 2-D maps, where each one is fitted with symmetric SNE. Multiview SNE [2] combines the data from t heterogeneous maps into a single embedding map. Each q_{ij} is a linear combination of input p_{ij}s, resulting in a single embedding of each datapoint.

4   Experiments

We demonstrate our approach with multipoint t-SNE. In experiments on small datasets, we compare multipoint t-SNE, multi-map t-SNE, mixture SNE and mixture t-SNE. The last one is created by switching Gaussian neighborhoods to Student's t neighborhoods in the distribution Q of mixture SNE. On larger datasets, we compare multipoint t-SNE with plain t-SNE, both using Barnes-Hut approximations.

In all experiments the datasets were reduced to $d = 50$ dimensions with Principal Component Analysis, and perplexity of Gaussians in the data space was set to 50. We optimize with Adam [8] using learning rate $\eta = 3.5$. In our experience it often works better than gradient descent with momentum. In subsequent stages of optimization, we reuse previous values of moments m, v. For mixture approaches, we found gradient descent with momentum and L-BFGS-B [3] to perform slightly better.

As multipoint t-SNE adds variables, it takes longer to converge, and we run the optimization for total of 2000 to 4000 gradient updates. In all experiments p_{ij} values were multiplied by 4 (or 12 in the case of large datasets [9]) for the first 250 gradient updates. In all experiments the datapoints were replicated in four stages, with replication of the top scoring 25% during each and an immediate cleanup of those $y_k \in Y_i$ with low probability mass $r_{ki} < 0.2$.

4.1   Pairwise Relationships Datasets

We adopt the word similarities dataset [13] (size 5000×5000) and the NIPS co-authorships dataset (size 1418×1418) from studies on multiple maps t-SNE and aspect maps [5,11]. Both datasets come in form of square matrices of pairwise relations. The former describes word similarities judged by human volunteers, the latter is a co-occurrence matrix of paper authors of two or more contributions accepted to NIPS in years 1988–2009. Quality of the embeddings can be scored with mean precision [11] or mean precision/recall [16] of k-nearest neighbors between $\mathbf{X}$ and $\mathbf{Y}$. For every datapoint in $\mathbf{X}$, k closest datapoints in $\mathbf{X}$ and $\mathbf{Y}$ are selected based on their Euclidean distances. For fairness of comparison, we select them based on their P and Q values. Both approaches are equivalent for SNE and t-SNE. However, the latter scores higher methods where each q is composed of many datapoints in Y, for instance through weights in mixture SNE.

Multipoint t-SNE achieves higher mean precision for $k < 15$ (Figs. 2a and b). This is due to better modeling of small neighborhoods. Moreover, 2-D visualizations with multipoint t-SNE are better in this regard than 3-D visualizations with t-SNE. Figure 3 shows a portion of the word association embedding, namely selected copies of datapoints and their 3-NNs. Words like *move, break, free* end up in different, natural contexts, or even among different parts of speech like *reason*.

4.2   Vector Datasets

Figure 4 shows embeddings of the COIL-20 dataset [14] (size 1440 × 16384). The dataset consists of B&W photos of a number of small objects, taken under different angles. Typically t-SNE embeds the images as closed, separated loops, sometimes torn or incomplete. In multipoint t-SNE, the lines between replicas connect certain areas of the embedding and pointing out imperfections, making for a more informative visualization.

Next, we visualize 10000 Google News dataset word embeddings[2] [12] of the most frequent words[3] and compare mean precision of multipoint t-SNE and t-SNE ($d = 2$ and $d = 3$), all using Barnes-Hut approximation with $\theta = 0.2$. Figure 2c shows the results. Multipoint t-SNE better preserves small neighborhoods in 2-D embeddings than t-SNE in 2-D and 3-D, outperforming them by a large margin.

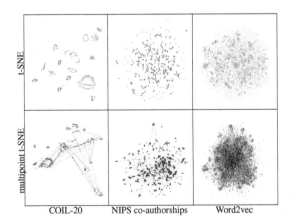

Fig. 4. Embeddings constructed with t-SNE and multipoint t-SNE. Copies of datapoints are shown in red. Lines connect copies of the same datapoints (only 10% of lines shown to avoid clutter). Copies of datapoints allow more clusters to form and reduce the number of satellite datapoints on the edges of embedding, by placing them in denser areas. Best viewed in color.

---

[2] Taken from https://code.google.com/archive/p/word2vec/.

[3] We exclude top 100 as mostly stop words or containing non-letter symbols.

Figure 4 compares t-SNE and multipoint t-SNE embeddings. Lines connecting datapoints with their copies form a dense network which connects remote areas of the embedding. Additional copies of datapoints allow small, isolated groups to form. They also seem to reduce the number of evenly spaced satellite datapoints on the verge of the embedding, which typically do not fit elsewhere.

5   Discussion

We have introduced the multipoint extension to NE methods and showed its effectiveness with multipoint t-SNE. It naturally extends NE methods through sensible addition of copies of certain datapoints during optimization. Moreover, the extension does not raise the complexity, and we have implemented the Barnes-Hut approximation.[4] The approach allows to preserve small neighborhoods better than multiple maps and mixture approaches, and even better than 3-D t-SNE. Embeddings constructed with multipoint t-SNE have fewer misplaced datapoints on the edge of the embedding, as well as crisper clusters datapoints inside. Datapoints are not weighted, but weights may be derived through analysis of their neighborhoods. Natural directions of future work include analysis of optimization stages and out-of-sample extension, i.e., embedding points not seen during training.

Acknowledgments. Adrian Lancucki was supported by local grant 0420/1710/16, and National Center for Research and Development (Poland) grant Audioscope (Applied Research Program, 3rd contest, submission no. 245755). Jan Chorowski was supported by National Science Center (Poland) grant Sonata 8 2014/15/D/ST6/04402. The authors also thank WCSS for computing power.

References

1. Barnes, J., Hut, P.: A hierarchical O(N log N) force-calculation algorithm. Nature **324**(6096), 446–449 (1986)
2. Xie, B., Yang, M., Tao, D., Huang, K.: m-SNE multiview stochastic neighbor embedding. IEEE Trans. Syst. Man Cybern. Part B (Cybern.) **41**(4), 1088–1096 (2011)
3. Byrd, R.H., Lu, P., Nocedal, J., Zhu, C.: A limited memory algorithm for bound constrained optimization. SIAM J. Sci. Comput. **16**(5), 1190–1208 (1995)
4. Carreira-Perpinán, M.A.: The elastic embedding algorithm for dimensionality reduction. In: ICML, vol. 10, pp. 167–174 (2010)
5. Cook, J., Sutskever, I., Mnih, A., Hinton, G.E.: Visualizing similarity data with a mixture of maps. In: International Conference on Artificial Intelligence and Statistics, pp. 67–74 (2007)
6. Hadsell, R., Chopra, S., LeCun, Y.: Dimensionality reduction by learning an invariant mapping. In: CVPR. vol. 2, pp. 1735–1742. IEEE (2006)

---

[4] Source code to reproduce the experiments available publicly at https://github.com/alancucki/multipoint_tsne.

7. Hinton, G.E., Roweis, S.T.: Stochastic neighbor embedding. In: Advances in Neural Information Processing Systems, pp. 833–840 (2002)

8. Kingma, D., Ba, J.: Adam: a method for stochastic optimization. arXiv preprint arXiv:1412.6980 (2014)

9. van der Maaten, L.: Accelerating t-SNE using tree-based algorithms. J. Mach. Learn. Res. **15**(1), 3221–3245 (2014)

10. van der Maaten, L., Hinton, G.: Visualizing data using t-SNE. J. Mach. Learn. Res. **9**(11), 2579–2605 (2008)

11. van der Maaten, L., Hinton, G.: Visualizing non-metric similarities in multiple maps. Mach. Learn. **87**(1), 33–55 (2012)

12. Mikolov, T., Chen, K., Corrado, G., Dean, J.: Efficient estimation of word representations in vector space. arXiv preprint arXiv:1301.3781 (2013)

13. Nelson, D.L., Mcevoy, C.L., Schreiber, T.A.: The University of South Florida Word Association, Rhyme, and Word Fragment Norms (1998)

14. Nene, S.A., Nayar, S.K., Murase, H.: Columbia Object Image Library (COIL-20). Technical report CUCS-005-96 (1996)

15. Yang, Z., Peltonen, J., Kaski, S.: Scalable optimization of neighbor embedding for visualization. In: ICML, vol. 28, pp. 127–135 (2013)

16. Yang, Z., Peltonen, J., Kaski, S.: Optimization equivalence of divergences improves neighbor embedding. In: ICML, pp. 460–468 (2014)

Significance of Interaction Parameter Levels in Interaction Quality Modelling for Human-Human Conversation

Anastasiia Spirina[1]([✉]), Alina Skorokhod[2], Tatiana Karaseva[2], Iana Polonskaia[2], and Maxim Sidorov[1]

[1] Ulm University, Ulm, Germany
{anastasiia.spirina,maxim.sidorov}@uni-ulm.de
[2] Reshetnev Siberian State University of Science and Technology,
Krasnoyarsk, Russia
{alina.skorokhod,tatiana.karaseva,iana.polonskaia}@sibsau.ru

Abstract. The Interaction Quality (IQ) metric, which originally was designed for spoken dialogue systems (SDSs) to assess human-computer spoken interaction (HCSI) and then adapted to human-human conversation (HHC), is based on features from three interaction parameter levels: an exchange, a window, and a dialogue level. To determine the significance of the window and dialogue interaction parameter levels, as well as their combination, computations, based on different data sets, have been performed using several classification algorithms. The obtained results may be used for further improvement of the IQ model for HHC in terms of the computational complexity.

Keywords: Human-human interaction · Task-oriented dialogues · Performance

1 Introduction

For improving of SDSs, which allow human to communicate with different computer systems via speech, the IQ metric was designed. This metric, proposed in [15,17], was considered as an indicator for SDSs, which might reflect some problematic situations during the interaction. Later, this metric was adapted to HHC [21], assuming that HHC and HCSI have a resemblance and, consequently, the results of such an adaptation may be used for further improvement of SDSs.

The IQ model for HHC is based on more than 1200 features, describing an agent's/ customer's/ overlapping speech and the dialogue itself [22]. All these features may be extracted automatically. The features for IQ modelling can be subdivided into the three parameter levels: exchange, window and dialogue [17,22].

To reduce the computational complexity in terms of the total feature extraction time and algorithm speed (for IQ modelling) we have tried to reduce the

K. Ekštein and V. Matoušek (Eds.): TSD 2017, LNAI 10415, pp. 465–472, 2017.
DOI: 10.1007/978-3-319-64206-2_52

number of features by analysing the significance of each parameter level's contribution. For this research we have designed the IQ models applying several classification algorithms, implemented in *Rapidminer*[1] and *WEKA* [10] on the different data sets (e.g. the data set containing only the features from the exchange and window levels).

The remainder of this paper is structured as follows: a brief description of IQ and the results of its modelling for HCSI using the different interaction parameter levels and their combinations is given in Sect. 2. Brief description of the spoken corpus, which was used for conducting all computations, is provided in Sect. 3, which is followed by a description of the formulated classification problems and utilized algorithms in Sect. 4. Section 5 presents the obtained results, which are then discussed in Sect. 6. Finally, we finished our paper by conclusions and future work in Sect. 7.

2   Related Work

The IQ paradigm idea was introduced in [15,17] for assessing the SDS performance during an ongoing interaction. Originally this paradigm was derived from the concept of User/Customer Satisfaction (CS), which is widely used in different spheres. Usually CS is assessed manually at the end of calls/transactions by customers during various surveys. In contrast to CS, the IQ metric helps to evaluate an SDS performance at any point during the interaction. The IQ model for HCSI is based on the features from the three parameter levels. The first of them, the exchange level, consists of information about the current system-user-exchange. The next one, namely the window level comprises the features (some statistics) from the n last exchanges. The third one, the dialogue level, describes the complete dialogue up to the current exchange [17]. The complete list of features can be found in [16,17].

For better understanding of each level's contribution to the overall estimation performance, different experiments, which were based on the features from each parameter level and their combinations, were conducted [23].

The results, described in [23], shows, that the best result in terms of Unweighted Average Recall (UAR) [14], Cohen's Kappa [4] linearly weighted [5], and Spearman's Rho [20] was achieved using all parameters [23]. Also these experiments reveal, that the parameters from the window level have an important role in the overall performance.

3   Corpus Description

The experiments, described in this paper, have been conducted based on the spoken corpus [22], which consists of 53 task-oriented dialogues between employees and customers. Subsequently, after the manual diarization all dialogues were split into 1,165 agent-customer-exchanges.

---

[1] http://rapidminer.com/.

Each agent-customer exchange is described by more than 1,200 features, which reflect the different interaction parameter levels: exchange, window, dialogue. The exchange level features contain acoustic attributes, extracted by *OpenSMILE* [7] (a feature vector, used in InterSpeech 2009 Emotion Challenge, contains 384 attributes [18]) for agent/customer/overlapping speech, information about the speech and pause duration, emotions (manually annotated) and others.

In turn, the dialogue and window levels are presented in the corpus by such features, as:

- the total/mean duration of an exchange,
- the total/mean duration and the percent of the duration of an agent, customer, overlapping speech and pauses between turns,
- the total pause duration between exchanges,
- the total/mean number of the fragments with speech overlaps,
- the number of the exchanges, where the first speaker is agent/customer/ overlapping speech.

The window level covers the three last exchanges with respect to the current exchange.

3.1 Interaction Quality

Each observation in the corpus, i.e. agent-customer exchange, was annotated with two IQ score labels. Two types of the IQ assessment are based on the different IQ-labeling guidelines, which can be found in [21].

For the first approach an absolute scale, which is similar to the IQ score annotation guideline for HCSI [16], was used. Despite the fact, that this scale consists of five indicators (1-bad, 2-poor, 3-fair, 4-good, 5-excellent), only three classes are presented (with the IQ scores "3", "4", "5") in this corpus. The biggest part (96.39%) of all observations belongs to the class with the IQ score "5", while the smallest class (the IQ score "3") covers only four observations. We will denote the first approach as *IQ1*.

In comparison with the first approach, the second approach is relied on a scale of changes, which, subsequently, is transformed into an absolute scale. The scale of changes is presented by the following scores: "−2", "−1", "0", "1", "2", "1_abs" (the last score is in the absolute scale). Then with the assumption from the first approach, that all dialogues start with the IQ score "5"(in absolute scale), the obtained labels were converted into an absolute scale. As a result, we have received four scores: "6", "5", "4", "3". The majority class (with the IQ score "5") in this case covers 88.24% of all exchanges, while the second biggest class "6" consists of 8.24% of all observations. Concerning the minority class "3", it also contains four exchanges. We will refer to the second approach as *IQ2*.

3.2 Emotions

Each agent/customer turn was annotated with three different emotion labels. For this labeling we have chosen three sets from [19], which then were adapted for

IQ modelling. The set *em1* contains such categories as: angry, sad, neutral, and happy. The next set *em2* differs from the *em1* by the presence of such categories as: disgust/irritation and boredom. In the last emotion set *em3*, in comparison with the set *em2*, the category "boredom" was replaced by the category "surprise". It should be pointed out, that not all categories in each set were presented in this corpus.

Then each set (*em{1, 2, 3}*) was subdivided into neutral and other emotions (denote them as *em{1, 2, 3}2*) and into negative, neutral, and positive emotions (denote them as *em{1, 2, 3}3*). This decomposition was performed to understand the complexity of the emotion sets, which is required for better IQ prediction.

4   Experimental Setup

The IQ score estimation task can be formulated as a classification problem, in our case with three classes for *IQ1* and four classes for *IQ2*. For our research a total number of the different sets is eighteen. Each set is a combination of an IQ label (*IQ1* or *IQ2*) and an emotion set (nine sets: the three main sets and two sets derived from each of them). Hereinafter we call it tasks.

Instead of using all possible combinations of the different parameter levels, as in [23], we have carried out our computations on the four sets:

- the exchange, window, dialogue levels,
- the exchange and window levels,
- the exchange and dialogue levels,
- the exchange level.

The reason is that the features from the window and dialogue levels do not contain sufficient information for the interaction description in HHC.

For the experiments the following classification algorithms were chosen: Kernel Naive Bayes classifier (NBK) [11], *k*-Nearest Neighbours algorithm (*k*NN) [25], L2 Regularised Logistic Regression (LR) [3], Support Vector Machines [6,24] trained by Sequential Minimal Optimisation (SVM) [13].

For the classification performance assessment we have accomplished 10-fold cross-validation to obtain statistically reliable results. We have split our data on the training and testing sets, afterwards we have introduced one more inner 10-fold cross-validation on the training sets. Subsequently we used inner 10-fold cross-validation for the grid parameter optimisation of the classification algorithms, where F_1-score [9] was maximized.

Regarding dimensionality reduction we employed a data transformation technique Principal Component Analysis (PCA) [1] with the fixed cumulative variance value 0.99. The data were pre-processed: each column (attribute values) was statistically normalised, it means, that the mean of each column is equal to 0 and the variance is equal to 1.

Furthermore, all non-numeric attributes such as emotions, speaker gender, and "who starts an exchange" have been transformed into numeric type using dummy coding, which replaces a nominal attribute with m different categories by m new attributes, containing 0 or 1. These values reflect the absence and presence of the respective categories of a nominal attribute for each observation.

5   Results

An assessment of the classification algorithms performance is based on such classification performance measures as accuracy, Unweighted Average Recall [14], F_1-score, which were averaged over ten computations on different train-test splits. However, in this paper we provide results for F_1-score, as the main classification performance measure for this study.

Partly the results for the classification task with *em13* for the different combinations of the interaction parameter levels are depicted in Fig. 1. The same results, but for all classification problems, for both *IQ1* and *IQ2* are presented in figures in Table 1. The results depicted in Fig. 1 and figures in Table 1 were achieved with *k*NN algorithm.

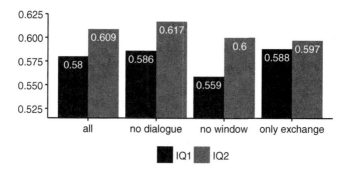

Fig. 1. *k*NN performance in F_1-score for the different combinations of the parameter level for the emotion set *em13*.

Regarding numerical evaluations in terms of accuracy the best results were obtained with *k*NN and LR.

6   Discussion

To define the statistically significant differences between the obtained results we have relied on the one-way analysis of variance (one-way ANOVA) [2] and the Tukey's honest significant difference (HSD) test [12] with the default settings, utilized in R programming language[2]. The one-way ANOVA determined, that the differences between means are statistically significant for *IQ1* and *IQ2* almost through all classification performance measures and all classification problems. To find out what algorithms gave statistically significant different results we have

---

[2] http://r-project.org/.

Table 1. kNN performance in F_1-score for the different combinations of the parameter levels.

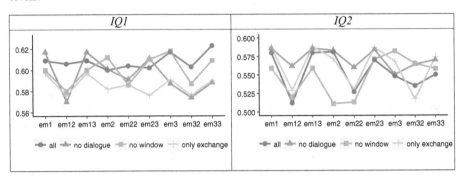

used the Tukey's HSD test. This test revealed that almost in all the cases there are statistically significant differences between the results of NBK and other algorithms.

Moreover, we have applied these tests to determine the statistically significant differences between results, which had been obtained with kNN algorithm based on the different combinations of the interaction parameter levels. The one-way ANOVA test found out, that there are no any statistically significant differences between the results.

However, it should be mentioned, that for *IQ1* almost for all classification problems the results, which are based on the data excluded the dialogue level, showed better results, than with the use of the dialogue parameter level. In turn, for *IQ2* almost for all classification problems the exclusion of the window and dialogue level simultaneously led to the result decreasing.

The baseline accuracy (classifier always predicts the majority class) for *IQ1* and *IQ2* are 0.964 and 0.882, correspondingly. For F_1-score the baselines are 0.327 and 0.234, respectively.

Given the fact that the data is highly unbalanced, the achieved results are not reasonable enough, although almost in all classification performance measures in all classification problems the obtained results outperform the baselines. For some algorithms, however, the results do not outperform the baseline in terms of accuracy.

Interestingly, that the best results in terms of F_1-score were achieved in all the cases with kNN algorithm. To determine whether the results, obtained with kNN and LR, in terms of accuracy statistically significant differ from the baselines, the Student's t-test [8] was employed. In the case of *IQ2* for all tasks and for both algorithms p-value is less than 0.007. Regarding the kNN model for *IQ1* p-value exceeds 0.15. It should be noted, that concerning the LR-based model results for *IQ1*, for some tasks p-values outperform 0.05.

Hence, from the results of the Student's t-test between the obtained results (in terms of accuracy) and the baseline (0.964) for *IQ1* we have concluded, that statistically significant results have been achieved with LR, but not for all tasks.

Almost in all the cases the use of only exchange parameter level is not enough. The use of the datasets with emotions dividing into two classes did not result in significantly better results.

In turn, for all classification problems the obtained results in terms of accuracy statistically significantly outperform the baseline for both algorithms, namely kNN and LR for *IQ2*.

It should be mentioned, that the use of PCA reduced the number of features approximately by factor of 2.5 (approx. from 1200 to 470), and consequently increased the computational speed in terms of execution time.

7   Conclusions and Future Work

In this paper we have analysed the significance of the different interaction parameter levels in the task of IQ modelling for HHC. Our research has revealed the impact of the different interaction parameter levels and their combination for IQ modelling for HHC. The differences between the feature lists for IQ modelling for HCSI an HHC did not lead to the same results. Partially it might be explained by the fact, that the used corpus is highly unbalanced and all labels were annotated only by one expert rater.

As a future direction we plan to apply ensemble-based classifiers for predicting an IQ score. Taking into account a rather high dimensionality of the feature space, other dimensionality reduction methods might be helpful. Moreover, considering the highly imbalanced classes, approaches for multiclass imbalanced data should be performed.

Acknowledgments. The work presented in this paper was partially supported by the DAAD (German Academic Exchange Service), the Ministry of Education and Science of Russian Federation within project 28.697.2016/2.2, and the Transregional Collaborative Research Centre SFB/TRR 62 "Companion-Technology for Cognitive Technical Systems" which is funded by the German Research Foundation (DFG).

References

1. Abdi, H., Williams, L.: Principal component analysis. WIREs Comput. Stat. **2**, 433–459 (2010)
2. Bailey, R.A.: Design of Comparative Experiments. Cambridge University Press, Cambridge (2008)
3. le Cessie, S., Houwelingen, J.C.: Ridge estimators in logistic regression. Appl. Stat. **41**(1), 191–201 (1992)
4. Cohen, J.: A coefficient of agreement for nominal scales. Educ. Psychol. Measur. **20**, 37–46 (1960)
5. Cohen, J.: Weighted kappa: nominal scale agreement provision for scaled disagreement or partial credit. Psychol. Bull. **70**(4), 213–220 (1968)
6. Cristianini, N., Shawe-Taylor, J.: An Introduction to Support Vector Machines and Other Kernel-based Learning Methods. Cambridge University Press, Cambridge (2000)

7. Eyben, F., Weninger, F., Gross, F., Schuller, B.: Recent developments in opensmile, the Munich open-source multimedia feature extractor. In: Proceedings of ACM Multimedia (MM), pp. 835–838 (2013)
8. Fay, M.P., Proschan, M.A.: Wilcoxon-Mann-Whitney or t-test? On assumptions for hypothesis tests and multiple interpretations of decision rules. Stat. Surv. **4**, 1–39 (2010)
9. Goutte, C., Gaussier, E.: A probabilistic interpretation of precision, recall and F-Score, with implication for evaluation. In: Losada, D.E., Fernández-Luna, J.M. (eds.) ECIR 2005. LNCS, vol. 3408, pp. 345–359. Springer, Heidelberg (2005). doi:10.1007/978-3-540-31865-1_25
10. Hall, M., Frank, E., Holmes, G., Pfahringer, B., Reutmann, P., Witten, I.H.: The WEKA data mining software: an update. SIGKDD Explor. **11**(1), 10–18 (2009)
11. John, G.H., Langley, P.: Estimating continuous distribution in Bayesian classifiers. In: Eleventh Conference on Uncertainty in Artificial Intelligence, pp. 338–345 (1995)
12. Kennedy, J.J., Bush, A.J.: An Introduction to the Design and Analysis of Experiments in Behavioural Research. University Press of America, Lanham (1985)
13. Platt, J.: Fast training of support vector machines using sequential minimal optimization. In: Schoelkopf, B., Burges, C., Smola, A. (eds.) Advances in Kernel Methods: Support Vector Learning. MIT Press, Cambridge (1999)
14. Rosenberg, A.: Classifying skewed data: importance to optimize average recall. In: Proceedings of INTERSPEECH 2012, pp. 2242–2245 (2012)
15. Schmitt, A., Schatz, B., Minker, W.: Modeling and predicting quality in spoken human-computer interaction. In: Proceedings of the SIGDIAL 2011 Conference, pp. 173–184. Association for Computational Linguistics (2011)
16. Schmitt, A., Ultes, S.: Interaction quality: assessing the quality of ongoing spoken dialog interaction by experts and how it relates to user satisfaction. Speech Commun. **74**, 12–36 (2015)
17. Schmitt, A., Ultes, S., Minker, W.: A parameterized and annotated corpus of the CMU lets go bus information system. In: International Conference on Language Resources and Evaluation (LREC), pp. 3369–3373 (2012)
18. Schuller, B., Steidl, S., Batliner, A.: The interspeech 2009 emotion challenge. In: Proceedings of INTERSPEECH 2009, pp. 312–315 (2009)
19. Sidorov, M., Brester, C., Schmitt, A.: Contemporary stochastic feature selection algorithms for speech-based emotion recognition. In: Proceedings of INTERSPEECH 2015, pp. 2699–2703 (2015)
20. Spearman, C.: The proof and measurement of association between two things. Am. J. Psychol. **15**(1), 72–101 (1904)
21. Spirina, A., Sidorov, M., Sergienko, R., Schmitt, A.: First experiments on interaction quality modelling for human-human conversation. In: Proceedings of the 13th International Conference on Informatics in Control, Automation and Robotics (ICINCO), vol. 2, pp. 374–380 (2016)
22. Spirina, A.V., Sidorov, M.Y., Sergienko, R.B., Semenkin, E.S., Minker, W.: Human-human task-oriented conversations corpus for interaction quality modelling. Vestnik SibSAU **17**(1), 84–90 (2016)
23. Ultes, S., Schmitt, A., Minker, W.: Analysis of temporal features for interaction quality estimation. In: Proceedings of the 7th International Workshop on Spoken Dialogue Systems (IWSDS) (2016)
24. Vapnik, V.N.: The Nature of Statistical Learning Theory. Springer, New York (1995)
25. Witten, I.H., Frank, E., Hall, M.A.: Data Mining: Practical Machine Learning Tools and Techniques. Morgan Kaufmann, San Francisco (2011)

Ship-LemmaTagger: Building an NLP Toolkit for a Peruvian Native Language

José Pereira-Noriega[1], Rodolfo Mercado-Gonzales[2], Andrés Melgar[2],
Marco Sobrevilla-Cabezudo[2], and Arturo Oncevay-Marcos[2(✉)]

[1] Facultad de Ciencias e Ingeniería,
Pontificia Universidad Católica del Perú, Lima, Peru
jpereira@pucp.pe
[2] Research Group on Pattern Recognition and Applied Artificial Intelligence,
Departamento de Ingeniería, Pontificia Universidad Católica del Perú, Lima, Peru
{rmercado,amelgar,msobrevilla,arturo.oncevay}@pucp.edu.pe

Abstract. Natural Language Processing deals with the understanding and generation of texts through computer programs. There are many different functionalities used in this area, but among them there are some functions that are the support of the remaining ones. These methods are related to the core processing of the morphology of the language (such as lemmatization) and automatic identification of the part-of-speech tag. Thereby, this paper describes the implementation of a basic NLP toolkit for a new language, focusing in the features mentioned before, and testing them in an own corpus built for the occasion. The obtained results exceeded the expected results and could be used for more complex tasks such as machine translation.

Keywords: Part-of-speech tagging · Lemmatization · Low resource language · Shipibo-konibo

1 Introduction

Traditionally, both Part-of-speech tagging (POS-tagging) and Lemmatization were made by the use of hand-crafted rules [6]. However, there are several recent studies showing that machine learning approaches are suitable to solve these tasks without taking any effort in defining all the rules and exceptions needed for a particular language.

Specifically, in the case of an agglutinative language like Shipibo-konibo, the labor of building rules is not feasible due to all of the possible combinations of affixes. Also, due to the lack of an established order in the words of a sentence in this language, the labor of developing rules for POS-tagging is particularly time-consuming.

Nevertheless, in order to use machine learning approaches, it is necessary to have an annotated corpus. In this way, since it is easier to build those datasets than the rules, it was decided to follow this learning approach for the develop of our NLP tools for this low resource language.

© Springer International Publishing AG 2017
K. Ekštein and V. Matoušek (Eds.): TSD 2017, LNAI 10415, pp. 473–481, 2017.
DOI: 10.1007/978-3-319-64206-2_53

The paper is organized as follows: in Sect. 2 are presented some works related to lemmatization and POS-tagging for agglutinative and low-resourced languages. After that, Sect. 3 describes the case study of this research: the Shipibo-konibo language. Later, the corpus annotation process is presented in Sect. 4. Then, Sect. 5 explains the functionalities developed in this work. Finally, the experiments performed are included in Sect. 6, and Sect. 7 presents some conclusions and potential future works.

2   Related Works

In the case of the Shipibo-konibo language, there have not been any direct attempts to solve the problem of POS-tagging or lemmatization. Moreover, this language does not even have an annotated corpus or any computational tool. However, there are some studies for similar agglutinative and low-resourced languages that show some progress in solving these tasks.

For the POS-Tagging task, in languages like Bhojpuri [11] and Bengali [3], the supervised learning approach had a great performance (between 86% and 90% of accuracy). The experiments made for these languages were performed with Support Vector Machines trained models. Also in similar languages like Nepali, approaches based in Hidden Markov Models were used with a little lower results [10].

Regarding the lemmatization task, in languages like Urdu [4] or Mongol [7], it is shown that a rule-based approach can be really effective in solving this problem. However, these studies used manually generated rules, a big corpus, and dictionaries of words to deal correctly with exceptions.

Although, due to the particular agglutinative characteristics of the Shipibo-konibo language, the labor of making manual derivation rules is not feasible. Therefore, it is also possible to develop rules automatically, like it is shown for the Afrikaans [2] and for some European languages [6]. However, since the corpus built for this study is currently smaller than the ones used for those languages, lower results were expected for this work.

3   The Shipibo-Konibo Language

Shipibo-konibo (SHP) is the sixth language with highest number of native speakers in Peru. It is a language spoken by about 150 communities (mainly in the Amazon region) and is taught in almost 300 public schools in Peru (schools with a bilingual education program) [1,8]. However, it does not have any own computational-linguistic resources yet, and this is the reason why it is considered a low-resourced language from a computational perspective, like most of the Peruvian native languages.

SHP is an agglutinative language which relies in the use of around 114 suffixes plus 31 prefixes [12] and their combinations for word derivation. However, there is not an official grammar established, so, in order to develop computational-linguistic resources it was a must to relied on the assistance of linguistic experts and bilingual speakers.

4 Corpus: Building and Annotation

Because there is no annotated corpus for SHP, a new one was built with the required data for the job. This task was achieved with the development of an annotation tool called ChAnot[1], the help of linguists with a vast knowledge of the language and some native speakers. It is important to note that they had no experience in annotation tasks. The final corpus for this study is available in a project site[2].

ChAnot is an annotation tool that allows to process a text by sentences and perform morphological (lemma and affixes), morpho-syntactic (POS-tag) and named-entity annotation. This tool was developed to make easier the process of creating an annotated corpus for peruvian low-resourced languages. Unlike annotators tools that allow highlighting parts of the document to annotate some information, the focus of this tool is to process a sentence sequentially word for word, allowing the splitting of its affixes and an specific annotation.

On the other side, a suitable tagset for the language was needed, and since Shipibo-konibo and most native languages in Peru do not have an official tagset, a linguist team defined a new one based on the standard tagset of Universal Dependencies [9] and linguistic studies regarding the language [12]. The tagset match with the UD standard names can be seen in the tool website. With the help of this tool, it was possible to develop a corpus of 219 annotated sentences, where each word of the sentence contains: an annotation of the lemma, POS-tag, sub-POS-tag, and a list of all the affixes that appears in the word.

This corpus was used entirely for the training of the POS-tagger. The distribution of the amount of words per tag in the Shipibo-konibo tagset is shown in Table 1.

Furthermore, with the help of a Shipibo-konibo dictionary (which entries included POS-tags information), it was possible to identify the derived word-forms of lemmas that were presented in the examples of the use of each entry. In that way, the corpus of the lemmatization task could get more annotated examples. At the end, the corpus achieved a total of 3544 unique input words (with their correspondent lemma and POS-tag) distributed by word category as it is shown in Table 2.

---

[1] Available in: chana.inf.pucp.edu.pe/chanot.
[2] Available in: chana.inf.pucp.edu.pe/resources.

Table 1. Structure of the corpus used in the POS-tagging task

POS category	Quantity
Adjective	66
Adverb	40
Particle	1
Conjunction	38
Determiner	53
Interjection	6
Noun	368
Proper noun	15
Numeral	6
Interrogative word	59
Adposition	14
Pronoun	65
Punctuation	311
Verb	361
Auxiliary	95

Table 2. Structure of the corpus used in the lemmatization task

POS category	Quantity
Adjective	309
Adverb	130
Particle	1
Conjunction	29
Determiner	4
Interjection	11
Noun	1474
Proper noun	0
Numeral	6
Interrogative word	30
Adposition	22
Pronoun	32
Verb	1490
Auxiliary	6

5   Ship-LemmaTagger

5.1   Part-of-Speech Tagging

Part-of-speech (POS) tagging is the process of assigning a part of speech to each word in a corpus [5]. For this process, it was defined a tagset aligned to the standard tagset of Universal Dependencies, and after that a supervised learning approach was considered.

The workflow for this step is shown in Fig. 1. Firstly, a sentence is received as an input. Then, a tokenization step is performed, where the sentence is split in tokens (words, numbers, symbols or signs). Each token in a sentence is checked to observe whether it is a numeral, a symbol or a punctuation sign. If the token is one of the three possibilities mentioned before, the POS-tag is assigned directly, otherwise the trained supervised model comes into action.

For the classification task, the approach of Ekbal et al. [3] was taken in consideration, since they trained a Support Vector Machine (SVM) algorithm using different features such as word information (initial and final substring of the word, which could be called prefixes and suffixes) and contextual information (previous word, previous POS-tags and next word). The complete list of the generated features is as follows:

- Current token
- Previous token

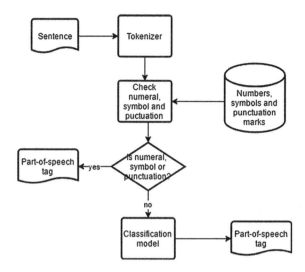

Fig. 1. Part-of-speech tagging process

- Next token
- A binary value that indicates whether it is the first token of sentence or not.
- A binary value that indicates whether it is the last token of sentence or not.
- A binary value that indicates whether the first character of the token is capitalized.
- A binary value that indicates whether all characters of the token are capitalized.
- Prefixes (initial characters) of length 1, 2, 3 and 4.
- Suffixes (last characters) of length 1, 2, 3 and 4.
- Two previous POS-tags.

For instance, in the sentence "Manaxawe betan chaxo iní?" (that means "The motelo and the deer"), the features regarding the information of word "Manaxawe" are 1 (first token), 0 (not last token), 1 (first character capitalized), 0 (some characters are not capitalized), "m", "ma", "man", "mana" (prefixes) and "e", "we", "awe", "xawe" (suffixes). Meanwhile, the features for the contextual information of the word "chaxo" are "betan" (previous token), "iní?" (next token), conjunction (previous POS-tag) and noun (POS-tag before previous POS-tag).

5.2   Lemmatization

The lemmatization process follows the workflow shown in Fig. 2. First, an individual input token is analyzed in order to determine if it could possess a suffix. This is done by contrasting the end of the word versus a list of all the existent suffixes identified in the Shipibo-konibo language.

In case there is a potential suffix present in the token, a possible rule is inferred with the use of a trained classification model. Once the potential rule is obtained, it is analyzed whether it could be applied for the input token to get the lemma. If the rule could no be used (there is no match) we retrieve the same word as the final lemma.

Regarding the rule prediction task, the approach of W. Daelemans [2] was followed, training a K-NN classification model using a number of features corresponding to the size of the biggest word of the corpus. In this feature vector, each character of the word is mapped to a dimension according to the position of the character in the word. Furthermore, since the language is highly agglutinative on the side of the suffixes, it was decided to reverse the order of the characters in a word to get an alignment between suffixes. On the other side, the derived rules of transformation were considered as the classes of the model.

The lemmatization rules are composed similarly to the ones shown in the previous work: two-elements tuples with (1) the string to be removed and (2) the string to be added to get the final lemma. In both cases, if there is no need to add or remove a string from the input word, the corresponding part of the tuple is left blank.

Additionally, since there are some particular suffixes that only appear in certain words categories, it was decided later to include the POS-tag as an additional feature.

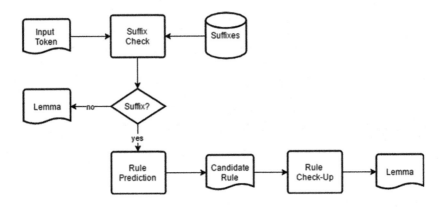

Fig. 2. Lemmatization process

For instance, for the word "ainbobo", that means "women" in Shipibo-konibo, and its lemma "ainbo", the features vector would be: ['o', 'b', 'o', 'b', 'n', 'i', 'a', 'noun'] and the rule of transformation would be [bo -] because we need to remove the substring "bo" from the input word and add a null string (" ") to get the lemma.

6   Experimentation and Results

Regarding the POS-tagger experiments, two methods were used: an SVM and a Decision Tree model. For the validation step, the corpus was split in sub-datasets for training (70%) and testing (30%). After repeating this split process 10, 50 and 100 times, the average accuracy of our part-of-speech tagger was obtained. The results are presented in Table 3.

Additionally, experiments with ensemble learning methods were tested, but the scores were lower than the expected. Finally, the best overall accuracy was 0.848, obtained with the SVM algorithm (kernel = RBF, C = 1, gamma = 0.1).

Table 3. Accuracy for part-of-speech tagging experiments

Algorithm/Iterations	10	50	100
SVM	0.847	0.847	**0.848**
Decision tree	0.808	0.810	0.811

For the lemmatization task using the K-NN algorithm, the performance was validated by splitting the corpus in two equal parts for training and testing (50-50). This division was made by stratifying every class of the corpus in two parts, in order to avoid the disproportion of some word categories with little data. This process was performed 100 times with random divisions each time, and the average accuracy obtained is presented in Table 4.

The experiment was fulfilled using different numbers of neighbors and distance metrics in order to find the optimal result. In this way, the best parameters configuration (neighbors = 5, distance = Manhattan) achieved an overall accuracy of 0.593. This is caused by the presence of high number of features for this task, and with the Manhattan measure, the relation between near features is isolated and the alignment of the characters obtained more relevance. Also, it is important to notice that the number of neighbors needed for the optimal result should not be too high, since that configuration could bias the results towards the rules with higher appearances in the corpus.

However, this result was not completely satisfactory in itself, but considering that only half of the corpus was used for training, it was a good step to then test it together with the POS-tagger.

Table 4. Accuracy results for the lemmatizer

Metric # of k	1	3	5	7	9	11	13
Euclidean	0.482	0.531	0.558	0.536	0.557	0.534	0.521
Chebyshev	0.486	0.514	0.543	0.539	0.558	0.539	0.541
Manhattan	0.502	0.539	**0.593**	0.562	0.547	0.556	0.551

Finally, both procedures were merged by using the best trained model of the POS-tagging step as an additional feature for the lemmatization. This new lemmatizer model was trained with the whole corpus obtained from the dictionary, and it was tested on the annotated sentences with ChAnot, that include a set of different words. With this procedure, it was obtained a new accuracy value of 81.4% for the trained lemmatizer as it is shown in Table 5.

Table 5. Accuracy results for the joint process

Metric # of k	1	3	5	7	9	11	13
Euclidean	0.805	0.565	0.507	0.486	0.474	0.483	0.476
Chebyshev	0.762	0.531	0.498	0.492	0.474	0.471	0.465
Manhattan	**0.814**	0.574	0.525	0.492	0.489	0.492	0.487

7   Conclusions and Future Work

This study focus on the developing of a basic NLP toolkit for a new language. As this language (SHP) is an agglutinative one, some approaches in similar contexts were taken in consideration in order to build a solid feature vector to fit learning models for the POS-tagger and Lemmatizer tasks.

The first results were uneven, highlighting the good performance of the POS-tagger. However, despite having achieved an individual low result for the lemmatization task, the integration with the POS-tagging process (as an input feature) led to very promising results in general. Likewise, since the approach used was a corpus-based, the continuous growth of the annotated corpus could lead to better accuracy results for both tasks.

As future work, semi-supervised learning methods will be considered for upcoming experiments. This approach could take advantage of the large unannotated corpus available and, with the integration of the predictive models in the annotation tool, it could support the development of more linguistic resources for this language.

Acknowledgments. For this study, the authors appreciate the linguistic team effort that made possible the corpus annotation, and also acknowledge the support of the "Consejo Nacional de Ciencia, Tecnología e Innovación Tecnológica" (CONCYTEC Perú) under the contract 225-2015-FONDECYT.

References

1. Acosta, S., Natalia, K., Huamancayo Curi, E., Mori Clement, M., Carbajal Solis, V.: Documento nacional de lenguas originarias del Perú (2013)
2. Daelemans, W., Groenewald, H.J., Van Huyssteen, G.B.: Prototype-Based Active Learning for Lemmatization (2009)
3. Ekbal, A., Bandyopadhyay, S.: Part of speech tagging in Bengali using support vector machine. In: International Conference on Information Technology 2008, ICIT 2008, pp. 106–111. IEEE (2008)

4. Gupta, V., Joshi, N., Mathur, I.: Design and development of a rule-based Urdu lemmatizer. In: Proceedings of International Conference on ICT for Sustainable Development, pp. 161–169. Springer (2016)

5. Jurafsky, D., Martin, J.H.: Speech and Language Processing, vol. 3. Pearson, London (2014)

6. Juršic, M., Mozetic, I., Erjavec, T., Lavrac, N.: Lemmagen: multilingual lemmatisation with induced ripple-down rules. J. Univers. Comput. Sci. **16**(9), 1190–1214 (2010)

7. Khaltar, B.O., Fujii, A.: A lemmatization method for mongolian and its application to indexing for information retrieval. Inf. Process. Manag. **45**(4), 438–451 (2009)

8. Ministerio de Educación del Perú: Minedu oficializa alfabetos de 24 lenguas originarias a ser utilizados por todas las entidades públicas. http://www.minedu.gob.pe/n/noticia.php?id=33082. Accessed 31 Mar 2016

9. Nivre, J., de Marneffe, M.C., Ginter, F., Goldberg, Y., Hajic, J., Manning, C.D., McDonald, R., Petrov, S., Pyysalo, S., Silveira, N., et al.: Universal dependencies v1: A multilingual treebank collection. In: Proceedings of the 10th International Conference on Language Resources and Evaluation (LREC 2016), pp. 1659–1666 (2016)

10. Paul, A., Purkayastha, B.S., Sarkar, S.: Hidden Markov model based part of speech tagging for Nepali language. In: 2015 International Symposium on Advanced Computing and Communication (ISACC), pp. 149–156. IEEE (2015)

11. Singh, S., Jha, G.N.: Statistical tagger for Bhojpuri (employing support vector machine). In: 2015 International Conference on Advances in Computing, Communications and Informatics (ICACCI), pp. 1524–1529. IEEE (2015)

12. Valenzuela, P.: Transitivity in Shipibo-Konibo grammar. Ph.D. thesis, University of Oregon (2003)

A Study of Abstractive Summarization Using Semantic Representations and Discourse Level Information

Gregory César Valderrama Vilca[1] and Marco Antonio Sobrevilla Cabezudo[2(✉)]

[1] Escuela de Posgrado, Maestría en Informática,
Pontificia Universidad Católica del Perú, Lima, Perú
a20133303@pucp.edu.pe
[2] Grupo de Reconocimiento de Patrones e Inteligencia Artificial Aplicada,
Departamento de Ingeniería, Pontificia Universidad Católica del Perú, Lima, Perú
msobrevilla@pucp.edu.pe

Abstract. The present work proposes an exploratory study of abstractive summarization integrating semantic analysis and discursive information. Firstly, we built a conceptual graph using some lexical resources and Abstract Meaning Representation (AMR). Secondly, we applied PageRank algorithm to get the most relevant concepts. Also, we incorporated discursive information of Rethorical Structure Theory (RST) into the PageRank to improve the relevant concepts identification. Finally, we made some rules over the relevant concepts and applied SimpleNLG to make the summaries. This study was performed on the corpus of DUC 2002 and the results showed a F1-measure of 24% in Rouge-1 when AMR and RST were used, proving their usefulness in this task.

Keywords: Abstractive summarization · Abstract Meaning Representation · Rethorical Structure Theory

1 Introduction

The web is a giant resource of data and information that has great utility for people. However, getting an abstract about one or many documents is an expensive labor, which with manual process might be impossible to complete due to the huge amount of data.

Automatic Summarization [12] is a challenging task, because it involves analysis and comprehension of the written text in non-structural natural language and it is dependent of a context that must describe an event synthesis or knowledge in a simple form, becoming natural for any reader. There are diverse approaches to summarize text and categorize into extractive or abstractive.

Abstractive summaries regenerate the content extracted from source text by terms fusion, compression or suppression processes. Thus, paraphrased sentences are obtained and these are not in the original text. This approach has a major

© Springer International Publishing AG 2017
K. Ekštein and V. Matoušek (Eds.): TSD 2017, LNAI 10415, pp. 482–490, 2017.
DOI: 10.1007/978-3-319-64206-2_54

probability to reach coherence and smoothness like one generated or made by a human beings.

Previous work has shown progress using semantic representations such as Abstract Meaning Representation (AMR) presented in [11], Discursive Analysis with Rhetorical Structure Theory (RST) present in [6] and conceptual models using linguistic resources such as WordNet present in [14].

This work presents an exploratory study of how to integrate semantic (AMR annotator) and discursive (RST annotator) information into an abstractive summarization method produces better results. In a first phase, the method generated a conceptual graph using AMR parsing and other lexical resources like WordNet and PropBank [16]. Thus, to find the most relevant concepts we use PageRank, considering all discursive information given by the O'Donell method application. Then, sentence candidates are built with the most important concepts and semantic roles information. Finally, an abstractive summary is generated using SimpleNLG, as Natural Language Generation tool, over the sentence candidates. This shows that using these techniques are workable and even more profitable, recommended configurations and useful tools for this task.

This paper organization is: first, the Sect. 2 presents the related works, Sect. 3 presents the proposed method, Sect. 4 presents experiments and results. Finally, Sect. 5 presents some conclusions and future works.

2   Related Works

The performance of extractive and abstract techniques was tested in [2], not only the automatic methods but also the summaries made by people. It conclude that in the linguistic-grammatical aspect, and in the quality of the content, summaries generated by humans are far superior to those generated automatically, and the abstractive methods have more possibilities to achieve results more similar to their human counterparts.

In [14] we can observe an intermediate representation models and the use of knowledge sources presented on the Web. The authors generated summaries of a single document using a semantic representation of texts through conceptual graphs, in which, the weights are associated with the edges linking concept nodes, creating a flow called "semantic flow". A semantic flow is the weight accumulated by the nodes and that transmit to other nodes increasing or decreasing its value when passing through any conceptual relation. For the graph generation, the authors used the semantic information from external sources as *WordNet* [8] and *VerbNet* [3] that rule the structural coherence of the graphs. In the synthesis stage, the graphs were reduced according to a set of generalization, union, weighting and pruning operations shown in [7]. In [6], the authors present opinion summarization by an abstractive method based on the analysis of the structures and relations of the discourse, and also they proposed a method to generate new sentences that uses the PageRank algorithm to identify the most important content.

In [11], the authors used AMR [10] for the representation and generation of abstract summaries for a single document. The authors generated an AMR graph for each document sentence using JAMR parser [4]. The AMR graphs are merged based on the concepts common between them. Thus, generate an unique graph for a document that reduced its concept redundancy.

3   Abstractive Summarization Method

We used the architecture proposed by [12] which comprises three stages: (1) in the analysis phase, input text are interpreted and represented in a computational format; (2) in the transformation phase, representation mentioned in first phase, is processed to identify and select the content more relevant and as a result a condensed computational representation of texts is got, and (3) in the synthesis phase, a natural language text is generated. In the Fig. 1, we may see the pipeline of the proposed Abstractive Summarization method.

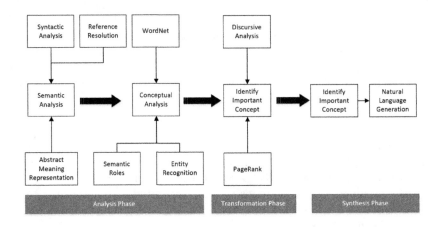

Fig. 1. Pipeline of the abstractive summarization method

3.1   Analysis Phase

This phase aimed at building the representation of a text as a graph. Given the abstractive approach, we had to change the original text using techniques of reference resolution to expand it and increase the amount of information in each sentence. This process helped the conceptual analysis.

In this work, we used the Natural Language Processing tools of theStanford[1], that includes also the syntactic tree generation, the part-of-speech tagging, delimitation of sentences and reference resolution among other tasks.

Because of the complexity of the reference resolution task, we only considered to exploit the references of pronouns to entities recognized by the annotations (*NN, NNS, NNP, NNPS*) into the *Part-of-Speech* Tagger. For example, in the

---

[1] Available at https://nlp.stanford.edu/software/. Accessed on February 2017.

following sentences we may see how this process increases the information contained when replacing the pronoun "it" with the full text of the organization that references:

*"**The United Nations Food and Agriculture** organization said hot and dry conditions in January and February were expected to reduce the total cereal harvest in 11 southern African countries to 16 m tonnes, 25% down on the average. [**It (PRP)—The United Nations Food and Agriculture (NNP)**]said Zimbabwe and South Africa, which normally offset shortages in the area with their own surpluses, would themselves have to import food"*

After these steps, we generated the knowledge graph that represents the document. To do this, we used an Abstract Meaning Representation parser called CAMR parser [17]. This parser has taken part in SemEval-2016[2] reaching an average F1 of 66.5% over the corpus of the competition.

Once generated the AMR graph for each sentence in the document, we needed to join all sentences via some analysis to generate a knowledge graph.

In the same line of work as used in [14], a model was necessary to take the analysis to a higher level of abstraction, which we called "Conceptual", due we needed to abstract the concepts to merge them and generate new sentences.

Unlike the work presented in [14], where VerbNet [9] was used through manual work to align concepts and semantic relationships, we generated conceptual graphs automatically based on the AMR output (and its features) that is already aligned with a linguistic resource such as Propbank[3], a corpus annotated with information related to syntactic and semantics of verbs.

In order to generate the conceptual graph, we used some and procedures and criteria to merge terms or expressions into a concept, which are shown as below:

- Semantic Roles: In AMR, the relationships between concepts have identifiers like `Arg0...Arg5` which are associated with a semantic role such as agent, patience, among others. In our work we used the relationship that exists between Propbank and VerbNet to identify the semantic relationships and semantic roles of each concept. Thus, we may find that `Arg0` usually represents the "Agent" for a verb. However, in case of ambiguity or when no exists information in PropBank, we associated semantic roles's PropBank with semantic role's VerbNet by default. This association may be seen in Table 1.
- Fusion by entities: AMR has ability to recognize entities like *Person, Organization, Location*, among others, which contains several subtypes. For example, *organization* contains *company, government, military, criminal organization*, among others. In our work, we merged entities (with the same name) which are considered "Agent", "Patient", "Goal" and "Theme" in the Semantic Role Labeling avoiding fusion of graphs by verbs because this generates confusion and ambiguity in the graph.

---

[2] Available at http://alt.qcri.org/semeval2016/. Accessed in February 2017.
[3] Proposition Bank Available at https://verbs.colorado.edu/mpalmer/projects/ace.html. Accessed on March 2017.

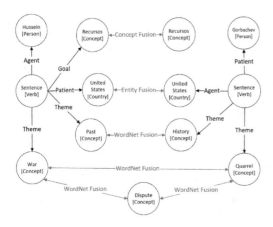

Table 1. Default relation between AMR and semantic roles

AMR relation	Semantic role
Arg0	Agent
Arg1	Patient
Arg2	Goal
Arg3	Start
Arg4	End

Fig. 2. Semantic graph fusion sample

- Fusion by WordNet concepts: Other criterion to merge terms/expressions into concepts was related to the measure got between two terms in the WordNet. To merge terms, we used the similarity measure Wu, proposed in [18]. In experiments, we determined that similarity measure must be greater than 0.9 to merge two terms into a concept.

Figure 2 shows an example of the fusion method using where we may appreciate that (1) some concepts that have been identified as *Agent* or *Patient* in different sentences, (2) entities recognized such as countries or persons, and (3) similar concepts in the WordNet (such as *Past* and *History*) may be merged.

3.2   Transformation Phase

In this phase, we needed to identify the most relevant concepts in the graph to create a summarization graph which includes them. To perform this, we executed the PageRank algorithm [1] over the conceptual graph. This algorithm is useful to identify relevant concepts considering the number of relations between different concepts and a possibility to do a random jump in a concept. In Eq. 1 we may see the formula where "M" represents the transition matrix (related to the number of relations), "v" represents the random jump vector, "c" represents a dumping factor and "Pr" represents the PageRank vector. In PageRank execution, the best results were obtained using a damping factor value of 0.65 and 30 iterations.

$$Pr = cMPr + (1 - c)v \qquad (1)$$

Once the PageRank was executed, we perceived that some nodes with many relations received higher weights (although the related nodes were less important) generating noise. To solve this problem, we incorporated discourse-level information into the PageRank, since this information has proven to be useful in extractive automatic summarization task [15]. We decided for this algorithm because use the nuclear-satellite information and take in consideration the relation type between the EDUs[4] to assign importance.

Thus, we applied a method proposed in [15] (called O'Donell method), which calculates the importance of each EDU according to the relations found in the Rhetorical Structure Theory [13]. The results got from O'Donell method were incorporated into the random jump vector in the PageRank algorithm. This made that concepts with in high score in O'Donell method benefit others around itself and unimportant concepts with many relations have low scores.

3.3 Synthesis Phase

Once the concepts in the graph have been weighted, the model iterated the conceptual graph to extract information about the actions done (*Verbs*), who has made those actions (*Agents*), who is affected by them (*Patients*), what is the theme (*Themes*) and what is the aim (*Goals*).

Then, our algorithm started in the verb nodes and as from there attempted to extract the nodes attached to it with the semantic relationship of Agent and thus for the semantic roles of *Patient, Theme* and *Goal*. Once these subgraphs were identified, it was the basis of a new sentence whose importance was given by:

$$Sentence\_Relevance = Sum(P(Agents) + P(Verbs) + P(Themes) + P(Goals))$$

These total values represented the final relevance of the expression. Then, we applied a descendant sorting over the sentence relevance to generate a summary with the most important expressions until up to a compression rate. To generate a sentence that has the synthesis of the document ideas a similar form as human production, we used `SimpleNLG` [5][5] as a tool for Natural Language Generation.

4 Evaluation

To evaluate the use of Abstract Meaning Representation and Discourse-level information into Automatic Summarization, we conducted experiments for each case. All experiments were performed on *Document Understanding Conference* (DUC) corpus[6].

Table 2 shows results of each experiment in the training corpus and test corpus, i.e., when only used the expanded conceptual graph with the reference resolution (conceptual + RR), when used a conceptual graph with reference resolution and includes discursive information (RST) (Conceptual + RR + RST) and when used a conceptual graph with reference resolution, includes discursive information (RST) and Natural Language Generation (Conceptual + RR + RST + NLG). Furthermore, the results improved in each experiment, i.e., Conceptual + RR + RST

---

[4] Elementary Discourse Unit is the basic unit in discourse-level.

[5] https://github.com/simplenlg/simplenlg. Last visited in February 2017.

[6] http://duc.nist.gov/data.html. Last visited in February 2017.

model was better than Conceptual + RR model and Conceptual + RR + RST + NLG model was better than Conceptual + RR + RST model.

We noted in our experiments with NLG that best combination was the use of *"with objective of"* like connector when we detected the *goal* semantic relation. For example, the sentence *"We agree possible international peaceful order devour large state and Gorbachev neighbor"* was transformed in *"We agreed with objective of possible international peaceful order devour large state and Gorbachev neighbor"*.

Another point to note is generated sentences had a correct use of the pronoun *We*, also we can identify the verb and expression goal. Table 2 shows a significant improvement in Rouge-1 and Rouge-L metrics and an important enhancement in Rouge-SU4 metric. This means a much better coherence in the generated text. In particular, the use of connectors like *And* and the correctness in the person and number over the generated expression improve the result.

In relation to incorporating discursive information into the original method, we may note an increment between conceptual and conceptual with discursive information. It based on the myopia of pure conceptual model to include additional concepts (*Agents, Patients, Goal, Themes*), because it only uses semantic relations at sentence level. For example, in a specific document the application of Conceptual + RR model produces 6 sentences, where four of them talk about the same subject. The discursive information incremented the possibility to detect expressions that can produce more valuable sentences to the summary. For example, in a same document, when applying the conceptual + RR + RST, it got 9 sentences where only two of them mentioned the same subject and the F1 Rouge-1 score was increment by 15%.

Table 2. F1 metric between Conceptual, Conceptual-RST and Conceptual-RST-NLG

Corpus	Training corpus			Test corpus		
F1/Rouge	R-1	R-L	R-SU4	R-1	R-L	R-SU4
Conceptual + RR	0.199	0.187	0.024	0.224	0.211	0.029
Conceptual + RR + RST	0.212	0.200	0.027	0.228	0.217	0.029
Conceptual + RR + RST + NLG	**0.230**	**0.216**	**0.031**	**0.244**	**0.231**	**0.033**

Finally, we may highlight that our experiments used none algorithm that may present an over-fitting to the specific data, so the goodness of the method only depends on the text in a document. Also, is important to note, the model never show a negative effect. However, the increment of the performance was not statistically significant according to the Wilcoxon Test.

5   Conclusions and Future Work

This work presented an automatic abstractive summarization using semantic representations and discourse-level information. The analysis phase used information from semantic analysis, got from use AMR parser for each sentence in a document. Then, we generated a conceptual graph by merging concepts with help of WordNet and Semantic Roles got from AMR. During the transformation phase, Discourse-level information was incorporated into PageRank algorithm to identify the most important concepts, resulting in an improvement on the concept identification.

In the synthesis phase, we implemented a navigation method to generate expressions from the ranked conceptual graph using hand-crafted rules based on semantic roles. With these rules we extracted many expressions that have a final score equal to the amount of their parts. After that, we sorted these expressions based on the amount score and take the most valuables for the natural language generation task, in our experiment we have worked with a compression rate of 20% more and less 100 words.

At last, the got expressions were used with SimpleNLG to generate a much natural expressions. In this work, we configured the tool to generate the sentence in a past form to get a coherent expression in tempo and number. The proposed method was evaluated on Document Understanding Conference (DUC) 2002 Corpus showing a F1 score of 24% on the Rouge-1 metric and outperformed the other variations of our method.

One limit related with the abstraction model is related with AMR. Although AMR is an important player in Semantic Analysis, in its current form is not enough to support the discovery and manipulation of the principal concepts, because it is too influenced by the syntax. We found evidence that different representations of the same idea are got, depending if these are written in active or passive voice.

One future work is related to the way of navigation or iteration over the Ranked Conceptual Graph with score information on its nodes to generate the candidate sentences. As a future work, we would like to explore other ways to navigate this graph to improve selection of concepts and generate better sentences.

References

1. Brin, S., Page, L.: The anatomy of a large-scale hypertextual web search engine. Comput. Netw. ISDN Syst. **30**(1), 107–117 (1998)
2. Carenini, G., Ng, R., Pauls, A.: Multi-document summarization of evaluative text. In: Proceedings of the Conference of the European Chapter of the Association for Computational Linguistics (2006)
3. Dang, H.T., Kipper, K., Palmer, M.: Integrating compositional semantics into a verb lexicon. In: Proceedings of the 18th conference on Computational linguistics, vol. 2, pp. 1011–1015. Association for Computational Linguistics (2000)
4. Flanigan, J., Thomson, S., Carbonell, J., Dyer, C., Smith, N.A.: A Discriminative Graph-Based Parser for the Abstract Meaning Representation, pp. 1426–1436. ACL (2014)
5. Gatt, A., Reiter, E.: SimpleNLG: a realisation engine for practical applications. In: Proceedings of the 12th European Workshop on Natural Language Generation, pp. 90–93. Association for Computational Linguistics (2009)

6. Gerani, S., Mehdad, Y., Carenini, G., Ng, R.T., Nejat, B.: Abstractive summarization of product reviews using discourse structure. In: EMNLP, pp. 1602–1613 (2014)
7. Montes-y-Gómez, M., Gelbukh, A., López-López, A., Baeza-Yates, R.: Flexible comparison of conceptual graphs. In: Mayr, H.C., Lazansky, J., Quirchmayr, G., Vogel, P. (eds.) DEXA 2001. LNCS, vol. 2113, pp. 102–111. Springer, Heidelberg (2001). doi:10.1007/3-540-44759-8_12
8. Kilgarriff, A., Fellbaum, C.: WordNet: An Electronic Lexical Database. MIT Press, Cambridge (2000)
9. Kipper, K., Dang, H.T., Palmer, M., et al.: Class-based construction of a verb lexicon. In: AAAI/IAAI, pp. 691–696 (2000)
10. Knight, K., Baranescu, L., Bonial, C., Georgescu, M., Griffitt, K., Hermjakob, U., Marcu, D., Palmer, M., Schneifer, N.: Abstract Meaning Representation (AMR) Annotation Release 1.0. Web download (2014)
11. Liu, F., Flanigan, J., Thomson, S., Sadeh, N., Smith, N.A.: Toward Abstractive Summarization Using Semantic Representations (2015)
12. Mani, I.: Automatic Summarization, vol. 3. John Benjamins Publishing, Amsterdam (2001)
13. Mann, W.C., Thompson, S.A.: Rhetorical structure theory: toward a functional theory of text organization. Text-Interdiscipl. J. Study Discourse 8(3), 243–281 (1988)
14. Miranda-Jiménez, S., Gelbukh, A., Sidorov, G.: Conceptual graphs as framework for summarizing short texts. Int. J. Concept. Struct. Smart Appl. (IJCSSA) 2(2), 55–75 (2014)
15. O Donnell, M.: Variable-length on-line document generation. In: the Proceedings of the 6th European Workshop on Natural Language Generation, Gerhard-Mercator University, Duisburg, Germany (1997)
16. Palmer, M., Gildea, D., Kingsbury, P.: The proposition bank: an annotated corpus of semantic roles. Comput. Linguist. 31(1), 71–106 (2005)
17. Wang, C., Pradhan, S., Pan, X., Ji, H., Xue, N.: CAMR at SemEval-2016 task 8: an extended transition-based AMR parser. In: Proceedings of the 10th International Workshop on Semantic Evaluation (SemEval-2016), pp. 1173–1178. Association for Computational Linguistics, San Diego (June 2016)
18. Wu, Z., Palmer, M.: Verbs semantics and lexical selection. In: Proceedings of the 32nd Annual Meeting on Association for Computational Linguistics, pp. 133–138. Association for Computational Linguistics (1994)

Development and Integration of Natural Brazilian Portuguese Synthetic Voices to Framework FIVE

Danilo S. Barbosa$^{(\boxtimes)}$, Byron L.D. Bezerra, and Alexandre M.A. Maciel

Polytechnic School of Pernambuco, University of Pernambuco, Recife, Brazil
{dsb3,byronleite,amam}@ecomp.poli.br
http://mestrado.ecomp.poli.br/

Abstract. The Framework FIVE is a multiplatform tool that assists the development of voice user interfaces applied in different technological environments. Several works have been carried out in order to provide increasingly in natural synthetic voices to FIVE, however, experiments realized with users has been reported the need for more friendly voices integrated to the framework. This paper describes the development and integration of natural synthetic voices in Brazilian Portuguese to the Framework FIVE. For this, a private audio and phonetics database were used on development of two voices (male and female) using the Unit Selection technique available on MaryTTS platform. For the integration process it was developed a specific web service. For comparison purposes, it was realized experiments to evaluate the naturalness and intelligibility of the voices, and the results obtained show that the constructed voices are more friendly, however, there is not a great difference when compared with HMM-based technique.

Keywords: Speech synthesis · Unit selection · HMM-based · Web service

1 Introduction

In recent years, the voice user interface area has received considerable attention from academics due to two reasons. First, due to improvements in the performance of automatic systems for speech processing, including speech synthesis, translation of spoken idioms and speech recognition. Second, due to the convergence of devices and the massive production of multimedia content that requires faster and more efficient means of interaction with the user. Besides those advancements, the construction of applications with voice user interface that can recognize speech, understand voice commands from its users and provide answers to users still is a challenge, due to its interdisciplinary and complexity of development [1].

The Portuguese language is the fifth most spoken language in the world with 220 million native speakers in four continents Africa, America, Asia and Europe.

K. Ekštein and V. Matoušek (Eds.): TSD 2017, LNAI 10415, pp. 491–499, 2017.
DOI: 10.1007/978-3-319-64206-2_55

According to Branco et al. [2], considering the new challenges encountered in the information society in a globalized world, there is a great need to direct efforts towards the creation of linguistic resources, development of tools and applications for voice processing in Portuguese, with the objective of attend the market's demand. For the Brazilian Portuguese language, several initiatives have been carried out in order to build a high-quality speech synthesis mechanism. Whether in the industry, such as the voices Felipe and Fernanda of Nuance[1] and the voices Tele and Heloisa of Microsoft[2], and on the academy with open source resources: the TextAnalysis4BP tool [3], the platform MaryTTS [4] and the Framework FIVE [5].

The Framework FIVE (*Framework for an Integrated Voice Environment*) is a tool built with the purpose to assist the process of construction and instantiation of speech engines in different technological environments (telephone, mobile, digital TV), in an integrated manner, scalable for various techniques and portable [5]. In Maciel et al. [6] was developed and integrated to Framework FIVE synthetic voices in Brazilian Portuguese using the HMM-based Speech Synthesis System (HTS), and in Souza et al. [7] was used Modular Architecture for Research on Speech Synthesis (MaryTTS) platform for the development of two new voices. These works were evaluated regarding audio transcription, appropriate voice quality criteria, and its portability capacity on operating system android.

Despite the good results obtained, the users reported the need to improve the naturalness and intelligibility of voices in Brazilian Portuguese. Another requirement mentioned by then was the possibility to integrate these voices into to Framework FIVE for use in several real applications. This article aims to present the development and integration of natural synthetic voices in Brazilian Portuguese to the Framework FIVE. For this, it was performed the preparation of the database (audio and phonetics) (Sect. 2). It was built a set of new voices using the Unit Selection technique available on MaryTTS platform (Sect. 3). Then, these voices were integrated to the Framework FIVE through a web service (Sect. 4). The voices were evaluated for their naturalness and intelligibility (Sect. 5), and Sect. 6 presents the conclusions and future works.

2 Preparing the Database

2.1 Audio Database

The audio base used in this work included the selection of 800 phonetically balanced phrases, containing between seven and 15 words, interrogative and affirmative, that were obtained from the journalistic context, historical texts, biographies, novels and poems. Examples in the Brazilian Portuguese: "Debaixo de uma pipa que pinga, há um pinto que pia", "A Amazônia é a reserva ecológica do globo", "Não tenho dúvida de quanto fui amado" and "Se os juízes ganham com o Tribunal de Pequenas Causas, por que os advogados não podem ganhar também?".

---

[1] Nuance Language: http://www.nuance.com.

[2] Microsoft Language: https://www.microsoft.com/en-us/download.

The editing of the audio database was performed at the ProTools tool[3], through the recording in the studio of four hours of audio, by two professional speakers, male and female. The editing consisted in segmenting the audio base, separation of audios and in the removal of noise which does not represent a voice signal. The characteristics of these audio files: sampling frequency of 16,000 kHz, with 16 bits per sample and stereo mode.

2.2 Phonetic Database

The words database used in this work was made available by VERO spelling checker of LibreOffice[4], with 140 thousand words in Brazilian Portuguese, containing substantive, adjective, verbs, and numerals. The phonetic alphabet used was based in SAMPA for Brazilian Portuguese with its 38 phonemes, and for generating the phonetic database it was used TextAnalysis4BP tool.

The process of generating the phonetic base consists of three phases. The Syllabification stage realizes the syllable division using an algorithm that identifies the vowels in words and applies a set of rules for syllable identification. The Grapheme-to-Phoneme (G2P) performs the conversion of the grapheme into its respective phoneme, through a set of rules defined for the Brazilian Portuguese language. The tonicity stage identified the tonic syllable of words [3]. Figure 1 shows the execution flowchart of the word house using the TextAnalysis4BP tool.

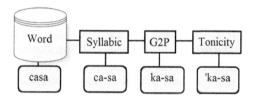

Fig. 1. Flowchart TextAnalysis4BP

3 Building Voices Using MaryTTS

The MaryTTS platform is composed of the client-server architecture, in which, a client sends an input text with a set of parameters to the server and receives an appropriate synthetic voice as results. The multi-threaded server allows the client to send multiple requests in this process and the client interface allows the users to choose different types of input (e.g. plain text, phonemes) and output types (e.g. audio, phonemes) [4]. The modular architecture of MaryTTS has support for adding several languages and build voice signal with the HMM-Based and Unit Selection techniques [8]. Recently this architecture has performed a

---

[3] ProTools: http://www.avid.com/pro-tools.
[4] Vero: https://pt-br.libreoffice.org/projetos/vero/.

modification of its modules with the use of Apache Maven [9] and of Gradle [10] to improve management of the modules. Thus, the MaryTTS platform modules have available the following parts: Core Client-Server responsible for the execution of web server. New Language Support Tools (NLST)[5] responsible for Natural Language Processing (NLP). Voice Creation Tools (VCT)[6] responsible for building the voice. And the Interface Installer responsible for installation of the synthetic voices [11].

The process of constructing voice in the MaryTTS platform realized in this work started with the execution of Client-Server module, a necessary condition for execution of the other modules. Then, the NLST was used to adapt the phonetic base to the MaryTTS input pattern. Although the Brazilian Portuguese language has NLST support [3], it was necessary to create a new support using the architecture Apache Maven [8] and Gradle [12]. The development of this support is required because the Transcription Tool (TT) receive as input the files generated by the phonetic base and obtain as output the grapheme-phoneme conversion files of words, rudimentary labeling, and phonetic transcriptions. In the VCT module, the Voice Import Tool (VIT) was used to construct two voices (male and female) through unit selection technique. This tool received as input the audio database with its corresponding transcriptions and the phonetic database with the phonetic segmentation. As output, it generated a set of the files to execute the synthetic voices [13].

4   Integration with FIVE

The FIVE architecture is composed five modules. FIVE CORE responsible for executing the feature extraction algorithms and pattern classification required for the construction of the speech engines. FIVE CONTROLLER responsible for handling all user requests between the layers that generate the engines and applications layer. The FIVE API consists of a set of routines and programming standards for access to the engines both by the applications (APP) and through the web service (FIVE WS). FIVE GUI consists of a front end, in wizard format, which assists developers in building the engines. Figure 2 shows the modules of the Framework FIVE.

In order to integrate the built voices on Mary TTS to framework FIVE it was necessary to adapt the *Synthesizer* and *Vocoder* classes contained in the FIVE API module, as well as, to build a particular package within FIVE CORE module necessary to access the *Web Client* module from MaryTTS. Figure 3 shows the integration modules between the Framework FIVE and MaryTTS.

The integration process occurred in three phases. The first phase involved importing the *maryttsTool, transcriptionTool*, and *voiceImport* libraries for the Framework FIVE. The *maryttsTool* library is responsible for communication between the client and server modules of MaryTTS. The *transcriptionTool*

---

[5] NLST: https://github.com/marytts/marytts/wiki/New-Language-Support.

[6] VCT: https://github.com/marytts/marytts/wiki/VoiceImportToolsTutorial.

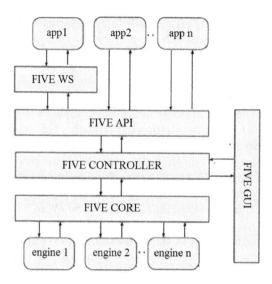

Fig. 2. Framework FIVE architecture

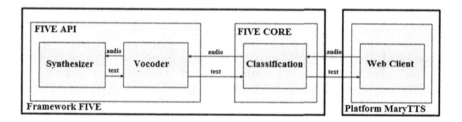

Fig. 3. Module between the framework FIVE and MaryTTS

library is responsible for running the transcription tool (TT), and the *voiceImport* library is responsible for running the voice import tool (VIT).

The second phase consisted of adapting the *Synthesizer* and *Vocoder* classes contained in the FIVE API module. In the *Synthesizer* class was added an *engineType* parameter to make easy the choice os which engine must be executed. In the *Vocoder* class, two additional parameters were added beyond *engineType*: the voice gender (male or female), and synthesis technique (unit selection or HMM-based).

The third phase consisted of creating the classes *MaryParameter*, *MaryCommands* and *MaryProcess* in the FIVE CORE module. The *MaryParameter* class is a model class composed of the necessary parameters for client-server execution e.g. host and server port parameters are required for running the server. The *MaryCommands* class utilize the *executeMaryTTS* method with the *phrase*, *speaker*, and *wavFileName* parameters for running the client request. And finally, in the *MaryProcess* class, a *test* method was structured with the *phrase, speaker* and *fileNamef* parameters for generating the automatic alignment file (lab) and the audio file (Wav).

5   Experiments

In order to evaluate the naturalness and intelligibility of the newly generated voices, it was realized a user evaluation experiment. The experiment used the voices constructed with the Unit Selection MaryTTS, HMM MaryTTS [7] and HMM HTS [6]. 400 audio samples were generated with this synthetic voices, containing between 10 and 15 words, not include in the database training. The audio samples were evaluated with Mean Opinion Score (MOS) metric, scale of 1 to 5 (1-bad, 2-poor, 3-fair, 4-Good and 5-Great). The Word Error Rates (WER) metric was utilized to identify the percentage of wrong words that have been replaced, deleted, and included in the transcription of the sentences.

The experiment selected 100 volunteers who heard a sequence with male and female audio. Each volunteer evaluated 40 audio samples using MOS and transcripted all sentences. The samples were divided into four parts: ten samples using Unit Selection MaryTTS voices, ten samples using HMM-Based MaryTTS voice, ten samples using HMM-Based HTS and ten samples using the natural audio base. Each of these parts was contemplated with five male and five female samples.

The mean time to perform this evaluation was 20 min, considering that each volunteer had 40 audio samples and each sample was evaluated in approximately 30 s. The evaluation of these voices was organized as follows: The Unit Selection MaryTTS receive the labels UM (Unit Selection-Male) and UF (Unit Selection-Female). The HMM MaryTTS receive the labels HM (HMM-Male) and HF (HMM-Female). The HMM HTS voices receive the labels TM (HTS-Male) and TF (HTS-Female). The natural voices receive the labels NM (Natural-Male) and NF (Natural-Female). Figures 4 and 5 show the results of boxplots for the MOS of naturalness and intelligibility.

The evaluation using the MOS scale has the following results. The male and female voices of the natural audio base (NM and NF) have better results, compared to the voices generated by synthesis techniques. The HMM HTS (TM and TF) has a result worse, compared to the methods created using MaryTTS. The naturalness assessment shows that the HMM Mary TTS (HM and HF) has better results compared to the Unit Selection MaryTTS (UM and UF). The intelligibility assessment indicates that HMM Mary TTS (HM and HF) has better results compared to the HMM HTS (TM and TF). And finally, the Unit Selection Mary TTS (UM and UF) has similar results to HMM Mary TTS (HM and HF).

The intelligibility evaluation using WER scale has the following results. The male and female voices of HMM HTS obtained the biggest error rate (TM 9.3%) and (TF 9.8%), and the natural voices got a lower error rate (NM 3.1%) and (NF 3.4%). Regarding the voices of Mary TTS platform, the male and female voices of based on HMM obtained a lower error rate (HM 4.6%, HF 4.5%), then based

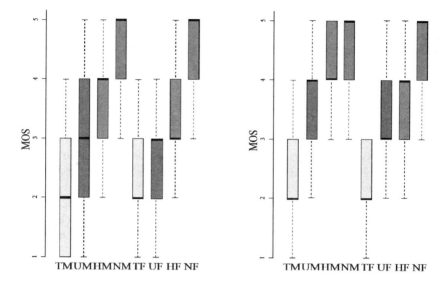

Fig. 4. Results of the naturalness   **Fig. 5.** Results of the intelligibility

on Unit Selection (UF 6.6%, UM 5.2%). Thus, it was observed that the HMM-based technique has a better intelligibility when compared with Unit Selection technique. Figures 6 and 7 shows the WER result male and female.

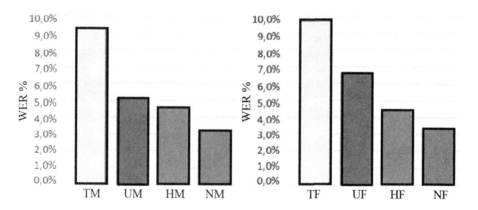

Fig. 6. Results WER male   **Fig. 7.** Results WER female

A detailed evaluation of the transcriptions shows that HMM HTS (TM and TF) has substitution problem in words with near punctuation. Confirming some punctuation problems noted in previous evaluations [6]. The MaryTTS unit selection technique has problems related the concatenating of some words. This problem occurred due to the small evidence of punctuation sentences (e.g. commas and dots). The audio database has 800 phrases, but only 200 phrases have any punctuation. The HMM Mary TTS voices presented problems related to the substitution of words with errors in verbal conjugation and plural identification. Examples in Portuguese: "As coisas tem tudo para darem certo mas é preciso batalhar por elas" replaced with "As coisas tem tudo para darem certo mas é preciso batalhar por ela".

6   Conclusions

This article presents the development of natural synthetic voices in Brazilian Portuguese, and the integration of this then to Framework FIVE. The process of constructing the voices, taking as reference the MaryTTS tools was successful. In addition, using this platform has optimized the construction of voices and made the learning curve faster. Regarding integration to Framework FIVE, a small set of changes in architecture optimized the extensibility for new techniques and portability for other platforms.

About voice quality, the construction of the new database with audio and phonetics contributed to improving the quality of voices. The voices built with MaryTTS platform have achieved better naturalness and intelligibility when compared with HTS tool. In the WER scale evaluation, any representative differences were detected when comparing Unit Selection technique and HMM technique.

Thus, this work managed to reach its objectives and contribute to the availability of more natural synthetic voices for Brazilian Portuguese at the FIVE. As future works, we intend to expand our audio database, specifically with samples with punctuation. In the integration process, we intend to test this new architecture in a real application.

Acknowledgments. The authors would like to thank the support of this work through the research projects granted by: "CNPQ-Bolsa de Produtividade DT" (Process 310752/ 2015-9), "CNPQ Edital Universal" (Process 444745/2014-9) and "FACEPE - Edital PRONEX" (Process APQ 0880-1.03/14).

References

1. Farinazzo, V., Kawamoto, A.L.S., de Oliveira Neto, J.S., Salvador, M.: An Empirical Approach for the Evaluation of Voice User Interfaces. INTECH Open Access Publisher, Rijeka (2010)
2. Branco, A., Mendes, A., Pereira, S., Henriques, P., Pellegrini, T., Meinedo, H., Trancoso, I., Quaresma, P., de Lima, V.L.S.: The Portuguese language in the digital age (2012)
3. Couto, I., Neto, N., Tadaiesky, V., Klautau, A., Maia, R.: An open source HMM-based text-to-speech system for Brazilian Portuguese. In: 7th International Telecommunications Symposium (2010)
4. Schröder, M., Trouvain, J.: The German text-to-speech synthesis system MARY: a tool for research, development and teaching. Int. J. Speech Technol. 6(4), 365–377 (2003)

5. Maciel, A., Carvalho, E.: FIVE-framework for an integrated voice environment. In: Proceedings of International Conference on Systems, Signal and Image Processing, Rio de Janeiro (2010)
6. Maciel, A., Carvalho Filho, E.: Integration and evaluation of an HMM-based text-to-speech system to FIVE. In: 2012 19th International Conference on Systems, Signals and Image Processing (IWSSIP), pp. 633–636. IEEE (2012)
7. Souza, D., Saturnino, L., Maciel, A.M.: A portability evaluation of Brazilian Portuguese voices produced with MARY TTS. In: 2014 International Conference on Systems, Signals and Image Processing (IWSSIP), pp. 95–98. IEEE (2014)
8. Charfuelan, M., Pammi, S., Steiner, I.: MARY TTS unit selection and HMM-based voices for the blizzard challenge 2013. In: Blizzard Challenge Workshop (2013)
9. Maven: Software project. https://maven.apache.org
10. Gradle: Build tool. https://gradle.org/
11. Le Maguer, S., Steiner, I.: The MaryTTS entry for the blizzard challenge 2016 (2016)
12. Steiner, I., Le Maguer, S., Manzoni, J., Gilles, P., Trouvain, J.: Developing new language tools for MaryTTS: the case of Luxembourgish. In: 28th Conference on Electronic Speech Signal Processing (ESSV), Saarbrücken, Germany (2017)
13. Pammi, S., Charfuelan, M., Schröder, M.: Multilingual voice creation toolkit for the MARY TTS platform. In: LREC. Citeseer (2010)

Fine-Tuning Word Embeddings
for Aspect-Based Sentiment Analysis

Duc-Hong Pham[1,2], Thi-Thanh-Tan Nguyen[2], and Anh-Cuong Le[3(✉)]

[1] Faculty of Information Technology, University of Engineering and Technology,
Vietnam National University, Hanoi, Vietnam
[2] Faculty of Information Technology, Electric Power University, Hanoi, Vietnam
{hongpd,tanntt}@epu.edu.vn
[3] Faculty of Information Technology, Ton Duc Thang University,
Ho Chi Minh, Vietnam
leanhcuong@tdt.edu.vn

Abstract. Nowadays word embeddings, also known as word vectors, play an important role for many natural language processing (NLP) tasks. In general, these word embeddings are learned from unsupervised learning models (e.g. Word2Vec, GloVe) with a large unannotated corpus and they are independent with the task of their application. In this paper we aim to enrich word embeddings by adding more information from a specific task that is the aspect based sentiment analysis. We propose a model using a convolutional neural network that takes a labeled data set, the learned word embeddings from an unsupervised learning model (e.g. Word2Vec) as input and fine-tunes word embeddings to capture aspect category and sentiment information. We conduct experiments on restaurant review data (http://spidr-ursa.rutgers.edu/datasets/). Experimental results show that fine-tuned word embeddings outperform unsupervisedly learned word embeddings.

1 Introduction

Word representations have now become a critical component of many natural language processing systems. Alternatively, words can be represented as vectors in a semantic space. This approach is also called as distributed representations of words, or word embeddings. The simplest way for this approach is to represent each word in the vocabulary as a one-hot vector, which contains only a one in a position and zeros in all other positions. The dimensionality of these vectors equals to the vocabulary size. Using this technique can overcome the "hand-crafted" drawback of semantic networks (e.g. WordNet), but its vectors still cannot provide useful evidence to evaluate similarity between words. To overcome this limitation, some studies such as [3, 8, 9] learn word representations based on contexts of words and the achieved results are words with similar grammatical usages and semantic meanings. As the result, each word is represented as a real-valued vector with a low-dimensional and continuous, this known as a word embedding. By this way each word will carry more semantic and grammatical

© Springer International Publishing AG 2017
K. Ekštein and V. Matoušek (Eds.): TSD 2017, LNAI 10415, pp. 500–508, 2017.
DOI: 10.1007/978-3-319-64206-2_56

information, expressing the relationship between the words through the measurement between the vectors, for example, two words "excellent" and "good" are mapped into neighboring vectors in the embedding space. Since word embeddings capture rich linguistic information, they have been used as inputs or a representational basis (i.e. features) for many NLP tasks, such as text classification, document classification, information retrieval, question answering, name entity recognition, sentiment analysis, and so on.

Aspect-based sentiment analysis is a special type of sentiment analysis, it aims to identify the different aspects of entities in textual reviews and the corresponding sentiment toward them. There are some related studies such as aspect term extraction [1,10], aspect category detection and aspect sentiment classification [1,4,5]. Although the context-based word embeddings have been proven useful in many NLP tasks (Collobert et al. [2]), but they can ignore the aspect category and sentiment information when applying to the aspect-based sentiment analysis tasks. The problem here is how to utilize these kinds of information and integrate them into the general word embeddings. We will consider this problem as the task of tuning the general word embeddings towards the objective of aspect category detection and sentiment analysis classification.

In this paper we consider convolutional neural networks which have achieved remarkably strong results in many studies, such as sentence classification [6], aspect extraction for opinion mining [10]. Inspiring from that, we will propose a model using a convolutional neural network that takes a labeled data set and the learned word embeddings from an unsupervised learning model (e.g. Word2Vec) as input, and fine-tunes the word embeddings to capture aspect category and sentiment information in labeled sentences. In experiment, we evaluate the effectiveness of word embeddings by applying them to two tasks of aspect-based sentiment analysis including aspect category detection and aspect sentiment classification.

2   Related Work

In this section we review existing works closely related to our work. It includes word embedding techniques, word embeddings in sentiment analysis, and aspect-based sentiment analysis.

Word embedding techniques began development in 2003 with the neural network language model (NNLM) in (Bengio et al. [3]). In 2013, Mikolov et al. [8] proposed two models, namely skip-gram and continuous bag-of-words (CBOW). They improve training efficiency and learn high-quality word representation by simplifying the internal structure of the NNLM. In 2014, Pennington et al. [9] proposed the GloVe model using a global log-bilinear regression model that outperforms the original models of Skip-gram and CBOW.

Many studies learned sentiment-specific word embeddings for sentiment analysis, they attempt to capture more information into word embeddings, such as Maas et al. [7] extended the unsupervised probabilistic model to incorporate sentiment information. Tang et al. [12] proposed three neural network models to learn word vectors from tweets containing positive and negative emoticons. Ren et al. [11] improved the models of Tang et al. [12] by incorporating topic information and learning topic-enriched multiple prototype embeddings for each word. Zhou et al. [15] proposed a semi-supervised word embedding based on the skip-gram model of Word2Vec algorithm, the word embeddings in their model can capture some information as semantic, the relations between sentiment words and aspects.

Some tasks of aspect-based sentiment analysis have been done by using lexical information and classification methods, such as aspect sentiment classifier (Wagner et al. [14]; Ganu et al. [4]), aspect category detection (Ganu et al. [4]; Kiritchenko et al. [5]). They have a limitation that they cannot capture semantic interactions between different words in sentences. Recently, some studies have been used word embeddings as input, such as Alghunaim et al. [1] using CRF-suite or SVM-HMM with word embeddings as features. Poria et al. [10] used word embeddings as input, they then use a deep convolutional neural network to produce local features around each word in a sentence and combine these features into a global feature vector. However, most these studies have been used the learned word embeddings from Word2Vec model (Mikolov et al. [8]) which only capture word semantic relations and ignore supervised information such as aspect category and aspect sentiment, that is our objective in this paper.

3   Proposed Method

Given a set of labeled sentences $D = \{d_1, d_2, ..., d_{|D|}\}$ extracting from a collection of reviews in a particular domain (e.g. restaurant), each sentence $d \in D$ is assigned two labels, i.e. aspect sentiment and aspect category. Let k be the number of aspect category labels and m be the number of aspect sentiment labels in D. We denote $a_d \in \mathrm{R}^k$ to be a binary label vector of the aspect categories in the sentence d. Each value in a_d indicates that the sentence d is discussing an aspect category or not. Denoted $o_d \in \mathrm{R}^m$ to be a binary label vector of the aspect sentiments in the sentence d. Each value in o_d indicates that the sentence d is discussing an aspect sentiment or not. Let $V = \{\omega_1, \omega_2, ..., \omega_{|V|}\}$ be a vocabulary and $W \in \mathrm{R}^{n \times |V|}$ be a word embedding matrix, where the i-th column of W is the n-dimensional embedded vector for word i-th, w_i in V. We assume that the matrix W is learned from an unsupervised learning model (e.g. Word2Vec) with a large number of unlabeled sentences. We need to fine-tune word embeddings in matrix W.

In the following, we will present a model using a convolutional neural network to fine-tune word embeddings. The word embeddings are initialized by the learned word embeddings from an unsupervised learning model (e.g. Word2Vec). They then are re-computed to capture aspect category and sentiment information. We named this model as Word Embedding Fine-Tuning (WEFT) model, its architecture is shown in Fig. 1. Our model is similar to the CNN-non-static model of (Kim et al. [6]), but different from it, our model has two output vectors corresponding to aspect category and aspect sentiment information.

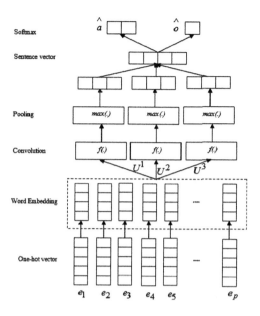

Fig. 1. An illustration of the WEFT model

For a sentence $d \in D$ containing p words, we present each layer with necessary formulations and notations as follows

Word Embedding Layer. The word embedding of word i-th is computed as:

$$x_i = W.e_i \tag{1}$$

where e_i is a one-hot vector of the word i-th in the sentence d.

Convolution Layer. This layer receives $x_1, x_2, ..., x_p$ as input and we use three convolutional filters with widths as 1, 2 and 3 to encode the semantics of unigrams, bigrams and trigrams in the sentence d. For the filter t-th, $1 \leq t \leq 3$, we use the convolution operation to obtain a new vector sequence $y_1^{1t}, y_2^{1t}, ..., y_p^{1t}$ according to the following equation:

$$y_i^{1t} = f(U^t.x_{i:i+h^t-1} + u^t) \tag{2}$$

where $x_{i:i+h^t-1}$ denotes the concatenation of words $x_i, x_{i+1}, ..., x_{i+h^t-1}$, h^t is a filter window size to combine embedding vectors and $f(.)$ is a non-linear function (i.e. $f(y) = \tanh(y) = \frac{e^y - e^{-y}}{e^y + e^{-y}}$). $U^t \in R^{C \times h^t.n}$ and $u^t \in R^C$ are model parameters, which are learned during training, C is the output dimension.

Pooling Layer. We apply the max pooling operation [2] to mix the varying number of features from the convolution layer into one vector with fixed dimension

$$y^{2t} = [max(y_{i1}^{1t}), max(y_{i2}^{1t}), ..., max(y_{iC}^{1t})] \tag{3}$$

where y_{ij}^{1t} denotes the j-th dimension of y_i^1, $y^2 \in R^C$ is the output of the filter t-th.

Sentence Vector Layer. In order to obtain the sentence representation vector, this layer concatenates the output vectors of three convolutional filters into one vector as follows $v(d) = [y^{21}, y^{22}, y^{23}]$, $v(d) \in R^{3C}$.

Softmax Layer. The aspect category vector $\hat{a}_d$ and the sentiment vector $\hat{o}_d$ are computed as follows: $\hat{a}_d = g(V^1.v(d) + b^1)$ and $\hat{o}_d = g(V^2.v(d) + b^2)$, where $V^1 \in R^{k \times 3C}$ and $V^2 \in R^{m \times 3C}$ are weight matrices from the sentence vector layer to the softmax layer, $b^1 \in R^k$ and $b^2 \in R^m$ are bias vectors, g is the softmax function.

Model Learning. The cross entropy cost function for the data set D is,

$$E(\theta) = -\sum_{d \in D}\left(\sum_{i=1}^{k} a_{di} \log \hat{a}_{di} + \sum_{i=1}^{m} o_{di} \log \hat{o}_{di}\right) + \frac{1}{2}\lambda_\theta \|\theta\|^2 \qquad (4)$$

where $\theta = [U^1, U^2, U^3, W, V^1, V^2, u^1, u^2, u^3, b^1, b^2]$ and $\|\theta\|^2 = \sum_i \theta_i^2$ is a norm regularization term. In order to compute the parameters θ, we apply back-propagation algorithm with stochastic gradient descent to minimize this cost function.

4 Experiments

4.1 Experimental Data

We use two data sets on the domain of restaurant products: the first data set contains 3,111,239 unlabeled sentences which is extracted from 229,907 reviews[1]. This data set will be used to learn word embeddings. The second data set contains 190,655 labeled sentences, is extracted from 52,574 reviews[2], this data set used in previous work (Ganu et al. [4]; Wang et al. [13]). It contains six aspect category labels as *Price, Food, Service, Ambience, Anecdotes,* and *Miscellaneous.* The four aspect sentiment labels include *Positive, Negative, Neutral* and *Conflict.* Each sentence is labeled for both aspect category and sentiment labels. We randomly get 75% of the given sentences to fine-tune word embeddings, the remaining 25% of given sentences to evaluate the quality of the WEFT model. Some statistics of the second data set are shown in Table 1.

4.2 Implementation of the WEFT Model

We develop a back-propagation algorithm with stochastic gradient descent to minimize the cost function in Eq. (4). We then use it to fine-tune word embeddings with the parameters: the filter window sizes $h^1 = 1$, $h^2 = 2$ and $h^3 = 3$;

---

[1] https://www.yelp.com/datasetchallenge/.
[2] http://spidr-ursa.rutgers.edu/datasets/.

Table 1. Statistics of the second data set

Aspect	Number of sentences used for	
	Fine-tuning word embeddings	Evaluating word embeddings
Price	4,386	1,462
Food	44,912	14,970
Service	22,470	7,489
Ambience	17,729	5,909
Anecdotes	18,396	6,132
Miscellaneous	35,100	11,700
Total	142,993	47,662

the output dimension C of each filter is 100; the mini-batch size is 60; the regularizations $\lambda_W = \lambda_{U^1} = \lambda_{U^2} = \lambda_{U^3} = 10^{-4}$, $\lambda_{u^1} = \lambda_{u^2} = \lambda_{u^3} = 10^{-5}$, $\lambda_{V^1} = \lambda_{b^1} = \lambda_{V^2} = \lambda_{b^2} = 10^{-3}$; the weight matrices U^1, U^2, U^3, V^1, V^2 are randomly initialized in the range of $[-1, 1]$; the bias vectors u^1, u^2, u^3, b^1, b^2 are initialized to be zero; the learning rate η is 0.025; the iterative threshold $I = 50$. Matrix W is initialized by the learned word embeddings from an unsupervised learning model (e.g. Word2Vec).

4.3 Evaluation

We evaluate our WEFT model through fine-tuning the learned word embeddings from the models: CBOW, skip-gram of Word2Vec (Mikolov et al. [8]) and GloVe (Pennington et al. [9]). We denote variants of the WEFT model as follows: WEFT-rand: Using randomly initialized word embeddings and then modified during training. WEFT-SG, WEFT-CB and WEFT-GV are three models that fine-tune the learned word embeddings from skip-gram, CBOW and GloVe model. Note that in our work, we use the tools of Word2Vec[3] and GloVe[4] to learn word embeddings, the size of dimensional word embedding is 300 and the window size of context is 4.

In fact, we can not evaluate the word embeddings directly because we can not obtain the ground-truth word embeddings from the given dataset. As a result, we choose to evaluate the word embeddings indirectly through using them as input of a prediction model in two tasks of aspect-based sentiment analysis, i.e. aspect category detection and aspect sentiment classification. Specifically, we use the CNN model (Kim et al. [6]) as the prediction model for these tasks. The lower prediction results in any experiment mean that the word embeddings used in that case are low. In each experiment case, the CNN model performs 4-fold cross validation with 47,662 sentences. The metrics are used to measure the prediction results as F1 score and Accuracy.

---

[3] https://github.com/piskvorky/gensim/.
[4] https://nlp.stanford.edu/projects/glove/.

Table 2. Results for ACD   **Table 3.** Results for ASP

Method	F1 score
SG	77.87
CB	78.54
GV	79.19
WEFT-rand	81.43
WEFT-SG	81.50
WEFT-CB	81.76
WEFT-GV	**82.09**

Method	Pos-F1	Neg-F1	Neu-F1	Con-F1	Accuracy
SG	87.05	52.03	65.74	55.46	78.77
CB	86.93	52.25	66.60	55.93	79.22
GV	87.10	51.07	71.02	57.85	80.35
WEFT-rand	88.65	64.18	74.13	56.40	82.15
WEFT-SG	90.87	64.63	73.82	60.23	83.82
WEFT-CB	93.12	64.70	77.03	61.17	84.05
WEFT-GV	93.61	64.77	77.11	61.43	**84.23**

Aspect Category Detection (ACD). In Table 2, we show the achieved F1-score of aspect category detection task for each method. In a general observation, using the fine-tuned word embeddings from the WEFT model gives the better results. This indicates that the fine-tuned word embeddings from the WEFT model is better than other models. Additionally, although the WEFT-rand only captures aspect category and sentiment information in each labeled sentence, but it outperforms CBOW, skip-gram and GloVe model. This shows that the aspect category and sentiment information play an important role in word embeddings.

Aspect Sentiment Prediction (ASP). In Table 3, we show the achieved results of each method. In most cases, using the fine-tuned word embeddings from the WEFT model give the better results. This indicates that the fine-tuned word embeddings from our model helped improve the aspect sentiment predicted results.

Table 4. Each target word is given with its four most similar words using cosine similarity of the vectors determined by each model

	good	bad	food	price
GloVe model	excellent	poor	props	prices
	decent	awful	postings	pricing
	fantastic	horrible	vary	penny
	costco	terrible	gosh	albeit
WEFT-rand model	delight	horrible	snacks	prices
	goodness	alien	restaurant	pricey
	millionaires	calling	hospitality	pricing
	paycheck	poor	fashion	350
WEFT-GV model	excellent	terrible	snacks	prices
	great	lousy	foods	pricing
	wonderful	worse	meal	bills
	bron	poor	variation	pricey

Querying Words. We also evaluate the quality of the word embeddings through querying words. Given some sentiment words or aspect words, we can find out the most similar words with them. In Table 4, we show the interesting results of the most similar words to four given words "good", "bad", "food", and "price". Two models GloVe and WEFT-GV capture broad semantic similarities. The WEFT-rand model captures the true semantic of two aspect words "food" and "price", but it does not capture the good meaning of the two sentiment words "good" and "bad", this is because the WEFT-rand model does not use the semantic of input text. The WEFT-GV model captures the three information kinds, i.e. semantic, aspect category and aspect sentiment, thus it seems to be better than GloVe. This comparison again indicate that adding aspect category and sentiment information into word embeddings is very important.

5   Conclusion

In this paper, we have proposed a model using a convolutional neural network to fine-tune word embeddings for aspect-based sentiment analysis. Word embeddings in our model are initialized by the learned word embeddings from unsupervised learning models and then recomputed to capture aspect category and sentiment information. From experimental results, we have demonstrated that the fine-tuned word embeddings from our WEFT model outperform other word embeddings, which are learned from CBOW, skip-gram and GloVe model.

Acknowledgement. This paper is supported by The Vietnam National Foundation for Science and Technology Development (NAFOSTED) under grant number 102.01-2014.22.

References

1. Alghunaim, A., Mohtarami, M., Cyphers, S., Glass, J.: A vector space approach for aspect based sentiment analysis. In: Proceedings of NAACL-HLT, pp. 116–122 (2015)
2. Collobert, R., Weston, J.: A unified architecture for natural language processing. In: Proceedings of the ICML, pp. 160–167 (2008)
3. Bengio, Y., Ducharme, R., Vincent, P., Janvin, C.: A neural probabilistic language model. J. Mach. Learn. Res. **3**, 1137–1155 (2003)
4. Ganu, G., Elhadad, N., Marian, A.: Beyond the stars: improving rating predictions using review text content. In: Proceedings of WebDB, pp. 1–6 (2009)
5. Kiritchenko, S., Zhu, X., Cherry, C., Mohammad, S.M.: NRC-Canada-2014: detecting aspects and sentiment in customer reviews. In: Proceedings of SemEval, pp. 437–442 (2014)
6. Kim, Y.: Convolutional neural networks for sentence classification. In: Proceedings of EMNLP, pp. 1746–1751 (2014)
7. Maas, A.L., Daly, R.E., Pham, P.T., Huang, D., Nguyen, A.Y., Potts, C.: Learning word vectors for sentiment analysis. In: Proceedings of ACL, pp. 142–150 (2011)
8. Mikolov, T., Chen, K., Corrado, G., Dean, J.: Efficient estimation of word representations in vector space. In: Proceedings of Workshop at ICLR (2013)

9. Pennington, J., Socher, R., Manning, C.D.: GloVe global vectors for word representation. In: Proceedings of EMNLP, pp. 1532–1543 (2014)

10. Poria, S., Cambria, E., Gelbukh, A.: Aspect extraction for opinion mining with a deep convolutional neural network. Knowl. Based Syst. **108**, 42–49 (2016)

11. Ren, Y., Zhang, Y., Zhang, M., Ji, D.: Improving twitter sentiment classification using topic-enriched multi-prototype word embeddings. In: AAAI, pp. 3038–3044 (2016)

12. Tang, D., Qin, B., Liu, T.: Learning sentiment-specific word embedding for Twitter sentiment classification. In: Proceedings of ACL, pp. 1555–1565 (2014)

13. Wang, L., Liu, K., Cao, Z., Zhao, J., de Melo, G.: Sentiment-aspect extraction based on restricted boltzmann machines. In: Proceedings of ACL, pp. 616–625 (2015)

14. Wagner, J., Arora, P., Cortes, S., Barman, U., Bogdanova, D., Foster, J., Tounsi, L.: Aspect based polarity classification for semeval task 4. In: Proceedings of SemEval, pp. 223–229 (2014)

15. Zhou, X., Wan, X., Xiao, J.: Representation learning for aspect category detection in online reviews. In: Proceedings of AAAI, pp. 417–423 (2015)

Recognition of the Electrolaryngeal Speech: Comparison Between Human and Machine

Petr Stanislav[⊠], Josef V. Psutka, and Josef Psutka

Department of Cybernetics, Faculty of Applied Sciences,
University of West Bohemia, Pilsen, Czech Republic
{pstanisl,psutka_j,psutka}@kky.zcu.cz
http://www.kky.zcu.cz/en

Abstract. Automatic recognition of an electrolaryngeal speech is usually a hard task due to the fact that all phonemes tend to be voiced. However, using a strong language model (LM) for continuous speech recognition task, we can achieve satisfactory recognition accuracy. On the other hand, the recognition of isolated words or phrase sentences containing only several words poses a problem, as in this case, the LM does not have a chance to properly support the recognition. At the same time, the recognition of short phrases has a great practical potential. In this paper, we would like to discuss poor performance of the electrolaryngeal speech automatic speech recognition (ASR), especially for isolated words. By comparing the results achieved by humans and the ASR system, we will attempt to show that even humans are unable to distinguish the identity of the word, differing only in voicing, always correctly. We describe three experiments: the one represents blind recognition, i.e., the ability to correctly recognize an isolated word selected from a vocabulary of more than a million words. The second experiment shows results achieved when there is some additional knowledge about the task, specifically, when the recognition vocabulary is reduced only to words that actually are included in the test. And the third test evaluates the ability to distinguish two similar words (differing only in voicing) for both the human and the ASR system.

Keywords: Electrolaryngeal speech · ASR · Listening tests

1 Introduction

Surgical removal of the vocal cords during total laryngectomy has a major impact on the quality of life of a person. After such surgery people are unable to produce a voiced sound, their speech is called alaryngeal speech. There are several methods of restoring the speech after total laryngectomy (TL). Esophageal speech belongs to the most common methods used for speech restoration. The idea is based on releasing gases from esophagus instead of lungs. But this method has a low acquisition rate (only ~6% of the patients are able to learn it [6]). Another

© Springer International Publishing AG 2017
K. Ekštein and V. Matoušek (Eds.): TSD 2017, LNAI 10415, pp. 509–517, 2017.
DOI: 10.1007/978-3-319-64206-2_57

method uses a tracheoesophageal prosthesis that connects the larynx with pharynx. The air passing into the pharynx causes the required vibrations and the voiced sound can be created [5].

Another option to produce the excitation necessary for proper voicing is using an external device - the electrolarynx (EL). EL is a battery-powered device that mechanically generates sound source signals which are conducted into the oral cavity from either the soft parts of the neck or from the lower jaw [9]. In most cases, a monotone pitch is generated and the produced speech sounds very mechanical. Moreover, in order to make the resulting speech sufficiently audible, the EL needs to generate rather strong vibrations that are also emitted outside of the speaker's body and could be perceived as noise by other people. Unfortunately, neither of these methods of rehabilitation can guarantee the natural sound of the resulting voice.

The great disadvantage of using EL is that all pronounced phonemes tend to be voiced. That's why the automatic recognition of the EL speech is usually a very hard task. Nevertheless, we can achieve good results but only in cases where we can use a strong language model (LM) during recognition. Unfortunately, the LM can fully unfold its strength only when recognizing longer sentence segments where it can make use of the word context. But the problem is that speakers with TL almost do not communicate. And if they do, they do so only in short phrases or instructions. This is mainly due to the unnatural sound of their voice which they are often ashamed of. The second reason is that shorter sentences are less exhausting for EL speakers. In this case, the benefit of the LM seems to be negligible (it is hard to determine the identity of words without the necessary context). However, such task has a great practical potential.

In this paper, we would like to compare the ability of humans and machines (equipped with ASR system) to distinguish the identity of acoustically very similar words and word bigrams. For this purpose, we created two listening tests. The first one contains 320 isolated words and the second one contains 333 word bigrams. Both tests are described in detail (including the results) in Sect. 3. In Sect. 4, we present the process of the ASR system training and ASR results on test sentences. And finally in Sect. 5, we offer the comparison between human and machine recognition results.

2   Database Description

Although there are various ways to rehabilitate the voice [1], none of them is perfect. One possible way of improving the patient's ability to communicate is to use ASR and speech synthesis (TTS) technology [4] in some means of communication, e.g., for phone calls, but the biggest issue is the lack of speech data for training the ASR acoustic models. In comparison with healthy speakers, speech recording is naturally significantly more exhausting for people with TL using the electrolarynx and it is therefore complicated to obtain the relevant amount of data.

For the purpose of our research, we recorded a speech corpus which contains 5589 sentences and 320 isolated words from one speaker. It consists of about

14 h of annotated speech. The speaker is an older woman who underwent TL more than 10 years ago and uses electrolarynx (EL) on a daily basis. All the sentences and words are in the Czech language. The database of text prompts from which the sentences were selected was obtained in an electronic form from the web pages of Czech newspaper publishers [8].

The corpus consists of two parts. The major part contains 40 phonetically rich and 5036 phonetically balanced sentences. The other part contains 419 sentences and 320 isolated words. Separation of the corpus into two parts was necessary due to the large time span between the recordings. During that period, the speaker changed the type of the electrolarynx, which plays an important role in the quality of the speech, and also the recording technology was upgraded.

Part of the corpus with isolated words contains 160 pairs of specifically selected words that have different meaning but they acoustically differ only in voicing. An example is a pair of Czech words "kosa[1]" - "koza[2]". For each word, we recorded at least one sentence containing the word.

All the speech utterances are sampled at 48 kHz and 16 bit amplitude resolution and resampled to a sampling frequency of 16 kHz for the speech recognition task.

3   Listening Tests

For our human vs. machine comparison, it is crucial to obtain relevant results from as large group of listeners as possible. For this purpose, we created two listening tests. The first one contained recordings of isolated words from the corpus mentioned above, the second one the combination of word pairs that differ only in the voicing of exactly one phoneme. The task was to correctly recognize the identity of recorded words. In both cases, participants were asked to choose one of the prepared answers through a web-based interface. Subjects were allowed to replay the speech recording as many times as they wanted. The participation in the test was voluntary and the subjects were predominantly fellow researchers from our lab.

The listening tests have been designed to verify the human ability to distinguish two words having various meanings but differing only in the voicing of a single phoneme.

Such task is rather easy when listening to "natural" speech but in the case of electrolaryngeal speech, the situation is quite different. The reader should bear in mind that the EL produces the excitation signal constantly and therefore even the originally voiceless phonemes should theoretically become voiced. The form of isolated words and word bigrams was chosen intentionally because it is much harder to determine the meaning – and thus correctly recognize the word identity – when there is no context.

---

[1] Scythe.
[2] Goat.

3.1   Isolated Words

In the test, participants were asked to listen to 320 recordings of isolated words and select one of the prepared options which were in the form: (a) *word A (e.g. kosa)*, (b) *word B (e.g. koza)*, (c) *cannot decide*. The first two options represent a combination of the word that was really uttered in the recording and a complementary word differing in a single phoneme voicing. These options were in all cases presented in the alphabetical order. If the participants had no clue about the identity of the replayed word, they were instructed to choose the third option (cannot decide). The samples were presented to each listener in random order. The test was completed by 19 subjects.

The output of the test is a table with percentages of responses for each option. Table 1 shows an excerpt from the results. The percentages of the correct answers are highlighted by boldface. The first example represents situations where participants were unable to make a clear decision about the word identity. The second example represents a pair of complementary words where all the participants selected in all cases the same option, regardless of the word that was actually uttered. It could be interpreted as that the acoustic form of the Czech word "kosa" is exactly the same as the word "koza" if the speaker uses the EL. The last example shows a pair of words for which the participants were able to identify the word correctly in almost all cases. The overall accuracy of answers was $Acc_w^{human} = 70,47\%$ and it was counted as follows

$$Acc_w^{human} = \frac{1}{n} \sum_{i=1}^{n} f_i * 100, \tag{1}$$

where $n = 320$ and f_i is equal to a relative frequency of the correct answer to the question i in the listening test.

Table 1. Sample results of the listening test with the isolated words with percentage of selected options. The percentages of the correct answers are highlighted by boldface.

Word	Option 1	Option 2	Option 3
borce	**57.90**	36.84	5.26
porce	21.05	**52.63**	26.32
kosa	**0.00**	100.00	0.00
koza	0.00	**100.00**	0.00
přibít	**94.74**	5.26	0.00
připít	10.52	**89.48**	0.00

3.2   Word Bigrams

In the test, participants were asked to listen 333 recordings of the word bigrams (two consecutively uttered words) and select one of the prepared options that have the forms (a) *word A + word A (e.g. kosa + kosa)*, (b) *word A + word B (e.g. kosa + koza)*, (c) *word B + word A (e.g. koza + kosa)*, (d) *word B + word B (e.g. koza + koza)*. It can be seen that this represents all the combinations of the words from the bigram and also that the bigrams were always assembled using the words from the complementary pairs differing only in voicing of a single phoneme. An audio file containing the bigram is a concatenation of two recordings of the isolated words with a short pause between them. Higher number of questions in the test is due to the fact that for some words there exists more than one possible combination with another word. In an effort to shorten the time of the already long test, we generated only the combinations of words corresponding to the second and third option in the test; the participants were not informed about this fact. However, the listening test was completed only by 12 subjects.

The output of the test is again a table with percentages of responses for each option. Table 2 shows an excerpt from the results. As in the previous case, the percentages of the correct answers are highlighted by boldface. It depicts the results for the same words that were shown for isolated word scenario in Table 1. Although now the test deals with word bigrams, the results from both tests naturally exhibit a correlation. For the first combination of words *"borci"* and *"porci"*, the participants were unable to convincingly decide about the content of the recording, just as they were not able to clearly identify those words in the first listening test. In the second example, all the participants except one voted for the combination of words *"koza + koza"* even though the bigrams contained both words from this complementary pair. The last example is an illustration of the situation where the participants were able to clearly distinguish the words. The overall accuracy of correct answers is $Acc_p^{human} = 66,24\%$ and it was computed using Eq. 1.

Table 2. Sample results of the listening test with the word bigrams with percentage of selected options. The percentages of the correct answers are highlighted by boldface.

Word bigram	Option 1	Option 2	Option 3	Option 4
borce + porce	16.67	**50.00**	0.00	33.33
porce + borce	8.33	0.00	**66.67**	25.00
kosa + koza	0.00	**8.33**	0.00	91.67
koza + kosa	0.00	0.00	**0.00**	100.00
přibít + připít	0.00	**100.00**	0.00	0.00
připít + přibít	0.00	0.00	**100.00**	0.00

4   Acoustic Modeling and Recognition Results

We followed the typical Kaldi [7] training recipe S5 [3] for a deep neural network (DNN) acoustic model (AM) training. In all our experiments we used features based on PLP parameterization (19 band pass filters, 12 cepstral coefficients with delta and delta-delta features with CMN). The first step of the Kaldi training recipe is a monophone acoustic model. A monophone AM is trained from the flat start using the PLPs features (static + delta + delta delta) and is treated as a special case of a context-dependent system, without any left or right context. Secondly, we train the triphone GMM AM. As the number of triphones is too large, we used decision trees to tie their states. The questions can be based on linguistic knowledge, but in our case, the questions were generated fully automatically (based on a tree-clustering of the phones). Due to a lack of acoustic training data, we performed several experiments to determine the best WER according to the number of clustered states. The best results were obtained for 2048 states.

We also applied linear discriminant analysis (LDA) and Maximum Likelihood Linear Transform (MLLT) over a central frame spliced across 3 frames. LDA+MLLT project the concatenated frames into 40 dimensions space. Despite the fact that we have only one speaker, we used feature space Maximum Likelihood Linear Regression (fMLLR) [2] because the recordings were taken in two separate series; the individual series differ not only in the recording equipment but also in speaker's EL.

The 40-dimensional features from GMM are spliced across 5 frames of context and used as input to the DNN. The resulting dimension of the input feature vector of the DNN is therefore 440. In this framework, we used the standard 6 layers topology (5 hidden layers, each with 2048 neurons). The output layer was a softmax layer with dimension equal to the number of clustered context-dependent states (the number of states was one of the task's parameters that were optimized).

The S5 recipe supports layer-wise restricted Boltzmann machine (RBM) pre-training, stochastic gradient descent (SGD) training and sequence-discriminative training, using lattice framework, optimizing state-level minimum Bayes risk (sMBR) criterion. The whole DNN training is running on a single GPU using CUDA. The training dataset consists of a total of 5000 sentences from both parts of the corpus.

For all recognition experiments, we used our own RT decoder. This in-house LVCSR system is optimized for low latency in the real-time operation with very large vocabularies. To enable recognition with a vocabulary containing more than one million words in real-time, we accelerated the decoding using parallel approach (Viterbi search on CPU and DNN segments scores on GPU). We used trigram back-off language models (LM) with mixed-case vocabularies with more than 1.2M words. Our text corpus containing the data from newspapers (520 million tokens), web news (350 million tokens), subtitles (200 million tokens) and transcriptions of some TV programs (175 million tokens) (details can be found in [10]). The overall accuracy on the test dataset containing 580 of randomly selected sentences is $Acc_w^{machine} = 86.10\%$ and it was computed as follows

$$Acc_w^{machine} = \frac{N - S - D - I}{N} * 100, \tag{2}$$

where N is the number of words, S is the number of substitutions, D is the number of deletions and I is the number of insertions.

5 Experiments and Evaluation

As mentioned above, the goal is to compare the ability of the human and the machine to distinguish the identity of words, differing only in voicing. In this section, we describe three experiments designed for such comparison and present the evaluation of the results.

From the listening tests described above, we obtained two sets of results. The first one represents the ability to determine the identity of the word pronounced in isolation, and the second one the ability to distinguish two similar words pronounced together. We created three ASR experiments that should emulate the same scenarios, using the Kaldi model described in Sect. 4.

The first experiment with isolated words and the zerogram LM[3] containing more than a million words should correspond to what we called a blind test. In the real situation, the listener does not know the word in advance, which is the reason why such a large vocabulary is used. We named the experiment *"onemil"*. However, in the actual listening test, the participants knew the list of included words and therefore had some advantage over the *"onemil"* machine setting. To compensate for this, we reduced the size of ASR vocabulary and included only the words that actually occurred in the test. We named it *"reduced"*. Both experiments are compared with the first listening test. Note that another possible solution would be to create a special vocabulary with just the specific complementary word pair. But since the listening test contains three possible answers (including the "cannot decide" option), and the ASR system with a two-word vocabulary would always select one of the words from the pair, the results would not correspond to the listening test. The last experiment corresponds to the second listening test. To obtain comparable results, we generated special LM for each word bigram. The LM contains only all four combinations of the words, same as the listening test. The output is the best-matched combination. We named the experiment *"bigrams"*.

The results from the recognizer are not directly comparable with those of listening tests. For resolving the issue we rated correct hypothesis with 1, otherwise 0, and computed the average. The overall results are presented in Table 3.

Table 3. Average accuracy results in percent for human and machine.

	onemil	reduced	bigrams
human	70.47	70.47	66.24
machine	61.76	69.91	54.82

---

[3] Zerogram LM is the setting where all words from a fixed vocabulary have the same probability.

The results show that the presented tasks are challenging even for humans. In the case of the *"onemil"* experiment, the performance of the ASR is further hurt by the enormous perplexity of the language model (note that for zerogram LM the perplexity is directly equal to the vocabulary size). Reducing the size of the vocabulary in the *"reduced"* experiment lifted the ASR accuracy almost to the human level; let us stress out that even this scenario is unfair to ASR as the humans are effectively facing only the perplexity of 3.

Interesting are the results of the experiment with word bigrams. At first glance, it may seem easier because the task is to select well-defined combinations of words. However, words are acoustically very similar and this makes it very difficult to tell them apart. In many cases, the difference of rating of the ASR hypotheses is very small. It indicates similarity between the incriminated models of phonemes. It occurs mostly in bigrams where humans were not able to conclusively decide.

6   Conclusion

In this paper, we compared the ability of humans and machines to distinguish the identity of acoustically similar words.

One important conclusion is that the recognition of isolated words pronounced using the electrolarynx is a hard task even for humans and nowadays ASR systems are not better. In the ideal situation, we are able to achieve competitive results (i.e. 86.10% accuracy) if there is sufficient context. However, with a smaller (or non-existent) context, the recognition quality dramatically decreases (both for humans and machines). An illustrative example is the performance drop in an isolated word scenario. On the other hand, after the experience with corpus recording, we should suppose that the communication using short phrases (and the resulting limited word context) is a realistic scenario when dealing with EL speakers – most of the recorded sentences were split into multiple parts during the post-processing since the speakers needed to make pauses for recovery in the middle of long sentences.

Some relatively significant differences between human and machine performance could be caused by the aforementioned lower effective perplexity in the human task, stemming from the listening tests design. Some participants may also unravel the fact that a test always contains only a subset of all possible answer variants. Also, the human participants could listen to the recording several times and then select an answer. However, we must honestly say that the ASR system as prepared for our experiments can hardly benefit from repeated processing of the same audio file.

All those drawbacks could be solved by a more sophisticated listening test, but the problem is the difficulty of even the existing test. The slightly more challenging listening test with word bigrams was completed only by two-thirds of the participants of the first test.

At the beginning of the research, we assumed all phonemes are voiced in EL speech. Experiments have shown that this is not always true. The context of the phoneme plays an important role in voicing. The human and the machine were able, in many cases, to distinguish voiced and unvoiced phonemes in the EL speech. The processing of these phonemes will be one of the subjects of further research.

Achieved results with an ASR system lead to the conclusion that if some preprocessing is done or special models are created, EL users can have access to ASR technologies.

Acknowledgements. The work has been supported by the grant of the University of West Bohemia, project No. SGS-2016-039 and by the Ministry of Education, Youth and Sports of the Czech Republic project No. LO1506.

References

1. Brown, D.H., Hilgers, F.J., Irish, J.C., Balm, A.J.: Postlaryngectomy voice rehabilitation: state of the art at the Millennium. World J. Surg. **27**(7), 824–831 (2003)
2. Fuchs, A.K., Morales-Cordovilla, J.A., Hagmller, M.: ASR for electro-laryngeal speech. In: 2013 IEEE Workshop on Automatic Speech Recognition and Understanding, pp. 234–238, December 2013
3. Github Kaldi: https://github.com/kaldi-asr/kaldi/tree/master/egs/wsj/s5
4. Jůzová, M., Romportl, J., Tihelka, D.: Speech Corpus Preparation for Voice Banking of Laryngectomised Patients, pp. 282–290. Springer, Cham (2015)
5. Kramp, B., Dommerich, S.: Tracheostomy cannulas and voice prosthesis. GMS Curr. Top. Otorhinolaryngol. Head Neck Surg. **8**, Doc05 (2009)
6. Liu, H., Ng, M.L.: Electrolarynx in voice rehabilitation. Auris Nasus Larynx **34**(3), 327–332 (2007)
7. Povey, D., Ghoshal, A., Boulianne, G., Burget, L., Glembek, O., Goel, N., Hannemann, M., Motlicek, P., Qian, Y., Schwarz, P., et al.: The Kaldi speech recognition toolkit. In: IEEE 2011 Workshop on Automatic Speech Recognition and Understanding. No. EPFL-CONF-192584. IEEE Signal Processing Society (2011)
8. Radová, V., Psutka, J.: UWB-S01 corpus: a Czech read-speech corpus. In: Proceedings of the 6th International Conference on Spoken Language Processing, ICSLP 2000, pp. 732–735 (2000)
9. Stanislav, P., Psutka, J.V.: Influence of different phoneme mappings on the recognition accuracy of electrolaryngeal speech. In: Proceedings of the International Conference on Signal Processing and Multimedia Applications and Wireless Information Networks and Systems, SIGMAP (ICETE 2012), vol. 1, pp. 204–207 (2012)
10. Švec, J., Hoidekr, J., Soutner, D., Vavruška, J.: Web Text Data Mining for Building Large Scale Language Modelling Corpus, pp. 356–363. Springer, Heidelberg (2011)

Author Index

Printed in the United States
By Bookmasters